Student Solutions Guide for

CALCULUS:
AN APPLIED APPROACH

SEVENTH EDITION

Larson / Edwards

Bruce H. Edwards
University of Florida

Houghton Mifflin Company   Boston   New York

Publisher: Jack Shira
Associate Sponsoring Editor: Cathy Cantin
Development Manager: Maureen Ross
Editorial Assistant: Elizabeth Kassab
Supervising Editor: Karen Carter
Senior Project Editor: Patty Bergin
Editorial Assistant: Julia Keller
Production Technology Supervisor: Gary Crespo
Executive Marketing Manager: Michael Busnach
Senior Marketing Manager: Danielle Potvin Curran
Marketing Coordinator: Nicole Mollica

Printed in the United States of America

ISBN: 0-618-54720-7

6789-EB-09 08 07

PREFACE

This *Student Solutions Guide* is designed as a supplement to *Calculus: An Applied Approach*, Seventh Edition, and *Brief Calculus: An Applied Approach*, Seventh Edition, by Ron Larson and Bruce H. Edwards. All references to chapters, theorems, and exercises relate to the main text. Solutions to every odd-numbered exercise in the text are given with all essential algebraic steps included. Although this supplement is not a substitute for good study habits, it can be valuable when incorporated into a well-planned course of study. For suggestions that may assist you in the use of this text, your lecture notes, and this *Guide*, please refer to the student web site for your text at *college.hmco.com*.

I have made every effort to see that the solutions are correct. However, I would appreciate hearing about any errors or other suggestions for improvement. Good Luck with your study of calculus.

Bruce H. Edwards
Department of Mathematics
University of Florida

CONTENTS

C H A P T E R 0
A Precalculus Review

CHAPTER 0
A Precalculus Review

Section 0.1   The Real Number Line and Order

Solutions to Odd-Numbered Exercises

1. Since $0.7 = \dfrac{7}{10}$, it is rational.

3. $\dfrac{3\pi}{2}$ is irrational because π is irrational.

5. $4.3\overline{451}$ is rational because it has a repeating decimal expansion.

7. Since $\sqrt[3]{64} = 4$, it is rational.

9. $\sqrt[3]{60}$ is irrational, since 60 is not the cube of a rational number.

11. (a) Yes, if $x = 3$, then $5(3) - 12 = 3 > 0$.

(b) No, if $x = -3$, then $5(-3) - 12 = -27 < 0$.

(c) Yes, if $x = \dfrac{5}{2}$, then $5\left(\dfrac{5}{2}\right) - 12 = \dfrac{1}{2} > 0$.

(d) No, if $x = \dfrac{3}{2}$, then $5\left(\dfrac{3}{2}\right) - 12 = -\dfrac{9}{2} < 0$.

13. $0 < \dfrac{x-2}{4} < 2$

$0 < x - 2 < 8$

$2 < x < 10$

(a) Yes, if $x = 4$, then $2 < x < 10$.

(b) No, if $x = 10$, then x is not less than 10.

(c) No, if $x = 0$, then x is not greater than 2.

(d) Yes, if $x = \dfrac{7}{2}$, then $2 < x < 10$.

15. $x - 5 \geq 7$

$x - 5 + 5 \geq 7 + 5$

$x \geq 12$

17. $4x + 1 < 2x$

$2x < -1$

$x < -\dfrac{1}{2}$

19. $4 - 2x < 3x - 1$

$4 - 5x < -1$

$-5x < -5$

$x > 1$

21. $-4 < 2x - 3 < 4$

$-4 + 3 < 2x - 3 + 3 < 4 + 3$

$-1 < 2x < 7$

$-\dfrac{1}{2} < x < \dfrac{7}{2}$

23. $\dfrac{3}{4} > x + 1 > \dfrac{1}{4}$

$-\dfrac{1}{4} > x > -\dfrac{3}{4}$

$-\dfrac{3}{4} < x < -\dfrac{1}{4}$

25. $\dfrac{x}{2} + \dfrac{x}{3} > 5$

$3x + 2x > 30$

$5x > 30$

$x > 6$

27.
$$2x^2 - x < 6$$
$$2x^2 - x - 6 < 0$$
$$(2x + 3)(x - 2) < 0$$

Zeros of the polynomial $(2x + 3)(x - 2)$ are $x = -\frac{3}{2}$ and $x = 2$. Testing the intervals $\left(-\infty, -\frac{3}{2}\right)$, $\left(-\frac{3}{2}, 2\right)$, and $(2, \infty)$, you see that the solution set is $-\frac{3}{2} < x < 2$.

29.

31. $R = 115.95x$ and $C = 95x + 750$ and we have:

$$R > C$$
$$115.95x > 95x + 750$$
$$20.95x > 750$$
$$x > \frac{750}{20.95} = 35.7995 \ldots$$

Therefore, $x \geq 36$ units.

33. Let x be the number of miles driven each week. Then the company reimburses according to the formula

$$C = 100 + 0.35x$$

According to the allocations, you have

$$200 \leq C \leq 250$$
$$200 \leq 100 + 0.35x \leq 250$$
$$100 \leq 0.35x \leq 150$$
$$\frac{100}{0.35} \leq x \leq \frac{150}{0.35}$$
$$285.71 < x < 428.57$$

Thus, the minimum is approximately 285.7 miles per week, and the maximum is approximately 428.6.

35. Given $a < b$,

(a) $-2a < -2b$ is false.

(b) $a + 2 < b + 2$ is true.

(c) $6a < 6b$ is true.

(d) $\dfrac{1}{a} < \dfrac{1}{b}$ is false if $ab > 0$ and true if $ab < 0$.

Section 0.2   Absolute Value and Distance on the Real Number Line

1. (a) The directed distance from a to b is $75 - 126 = -51$.

(b) The directed distance from b to a is $126 - 75 = 51$.

(c) The distance between a and b is $|75 - 126| = 51$.

3. (a) The directed distance from a to b is $-5.65 - 9.34 = -14.99$.

(b) The directed distance from b to a is $9.34 - (-5.65) = 14.99$.

(c) The distance between a and b is $|-5.65 - 9.34| = 14.99$.

5. (a) The directed distance from a to b is $\frac{112}{75} - \frac{16}{5} = -\frac{128}{75}$.

(b) The directed distance from b to a is $\frac{16}{5} - \frac{112}{75} = \frac{128}{75}$.

(c) The distance between a and b is $\left|\frac{112}{75} - \frac{16}{5}\right| = \frac{128}{75}$.

7. $|x| \leq 2$ **9.** $|x| > 2$ **11.** $|x - 4| \leq 2$

13. $|x - 2| > 2$

15. $|x - 4| < 2$

17. $|y - a| \leq 2$

19. $-5 < x < 5$

21. $\dfrac{x}{2} < -3$ or $\dfrac{x}{2} > 3$

$x < -6$ $x > 6$

23. $-5 < x + 2 < 5$

$-7 < x < 3$

25. $\dfrac{x - 3}{2} \leq -5$ or $\dfrac{x - 3}{2} \geq 5$

$\dfrac{x - 3}{2}(2) \leq -5(2)$ $\dfrac{x - 3}{2}(2) \geq 5(2)$

$x - 3 \leq -10$ $x - 3 \geq 10$

$x - 3 + 3 \leq -10 + 3$ $x - 3 + 3 \geq 10 + 3$

$x \leq -7$ $x \geq 13$

27. $10 - x < -4$ or $10 - x > 4$

$-x < -14$ $-x > -6$

$x > 14$ $x < 6$

29. $-1 < 9 - 2x < 1$

$-10 < -2x < -8$

$5 > x > 4$

$4 < x < 5$

31. $-b \leq x - a \leq b$

$a - b \leq x \leq a + b$

33. $-2b < \dfrac{3x - a}{4} < 2b$

$-8b < 3x - a < 8b$

$-8b + a < 3x < 8b + a$

$\dfrac{1}{3}(a - 8b) < x < \dfrac{1}{3}(a + 8b)$

35. Midpoint $= \dfrac{7 + 21}{2} = 14$

37. Midpoint $= \dfrac{-6.85 + 9.35}{2} = 1.25$

39. Midpoint $= \dfrac{-\frac{1}{2} + \frac{3}{4}}{2} = \dfrac{\frac{1}{4}}{2} = \dfrac{1}{8}$

41. $1083.4 - 0.2 \leq M \leq 1083.4 + 0.2$

$1083.2 \leq M \leq 1083.6$

$|M - 1083.4| \leq 0.2$

43. $\left|\dfrac{h - 68.5}{2.7}\right| \leq 1$

$-1 \leq \dfrac{h - 68.5}{2.7} \leq 1$

$-2.7 \leq h - 68.5 \leq 2.7$

$65.8 \leq h \leq 71.2$

45. $|x - 200{,}000| \leq 25{,}000$

$-25{,}000 \leq x - 200{,}000 \leq 25{,}000$

$175{,}000 \leq x \leq 225{,}000$

47. (a) $|E - 4750| \leq 500 \Rightarrow 4250 \leq E \leq 5250$

$0.05(4750) = 237.50$

$|E - 4750| \leq 237.50 \Rightarrow 4512.50 \leq E \leq 4987.50$

(b) $5116.37 is not within 5% of the specified budgeted amount; at variance.

49. (a) $|E - 20,000| \leq 500 \Rightarrow 19,500 \leq E \leq 20,500$

$0.05(20,000) = 1000$

$|E - 20,000| \leq 1000 \Rightarrow 19,000 \leq E \leq 21,000$

(b) \$22,718.35 is at variance with both budget restrictions.

Section 0.3   Exponents and Radicals

1. $-3(2)^3 = -3(8) = -24$

3. $4(2)^{-3} = 4\left(\frac{1}{8}\right) = \frac{1}{2}$

5. $\dfrac{1 + (2)^{-1}}{(2)^{-1}} = \dfrac{1 + (1/2)}{1/2} = \dfrac{3/2}{1/2} = 3$

7. $3(-2)^2 - 4(-2)^3 = 3(4) - 4(-8) = 12 + 32 = 44$

9. $6(10)^0 - [6(10)]^0 = 6(1) - 1 = 5$

11. $\sqrt[3]{27^2} = \left(\sqrt[3]{27}\right)^2 = 3^2 = 9$

13. $4^{-1/2} = \dfrac{1}{\sqrt{4}} = \dfrac{1}{2}$

15. $(-32)^{-2/5} = \dfrac{1}{\left(\sqrt[5]{-32}\right)^2} = \dfrac{1}{(-2)^2} = \dfrac{1}{4}$

17. $500(1.01)^{60} \approx 908.3483$

19. $\sqrt[3]{-154} \approx -5.3601$

21. $6y^{-2}(2y^4)^{-3} = 6y^{-2}(2^{-3}y^{-12}) = 6\left(\dfrac{1}{8}\right)y^{-14} = \dfrac{3}{4y^{14}}$

23. $10(x^2)^2 = 10x^4$

25. $\dfrac{7x^2}{x^{-3}} = 7x^5$

27. $\dfrac{12(x + y)^3}{9(x + y)^{-2}} = \dfrac{4}{3}(x + y)^5, \quad x + y \neq 0$

29. $\dfrac{3x\sqrt{x}}{x^{1/2}} = \dfrac{3x\sqrt{x}}{\sqrt{x}} = 3x, \quad x > 0$

31. (a) $\sqrt{8} = \sqrt{4 \cdot 2} = \sqrt{4}\sqrt{2} = 2\sqrt{2}$

(b) $\sqrt{18} = \sqrt{9 \cdot 2} = \sqrt{9}\sqrt{2} = 3\sqrt{2}$

33. (a) $\sqrt[3]{16x^5} = \sqrt[3]{(8x^3)(2x^2)} = \sqrt[3]{8x^3}\sqrt[3]{2x^2} = 2x\sqrt[3]{2x^2}$

(b) $\sqrt[4]{32x^4z^5} = \sqrt[4]{16x^4z^42z} = \sqrt[4]{16x^4z^4}\sqrt[4]{2z} = 2|x|z\sqrt[4]{2z}$

[*Note:* Since x^4 is under the radical, x could be positive or negative. For z^5 to be under the radical, z must be positive.]

35. (a) $\sqrt[3]{144x^9y^{-4}z^5} = \sqrt[3]{18(2^3)(x^3)^3(y^{-1})^3y^{-1}z^2z^3}$

$= 2x^3y^{-1}z\,\sqrt[3]{18y^{-1}z^2}$

$= \dfrac{2x^3z}{y}\sqrt[3]{\dfrac{18z^2}{y}}$

$= \dfrac{2x^3z}{y^2}\sqrt[3]{18z^2y^2}$

(b) $\sqrt{12(3x + 5)^7} = \sqrt{3(2^2)(3x + 5)^6(3x + 5)}$

$= 2(3x + 5)^3\sqrt{3(3x + 5)}$

$= 2(3x + 5)^3\sqrt{9x + 15}$

37. $4x^3 - 6x = 2x(2x^2 - 3)$

39. $2x^{5/2} + x^{-1/2} = x^{-1/2}(2x^3 + 1) = \dfrac{2x^3 + 1}{x^{1/2}}$

41. $3x(x + 1)^{3/2} - 6(x + 1)^{1/2} = 3(x + 1)^{1/2}(x(x + 1) - 2)$

$$= 3(x + 1)^{1/2}(x^2 + x - 2)$$

$$= 3(x + 1)^{1/2}(x - 1)(x + 2)$$

43. $\dfrac{(x + 1)(x - 1)^2 - (x - 1)^3}{(x + 1)^2} = \dfrac{(x - 1)^2}{(x + 1)^2}((x + 1) - (x - 1))$

$$= \dfrac{(x - 1)^2}{(x + 1)^2}(2)$$

$$= \dfrac{2(x - 1)^2}{(x + 1)^2}$$

45. $(x^2 + 1)^2(x - 1)^{-1/2} + 2x(x - 1)^{1/2}(x^2 + 1) = (x^2 + 1)(x - 1)^{-1/2}((x^2 + 1) + 2x(x - 1))$

$$= (x^2 + 1)(x - 1)^{-1/2}(3x^2 - 2x + 1)$$

$$= \dfrac{(x^2 + 1)(3x^2 - 2x + 1)}{(x - 1)^{1/2}}$$

47. $\sqrt{x - 1}$ is defined when $x \geq 1$. Therefore, the domain is $[1, \infty)$.

49. $\sqrt{x^2 + 3}$ is defined for all real numbers. Therefore, the domain is $(-\infty, \infty)$.

51. $\dfrac{1}{\sqrt[3]{x - 1}}$ is defined for all real numbers except $x = 1$. Therefore, the domain is $(-\infty, 1) \cup (1, \infty)$.

53. The numerator is defined for $x \geq -2$. The denominator is defined for all $x \neq 4$. Therefore, the domain is $[-2, 4) \cup (4, \infty)$.

55. $\sqrt{x - 1}$ is defined when $x \geq 1$, and $\sqrt{5 - x}$ is defined when $x \leq 5$. Therefore, the domain of $\sqrt{x - 1} + \sqrt{5 - x}$ is $1 \leq x \leq 5$.

57. $A = 10{,}000\left(1 + \dfrac{0.065}{12}\right)^{120} \approx \$19{,}121.84$

59. $A = 5000\left(1 + \dfrac{0.055}{4}\right)^{60} \approx \$11{,}345.46$

61. $T = 2\pi\sqrt{\dfrac{L}{32}}$

$$= 2\pi\sqrt{\dfrac{4}{32}}$$

$$= 2\pi\sqrt{\dfrac{1}{8}}$$

$$= 2\pi\dfrac{1}{2\sqrt{2}}$$

$$= \dfrac{\pi}{\sqrt{2}} = \dfrac{\pi\sqrt{2}}{2} \approx 2.22 \text{ seconds}$$

Section 0.4   Factoring Polynomials

1. Since $a = 6$, $b = -1$, and $c = -1$, we have

$$x = \dfrac{1 \pm \sqrt{1 - (-24)}}{12} = \dfrac{1 \pm 5}{12}.$$

Thus, $x = \dfrac{1 + 5}{12} = \dfrac{1}{2}$ or $x = \dfrac{1 - 5}{12} = -\dfrac{1}{3}$.

3. Since $a = 4$, $b = -12$, and $c = 9$, we have

$$x = \dfrac{12 \pm \sqrt{144 - 144}}{8} = \dfrac{12}{8} = \dfrac{3}{2}.$$

5. Since $a = 1$, $b = 4$, and $c = 1$, we have

$$y = \frac{-4 \pm \sqrt{16 - 4}}{2} = \frac{-4 \pm 2\sqrt{3}}{2} = -2 \pm \sqrt{3}.$$

7. Since $a = 2$, $b = 3$, and $c = -4$, we have

$$x = \frac{-3 \pm \sqrt{9 - 4(2)(-4)}}{4} = \frac{-3 \pm \sqrt{41}}{4}$$

9. $x^2 - 4x + 4 = (x - 2)^2$

11. $4x^2 + 4x + 1 = (2x + 1)^2$

13. $x^2 + x - 2 = (x + 2)(x - 1)$

15. $3x^2 - 5x + 2 = (3x - 2)(x - 1)$

17. $x^2 - 4xy + 4y^2 = (x - 2y)^2$

19. $81 - y^4 = (9 + y^2)(9 - y^2)$
$$= (9 + y^2)(3 + y)(3 - y)$$

21. $x^3 - 8 = x^3 - 2^3$
$$= (x - 2)(x^2 + 2x + 4)$$

23. $y^3 + 64 = y^3 + 4^3$
$$= (y + 4)(y^2 - 4y + 16)$$

25. $x^3 - 27 = x^3 - 3^3$
$$= (x - 3)(x^2 + 3x + 9)$$

27. $x^3 - 4x^2 - x + 4 = x^2(x - 4) - (x - 4)$
$$= (x - 4)(x^2 - 1)$$
$$= (x - 4)(x + 1)(x - 1)$$

29. $2x^3 - 3x^2 + 4x - 6 = x^2(2x - 3) + 2(2x - 3)$
$$= (2x - 3)(x^2 + 2)$$

31. $2x^3 - 4x^2 - x + 2 = 2x^2(x - 2) - (x - 2)$
$$= (x - 2)(2x^2 - 1)$$

33. $x^4 - 15x^2 - 16 = (x^2 - 16)(x^2 + 1)$
$$= (x - 4)(x + 4)(x^2 + 1)$$

35. $x^2 - 5x = 0$
  $x(x - 5) = 0$
  $x = 0, 5$

37. $x^2 - 9 = 0$
  $(x + 3)(x - 3) = 0$
  $x = -3, 3$

39. $x^2 - 3 = 0$
  $(x + \sqrt{3})(x - \sqrt{3}) = 0$
  $x = \pm\sqrt{3}$

41. $(x - 3)^2 - 9 = 0$
  $x^2 - 6x + 9 - 9 = 0$
  $x(x - 6) = 0$
  $x = 0, 6$

43. $x^2 + x - 2 = 0$
  $(x + 2)(x - 1) = 0$
  $x = -2, 1$

45. $x^2 - 5x + 6 = 0$
  $(x - 2)(x - 3) = 0$
  $x = 2, 3$

47. $x^3 + 64 = 0$
  $x^3 = -64$
  $x = \sqrt[3]{-64} = -4$

49. $x^4 - 16 = 0$
  $x^4 = 16$
  $x = \pm\sqrt[4]{16} = \pm 2$

51. $x^3 - x^2 - 4x + 4 = 0$
  $x^2(x - 1) - 4(x - 1) = 0$
  $(x - 1)(x^2 - 4) = 0$
  $(x - 1)(x - 2)(x + 2) = 0$
  $x = 1, \pm 2$

53. Since $\sqrt{x^2 - 4} = \sqrt{(x + 2)(x - 2)}$, the roots are $x = \pm 2$. By testing points inside and outside the interval $[-2, 2]$, we find that the expression is defined when $x \leq -2$ or $x \geq 2$. Thus, the domain is $(-\infty, -2] \cup [2, \infty)$.

55. Since $\sqrt{x^2 - 7x + 12} = \sqrt{(x - 3)(x - 4)}$, the roots are $x = 3$ and $x = 4$. By testing points inside and outside the interval $[3, 4]$, we find that the expression is defined when $x \leq 3$ or $x \geq 4$. Thus, the domain is $(-\infty, 3] \cup [4, \infty)$.

57.
$$\begin{array}{r|rrrr} -1 & 1 & -3 & -6 & -2 \\ & & -1 & 4 & 2 \\ \hline & 1 & -4 & -2 & 0 \end{array}$$

Therefore, the factorization is

$$x^3 - 3x^2 - 6x - 2 = (x + 1)(x^2 - 4x - 2)$$

59.
$$\begin{array}{r|rrrr} 1 & 2 & -1 & -2 & 1 \\ & & 2 & 1 & -1 \\ \hline & 2 & 1 & -1 & 0 \end{array}$$

Therefore, the factorization is

$$2x^3 - x^2 - 2x + 1 = (x - 1)(2x^2 + x - 1).$$

61. Possible rational zeros: $\pm 8, \pm 4, \pm 2, \pm 1$
Using synthetic division for $x = -1$, we have

$$\begin{array}{r|rrrr} -1 & 1 & -1 & -10 & -8 \\ & & -1 & 2 & 8 \\ \hline & 1 & -2 & -8 & 0 \end{array}$$

Therefore,

$$x^3 - x^2 - 10x - 8 = 0$$
$$(x + 1)(x^2 - 2x - 8) = 0$$
$$(x + 1)(x + 2)(x - 4) = 0$$
$$x = -1, -2, 4$$

63. Possible rational roots: $\pm 1, \pm 2, \pm 3, \pm 6$
Using synthetic division for $x = 1$, we have the following.

$$\begin{array}{r|rrrr} 1 & 1 & -6 & 11 & -6 \\ & & 1 & -5 & 6 \\ \hline & 1 & -5 & 6 & 0 \end{array}$$

Therefore, we have

$$x^3 - 6x^2 + 11x - 6 = 0$$
$$(x - 1)(x^2 - 5x + 6) = 0$$
$$(x - 1)(x - 2)(x - 3) = 0$$
$$x = 1, 2, 3.$$

65. Possible rational zeros: $\pm 6, \pm 3, \pm 2, \pm 1, \pm \frac{3}{2}, \pm \frac{1}{2}, \pm \frac{2}{3}, \pm \frac{1}{3}, \pm \frac{1}{6}$
Using synthetic division for $x = 3$, we have

$$\begin{array}{r|rrrr} 3 & 6 & -11 & -19 & -6 \\ & & 18 & 21 & 6 \\ \hline & 6 & 7 & 2 & 0 \end{array}$$

Therefore,

$$6x^3 - 11x^2 - 19x - 6 = 0$$
$$(x - 3)(6x^2 + 7x + 2) = 0$$
$$(x - 3)(3x + 2)(2x + 1) = 0$$
$$x = 3, -\tfrac{2}{3}, -\tfrac{1}{2}$$

67. Possible rational roots: $\pm 1, \pm 2, \pm 4$
Using synthetic division for $x = 4$, we have the following.

$$\begin{array}{r|rrrr} 4 & 1 & -3 & -3 & -4 \\ & & 4 & 4 & 4 \\ \hline & 1 & 1 & 1 & 0 \end{array}$$

Therefore, we have

$$x^3 - 3x^2 - 3x - 4 = 0$$
$$(x - 4)(x^2 + x + 1) = 0.$$

Since $x^2 + x + 1$ has no real solutions, $x = 4$ is the only real solution.

69.
$$0.0003x^2 - 1200 = 0$$
$$0.0003x^2 = 1200$$
$$x^2 = 4{,}000{,}000$$
$$x = 2000 \text{ units}$$

71.
$$1.8 \times 10^{-5} = \frac{x^2}{1.0 \times 10^{-4} - x}$$
$$1.8 \times 10^{-9} - 1.8 \times 10^{-5}x = x^2$$
$$x^2 + 1.8 \times 10^{-5}x - 1.8 \times 10^{-9} = 0$$

By the Quadratic Formula:

$$x = \frac{-1.8 \times 10^{-5} \pm \sqrt{(1.8 \times 10^{-5})^2 + 4 \times 1.8 \times 10^{-9}}}{2}$$
$$\approx \frac{-1.8 \times 10^{-5} \pm \sqrt{7.524 \times 10^{-9}}}{2}$$
$$\approx 3.437 \times 10^{-5}[H^+]$$

Section 0.5   Fractions and Rationalization

1. $\dfrac{5}{x-1} + \dfrac{x}{x-1} = \dfrac{5+x}{x-1} = \dfrac{x+5}{x-1}$

3. $\dfrac{2x}{x^2+2} - \dfrac{1-3x}{x^2+2} = \dfrac{2x-(1-3x)}{x^2+2} = \dfrac{5x-1}{x^2+2}$

5. $\dfrac{2}{x^2-4} - \dfrac{1}{x-2} = \dfrac{2}{(x-2)(x+2)} - \dfrac{1}{x-2}\dfrac{(x+2)}{(x+2)}$

$\qquad\qquad = \dfrac{2-(x+2)}{(x-2)(x+2)}$

$\qquad\qquad = \dfrac{-x}{x^2-4}$

$\qquad\qquad = \dfrac{x}{4-x^2}$

7. $\dfrac{5}{x-3} + \dfrac{3}{3-x} = \dfrac{5}{x-3} + \dfrac{-3}{x-3}$

$\qquad\qquad = \dfrac{2}{x-3}$

9. $\dfrac{A}{x+1} + \dfrac{B}{(x+1)^2} + \dfrac{C}{x-2} = \dfrac{A(x+1)(x-2) + B(x-2) + C(x+1)^2}{(x+1)^2(x-2)}$

$\qquad\qquad = \dfrac{A(x^2-x-2) + B(x-2) + C(x^2+2x+1)}{(x+1)^2(x-2)}$

$\qquad\qquad = \dfrac{Ax^2 - Ax - 2A + Bx - 2B + Cx^2 + 2Cx + C}{(x+1)^2(x-2)}$

$\qquad\qquad = \dfrac{(A+C)x^2 - (A-B-2C)x - (2A+2B-C)}{(x+1)^2(x-2)}$

11. $\dfrac{A}{x-6} + \dfrac{Bx+C}{x^2+3} = \dfrac{A(x^2+3) + (Bx+C)(x-6)}{(x-6)(x^2+3)}$

$\qquad\qquad = \dfrac{(A+B)x^2 + (C-6B)x + 3A - 6C}{(x-6)(x^2+3)}$

13. $-\dfrac{1}{x} + \dfrac{2}{x^2+1} = \dfrac{-(x^2+1) + 2x}{x(x^2+1)}$

$\qquad\qquad = \dfrac{-x^2 + 2x - 1}{x(x^2+1)}$

$\qquad\qquad = \dfrac{-(x^2 - 2x + 1)}{x(x^2+1)}$

$\qquad\qquad = \dfrac{-(x-1)^2}{x(x^2+1)}$

15. $\dfrac{1}{x^2-x-2} - \dfrac{x}{x^2-5x+6} = \dfrac{1}{(x+1)(x-2)} - \dfrac{x}{(x-2)(x-3)}$

$\qquad\qquad = \dfrac{(x-3) - x(x+1)}{(x+1)(x-2)(x-3)}$

$\qquad\qquad = \dfrac{-x^2 - 3}{(x+1)(x-2)(x-3)}$

$\qquad\qquad = -\dfrac{x^2+3}{(x+1)(x-2)(x-3)}$

17. $\dfrac{-x}{(x+1)^{3/2}} + \dfrac{2}{(x+1)^{1/2}} = \dfrac{-x + 2(x+1)}{(x+1)^{3/2}}$

$\qquad\qquad = \dfrac{x+2}{(x+1)^{3/2}}$

19. $\dfrac{2-t}{2\sqrt{1+t}} - \sqrt{1+t} = \dfrac{2-t}{2\sqrt{1+t}} - \dfrac{\sqrt{1+t}}{1}\cdot\dfrac{2\sqrt{1+t}}{2\sqrt{1+t}}$

$\qquad\qquad = \dfrac{(2-t) - 2(1+t)}{2\sqrt{1+t}}$

$\qquad\qquad = \dfrac{-3t}{2\sqrt{1+t}}$

21. $\left(2x\sqrt{x^2+1} - \dfrac{x^3}{\sqrt{x^2+1}}\right) \div (x^2+1) = \dfrac{2x(x^2+1) - x^3}{\sqrt{x^2+1}} \cdot \dfrac{1}{x^2+1}$

$$= \dfrac{x^3 + 2x}{\sqrt{x^2+1}(x^2+1)}$$

$$= \dfrac{x(x^2+2)}{(x^2+1)^{3/2}}$$

23. $\dfrac{(x^2+2)^{1/2} - x^2(x^2+2)^{-1/2}}{x^2} = \dfrac{(x^2+2)^{-1/2}[(x^2+2) - x^2]}{x^2}$

$$= \dfrac{2}{x^2\sqrt{x^2+2}}$$

25. $\dfrac{\dfrac{\sqrt{x+1}}{\sqrt{x}} - \dfrac{\sqrt{x}}{\sqrt{x+1}}}{2(x+1)} = \dfrac{(x+1) - x}{\sqrt{x}\sqrt{x+1}} \cdot \dfrac{1}{2(x+1)}$

$$= \dfrac{1}{2\sqrt{x}(x+1)^{3/2}}$$

27. $\dfrac{x}{2(x+2)^{1/2}} + (x+2)^{1/2} = \dfrac{x + 2(x+2)}{2(x+2)^{1/2}}$

$$= \dfrac{3x+4}{2(x+2)^{1/2}}$$

29. $\dfrac{-x^2}{(2x+3)^{3/2}} + \dfrac{2x}{(2x+3)^{1/2}} = \dfrac{-x^2}{(2x+3)^{3/2}} + \dfrac{2x}{(2x+3)^{1/2}} \cdot \dfrac{2x+3}{2x+3}$

$$= \dfrac{-x^2 + 2x(2x+3)}{(2x+3)^{3/2}}$$

$$= \dfrac{3x^2 + 6x}{(2x+3)^{3/2}}$$

$$= \dfrac{3x(x+2)}{(2x+3)^{3/2}}$$

31. $\dfrac{3}{\sqrt{6}} = \dfrac{3}{\sqrt{6}} \cdot \dfrac{\sqrt{6}}{\sqrt{6}} = \dfrac{3\sqrt{6}}{6} = \dfrac{\sqrt{6}}{2}$

33. $\dfrac{x}{\sqrt{x-4}} = \dfrac{x}{\sqrt{x-4}} \cdot \dfrac{\sqrt{x-4}}{\sqrt{x-4}} = \dfrac{x\sqrt{x-4}}{x-4}$

35. $\dfrac{49(x-3)}{\sqrt{x^2-9}} = \dfrac{49(x-3)}{\sqrt{x^2-9}} \cdot \dfrac{\sqrt{x^2-9}}{\sqrt{x^2-9}}$

$$= \dfrac{49(x-3)\sqrt{x^2-9}}{(x+3)(x-3)}$$

$$= \dfrac{49\sqrt{x^2-9}}{x+3}, x \neq 3$$

37. $\dfrac{5}{\sqrt{14}-2} = \dfrac{5}{\sqrt{14}-2} \cdot \dfrac{\sqrt{14}+2}{\sqrt{14}+2}$

$$= \dfrac{5(\sqrt{14}+2)}{14-4}$$

$$= \dfrac{\sqrt{14}+2}{2}$$

39. $\dfrac{2x}{5-\sqrt{3}} = \dfrac{2x}{5-\sqrt{3}} \cdot \dfrac{5+\sqrt{3}}{5+\sqrt{3}}$

$$= \dfrac{2x(5+\sqrt{3})}{25-3}$$

$$= \dfrac{x(5+\sqrt{3})}{11}$$

41. $\dfrac{1}{\sqrt{6}+\sqrt{5}} = \dfrac{1}{\sqrt{6}+\sqrt{5}} \cdot \dfrac{\sqrt{6}-\sqrt{5}}{\sqrt{6}-\sqrt{5}}$

$$= \dfrac{\sqrt{6}-\sqrt{5}}{6-5}$$

$$= \sqrt{6}-\sqrt{5}$$

43. $\dfrac{2}{\sqrt{x} + \sqrt{x-2}} = \dfrac{2}{\sqrt{x} + \sqrt{x-2}} \cdot \dfrac{\sqrt{x} - \sqrt{x-2}}{\sqrt{x} - \sqrt{x-2}}$

$\qquad = \dfrac{2\left(\sqrt{x} - \sqrt{x-2}\right)}{x - (x-2)}$

$\qquad = \sqrt{x} - \sqrt{x-2}$

45. $\dfrac{\dfrac{\sqrt{4-x^2}}{x^4} - \dfrac{2}{x^2\sqrt{4-x^2}}}{4-x^2} = \dfrac{(4-x^2) - 2x^2}{x^4(4-x^2)^{3/2}}$

$\qquad = \dfrac{4-3x^2}{x^4(4-x^2)^{3/2}}$

47. $P = 10,000,\ r = 0.14,\ N = 60$

$M = 10,000\left[\dfrac{0.14/12}{1 - \left(\dfrac{1}{(0.14/12)+1}\right)^{60}}\right] \approx \$232.68$

Practice Test for Chapter 0

1. Determine whether $\sqrt[4]{81}$ is rational or irrational.

2. Determine whether the given value of x satisfies the inequality $3x + 4 \leq x/2$.

    (a) $x = -2$ (b) $x = 0$ (c) $x = -\frac{8}{5}$ (d) $x = -6$

3. Solve the inequality $3x + 4 \geq 13$.

4. Solve the inequality $x^2 < 6x + 7$.

5. Determine which of the two given real numbers is greater, $\sqrt{19}$ or $\frac{13}{3}$.

6. Given the interval $[-3, 7]$, find (a) the distance between -3 and 7 and (b) the midpoint of the interval.

7. Solve the inequality $|3x + 1| \leq 10$.

8. Solve the inequality $|4 - 5x| > 29$.

9. Solve the inequality $\left| 3 - \dfrac{2x}{5} \right| < 8$.

10. Use absolute values to describe the interval $[-3, 5]$.

11. Simplify $\dfrac{12x^3}{4x^{-2}}$.

12. Simplify $\left(\dfrac{\sqrt{3}\sqrt{x^3}}{x} \right)^0$, $x \neq 0$.

13. Remove all possible factors from the radical $\sqrt[3]{32x^4y^3}$.

14. Complete the factorization: $\frac{3}{2}(x + 1)^{-1/3} + \frac{1}{4}(x + 1)^{2/3} = \frac{1}{4}(x + 1)^{-1/3}(\quad)$

15. Find the domain: $\dfrac{1}{\sqrt{5 - x}}$

16. Factor completely: $3x^2 - 19x - 14$

17. Factor completely: $25x^2 - 81$

18. Factor completely: $x^3 + 8$

19. Use the Quadratic Formula to find all real roots of $x^2 + 6x - 2 = 0$.

20. Use the Rational Zero Theorem to find all real roots of $x^3 - 4x^2 + x + 6 = 0$.

21. Combine terms and simplify: $\dfrac{x}{x^2 + 2x - 3} - \dfrac{1}{x - 1}$

22. Combine terms and simplify: $\dfrac{3 - x}{2\sqrt{x + 5}} + \sqrt{x + 5}$

23. Combine terms and simplify: $\dfrac{\dfrac{\sqrt{x + 2}}{\sqrt{x}} - \dfrac{\sqrt{x}}{\sqrt{x + 2}}}{2(x + 2)}$

24. Rationalize the denominator: $\dfrac{3y}{\sqrt{y^2 + 9}}$

25. Rationalize the numerator: $\dfrac{\sqrt{x} + \sqrt{x + 7}}{14}$

Graphing Calculator Required

26. Use a graphing calculator to find the real solutions of $x^3 - 5x^2 + 2x + 8 = 0$ by graphing $y = x^3 - 5x^2 + 2x + 8$ and finding the x-intercepts.

CHAPTER 1
Functions, Graphs, and Limits

CHAPTER 1
Functions, Graphs, and Limits

Section 1.1 The Cartesian Plane and the Distance Formula

Solutions to Odd-Numbered Exercises

1. (a) $a = 4$

$b = 3$

$c = \sqrt{(4-0)^2 + (3-0)^2} = 5$

(b) $a^2 + b^2 = 16 + 9 = 25 = c^2$

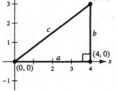

3. (a) $a = 10$

$b = 3$

$c = \sqrt{(7+3)^2 + (4-1)^2} = \sqrt{109}$

(b) $a^2 + b^2 = 100 + 9 = 109 = c^2$

5. (a) $a = 1 - (-3) = 4$

$b = 3 - (-2) = 5$

$c = \sqrt{(1-(-3))^2 + (-2-3)^2} = \sqrt{16 + 25} = \sqrt{41}$

(b) $a^2 + b^2 = 4^2 + 5^2 = 41 = c^2$

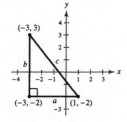

7. (a)

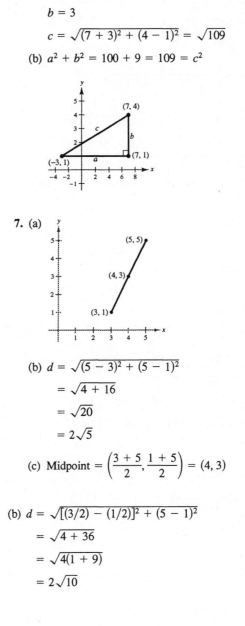

(b) $d = \sqrt{(5-3)^2 + (5-1)^2}$

$= \sqrt{4 + 16}$

$= \sqrt{20}$

$= 2\sqrt{5}$

(c) Midpoint $= \left(\dfrac{3+5}{2}, \dfrac{1+5}{2}\right) = (4, 3)$

9. (a)

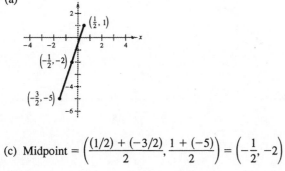

(b) $d = \sqrt{[(3/2) - (1/2)]^2 + (5-1)^2}$

$= \sqrt{4 + 36}$

$= \sqrt{4(1+9)}$

$= 2\sqrt{10}$

(c) Midpoint $= \left(\dfrac{(1/2) + (-3/2)}{2}, \dfrac{1 + (-5)}{2}\right) = \left(-\dfrac{1}{2}, -2\right)$

11. (a)

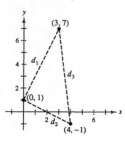

(b) $d = \sqrt{(4-2)^2 + (14-2)^2}$

$= \sqrt{4 + 144}$

$= \sqrt{4(1 + 36)}$

$= 2\sqrt{37}$

(c) Midpoint $= \left(\dfrac{2+4}{2}, \dfrac{2+14}{2}\right) = (3, 8)$

13. (a)

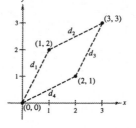

(b) $d = \sqrt{(-1-1)^2 + \left(1 - \sqrt{3}\right)^2}$

$= \sqrt{8 - 2\sqrt{3}}$

(c) Midpoint $= \left(\dfrac{1 + (-1)}{2}, \dfrac{\sqrt{3} + 1}{2}\right) = \left(0, \dfrac{\sqrt{3}+1}{2}\right)$

15. $d_1 = \sqrt{(3-0)^2 + (7-1)^2} = \sqrt{45} = 3\sqrt{5}$

$d_2 = \sqrt{(4-0)^2 + (-1-1)^2} = \sqrt{20} = 2\sqrt{5}$

$d_3 = \sqrt{(3-4)^2 + [7 - (-1)]^2} = \sqrt{65}$

Since $d_1{}^2 + d_2{}^2 = d_3{}^2$, the triangle is a right triangle.

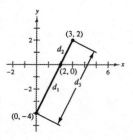

17. $d_1 = \sqrt{(1-0)^2 + (2-0)^2} = \sqrt{5}$

$d_2 = \sqrt{(3-1)^2 + (3-2)^2} = \sqrt{5}$

$d_3 = \sqrt{(2-3)^2 + (1-3)^2} = \sqrt{5}$

$d_4 = \sqrt{(0-2)^2 + (0-1)^2} = \sqrt{5}$

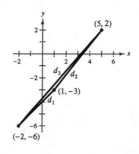

19. $d_1 = \sqrt{(2-0)^2 + (0+4)^2} = \sqrt{20} = 2\sqrt{5}$

$d_2 = \sqrt{(3-2)^2 + (2-0)^2} = \sqrt{5}$

$d_3 = \sqrt{(3-0)^2 + (2+4)^2} = \sqrt{45} = 3\sqrt{5}$

Since $d_1 + d_2 = d_3$, the points are collinear.

21. $d_1 = \sqrt{[1-(-2)]^2 + [-3-(-6)]^2} = \sqrt{18} = 3\sqrt{2}$

$d_2 = \sqrt{(5-1)^2 + [2-(-3)]^2} = \sqrt{41}$

$d_3 = \sqrt{[5-(-2)]^2 + [2-(-6)]^2} = \sqrt{113}$

Since $d_1 + d_2 \neq d_3$, the points are not collinear.

23. $d = \sqrt{(x-1)^2 + (-4-0)^2} = 5$

$\sqrt{x^2 - 2x + 17} = 5$

$x^2 - 2x + 17 = 25$

$x^2 - 2x - 8 = 0$

$(x-4)(x+2) = 0$

$x = 4, -2$

25. $d = \sqrt{(3-0)^2 + (y-0)^2} = 8$

$\sqrt{9 + y^2} = 8$

$9 + y^2 = 64$

$y^2 = 55$

$y = \pm\sqrt{55}$

27. Midpoint $= \left(\dfrac{x_1 + x_2}{2}, \dfrac{y_1 + y_2}{2} \right)$

The point one-fourth of the way between (x_1, y_1) and (x_2, y_2) is the midpoint of the line segment from (x_1, y_1) to

$$\left(\frac{x_1 + x_2}{2}, \frac{y_1 + y_2}{2} \right),$$

which is

$$\left(\frac{x_1 + \dfrac{x_1 + x_2}{2}}{2}, \frac{y_1 + \dfrac{y_1 + y_2}{2}}{2} \right) = \left(\frac{3x_1 + x_2}{4}, \frac{3y_1 + y_2}{4} \right).$$

The point three-fourths of the way between (x_1, y_1) and (x_2, y_2) is the midpoint of the line segment from

$$\left(\frac{x_1 + x_2}{2}, \frac{y_1 + y_2}{2} \right)$$

to (x_2, y_2), which is

$$\left(\frac{\dfrac{x_1 + x_2}{2} + x_2}{2}, \frac{\dfrac{y_1 + y_2}{2} + y_2}{2} \right) = \left(\frac{x_1 + 3x_2}{4}, \frac{y_1 + 3y_2}{4} \right).$$

Thus,

$$\left(\frac{3x_1 + x_2}{4}, \frac{3y_1 + y_2}{4} \right), \left(\frac{x_1 + x_2}{2}, \frac{y_1 + y_2}{2} \right), \text{ and } \left(\frac{x_1 + 3x_2}{4}, \frac{y_1 + 3y_2}{4} \right)$$

are the three points that divide the line segment joining (x_1, y_1) and (x_2, y_2) into four equal parts.

29. (a) $\left(\dfrac{3(1) + 4}{4}, \dfrac{3(-2) - 1}{4} \right) = \left(\dfrac{7}{4}, -\dfrac{7}{4} \right)$

$\left(\dfrac{1 + 4}{2}, \dfrac{-2 - 1}{2} \right) = \left(\dfrac{5}{2}, -\dfrac{3}{2} \right)$

$\left(\dfrac{1 + 3(4)}{4}, \dfrac{-2 + 3(-1)}{4} \right) = \left(\dfrac{13}{4}, -\dfrac{5}{4} \right)$

(b) $\left(\dfrac{3(-2) + 0}{4}, \dfrac{3(-3) + 0}{4} \right) = \left(-\dfrac{3}{2}, -\dfrac{9}{4} \right)$

$\left(\dfrac{-2 + 0}{2}, \dfrac{-3 + 0}{2} \right) = \left(-1, -\dfrac{3}{2} \right)$

$\left(\dfrac{-2 + 3(0)}{4}, \dfrac{-3 + 3(0)}{4} \right) = \left(-\dfrac{1}{2}, -\dfrac{3}{4} \right)$

31. (a) $d^2 = 16^2 + 5^2, \quad d > 0$

$d^2 = 281$

$d = \sqrt{281} \approx 16.76 \text{ feet}$

(b) $A = 2(40)\left(\sqrt{281} \right)$

$= 80\sqrt{281}$

$\approx 1341.04 \text{ square feet}$

33.

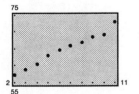

Answers will vary. The number of subscribers appears to be increasing linearly.

35. (a) 10,400

(b) 8,900

(c) 8,500

(d) 10,500

37. (a) $85 thousand

(b) $100 thousand

(c) $122 thousand

(d) $159 thousand

39. (a) Revenue midpoint $= \left(\dfrac{1999 + 2003}{2}, \dfrac{17{,}839 + 32{,}505}{2} \right)$

$$= (2001, 25{,}172)$$

Revenue estimate for 2001: $25,172 million

Profit midpoint $= \left(\dfrac{1999 + 2003}{2}, \dfrac{624.1 + 1157.3}{2} \right)$

$$= (2001, 890.7)$$

Profit estimate for 2001: $890.7 million

(b) Actual 2001 revenue: $24,623 million

   Actual 2001 profit: $885.6 million

(c) Yes, the increase in revenue is approximately $3667 million per year.

   Yes , the increase in profit is approximately $133 million per year.

(d) 1999 Expenses: $17{,}839 - 624.1 = \$17{,}214.9$ million

   2001 Expenses: $24{,}623 - 885.6 = \$23{,}737.4$ million

   2003 Expenses: $32{,}505 - 1157.3 = \$31{,}347.7$ million

(e) Answers will vary.

41. (a) $(0, 0)$ is translated to $(0 + 2, 0 + 3) = (2, 3)$.

   $(-3, -1)$ is translated to $(-3 + 2, -1 + 3) = (-1, 2)$.

   $(-1, -2)$ is translated to $(-1 + 2, -2 + 3) = (1, 1)$.

(b)

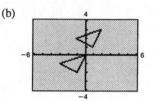

43. (a)

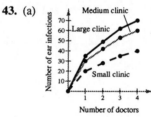

(b) The larger the clinic, the more patients a doctor can handle.

Section 1.2   Graphs of Equations

1. (a) This is not a solution point since $2x - y - 3 = 2(1) - 2 - 3 = -3 \neq 0$.

(b) This is a solution point since $2x - y - 3 = 2(1) - (-1) - 3 = 0$.

(c) This is a solution point since $2x - y - 3 = 2(4) - 5 - 3 = 0$.

3. (a) This is a solution point since $(1)^2 + \left(-\sqrt{3} \right)^2 = 4$.

(b) This is not a solution point since $\left(\tfrac{1}{2} \right)^2 + (-1)^2 = \tfrac{5}{4} \neq 4$.

(c) This is not a solution point since $\left(\tfrac{3}{2} \right)^2 + \left(\tfrac{7}{2} \right)^2 = \tfrac{29}{2} \neq 4$.

5. (a) This is not a solution point since $(0)^2 - (0)(2) + 4(2) = 8 \neq 3$.

(b) This is a solution point since $(-2)^2 - (-2)\left(-\tfrac{1}{6} \right) + 4\left(-\tfrac{1}{6} \right) = 3$.

(c) This is a solution point since $(3)^2 - (3)(-6) + 4(-6) = 3$.

7. The graph of $y = x - 2$ is a straight line with y-intercept at $(0, -2)$. Thus, it matches (e).

9. The graph of $y = x^2 + 2x$ is a parabola opening up with vertex at $(-1, -1)$. Thus, it matches (c).

11. The graph of $y = |x| - 2$ has a y-intercept at $(0, -2)$ and has x-intercepts at $(-2, 0)$ and $(2, 0)$. Thus, it matches (a).

13. To find the y-intercept, let $x = 0$ to obtain

$$2(0) - y - 3 = 0$$
$$y = -3.$$

Thus, the y-intercept is $(0, -3)$. To find the x-intercept, let $y = 0$ to obtain

$$2x - (0) - 3 = 0$$
$$x = \tfrac{3}{2}.$$

Thus, the x-intercept is $\left(\tfrac{3}{2}, 0\right)$.

15. The y-intercept occurs at $(0, -2)$. To find the x-intercepts, let $y = 0$ to obtain

$$x^2 + x - 2 = 0$$
$$(x + 2)(x - 1) = 0$$
$$x = -2, 1.$$

Thus, the x-intercepts are $(-2, 0)$ and $(1, 0)$.

17. The y-intercept occurs at $(0, 0)$. To find the x-intercepts, let $y = 0$ to obtain

$$x^2\sqrt{9 - x^2} = 0$$
$$x = 0, \pm 3.$$

Thus, the x-intercepts are $(0, 0)$, $(-3, 0)$, and $(3, 0)$.

19. The y-intercept occurs at $(0, 2)$. The x-intercept occurs when the numerator equals zero and the denominator does not equal zero. Thus, $(-2, 0)$ is the only x-intercept.

21. Let $x = 0$. Then $4y = 0 \Longrightarrow y = 0$.

Let $y = 0$. Then $-x^2 = 0 \Longrightarrow x = 0$.

The x-intercept and y-intercept both occur at $(0, 0)$.

23. $y = 2x + 3$ is a line with intercepts $(0, 3)$ and $\left(-\tfrac{3}{2}, 0\right)$.

x	-2	$-\tfrac{3}{2}$	-1	0	1	2
y	-1	0	1	3	5	7

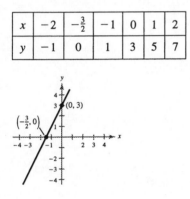

25. $y = x^2 - 3$

x	-2	-1	0	1	2	3
y	1	-2	-3	-2	1	6

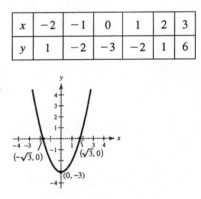

Intercepts: $(0, -3), \left(\pm\sqrt{3}, 0\right)$

27. $y = (x - 1)^2$

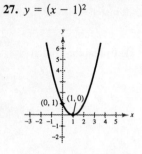

Intercepts: $(0, 1), (1, 0)$

29.

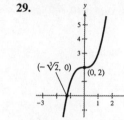

Intercepts: $\left(-\sqrt[3]{2}, 0\right)$ and $(0, 2)$

x	-2	-1	0	1	2
y	-6	1	2	3	10

31. $y = -\sqrt[3]{x + 1}$ has intercepts $(0, -1)$ and $(-1, 0)$.

x	-9	-1	0	7
y	2	0	-1	-2

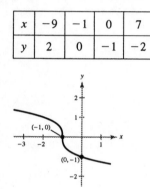

33. $y = |x + 1|$

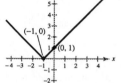

Intercepts: $(0, 1), (-1, 0)$

35.

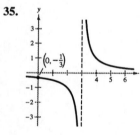

x	-1	0	1	2	2.5	3.5	4	5	6
y	$-\frac{1}{4}$	$-\frac{1}{3}$	$-\frac{1}{2}$	-1	-2	2	1	$\frac{1}{2}$	$\frac{1}{3}$

37.

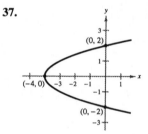

The graph is a parabola with vertex at $(-4, 0)$ and intercepts at $(-4, 0)$, $(0, 2)$, and $(0, -2)$.

x	5	0	-3	-4
y	± 3	± 2	± 1	0

39. $(x - 0)^2 + (y - 0)^2 = 3^2$

$$x^2 + y^2 = 9$$

$$x^2 + y^2 - 9 = 0$$

41. $(x - 2)^2 + (y + 1)^2 = 4^2$

$$x^2 - 4x + 4 + y^2 + 2y + 1 = 16$$

$$x^2 + y^2 - 4x + 2y - 11 = 0$$

43. Since the point $(0, 0)$ lies on the circle, the radius must be the distance between $(0, 0)$ and $(-1, 2)$.

$$\text{Radius} = \sqrt{(0 + 1)^2 + (0 - 2)^2} = \sqrt{5}$$

$$(x + 1)^2 + (y - 2)^2 = 5$$

$$x^2 + y^2 + 2x - 4y = 0$$

45. Center = midpoint = $(0, 3)$

Radius = distance from center to an endpoint

$$= \sqrt{(0 - 3)^2 + (3 - 3)^2} = 3$$

$$(x - 0)^2 + (y - 3)^2 = 3^2$$

$$x^2 + y^2 - 6y = 0$$

47. $(x^2 - 2x + 1) + (y^2 + 6y + 9) = -6 + 1 + 9$

$$(x - 1)^2 + (y + 3)^2 = 4$$

Center: $(1, -3)$

Radius: 2

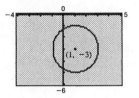

49. $(x^2 + 4x + 4) + (y^2 + 6y + 9) = 3 + 4 + 9$

$$(x + 2)^2 + (y + 3)^2 = 16$$

Center: $(-2, -3)$

Radius: 4

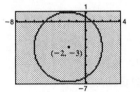

51.
$$x^2 + y^2 - x - y = \tfrac{3}{2}$$

$$\left(x^2 - x + \tfrac{1}{4}\right) + \left(y^2 - y + \tfrac{1}{4}\right) = \tfrac{3}{2} + \tfrac{1}{4} + \tfrac{1}{4}$$

$$\left(x - \tfrac{1}{2}\right)^2 + \left(y - \tfrac{1}{2}\right)^2 = 2$$

Center: $\left(\tfrac{1}{2}, \tfrac{1}{2}\right)$

Radius: $\sqrt{2}$

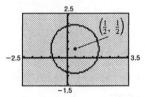

53.
$$x^2 + y^2 + x + \tfrac{5}{2}y = \tfrac{7}{16}$$

$$\left(x^2 + x + \tfrac{1}{4}\right) + \left(y^2 + \tfrac{5}{2}y + \tfrac{25}{16}\right) = \tfrac{7}{16} + \tfrac{1}{4} + \tfrac{25}{16}$$

$$\left(x + \tfrac{1}{2}\right)^2 + \left(y + \tfrac{5}{4}\right)^2 = \tfrac{9}{4}$$

Center: $\left(-\tfrac{1}{2}, -\tfrac{5}{4}\right)$

Radius: $\tfrac{3}{2}$

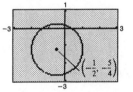

55. Solving for y in the equation $x + y = 2$ yields $y = 2 - x$, and solving for y in the equation $2x - y = 1$ yields $y = 2x - 1$. Then setting these two y-values equal to each other, we have

$$2 - x = 2x - 1$$

$$3 = 3x$$

$$x = 1.$$

The corresponding y-value is $y = 2 - 1 = 1$, so the point of intersection is $(1, 1)$.

57. Solving for y in the second equation yields $y = 10 - 2x$ and substituting this into the first equation gives

$$x^2 + (10 - 2x)^2 = 25$$

$$x^2 + 100 + 4x^2 - 40x = 25$$

$$5x^2 - 40x + 75 = 0$$

$$x^2 - 8x + 15 = 0$$

$$(x - 3)(x - 5) = 0$$

$$x = 3, 5.$$

The corresponding y-values are $y = 4$ and $y = 0$, so the points of intersection are $(3, 4)$ and $(5, 0)$.

59. By equating the y-values for the two equations, we have

$$x^3 = 2x$$

$$x^3 - 2x = 0$$

$$x(x^2 - 2) = 0$$

$$x = 0, \pm\sqrt{2}.$$

The corresponding y-values are $y = 0$, $y = -2\sqrt{2}$, and $y = 2\sqrt{2}$, so the points of intersection are $(0, 0)$, $\left(-\sqrt{2}, -2\sqrt{2}\right)$, and $\left(\sqrt{2}, 2\sqrt{2}\right)$.

61. By equating the y-values for the two equations, we have

$$x^4 - 2x^2 + 1 = 1 - x^2$$

$$x^4 - x^2 = 0$$

$$x^2(x + 1)(x - 1) = 0$$

$$x = 0, \pm 1.$$

The corresponding y-values are $y = 1$, 0, and 0, so the points of intersection are $(-1, 0)$, $(0, 1)$, and $(1, 0)$.

63. (a) $C = 11.8x + 15,000$

$R = 19.30x$

(b) $C = R$

$$11.8x + 15,000 = 19.30x$$

$$15,000 = 7.5x$$

$$x = 2000 \text{ units}$$

(c) $P = R - C$

$$1000 = 19.3x - (11.8x + 15,000)$$

$$16,000 = 7.5x$$

$$x \approx 2133.3$$

Hence, 2134 units.

65. $R = C$

$$1.55x = 0.85x + 35,000$$

$$0.7x = 35,000$$

$$x = \frac{35,000}{0.7} = 50,000 \text{ units}$$

67. $R = C$

$$9950x = 8650x + 250,000$$

$$1300x = 250,000$$

$$x = \frac{250,000}{1300} \approx 193 \text{ units}$$

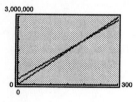

69. $p = 180 - 4x$

$p = 75 + 3x$

$$180 - 4x = 75 + 3x$$

$$105 = 7x$$

$$x = 15 \text{ units, in thousands}$$

$(15, 120)$

71. (a) Model: $y = 2.177t^3 - 41.99t^2 + 497.1 + 985$

($t = 5$ correponds to 1995)

t	5	6	7	8	9	10	11	12
Model	2693	2926	3154	3389	3645	3934	4270	4665
Exact	2708	2920	3110	3365	3773	3905	4187	4706

The model is a good fit. Answers will vary.

(b) For 2010, $t = 20$ and $y \approx \$11,547$ million.

73. Model: $y = \dfrac{292.48 + 37.72t}{1 + 0.02t}$

($t = 7$ corresponds to 1997)

(a)

Year	1997	1998	1999	2000	2001	2004
Salary	488.18	512.28	535.56	558.07	579.84	641.06

(b) Answers will vary.

(c) For 2006, $t = 16$ and $y \approx \$678.79$.

Yes, this prediction seems reasonable.

75.

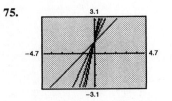

The greater the value of c, the steeper the line.

77. Intercept: $(0, 5.36)$

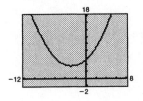

79. Intercepts: $(1.4780, 0), (12.8553, 0), (0, 2.3875)$

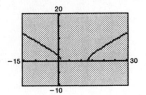

81. Intercepts: $\left(0, \frac{5}{12}\right) \approx (0, 0.4167)$

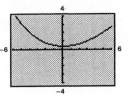

Section 1.3   Lines in the Plane and Slope

1. The slope is $m = 1$ since the line rises one unit vertically for each unit of horizontal change from left to right.

3. The slope is $m = 0$ since the line is horizontal.

5. The points are plotted in the accompanying graph and the slope is

$$m = \frac{2 - (-4)}{5 - 3} = 3.$$

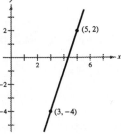

7. The points are plotted in the accompanying graph and the slope is

$$m = \frac{2 - 2}{6 - (1/2)} = 0.$$

Thus, the line is horizontal.

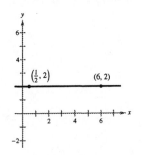

9. The points are plotted in the accompanying graph. The slope is undefined since

$$m = \frac{-5 - (-3)}{-8 - (-8)}. \quad \text{(undefined slope)}$$

Thus, the line is vertical.

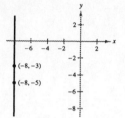

11. The points are plotted in the accompanying graph and the slope is

$$m = \frac{-3 - 1}{4 - (-2)} = \frac{-4}{6} = \frac{-2}{3}.$$

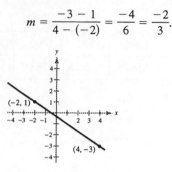

13. The points are plotted in the accompanying graph and the slope is

$$m = \frac{1 - (-2)}{-\frac{3}{8} - \frac{1}{4}} = \frac{3}{-\frac{5}{8}} = \frac{-24}{5}.$$

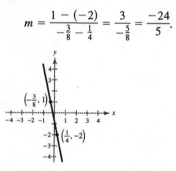

15. The points are plotted in the accompanying graph and the slope is

$$m = \frac{\frac{5}{2} - \left(-\frac{5}{6}\right)}{\frac{2}{3} - \frac{1}{4}} = \frac{\frac{10}{3}}{\frac{5}{12}} = \frac{10}{3} \cdot \frac{12}{5} = 8.$$

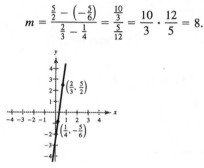

17. The equation of this horizontal line is $y = 1$. Therefore, three additional points are $(0, 1)$, $(1, 1)$, and $(3, 1)$.

19. The equation of this line is

$$y + 4 = \frac{2}{3}(x - 6)$$

$$y = \frac{2}{3}x - 8.$$

Therefore, three additional points are $(3, -6)$, $(9, -2)$, and $(12, 0)$.

21. The equation of the line is

$$y - 7 = -3(x - 1)$$

$$y = -3x + 10.$$

Therefore, three additional points are $(0, 10)$, $(2, 4)$, and $(3, 1)$.

23. The equation of this vertical line is $x = -8$. Therefore, three additional points are $(-8, 0)$, $(-8, 2)$, and $(-8, 3)$.

25. $x + 5y = 20$

$$y = -\frac{1}{5}x + 4$$

Therefore, the slope is $m = -\frac{1}{5}$ and the y-intercept is $(0, 4)$.

27. $7x - 5y = 15$

$$y = \frac{7}{5}x - 3$$

Therefore, the slope is $m = \frac{7}{5}$ and the y-intercept is $(0, -3)$.

29. $3x - y = 15$

$$y = 3x - 15$$

Therefore, the slope is 3 and the y-intercept is $(0, -15)$.

31. Since the line is vertical, the slope is undefined and there is no y-intercept.

33. $y - 4 = 0$

$$y = 4$$

Therefore, the slope is 0 and the y-intercept is $(0, 4)$.

35. The slope of the line is

$$m = \frac{3 - (-5)}{4 - 0} = 2.$$

Using the point-slope form, we have

$$y + 5 = 2(x - 0)$$
$$y = 2x - 5.$$

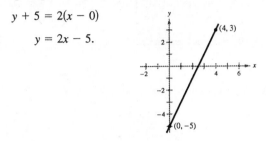

37. The slope of the line is

$$m = \frac{3 - 0}{-1 - 0} = -3.$$

Using the point-slope form, we have

$$y = -3x$$
$$3x + y = 0.$$

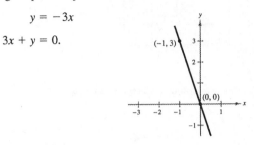

39. The slope of the line is undefined, so the line is vertical and its equation is

$$x = 2$$
$$x - 2 = 0.$$

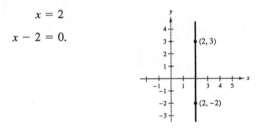

41. The slope of the line is $m = 0$, so the line is horizontal and its equation is

$$y = -1$$
$$y + 1 = 0.$$

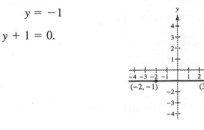

43. The slope of the line is

$$m = \frac{1 - 5/6}{(-1/3) + 2/3} = \frac{1/6}{1/3} = \frac{1}{2}$$

Using the point-slope form, we have

$$y - 1 = \frac{1}{2}\left(x + \frac{1}{3}\right)$$
$$y = \frac{1}{2}x + \frac{7}{6}$$
$$3x - 6y + 7 = 0.$$

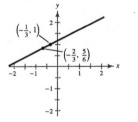

45. The slope of the line is

$$m = \frac{8 - 4}{1/2 + 1/2} = 4.$$

Using the point-slope form, we have

$$y - 8 = 4\left(x - \frac{1}{2}\right)$$
$$y = 4x + 6.$$

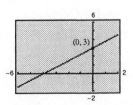

47. Using the slope-intercept form, we have

$$y = \tfrac{3}{4}x + 3$$
$$4y = 3x + 12$$
$$3x - 4y + 12 = 0.$$

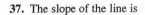

49. Since the slope is undefined, the line is vertical and its equation is $x = -1$.

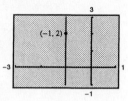

51. Since the slope is 0, the line is horizontal and its equation is $y = 7$.

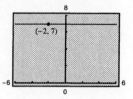

53. Using the point-slope form, we have

$$y + 2 = -4(x - 0)$$
$$y = -4x - 2$$
$$4x + y + 2 = 0.$$

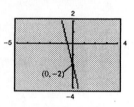

55. Using the slope-intercept form, we have

$$y = \tfrac{3}{4}x + \tfrac{2}{3}$$
$$12y = 9x + 8$$
$$9x - 12y + 8 = 0.$$

57. The slope of the line joining $(-2, 1)$ and $(-1, 0)$ is

$$\frac{1 - 0}{-2 - (-1)} = \frac{1}{-1} = -1.$$

The slope of the line joining $(-1, 0)$ and $(2, -2)$ is

$$\frac{0 - (-2)}{-1 - 2} = \frac{2}{-3} = -\frac{2}{3}.$$

Since the slopes are different, the points are not collinear.

$$d_1 = \sqrt{[-2 - (-1)]^2 + (1 - 0)^2} = \sqrt{1 + 1} = \sqrt{2} \approx 1.41421$$
$$d_2 = \sqrt{(-1 - 2)^2 + [0 - (-2)]^2} = \sqrt{9 + 4} = \sqrt{13} \approx 3.60555$$
$$d_3 = \sqrt{(-2 - 2)^2 + [1 - (-2)]^2} = \sqrt{16 + 9} = 5$$

Since $d_1 + d_2 \neq d_3$, the points are not collinear.

59. Since the line is vertical, it has an undefined slope and its equation is

$$x = 3$$
$$x - 3 = 0.$$

61. Slope $= 0$ and y-intercept at -10:

$$y = -10.$$

63. Given line: $y = -x + 7$, $m_1 = -1$

(a) Parallel; $m_1 = -1$

$$y - 2 = -1(x + 3)$$
$$x + y + 1 = 0$$

(b) Perpendicular; $m_2 = 1$

$$y - 2 = 1(x + 3)$$
$$x - y + 5 = 0$$

65. Given line: $y = -\frac{3}{4}x + \frac{7}{4}$, $m_1 = -\frac{3}{4}$

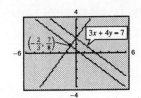

(a) Parallel; $m = -\frac{3}{4}$

$$y - \frac{7}{8} = -\frac{3}{4}\left(x + \frac{2}{3}\right) = -\frac{3}{4}x - \frac{1}{2}$$

$$8y - 7 = -6x - 4$$

$$8y + 6x - 3 = 0$$

(b) Perpendicular; $m = \frac{4}{3}$

$$y - \frac{7}{8} = \frac{4}{3}\left(x + \frac{2}{3}\right) = \frac{4}{3}x + \frac{8}{9}$$

$$72y - 63 = 96x + 64$$

$$96x - 72y + 127 = 0$$

67. Given line: $y = -3$ is horizontal and $m_1 = 0$

(a) Parallel: $y = 0$ or the x-axis

(b) Perpendicular: $x = -1$ or $x + 1 = 0$

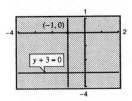

69. Given line: $x - 2 = 0$ is vertical

(a) Parallel: m is undefined, $x = 1$

(b) Perpendicular: $m = 0$, $y = 1$

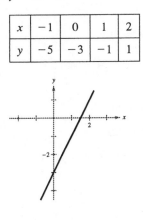

71. $y = -2$ is a horizontal line.

x	-2	-1	0	1
y	-2	-2	-2	-2

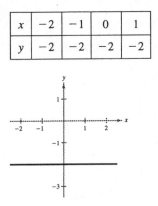

73. $y = 2x - 3$

x	-1	0	1	2
y	-5	-3	-1	1

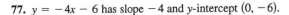

75.

x	-1	0	1	2
y	3	1	-1	-3

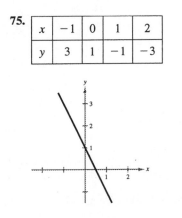

77. $y = -4x - 6$ has slope -4 and y-intercept $(0, -6)$.

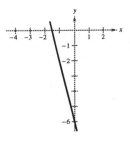

79. (a) $(1997, 3860), (2002, 4107)$

$$\frac{4107 - 3860}{2002 - 1997} = \frac{247}{5} = 49.4$$

$$4 - 3860 = 49.4(t - 7)$$

$$y = 49.4t + 3514.2$$

The slope $m = 49.4$ tells you that the population is increasing 49.4 thousand per year.

(b) For 1999, $t = 9$ and $y = 3958.8$ thousand (3,958,800).

(c) For 2001, $t = 11$ and $y = 4057.6$ thousand (4,057,600).

(d) 1999: 3975 thousand
2001: 4062 thousand

The estimates were close to the exact values.

(e) The model could possible be used to predict the population in 2006 if the population continues to grow at the same linear rate.

83. $C = 150 + 0.34x$

85. (a) $(0, -45), (100, 115)$

$$C + 45 = \frac{115 + 45}{100 - 0}(A - 0)$$

$$C = \tfrac{8}{5}A - 45 \text{ or } A = \tfrac{5}{8}(C + 45)$$

(b) $C = \tfrac{5}{9}(F - 32) = \tfrac{8}{5}A - 45$

$$25F - 800 = 72A - 2025$$

$$25F = 72A - 1225$$

$$F = \tfrac{72}{25}A - 49 \text{ or } A = \tfrac{25}{72}(F + 49)$$

(c) $C = \tfrac{8}{5}A - 45 = A$

$$\tfrac{3}{5}A = 45$$

$$A = 75°$$

(d) $C = \tfrac{8}{5}(86) - 45 = 92.6°$;

$$F = \tfrac{72}{25}(86) - 49 = 198.68°$$

(e) $A = \tfrac{5}{8}(45 + 45) = 56.25°$

89. (a) The slope is $\dfrac{47 - 50}{425 - 380} = \dfrac{-3}{45} = -\dfrac{1}{15}$. Hence,

$$x - 50 = -\frac{1}{15}(p - 380)$$

$$x = -\frac{1}{15}p + \frac{226}{3}.$$

(b) If $p = 455$, $x = -\dfrac{1}{15}(455) + \dfrac{226}{3} = 45$ units.

(c) If $p = 395$, $x = -\dfrac{1}{15}(395) + \dfrac{226}{3} = 49$ units.

81. $(0, 32), (100, 212)$

$$F - 32 = \frac{212 - 32}{100 - 0}(C - 0)$$

$$F = 1.8C + 32 = \tfrac{9}{5}C + 32$$

or

$$C = \tfrac{5}{9}(F - 32)$$

87. (a) The equipment depreciates

$$\frac{1025}{5} = \$205 \text{ per year,}$$

so the value is $y = 1025 - 205t$, where $0 \le t \le 5$.

(b)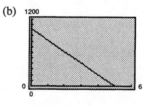

(c) When $t = 3$, the value is $410.00.

(d) The value is $600 when $t = 2.07$ years.

91. (a) $Y - 6937 = \dfrac{8685 - 6937}{11 - 7}(t - 7)$

$$Y = 437t + 3878$$

($t = 0$ corresponds to 1990)

(b) For 1999, $t = 9$ and $Y = \$7811$ billion.

(c) For 2002, $t = 12$ and $Y = \$9122$ billion.

(d) 1999: $7786.5 billion
2002: $8929.1 billion

93. $23,500 + 3100x \leq 100,000$

$3100x \leq 76,500$

$x \leq 24.677$

$x \leq 24$ units or 24.67 units if
fractional units are allowed.

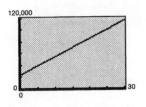

95. $C = 18,375 + 1150x \leq 100,000$

$1150x \leq 81,625$

$x \leq 70.978$

$x \leq 70$ units

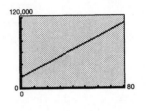

97. $C = 75,500 + 89x \leq 100,000$

$89x \leq 24,500$

$x \leq 275.28$

$x \leq 275$ units

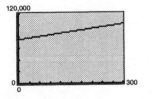

99. $C = 32,000 + 650x \leq 100,000$

$650x \leq 68,000$

$x \leq 104.62$

$x \leq 104$ units

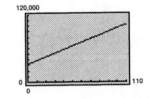

101. $C = 50,000 + 0.25x \leq 100,000$

$0.25x \leq 50,000$

$x \leq 200,000$ units

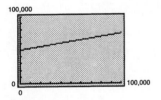

Section 1.4   Functions

1. $y = \pm\sqrt{4 - x^2}$

y is *not* a function of x since there are two values of y for some x.

3. $\frac{1}{2}x - 6y = -3$

$y = \frac{1}{12}x + \frac{1}{2}$

y *is* a function of x since there is only one value of y for each x.

5. $y = 4 - x^2$

y *is* a function of x since there is only one value of y for each x.

7. $y = \pm\sqrt{x^2 - 1}$

y is *not* a function of x since there are two values of y for some x.

9. Domain: $(-\infty, \infty)$

Range: $[-2.125, \infty)$

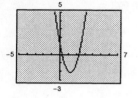

11. Domain: $(-\infty, 0) \cup (0, \infty)$

Range: $\{-1, 1\}$

13. Domain: $(4, \infty)$

Range: $[4, \infty)$

15. Domain: $(-\infty, -4) \cup (-4, \infty)$

Range: $(-\infty, 1) \cup (1, \infty)$

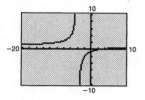

17. Domain: $(-\infty, \infty)$

Range: $(-\infty, \infty)$

19. Domain: $(-\infty, \infty)$

Range: $(-\infty, 4]$

21. $f(x) = 2x - 3$

(a) $f(0) = 2(0) - 3 = -3$

(b) $f(-3) = 2(-3) - 3 = -9$

(c) $f(x - 1) = 2(x - 1) - 3 = 2x - 5$

(d) $f(x + \Delta x) = 2(x + \Delta x) - 3 = 2x + 2\Delta x - 3$

23. $g(x) = \dfrac{1}{x}$

(a) $g(2) = \dfrac{1}{2}$

(b) $g\left(\dfrac{1}{4}\right) = \dfrac{1}{1/4} = 4$

(c) $g(x + 4) = \dfrac{1}{x + 4}$

(d) $g(x + \Delta x) - g(x) = \dfrac{1}{x + \Delta x} - \dfrac{1}{x}$

$$= \dfrac{x - (x + \Delta x)}{x(x + \Delta x)} = \dfrac{-\Delta x}{x(x + \Delta x)}$$

25. $f(x) = x^2 - 4x + 1$

$$\dfrac{f(x + \Delta x) - f(x)}{\Delta x} = \dfrac{[(x + \Delta x)^2 - 4(x + \Delta x) + 1] - [x^2 - 4x + 1]}{\Delta x}$$

$$= \dfrac{[x^2 + 2x\Delta x + (\Delta x)^2 - 4x - 4\Delta x - x^2 + 4x]}{\Delta x}$$

$$= \dfrac{2x\Delta x + (\Delta x)^2 - 4\Delta x}{\Delta x}$$

$$= 2x - 4 + \Delta x, \qquad \Delta x \neq 0$$

27. $\dfrac{g(x + \Delta x) - g(x)}{\Delta x} = \dfrac{\sqrt{x + \Delta x + 3} - \sqrt{x + 3}}{\Delta x} \cdot \dfrac{\sqrt{x + \Delta x + 3} + \sqrt{x + 3}}{\sqrt{x + \Delta x + 3} + \sqrt{x + 3}}$

$$= \dfrac{(x + \Delta x + 3) - (x + 3)}{\Delta x\left[\sqrt{x + \Delta x + 3} + \sqrt{x + 3}\right]}$$

$$= \dfrac{1}{\sqrt{x + \Delta x + 3} + \sqrt{x + 3}}, \qquad \Delta x \neq 0$$

29. $\dfrac{f(x + \Delta x) - f(x)}{\Delta x} = \dfrac{\dfrac{1}{x + \Delta x - 2} - \dfrac{1}{x - 2}}{\Delta x}$

$$= \dfrac{(x - 2) - (x + \Delta x - 2)}{(x + \Delta x - 2)(x - 2)\Delta x}$$

$$= \dfrac{-1}{(x + \Delta x - 2)(x - 2)}, \quad \Delta x \neq 0$$

31. *y* is *not* a function of *x*.

33. *y is* a function of *x*.

35. (a) $f(x) + g(x) = (2x - 5) + 5 = 2x$

   (b) $f(x) \cdot g(x) = (2x - 5)(5) = 10x - 25$

   (c) $\dfrac{f(x)}{g(x)} = \dfrac{2x - 5}{5} = \dfrac{2}{5}x - 1$

   (d) $f(g(x)) = f(5) = 2(5) - 5 = 5$

   (e) $g(f(x)) = g(2x - 5) = 5$

37. (a) $f(x) + g(x) = (x^2 + 1) + (x - 1) = x^2 + x$

   (b) $f(x) \cdot g(x) = (x^2 + 1)(x - 1) = x^3 - x^2 + x - 1$

   (c) $\dfrac{f(x)}{g(x)} = \dfrac{x^2 + 1}{x - 1}$

   (d) $f(g(x)) = f(x - 1) = (x - 1)^2 + 1 = x^2 - 2x + 2$

   (e) $g(f(x)) = g(x^2 + 1) = (x^2 + 1) - 1 = x^2$

39. (a) $f(x) + g(x) = \dfrac{1}{x} + \dfrac{1}{x^2} = \dfrac{x + 1}{x^2}$

   (b) $f(x) \cdot g(x) = \dfrac{1}{x} \cdot \dfrac{1}{x^2} = \dfrac{1}{x^3}$

   (c) $\dfrac{f(x)}{g(x)} = \dfrac{1/x}{1/x^2} = x, \quad x \neq 0$

   (d) $f(g(x)) = f(1/x^2) = x^2. \quad x \neq 0$

   (e) $g(f(x)) = g(1/x) = x^2, \quad x \neq 0$

41. $f(x) = \sqrt{x}, \quad g(x) = x^2 - 1$

   (a) $f(g(1)) = f(1^2 - 1) = f(0) = 0$

   (b) $g(f(1)) = g(\sqrt{1}) = g(1) = 0$

   (c) $g(f(0)) = g(0) = -1$

   (d) $f(g(-4)) = f(15) = \sqrt{15}$

   (e) $f(g(x)) = f(x^2 - 1) = \sqrt{x^2 - 1}$

   (f) $g(f(x)) = g(\sqrt{x}) = x - 1, \quad x \geq 0$

43. The data fits the function (b) $g(x) = -2x^2$ with $c = -2$.

45. The data fits the function (d) $r(x) = \dfrac{32}{x}$ with $c = 32$.

47. $f(g(x)) = f\left(\dfrac{x - 1}{5}\right) = 5\left(\dfrac{x - 1}{5}\right) + 1 = x$

$g(f(x)) = g(5x + 1) = \dfrac{(5x + 1) - 1}{5} = x$

49. $f(g(x)) = f(\sqrt{9 - x}) = 9 - (\sqrt{9 - x})^2 = x$

$g(f(x)) = g(9 - x^2)$

$$= \sqrt{9 - (9 - x^2)}$$

$$= \sqrt{x^2}$$

$$= x, \quad x \geq 0$$

51. $f(x) = 2x - 3 = y$

   $2y - 3 = x$

   $y = \dfrac{x + 3}{2}$

   $f^{-1}(x) = \dfrac{x + 3}{2}$

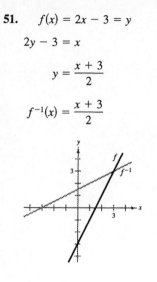

53. $f(x) = x^5 = y$

   $x = y^5$

   $y = \sqrt[5]{x}$

   $f^{-1}(x) = \sqrt[5]{x}$

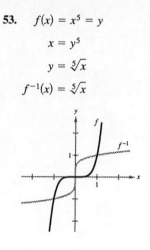

55. $f(x) = \sqrt{9 - x^2} = y,\quad 0 \le x \le 3$

   $x = \sqrt{9 - y^2}$

   $x^2 = 9 - y^2$

   $y^2 = 9 - x^2$

   $y = \sqrt{9 - x^2}$

   $f^{-1}(x) = \sqrt{9 - x^2},\quad 0 \le x \le 3$

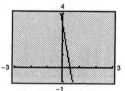

57. $f(x) = x^{2/3} = y,\quad x \ge 0$

   $x = y^{2/3}$

   $y = x^{3/2}$

   $f^{-1}(x) = x^{3/2}$

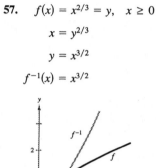

59. $f(x) = 3 - 7x$ is one-to-one.

   $y = 3 - 7x$

   $x = 3 - 7y$

   $y = \dfrac{3 - x}{7}$

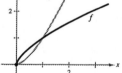

61. $f(x) = x^2 = y$

   f is *not* one-to-one since $f(1) = 1 = f(-1)$.

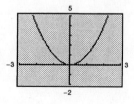

63. $f(x) = |x - 2| = y$

   f is *not* one-to-one since $f(0) = 2 = f(4)$.

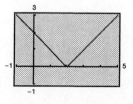

65. (a) $y = \sqrt{x} + 2$ (b) $y = -\sqrt{x}$ (c) $y = \sqrt{x - 2}$

(d) $y = \sqrt{x} + 3$ (e) $y = \sqrt{x - 4}$ (f) $y = 2\sqrt{x}$

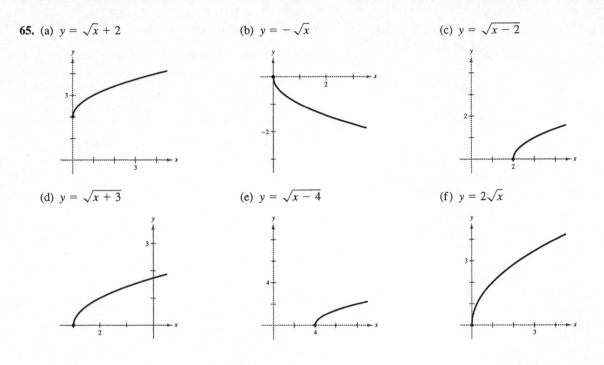

67. (a) Shifted three units to the left: $y = (x + 3)^2$

(b) Shifted three units upward: $y = x^2 + 3$

(c) Shifted three units to the right, six units upward, and reflected: $y = -(x - 3)^2 + 6$

(d) Shifted six units to the left, three units downward, and reflected: $y = -(x + 6)^2 - 3$

69. Total sales $= R = R_1 + R_2$

$$= (480 - 8t - 0.8t^2) + (254 + 0.78t)$$

$$= 734 - 7.22t - 0.8t^2, \quad t = 0, 1, 2, 3, 4, 5, 6$$

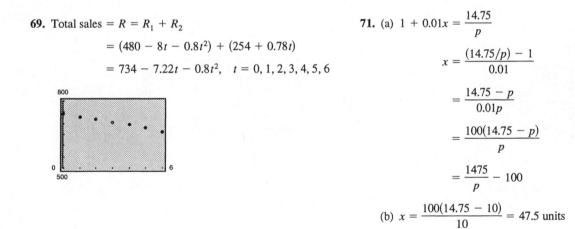

71. (a) $1 + 0.01x = \dfrac{14.75}{p}$

$$x = \frac{(14.75/p) - 1}{0.01}$$

$$= \frac{14.75 - p}{0.01p}$$

$$= \frac{100(14.75 - p)}{p}$$

$$= \frac{1475}{p} - 100$$

(b) $x = \dfrac{100(14.75 - 10)}{10} = 47.5$ units

73. $C(x) = 70x + 375$

$x(t) = 40t$

$C(x(t)) = C(40t) = 2800t + 375$

This is the weekly cost per t hours of production.

75. (a) If $0 \leq x \leq 100$, then $p = 90$. If $100 < x \leq 1600$, then $p = 90 - 0.01(x - 100) = 91 - 0.01x$. If $x > 1600$, then $p = 75$. Thus,

$$p = \begin{cases} 90, & 0 \leq x \leq 100 \\ 91 - 0.01x, & 100 < x \leq 1600 \\ 75, & x > 1600. \end{cases}$$

(b) $P = px - 60x$

$$P = \begin{cases} 90x - 60x, & 0 \leq x \leq 100 \\ (91 - 0.01x)x - 60x, & 100 < x \leq 1600 \\ 75x - 60x, & x > 1600 \end{cases}$$

$$= \begin{cases} 30x, & 0 \leq x \leq 100 \\ 31x - 0.01x^2, & 100 < x \leq 1600 \\ 15x, & x > 1600 \end{cases}$$

77. $r = 8 - 0.05(n - 80), \quad n \geq 80$

(a) Revenue $= R = rn = [8 - 0.05(n - 80)]n$

(b)

n	90	100	110	120	130	140	150
R	675	700	715	720	715	700	675

(c) The revenue begins to decrease for $n > 120$.

79. $f(x) = 9x - 4x^2$

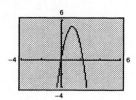

Zeros: $x(9 - 4x) = 0 \implies x = 0, \dfrac{9}{4}$

f is not one-to-one.

81. $g(t) = \dfrac{t + 3}{1 - t}$

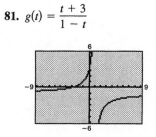

Zero: $t = -3$

The function is one-to-one.

83. $f(x) = \dfrac{4 - x^2}{x}$

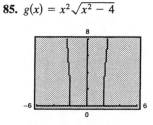

Zeros: $x = \pm 2$

The function is not one-to-one.

85. $g(x) = x^2\sqrt{x^2 - 4}$

Domain: $|x| \geq 2$

Zeros: $x = \pm 2$

g is not one-to-one.

87. Answers will vary.

Section 1.5   Limits

1.

x	1.9	1.99	1.999	2	2.001	2.01	2.1
$f(x)$	13.5	13.95	13.995	14	14.005	14.05	14.5

$\lim\limits_{x \to 2} (5x + 4) = 14$

3.

x	1.9	1.99	1.999	2	2.001	2.01	2.1
$f(x)$	0.2564	0.2506	0.2501	undefined	0.2499	0.2494	0.2439

$$\lim_{x \to 2} \frac{x - 2}{x^2 - 4} = \frac{1}{4}$$

5.

x	-0.1	-0.01	-0.001	0	0.001	0.01	0.1
$f(x)$	0.2911	0.2889	0.2887	undefined	0.2887	0.2884	0.2863

$$\lim_{x \to 0} \frac{\sqrt{x + 3} - \sqrt{3}}{x} = \frac{1}{2\sqrt{3}} \approx 0.289$$

7.

x	-0.5	-0.1	-0.01	-0.001	0
$f(x)$	-0.0714	-0.0641	-0.0627	-0.0625	undefined

$$\lim_{x \to 0^-} \frac{\left(\dfrac{1}{x + 4}\right) - \left(\dfrac{1}{4}\right)}{x} = -\frac{1}{16} = -0.0625$$

9. (a) $\lim_{x \to 0} f(x) = 1$ (b) $\lim_{x \to -1} f(x) = 3$

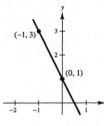

11. (a) $\lim_{x \to 0} g(x) = 1$ (b) $\lim_{x \to -1} g(x) = 3$

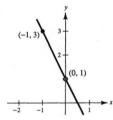

13. (a) $\lim_{x \to c} [f(x) + g(x)] = \lim_{x \to c} f(x) + \lim_{x \to c} g(x)$

$$= 3 + 9 = 12$$

(b) $\lim_{x \to c} [f(x)g(x)] = \left[\lim_{x \to c} f(x)\right]\left[\lim_{x \to c} g(x)\right] = 3 \cdot 9 = 27$

(c) $\lim_{x \to c} \dfrac{f(x)}{g(x)} = \dfrac{\lim_{x \to c} f(x)}{\lim_{x \to c} g(x)} = \dfrac{3}{9} = \dfrac{1}{3}$

15. $\lim_{x \to c} f(x) = 16$

(a) $\lim_{x \to c} \sqrt{f(x)} = \sqrt{16} = 4$

(b) $\lim_{x \to c} [3f(x)] = 3(16) = 48$

(c) $\lim_{x \to c} [f(x)]^2 = 16^2 = 256$

17. (a) $\lim_{x \to 3^+} f(x) = 1$

(b) $\lim_{x \to 3^-} f(x) = 1$

(c) $\lim_{x \to 3} f(x) = 1$

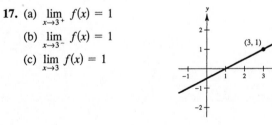

19. (a) $\lim_{x \to 3^+} f(x) = 0$

(b) $\lim_{x \to 3^-} f(x) = 0$

(c) $\lim_{x \to 3} f(x) = 0$

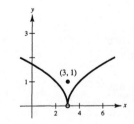

21. (a) $\displaystyle\lim_{x\to 3^+} f(x) = 3$

(b) $\displaystyle\lim_{x\to 3^-} f(x) = -3$

(c) $\displaystyle\lim_{x\to 3} f(x)$ does not exist.

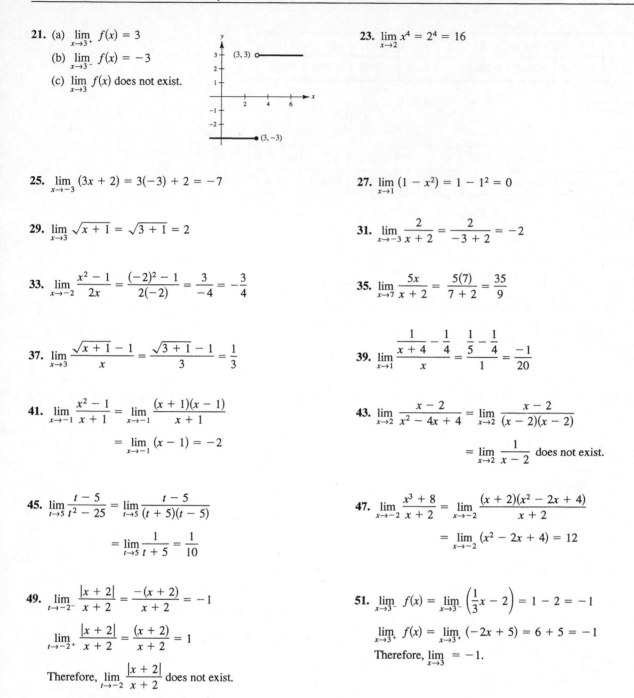

23. $\displaystyle\lim_{x\to 2} x^4 = 2^4 = 16$

25. $\displaystyle\lim_{x\to -3} (3x + 2) = 3(-3) + 2 = -7$

27. $\displaystyle\lim_{x\to 1} (1 - x^2) = 1 - 1^2 = 0$

29. $\displaystyle\lim_{x\to 3} \sqrt{x + 1} = \sqrt{3 + 1} = 2$

31. $\displaystyle\lim_{x\to -3} \frac{2}{x + 2} = \frac{2}{-3 + 2} = -2$

33. $\displaystyle\lim_{x\to -2} \frac{x^2 - 1}{2x} = \frac{(-2)^2 - 1}{2(-2)} = \frac{3}{-4} = -\frac{3}{4}$

35. $\displaystyle\lim_{x\to 7} \frac{5x}{x + 2} = \frac{5(7)}{7 + 2} = \frac{35}{9}$

37. $\displaystyle\lim_{x\to 3} \frac{\sqrt{x + 1} - 1}{x} = \frac{\sqrt{3 + 1} - 1}{3} = \frac{1}{3}$

39. $\displaystyle\lim_{x\to 1} \frac{\dfrac{1}{x + 4} - \dfrac{1}{4}}{x} = \frac{\dfrac{1}{5} - \dfrac{1}{4}}{1} = \frac{-1}{20}$

41. $\displaystyle\lim_{x\to -1} \frac{x^2 - 1}{x + 1} = \lim_{x\to -1} \frac{(x + 1)(x - 1)}{x + 1}$

$$= \lim_{x\to -1} (x - 1) = -2$$

43. $\displaystyle\lim_{x\to 2} \frac{x - 2}{x^2 - 4x + 4} = \lim_{x\to 2} \frac{x - 2}{(x - 2)(x - 2)}$

$$= \lim_{x\to 2} \frac{1}{x - 2} \text{ does not exist.}$$

45. $\displaystyle\lim_{t\to 5} \frac{t - 5}{t^2 - 25} = \lim_{t\to 5} \frac{t - 5}{(t + 5)(t - 5)}$

$$= \lim_{t\to 5} \frac{1}{t + 5} = \frac{1}{10}$$

47. $\displaystyle\lim_{x\to -2} \frac{x^3 + 8}{x + 2} = \lim_{x\to -2} \frac{(x + 2)(x^2 - 2x + 4)}{x + 2}$

$$= \lim_{x\to -2} (x^2 - 2x + 4) = 12$$

49. $\displaystyle\lim_{t\to -2^-} \frac{|x + 2|}{x + 2} = \frac{-(x + 2)}{x + 2} = -1$

$\displaystyle\lim_{t\to -2^+} \frac{|x + 2|}{x + 2} = \frac{(x + 2)}{x + 2} = 1$

Therefore, $\displaystyle\lim_{t\to -2} \frac{|x + 2|}{x + 2}$ does not exist.

51. $\displaystyle\lim_{x\to 3^-} f(x) = \lim_{x\to 3^-} \left(\frac{1}{3}x - 2\right) = 1 - 2 = -1$

$\displaystyle\lim_{x\to 3^+} f(x) = \lim_{x\to 3^+} (-2x + 5) = 6 + 5 = -1$

Therefore, $\displaystyle\lim_{x\to 3} = -1.$

53. $\displaystyle\lim_{\Delta x\to 0} \frac{2(x + \Delta x) - 2x}{\Delta x} = \lim_{\Delta x\to 0} \frac{2x + 2\Delta x - 2x}{\Delta x} = \lim_{\Delta x\to 0} 2 = 2$

55. $\displaystyle\lim_{\Delta x\to 0} \frac{\sqrt{x + 2 + \Delta x} - \sqrt{x + 2}}{\Delta x} = \lim_{\Delta x\to 0} \frac{\sqrt{x + 2 + \Delta x} - \sqrt{x + 2}}{\Delta x} \left(\frac{\sqrt{x + 2 + \Delta x} + \sqrt{x + 2}}{\sqrt{x + 2 + \Delta x} + \sqrt{x + 2}}\right)$

$$= \lim_{\Delta x\to 0} \frac{(x + 2 + \Delta x) - (x + 2)}{\Delta x\left[\sqrt{x + 2 + \Delta x} + \sqrt{x + 2}\right]}$$

$$= \lim_{\Delta x\to 0} \frac{1}{\sqrt{x + 2 + \Delta x} + \sqrt{x + 2}}$$

$$= \frac{1}{2\sqrt{x + 2}}$$

57. $\displaystyle\lim_{\Delta t \to 0} \frac{(t + \Delta t)^2 - 5(t + \Delta t) - (t^2 - 5t)}{\Delta t} = \lim_{\Delta t \to 0} \frac{t^2 + 2t(\Delta t) + (\Delta t)^2 - 5t - 5(\Delta t) - t^2 + 5t}{\Delta t}$

$$= \lim_{\Delta t \to 0} \frac{2t(\Delta t) + (\Delta t)^2 - 5(\Delta t)}{\Delta t}$$

$$= \lim_{\Delta t \to 0} 2t + (\Delta t) - 5$$

$$= 2t - 5$$

59. 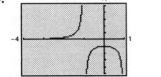 $\displaystyle\lim_{x \to 1^-} \frac{2}{x^2 - 1} = -\infty$

x	0	0.5	0.9	0.99	0.999	0.9999	1
$f(x)$	-2	-2.67	-10.53	-100.5	-1000.5	$-10{,}000.5$	undefined

Because $f(x) = \dfrac{2}{x^2 - 1}$ decreases without bound as x tends to 1 from the left, the limit does not exist.

$$\lim_{x \to 1^-} \frac{2}{x^2 - 1} = -\infty$$

61. 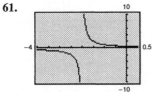 $\displaystyle\lim_{x \to -2^-} \frac{1}{x + 2} = -\infty$

x	-3	-2.5	-2.1	-2.01	-2.001	-2.0001	-2
$f(x)$	-1	-2	-10	-100	-1000	$-10{,}000$	undefined

Because $f(x) = \dfrac{1}{x + 2}$ decreases without bound as x tends to -2 from the left, the limit does not exist.

$$\lim_{x \to -2^-} \frac{1}{x + 2} = -\infty$$

63. $f(x) = \dfrac{x^2 - 5x + 6}{x^2 - 4x + 4} = \dfrac{(x - 2)(x - 3)}{(x - 2)(x - 2)}$

$\displaystyle\lim_{x \to 2} f(x) = \lim_{x \to 2} \frac{x - 3}{x - 2}$ does not exist.

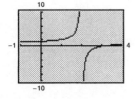

65. $\displaystyle\lim_{x \to -4} \frac{x^3 + 4x^2 + x + 4}{2x^2 + 7x - 4} = \lim_{x \to -4} \frac{(x + 4)(x^2 + 1)}{(x + 4)(2x - 1)}$

$$= -\frac{17}{9} \approx -1.889$$

67. (a)

x	-0.01	-0.001	-0.0001	0	0.0001	0.001	0.01
$f(x)$	2.732	2.720	2.718	undefined	2.718	2.717	2.705

$$\lim_{x \to 0} (1 + x)^{1/x} \approx 2.718$$

(b)

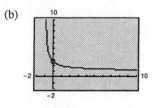

(c) Domain: $(-1, 0) \cup (0, \infty)$

Range: $(1, e) \cup (e, \infty)$

69. $C = \dfrac{25,000p}{100 - p}, \quad 0 \le p < 100$

(a) If $p = 50$, $C = \dfrac{25,000(50)}{100 - 50} = \$25,000.$

(b) If $C = 100,000 = \dfrac{25,000p}{100 - p}$, then

$$4(100 - p) = p$$

$$400 = 5p$$

$$p = 80 \Longrightarrow 80\%.$$

(c) $\displaystyle\lim_{p \to 100^-} C = \infty$

The cost function increases without bound as p approaches 100%.

71. (a)

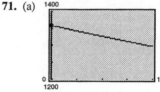

(b) For $x = 0.25$, $A = 1342.53.$

For $x = \frac{1}{365}$, $A = 1358.95.$

(c) $\displaystyle\lim_{x \to 0^+} 500(1 + 0.10x)^{10/x} = 500e \approx \$1359.14$

Continuous compounding

Section 1.6   Continuity

1. The polynomial $f(x) = 5x^3 - x^2 + 2$ is continuous on the entire real line.

3. The rational function $f(x) = \dfrac{1}{x^2 - 4}$ is not continuous on the entire real line. It is continuous at all $x \neq \pm 2$.

5. The rational function $f(x) = \dfrac{1}{4 + x^2}$ is continuous on the entire real line.

7. f is not continuous on the entire real line.

f is not defined at $x = 3, 5$.

9. g is not continuous on the entire real line.

g is not defined at $x = \pm 2$.

11. $f(x) = \dfrac{x^2 - 1}{x}$ is continuous on $(-\infty, 0)$ and $(0, \infty)$.

13. $f(x) = \dfrac{x^2 - 1}{x + 1}$ is continuous on $(-\infty, -1)$ and $(-1, \infty)$.

15. $f(x) = x^2 - 2x + 1$ is continuous on $(-\infty, \infty)$.

17. $f(x) = \dfrac{x}{x^2 - 1} = \dfrac{x}{(x + 1)(x - 1)}$

f is not defined at $x = \pm 1$.

Continuous on the intervals $(-\infty, -1)$, $(-1, 1)$, and $(1, \infty)$.

19. $f(x) = \dfrac{x}{x^2 + 1}$ is continuous on $(-\infty, \infty)$.

21. $f(x) = \dfrac{x - 5}{x^2 - 9x + 20} = \dfrac{x - 5}{(x - 5)(x - 4)}$ is continuous on $(-\infty, 4)$, $(4, 5)$ and $(5, \infty)$.

23. $f(x) = [\![2x]\!] + 1$ is continuous on all intervals of the form $\left(\frac{1}{2}c, \frac{1}{2}c + \frac{1}{2}\right)$, where c is an integer. That is, f is continuous on $\ldots, \left(-\frac{1}{2}, 0\right), \left(0, \frac{1}{2}\right), \left(\frac{1}{2}, 1\right), \ldots$.

f is not continuous at all points $\frac{1}{2}c$, where c is an integer.

25. $\lim\limits_{x \to 1^-} f(x) = \lim\limits_{x \to 1^-} (-2x + 3) = 1$

$\lim\limits_{x \to 1^+} f(x) = \lim\limits_{x \to 1^+} x^2 = 1$

f is continuous on $(-\infty, \infty)$.

27. $\lim\limits_{x \to 2^-} f(x) = 2$ and $\lim\limits_{x \to 2^+} f(x) = 1$. Hence

f is continuous on $(-\infty, 2]$ and $(2, \infty)$.

29. $\lim\limits_{x \to -1^-} \dfrac{|x + 1|}{x + 1} = \lim\limits_{x \to -1^-} \dfrac{-(x + 1)}{x + 1} = -1$

$\lim\limits_{x \to -1^+} \dfrac{|x + 1|}{x + 1} = \lim\limits_{x \to -1^+} \dfrac{x + 1}{x + 1} = 1$

Since $\lim\limits_{x \to -1} f(x)$ does not exist, f is continuous on $(-\infty, -1)$ and $(-1, \infty)$.

31. $\lim\limits_{x \to c^-} [\![x - 1]\!] = c - 2$, c is any integer.

$\lim\limits_{x \to c^+} [\![x - 1]\!] = c - 1$, c is any integer.

Since $\lim\limits_{x \to c} [\![x - 1]\!]$ does not exist, f is continuous on all intervals $(c, c + 1)$.

33. $h(x) = f(g(x)) = f(x - 1) = \dfrac{1}{\sqrt{x - 1}}$, $x > 1$

Thus, h is continuous on its entire domain $(1, \infty)$.

35. Since $f(x) = x^2 - 4x - 5$ is a polynomial, it is continuous on $[-1, 5]$.

37. Since $\lim\limits_{x \to 2^+} \dfrac{1}{x - 2} = \infty$, f has a nonremovable discontinuity at $x = 2$ on the closed interval $[1, 4]$.

39. $f(x) = \dfrac{x^2 - 16}{x - 4} = \dfrac{(x + 4)(x - 4)}{x - 4} = x + 4$, $x \neq 4$

Removable discontinuity at $x = 4$

Continuous on $(-\infty, 4)$ and $(4, \infty)$

41. $f(x) = \dfrac{x^3 + x}{x} = \dfrac{x(x^2 + 1)}{x} = x^2 + 1$, $x \neq 0$

Removable discontinuity at $x = 0$

Continuous on $(-\infty, 0)$ and $(0, \infty)$

43. $f(x) = \begin{cases} x^2 + 1, & x < 0 \\ x - 1, & x \geq 0 \end{cases}$

Nonremovable discontinuity at $x = 0$

Continuous on $(-\infty, 0), (0, \infty)$

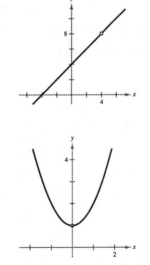

45. $\lim\limits_{x \to 2^-} f(x) = \lim\limits_{x \to 2^-} x^3 = 8$

$\lim\limits_{x \to 2^+} f(x) = \lim\limits_{x \to 2^+} ax^2 = 4a$

Therefore, $8 = 4a$ and $a = 2$.

47. $h(x) = \dfrac{1}{x^2 - x - 2} = \dfrac{1}{(x-2)(x+1)}$

h is not continuous at $x = 2$ and $x = -1$.

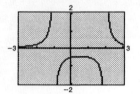

49. $f(x) = \begin{cases} 2x - 4, & x \le 3 \\ x^2 - 2x, & x > 3 \end{cases}$

f is not continuous at $x = 3$.

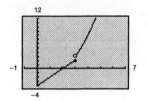

51.

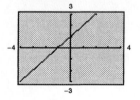

f is not continuous at all integers c.

53. $f(x) = \dfrac{x}{x^2 + 1}$ is continuous on $(-\infty, \infty)$.

55. $f(x) = \dfrac{1}{2}[\![2x]\!]$ is continuous on all intervals of the form $\left(\dfrac{c}{2}, \dfrac{c+1}{2} \right)$, where c is an integer.

57. $f(x) = \dfrac{x^2 + x}{x} = \dfrac{x(x+1)}{x}$ appears to be continuous on $[-4, 4]$. But it is not continuous at $x = 0$ (removable discontinuity).

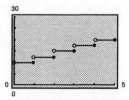

59. $A = 7500(1.015)^{[\![4t]\!]}, \quad t \ge 0$

(a)

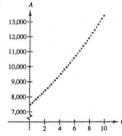

The graph has nonremovable discontinuities at $t = \dfrac{1}{4}, \dfrac{1}{2}, \dfrac{3}{4}, 1, \dfrac{5}{4}, \dots$ (every 3 months).

(b) For $t = 7$, $A = 7500(1.015)^{[\![4 \cdot 7]\!]} = \$11,379.17$.

61. $c(x) = 9.80 - 2.50[\![1 - x]\!] = \begin{cases} 9.80 + 2.50[\![x]\!], & x > 0, x \text{ not an integer} \\ 9.80 + 2.50[\![x - 1]\!], & x > 0, x \text{ an integer} \end{cases}$

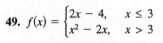

$c(x)$ is not continuous at $x = 1, 2, 3, \dots$

63. (a) $C(t) = \begin{cases} 1.04, & 0 < t \le 2 \\ 1.04 + 0.36[\![t - 1]\!], & t > 2, t \text{ is not an integer.} \\ 1.04 + 0.36(t - 2), & t > 2, t \text{ is an integer.} \end{cases}$

      C is not continuous at $t = 2, 3, 4, \ldots$.

    (b) $C(9) = 1.04 + 0.36[\![9 - 2]\!] = \$3.56$

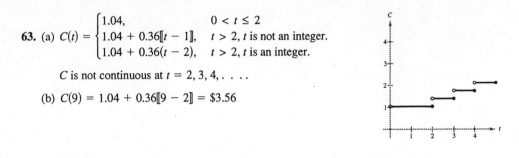

65. (a) Nonremovable discontinuities at $t = 2, 4, 6, 8, \ldots$

    (b) $N \to 0$ when $t \to 2^-, 4^-, 6^-, 8^-, \ldots$, so the inventory is replenished every two months.

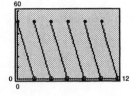

67. There are nonremovable discontinuities at $t = 1, 2, 3, 4, 5,$ and 6.

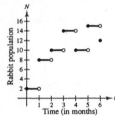

Review Exercises for Chapter 1

1. Matches (a)

3. Matches (b)

5. Distance $= \sqrt{(0 - 5)^2 + (0 - 2)^2}$
          $= \sqrt{25 + 4} = \sqrt{29}$

7. Distance $= \sqrt{[-1 - (-4)]^2 + (3 - 6)^2}$
           $= \sqrt{9 + 9} = \sqrt{18} = 3\sqrt{2}$

9. Midpoint $= \left(\dfrac{5 + 9}{2}, \dfrac{6 + 2}{2} \right) = (7, 4)$

11. Midpoint $= \left(\dfrac{-10 - 6}{2}, \dfrac{4 + 8}{2} \right) = (-8, 6)$

13. $P = R - C$. The tallest bars represent revenues. The middle bars represent costs. The bars on the left of each group represent profit, because $P = R - C$.

15. The translated vertices are $(1 + 3, 3 + 4) = (4, 7)$, $(2 + 3, 4 + 4) = (5, 8)$, and $(5 + 3, 6 + 4) = (8, 10)$.

17. Bar graph for data:

Mytilus	107
Gammarus	78
Littorina	65
Nassarius	112
Arbacia	6
Mya	18

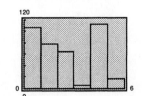

19. $y = 4 - 3x$

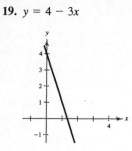

21. $y = 1 - x^2$

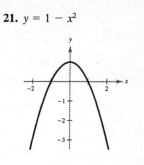

23. $y = |2x - 3|$

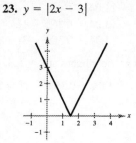

25. $y = x^3 + 2x^2 - x + 2$

x	-3	-2	-1	0	1	2	3
y	-4	4	4	2	4	16	44

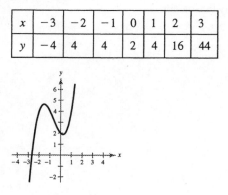

27. $y = \sqrt{2x}$

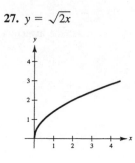

29. $y = (x - 1)^3 + 2(x - 1)^2$

$x = 0 \rightarrow y = (-1)^3 + 2(-1)^2 = 1 \rightarrow (0, 1)$

y-intercept

$y = 0 \rightarrow 0 = (x - 1)^3 + 2(x - 1)^2 = (x - 1)^2(x - 1 + 2)$

$\qquad = (x - 1)^2(x + 1) \rightarrow (1, 0), (-1, 0)$

x-intercepts

31. $\quad (x - 2)^2 + (y + 1)^2 = r^2$

$(-1 - 2)^2 + (7 + 1)^2 = r^2$

$9 + 64 = r^2$

$73 = r^2$

$(x - 2)^2 + (y + 1)^2 = 73$

33. $\qquad x^2 + y^2 + 10x + 4y - 7 = 0$

$(x^2 + 10x + 25) + (y^2 + 4y + 4) = 7 + 25 + 4$

$\qquad (x + 5)^2 + (y + 2)^2 = 36$

Center: $(-5, -2)$

Radius: 6

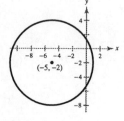

35. $x^2 + y^2 = 5$

$x - y = 1 \rightarrow y = x - 1$

$x^2 + (x - 1)^2 = 5$

$2x^2 - 2x + 1 = 5$

$2x^2 - 2x - 4 = 0$

$2(x - 2)(x + 1) = 0$

$x = -1, 2$

Points of intersection: $(-1, -2), (2, 1)$

37. $y = \sqrt{x} \qquad\qquad \sqrt{x} = x$

$\quad y = x \qquad\qquad\quad x = x^2$

$\qquad\qquad\qquad x^2 - x = 0$

$\qquad\qquad\qquad x(x - 1) = 0$

Points of intersection: $(0, 0), (1, 1)$

39. (a) $C = 6000 + 6.50x$

$R = 13.90x$

(b)
$$C = R$$
$$6000 + 6.5x = 13.9x$$
$$6000 = 7.4x$$
$$x \approx 810.81, \text{ or } 811 \text{ units}$$

41. $3x + y = -2$

$\qquad y = -3x - 2$

Slope: -3

y-intercept: $(0, -2)$

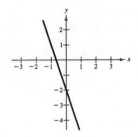

43. $y = -\frac{5}{3}$

Slope: 0 (horizontal line)

y-intercept: $\left(0, -\frac{5}{3}\right)$

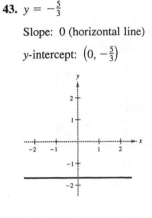

45. $-2x - 5y - 5 = 0$

$\qquad 5y = -2x - 5$

$\qquad y = -\frac{2}{5}x - 1$

Slope: $-\frac{2}{5}$

y-intercept: $(0, -1)$

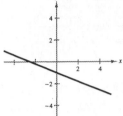

47. Slope $= \dfrac{6 - 0}{7 - 0} = \dfrac{6}{7}$

49. Slope $= \dfrac{17 - (-3)}{10 - (-11)} = \dfrac{20}{21}$

51. $y - (-1) = -2(x - 3)$

$\qquad y = -2x + 5$

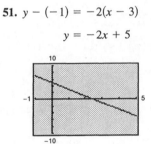

53. (a) $\qquad y - 6 = \frac{7}{8}[x - (-3)]$

$\qquad\qquad y = \frac{7}{8}x + \frac{69}{8}$

$\qquad 7x - 8y + 69 = 0$

(b) $4x + 2y = 7 \Longrightarrow y = -2x + \frac{7}{2}; \quad \text{slope} = -2$

$\qquad\qquad y - 6 = -2[x - (-3)]$

$\qquad\qquad\qquad y = -2x$

(c) The line through $(0, 0)$ and $(-3, 6)$ has slope

$\qquad \frac{6}{-3} = -2 \Longrightarrow y = -2x.$

(d) $3x - 2y = 2 \Longrightarrow y = \frac{3}{2}x - 1$

$\qquad$ Slope of perpendicular is $-\frac{2}{3}$.

$\qquad\qquad y - 6 = -\frac{2}{3}[x - (-3)]$

$\qquad\qquad\qquad y = -\frac{2}{3}x + 4$

$\qquad 2x + 3y - 12 = 0$

55. $(32, 750), (37, 700)$

$\qquad m = \dfrac{750 - 700}{32 - 37} = \dfrac{50}{-5} = -10$

(a) $x - 750 = -10(p - 32)$

$\qquad\qquad x = -10p + 1070$

(b) If $p = 34.50$, $x = -10(34.50) + 1070 = 725.$

(c) If $p = 42.00$, $x = -10(42.00) + 1070 = 650.$

57. Yes

59. No

61. $f(x) = 3x + 4$

(a) $f(1) = 3(1) + 4 = 7$

(b) $f(x + 1) = 3(x + 1) + 4 = 3x + 7$

(c) $f(2 + \Delta x) = 3(2 + \Delta x) + 4 = 10 + 3\Delta x$

63. $f(x) = x^2 + 3x + 2$

$= \left(x + \frac{3}{2}\right)^2 - \frac{1}{4}$

Domain: $(-\infty, \infty)$

Range: $\left[-\frac{1}{4}, \infty\right)$

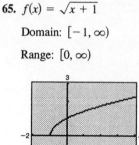

65. $f(x) = \sqrt{x + 1}$

Domain: $[-1, \infty)$

Range: $[0, \infty)$

67. $f(x) = -|x| + 3$

Domain: $(-\infty, \infty)$

Range: $(-\infty, 3]$

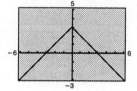

69. (a) $f(x) + g(x) = (1 + x^2) + (2x - 1) = x^2 + 2x$

(b) $f(x) - g(x) = (1 + x^2) - (2x - 1) = x^2 - 2x + 2$

(c) $f(x)g(x) = (1 + x^2)(2x - 1) = 2x^3 - x^2 + 2x - 1$

(d) $\dfrac{f(x)}{g(x)} = \dfrac{1 + x^2}{2x - 1}$

(e) $f(g(x)) = f(2x - 1) = 1 + (2x - 1)^2 = 4x^2 - 4x + 2$

(f) $g(f(x)) = g(1 + x^2) = 2(1 + x^2) - 1 = 2x^2 + 1$

71. $f(x) = \frac{3}{2}x$ has an inverse by the horizontal line test.

$y = \frac{3}{2}x$

$x = \frac{3}{2}y$

$y = \frac{2}{3}x$

$f^{-1}(x) = \frac{2}{3}x$

73. $f(x) = -x^2 + \frac{1}{2}$ does not have an inverse by the horizontal line test.

75. $\lim\limits_{x \to 2} (5x - 3) = 5(2) - 3 = 7$

77. $\lim\limits_{x \to 2} (5x - 3)(2x + 3) = [5(2) - 3][2(2) + 3] = 49$

79. $\lim\limits_{t \to 3} \dfrac{t^2 + 1}{t} = \dfrac{(3)^2 + 1}{3} = \dfrac{10}{3}$

81. $\lim\limits_{t \to 1} \dfrac{t + 1}{t - 2} = \dfrac{1 + 1}{1 - 2} = -2$

83. $\lim\limits_{x \to -2} \dfrac{x + 2}{x^2 - 4} = \lim\limits_{x \to -2} \dfrac{x + 2}{(x + 2)(x - 2)} = \lim\limits_{x \to -2} \dfrac{1}{x - 2} = -\dfrac{1}{4}$

85. $\lim\limits_{x \to 0^+} \left(x - \dfrac{1}{x}\right) = \lim\limits_{x \to 0^+} \dfrac{x^2 - 1}{x} = -\infty$

87. $\lim\limits_{x \to 0} \dfrac{[1/(x - 2)] - 1}{x} = \lim\limits_{x \to 0} \dfrac{1 - (x - 2)}{x(x - 2)} = \lim\limits_{x \to 0} \dfrac{3 - x}{x(x - 2)}$ does not exist.

89. $\lim\limits_{t \to 0} \dfrac{\dfrac{1}{\sqrt{t + 4}} - \dfrac{1}{2}}{t} \cdot \left(\dfrac{\dfrac{1}{\sqrt{t + 4}} + \dfrac{1}{2}}{\dfrac{1}{\sqrt{t + 4}} + \dfrac{1}{2}}\right) = \lim\limits_{t \to 0} \dfrac{\dfrac{1}{t + 4} - \dfrac{1}{4}}{t\left[\dfrac{1}{\sqrt{t + 4}} + \dfrac{1}{2}\right]} = \lim\limits_{t \to 0} \dfrac{4 - (t + 4)}{t\left[\dfrac{1}{\sqrt{t + 4}} + \dfrac{1}{2}\right][4(t + 4)]}$

$= \lim\limits_{t \to 0} \dfrac{-1}{\left[\dfrac{1}{\sqrt{t + 4}} + \dfrac{1}{2}\right][4t + 16]} = -\dfrac{1}{16}$

91. $\displaystyle\lim_{\Delta x \to 0} \frac{(x + \Delta x)^3 - (x + \Delta x) - (x^3 - x)}{\Delta x} = \lim_{\Delta x \to 0} \frac{x^3 + 3x^2\Delta x + 3x(\Delta x)^2 + (\Delta x)^3 - x - \Delta x - x^3 + x}{\Delta x}$

$$= \lim_{\Delta x \to 0} \frac{3x^2\Delta x + 3x(\Delta x)^2 + (\Delta x)^3 - \Delta x}{\Delta x}$$

$$= \lim_{\Delta x \to 0} [3x^2 + 3x\Delta x + (\Delta x)^2 - 1]$$

$$= 3x^2 - 1$$

93.

x	1.1	1.01	1.001	1.0001
$f(x)$	0.5680	0.5764	0.5773	0.5773

$$\lim_{x \to 1^+} \frac{\sqrt{2x + 1} - \sqrt{3}}{x - 1} = \frac{1}{\sqrt{3}} \approx 0.5774$$

95. The statement is false since $\displaystyle\lim_{x \to 0^-} \frac{|x|}{x} = -1.$

97. The statement is false since $\displaystyle\lim_{x \to 0^-} \sqrt{x}$ is undefined.

99. The statement is false since $\displaystyle\lim_{x \to 2^+} f(x) = \lim_{x \to 2^+} 0 = 0.$

101. $f(x) = \dfrac{1}{(x + 4)^2}$ is continuous on the intervals $(-\infty, -4)$ and $(-4, \infty).$

103. $f(x) = \dfrac{3}{x + 1}$ is continuous on the intervals $(-\infty, -1)$ and $(-1, \infty).$

105. $f(x) = [\![x + 3]\!]$ is continuous on all intervals of the form $(c, c + 1)$, where c is an integer.

107. $f(x)$ is continuous on the intervals $(-\infty, 0)$ and $(0, \infty).$

109. $\displaystyle\lim_{x \to 3^-} f(x) = \lim_{x \to 3^-} (-x + 1) = -2$

$$\lim_{x \to 3^+} f(x) = \lim_{x \to 3^+} (ax - 8) = 3a - 8$$

Thus, $-2 = 3a - 8$ and $a = 2.$

111. (a)

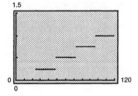

7000

0   13
3000

(b)

t	0	1	2	3	4
Debt	3206.3	3598.2	4001.8	4351.0	4643.3
Model	3103.6	3644.1	4078.9	4424.7	4697.8

t	5	6	7	8	9
Debt	4920.6	5181.5	5369.2	5478.2	5605.5
Model	4914.8	5092.2	5246.5	5394.1	5551.7

t	10	11	12	13
Debt	5628.7	5769.9	6198.4	6752.0
Model	5735.6	5962.4	6248.6	6610.8

(c) For 2008, $t = 18$ and $D \approx \$10{,}137.2$ billion

113. (a) Graphing utility graph of $A = (1/4)[\![x/24]\!]$ not continuous at $x = 24n$, n a positive integer. Graph as shown below.

(b) If $x = 1500$, $A = \$15.50.$

1.5

0   120
0

Practice Test for Chapter 1

1. Find the distance between $(3, 7)$ and $(4, -2)$.

2. Find the midpoint of the line segment joining $(0, 5)$ and $(2, 1)$.

3. Determine whether the points $(0, -3)$, $(2, 5)$, and $(-3, -15)$ are collinear.

4. Find x so that the distance between $(0, 3)$ and $(x, 5)$ is 7.

5. Sketch the graph of $y = 4 - x^2$.

6. Sketch the graph of $y = \sqrt{x - 2}$.

7. Sketch the graph of $y = |x - 3|$.

8. Write the equation of the circle in standard form and sketch its graph.

$$x^2 + y^2 - 8x + 2y + 8 = 0$$

9. Find the points of intersection of the graphs of $x^2 + y^2 = 25$ and $x - 2y = 10$.

10. Find the general equation of the line passing through the points $(7, 4)$ and $(6, -2)$.

11. Find the general equation of the line passing through the point $(-2, -1)$ with a slope of $m = \frac{2}{3}$.

12. Find the general equation of the line passing through the point $(6, -8)$ with undefined slope.

13. Find the general equation of the line passing through the point $(0, 3)$ and perpendicular to the line given by $2x - 5y = 7$.

14. Given $f(x) = x^2 - 5$, find the following.
    (a) $f(3)$ (b) $f(-6)$ (c) $f(x - 5)$ (d) $f(x + \Delta x)$

15. Find the domain and range of $f(x) = \sqrt{3 - x}$.

16. Given $f(x) = 2x + 3$ and $g(x) = x^2 - 1$, find the following.
    (a) $f(g(x))$ (b) $g(f(x))$

17. Given $f(x) = x^3 + 6$, find $f^{-1}(x)$.

18. Find $\lim\limits_{x \to -4} (2 - 5x)$.

19. Find $\lim\limits_{x \to 6} \dfrac{x^2 - 36}{x - 6}$.

20. Find $\lim\limits_{x \to -1} \dfrac{|x + 1|}{x + 1}$.

21. Find $\lim\limits_{x \to 0} \dfrac{\sqrt{x + 5} - \sqrt{5}}{x}$.

22. Find $\lim\limits_{x \to 1} f(x)$, where $f(x) = \begin{cases} 2x + 3, & x \le 1 \\ x^2 + 4, & x > 1. \end{cases}$

23. Find the discontinuities of $f(x) = \dfrac{x - 8}{x^2 - 64}$. Which are removable?

24. Find the discontinuities of $f(x) = \dfrac{|x - 3|}{x - 3}$. Which are removable?

25. Sketch the graph of $f(x) = \dfrac{x^2 - 5x + 6}{x - 3}$.

Graphing Calculator Required

26. Solve the equation for y and graph the resulting two equations on the same set of coordinate axes.

$$x^2 + y^2 + 6x + 5 = 0$$

27. Use a graphing calculator to graph $f(x) = \dfrac{x^2 - 9}{x - 3}$ and find $\lim\limits_{x \to 3} f(x)$. Is the graph displayed correctly at $x = 3$?

C H A P T E R 2
Differentiation

CHAPTER 2
Differentiation

Section 2.1 The Derivative and the Slope of a Graph

Solutions to Odd-Numbered Exercises

1. The tangent line at (x_1, y_1) has a positive slope. The tangent line at (x_2, y_2) has a negative slope.

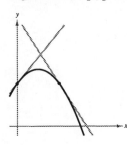

3. The tangent line at (x_1, y_1) has a positive slope. The tangent line at (x_2, y_2) has zero slope.

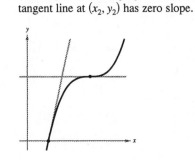

5. The slope is $m = 1$.

7. The slope is $m = 0$.

9. The slope is $m = -\frac{1}{3}$.

11. For 1997, $t = 7$ and $m \approx 250$.

For 2000, $t = 10$ and $m \approx 200$.

For 2002, $t = 12$ and $m \approx 0$.

13. For $t = 1$: $m \approx 65$

For $t = 8$: $m \approx 0$

For $t = 12$: $m \approx -1000$

15.
$$f(x) = 3$$
$$f(x + \Delta x) = 3$$
$$f(x + \Delta x) - f(x) = 0$$
$$\frac{f(x + \Delta x) - f(x)}{\Delta x} = 0$$
$$\lim_{\Delta x \to 0} \frac{f(x + \Delta x) - f(x)}{\Delta x} = 0$$

17.
$$f(x) = -5x + 3$$
$$f(x + \Delta x) = -5(x + \Delta x) + 3 = -5x - 5\Delta x + 3$$
$$f(x + \Delta x) - f(x) = -5\Delta x$$
$$\frac{f(x + \Delta x) - f(x)}{\Delta x} = -5$$
$$\lim_{\Delta x \to 0} \frac{f(x + \Delta x) - f(x)}{\Delta x} = -5$$

19.
$$f(x) = x^2 - 4$$
$$f(x + \Delta x) = (x + \Delta x)^2 - 4 = x^2 + 2x\Delta x + (\Delta x)^2 - 4$$
$$f(x + \Delta x) - f(x) = 2x\Delta x + (\Delta x)^2$$
$$\frac{f(x + \Delta x) - f(x)}{\Delta x} = 2x + \Delta x$$
$$\lim_{\Delta x \to 0} \frac{f(x + \Delta x) - f(x)}{\Delta x} = 2x$$

21.
$$h(t) = \sqrt{t - 1}$$
$$h(t + \Delta t) = \sqrt{t + \Delta t - 1}$$
$$h(t + \Delta t) - h(t) = \sqrt{t + \Delta t - 1} - \sqrt{t - 1}$$
$$= \frac{\sqrt{t + \Delta t - 1} - \sqrt{t - 1}}{1} \cdot \frac{\sqrt{t + \Delta t - 1} + \sqrt{t - 1}}{\sqrt{t + \Delta t - 1} + \sqrt{t - 1}}$$
$$= \frac{\Delta t}{\sqrt{t + \Delta t - 1} + \sqrt{t - 1}}$$
$$\frac{h(t + \Delta t) - h(t)}{\Delta t} = \frac{1}{\sqrt{t + \Delta t - 1} + \sqrt{t - 1}}$$
$$\lim_{\Delta t \to 0} \frac{h(t + \Delta t) - h(t)}{\Delta t} = \frac{1}{2\sqrt{t - 1}}$$

23.
$$f(t) = t^3 - 12t$$
$$f(t + \Delta t) = (t + \Delta t)^3 - 12(t + \Delta t)$$
$$= t^3 + 3t^2\Delta t + 3t(\Delta t)^2 + (\Delta t)^3 - 12t - 12\Delta t$$
$$f(t + \Delta t) - f(t) = 3t^2\Delta t + 3t(\Delta t)^2 + (\Delta t)^3 - 12\Delta t$$
$$\frac{f(t + \Delta t) - f(t)}{\Delta t} = 3t^2 + 3t\Delta t + (\Delta t)^2 - 12$$
$$\lim_{\Delta t \to 0} \frac{f(t + \Delta t) - f(t)}{\Delta t} = 3t^2 - 12$$

25.
$$f(x) = \frac{1}{x + 2}$$
$$f(x + \Delta x) = \frac{1}{x + \Delta x + 2}$$
$$f(x + \Delta x) - f(x) = \frac{1}{x + \Delta x + 2} - \frac{1}{x + 2}$$
$$= \frac{-\Delta x}{(x + \Delta x + 2)(x + 2)}$$
$$\frac{f(x + \Delta x) - f(x)}{\Delta x} = \frac{-1}{(x + \Delta x + 2)(x + 2)}$$
$$\lim_{\Delta x \to 0} \frac{f(x + \Delta x) - f(x)}{\Delta x} = \frac{-1}{(x + 2)^2}$$

27.
$$f(x) = 6 - 2x$$
$$f(x + \Delta x) = 6 - 2(x + \Delta x)$$
$$= 6 - 2x - 2\Delta x$$
$$f(x + \Delta x) - f(x) = -2\Delta x$$
$$\frac{f(x + \Delta x) - f(x)}{\Delta x} = -2$$
$$\lim_{\Delta x \to 0} \frac{f(x + \Delta x) - f(x)}{\Delta x} = -2 = f'(x)$$

At $(2, 2)$, the slope of the tangent line is $m = -2$. The figure shows the graph of f and the tangent line.

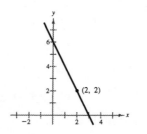

29.

$$f(x) = -1$$

$$f(x + \Delta x) = -1$$

$$f(x + \Delta x) - f(x) = -1 - (-1) = 0$$

$$\frac{f(x + \Delta x) - f(x)}{\Delta x} = 0$$

$$\lim_{\Delta x \to 0} \frac{f(x + \Delta x) - f(x)}{\Delta x} = 0 = f'(x)$$

At $(0, -1)$, the slope of the tangent line is 0.

31.

$$f(x + \Delta x) = (x + \Delta x)^2 - 2 = x^2 + 2x\Delta x + (\Delta x)^2 - 2$$

$$f(x + \Delta x) - f(x) = 2x\Delta x + (\Delta x)^2$$

$$\frac{f(x + \Delta x) - f(x)}{\Delta x} = 2x + \Delta x$$

$$\lim_{\Delta x \to 0} \frac{f(x + \Delta x) - f(x)}{\Delta x} = 2x = f'(x)$$

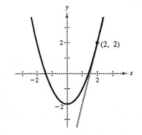

At $(2, 2)$, the slope of the tangent line is $m = 2(2) = 4$.
The figure shows the graph of f and the tangent line.

33.

$$f(x + \Delta x) = (x + \Delta x)^3 - (x + \Delta x) = x^3 + 3x^2\Delta x + 3x(\Delta x)^2 + (\Delta x)^3 - x - \Delta x$$

$$f(x + \Delta x) - f(x) = 3x^2\Delta x + 3x(\Delta x)^2 + (\Delta x)^3 - \Delta x$$

$$\frac{f(x + \Delta x) - f(x)}{\Delta x} = 3x^2 + 3x(\Delta x) + (\Delta x)^2 - 1$$

$$\lim_{\Delta x \to 0} \frac{f(x + \Delta x) - f(x)}{\Delta x} = 3x^2 - 1$$

At $(2, 6)$, the slope of the tangent line is $f'(2) = 3(2)^2 - 1 = 11$.

35.

$$f(x + \Delta x) = \sqrt{1 - 2(x + \Delta x)} = \sqrt{1 - 2x - 2\Delta x}$$

$$f(x + \Delta x) - f(x) = \sqrt{1 - 2x - 2\Delta x} - \sqrt{1 - 2x}$$

$$= \frac{\sqrt{1 - 2x - 2\Delta x} - \sqrt{1 - 2x}}{1} \cdot \frac{\sqrt{1 - 2x - 2\Delta x} + \sqrt{1 - 2x}}{\sqrt{1 - 2x - 2\Delta x} + \sqrt{1 - 2x}}$$

$$= \frac{(1 - 2x - 2\Delta x) - (1 - 2x)}{\sqrt{1 - 2x - 2\Delta x} + \sqrt{1 - 2x}}$$

$$= \frac{-2\Delta x}{\sqrt{1 - 2x - 2\Delta x} + \sqrt{1 - 2x}}$$

$$\frac{f(x + \Delta x) - f(x)}{\Delta x} = \frac{-2}{\sqrt{1 - 2x - 2\Delta x} + \sqrt{1 - 2x}}$$

$$\lim_{\Delta x \to 0} \frac{f(x + \Delta x) - f(x)}{\Delta x} = \frac{-2}{2\sqrt{1 - 2x}} = \frac{-1}{\sqrt{1 - 2x}}$$

At $(-4, 3)$, the slope of the tangent line is $f'(-4) = \dfrac{-1}{\sqrt{1 - 2(-4)}} = -\dfrac{1}{3}$.

37.

$$f(x + \Delta x) = \frac{1}{2}(x + \Delta x)^2 = \frac{1}{2}x^2 + x\Delta x + \frac{1}{2}\Delta x^2$$

$$f(x + \Delta x) - f(x) = x\Delta x + \frac{1}{2}(\Delta x)^2$$

$$\frac{f(x + \Delta x) - f(x)}{\Delta x} = x + \frac{1}{2}(\Delta x)$$

$$\lim_{\Delta x \to 0} \frac{f(x + \Delta x) - f(x)}{\Delta x} = x$$

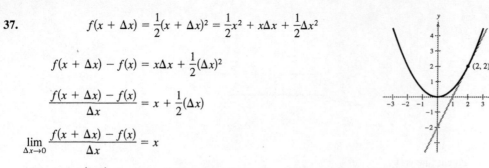

At the point $(2, 2)$, the slope of the tangent line is $m = 2$. The equation of the tangent line is

$$y - 2 = 2(x - 2)$$

$$y = 2x - 2.$$

39.

$$f(x + \Delta x) = [(x + \Delta x) - 1]^2$$

$$= x^2 + 2x\Delta x + \Delta x^2 - 2x - 2\Delta x + 1$$

$$f(x + \Delta x) - f(x) = 2x\Delta x + (\Delta x)^2 - 2\Delta x$$

$$\frac{f(x + \Delta x) - f(x)}{\Delta x} = 2x + \Delta x - 2$$

$$\lim_{\Delta x \to 0} \frac{f(x + \Delta x) - f(x)}{\Delta x} = 2x - 2$$

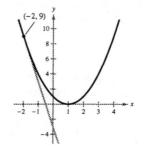

At the point $(-2, 9)$, the slope of the tangent line is $m = 2(-2) - 2 = -6$. The equation of the tangent line is

$$y - 9 = -6[x - (-2)]$$

$$y = -6x - 3.$$

41.

$$f(x + \Delta x) = \sqrt{x + \Delta x} + 1$$

$$f(x + \Delta x) - f(x) = \sqrt{x + \Delta x} - \sqrt{x}$$

$$= \frac{\sqrt{x + \Delta x} - \sqrt{x}}{1} \cdot \frac{\sqrt{x + \Delta x} + \sqrt{x}}{\sqrt{x + \Delta x} + \sqrt{x}}$$

$$= \frac{\Delta x}{\sqrt{x + \Delta x} + \sqrt{x}}$$

$$\frac{f(x + \Delta x) - f(x)}{\Delta x} = \frac{1}{\sqrt{x + \Delta x} + \sqrt{x}}$$

$$\lim_{\Delta x \to 0} \frac{f(x + \Delta x) - f(x)}{\Delta x} = \frac{1}{\sqrt{x} + \sqrt{x}} = \frac{1}{2\sqrt{x}}$$

At the point $(4, 3)$, $f'(4) = \frac{1}{2\sqrt{4}} = \frac{1}{4}$. The equation of the tangent line is

$$y - 3 = \frac{1}{4}(x - 4)$$

$$y = \frac{1}{4}x + 2.$$

43.
$$f(x + \Delta x) = \frac{1}{x + \Delta x}$$

$$f(x + \Delta x) - f(x) = \frac{1}{x + \Delta x} - \frac{1}{x}$$

$$= \frac{x - (x + \Delta x)}{(x + \Delta x)(x)}$$

$$= -\frac{\Delta x}{x(x + \Delta x)}$$

$$\frac{f(x + \Delta x) - f(x)}{\Delta x} = -\frac{1}{x(x + \Delta x)}$$

$$\lim_{\Delta x \to 0} \frac{f(x + \Delta x) - f(x)}{\Delta x} = -\frac{1}{x^2}$$

At the point $(1, 1)$, the slope of the tangent line is $m = -\dfrac{1}{(1)^2} = -1$. The equation of the tangent line is

$$y - 1 = -1(x - 1)$$

$$y - 1 = -x + 1$$

$$y = -x + 2$$

45.
$$f(x + \Delta x) = -\frac{1}{4}(x + \Delta x)^2 = -\frac{1}{4}[x^2 + 2x\Delta x + (\Delta x)^2]$$

$$f(x + \Delta x) - f(x) = -\frac{1}{4}[2x\Delta x + (\Delta x)^2]$$

$$\frac{f(x + \Delta x) - f(x)}{\Delta x} = -\frac{x}{2} - \frac{1}{4}\Delta x$$

$$\lim_{\Delta x \to 0} \frac{f(x + \Delta x) - f(x)}{\Delta x} = -\frac{x}{2} \quad \text{(slope of tangent line)}$$

Since the slope of the given line is -1, we have

$$-\frac{x}{2} = -1$$

$$x = 2 \text{ and } f(2) = -1.$$

Therefore, at the point $(2, -1)$, the tangent line parallel to $x + y = 0$ is

$$y - (-1) = -1(x - 2)$$

$$y = -x + 1.$$

47.
$$f(x + \Delta x) = -\frac{1}{2}(x + \Delta x)^3 = -\frac{1}{2}[x^3 + 3x^2\Delta x + 3x(\Delta x)^2 + (\Delta x)^3]$$

$$f(x + \Delta x) - f(x) = -\frac{1}{2}[3x^2\Delta x + 3x(\Delta x)^2 + (\Delta x)^3]$$

$$\frac{f(x + \Delta x) - f(x)}{\Delta x} = -\frac{1}{2}[3x^2 + 3x\Delta x + (\Delta x)^2]$$

$$\lim_{\Delta x \to 0} \frac{f(x + \Delta x) - f(x)}{\Delta x} = -\frac{1}{2}(3x^2) = -\frac{3}{2}x^2 \quad \text{(slope of tangent line)}$$

Since the slope of the given line is -6, we have

$$-\frac{3}{2}x^2 = -6$$

$$x^2 = 4$$

$$x = \pm 2 \text{ and } f(2) = -4 \text{ and } f(-2) = 4.$$

At the point $(2, -4)$, the tangent line is

$$y + 4 = -6(x - 2)$$

$$y = -6x + 8.$$

At the point $(-2, 4)$, the tangent line is

$$y - 4 = -6(x + 2)$$

$$y = -6x - 8.$$

49. y is not differentiable when $x = -3$. At $(-3, 0)$, the graph has a node. y is differentiable for all $x \neq -3$.

51. y is differentiable everywhere except at $x = 3$. At $(3, 0)$, the graph has a cusp.

53. f is differentiable on the open interval $(1, \infty)$.

55. f is differentiable everywhere except at $x = 0$, which is a nonremovable discontinuity.

57. $f(x) = \frac{1}{4}x^3$

x	-2	$-\frac{3}{2}$	-1	$-\frac{1}{2}$	0	$\frac{1}{2}$	1	$\frac{3}{2}$	2
$f(x)$	-2	-0.8438	-0.25	-0.0313	0	0.0313	0.25	0.8438	2
$f'(x)$	3	1.6875	0.75	0.1875	0	0.1875	0.75	1.6875	3

Analytically, the slope of $f(x) = \frac{1}{4}x^3$ is

$$m = \lim_{\Delta x \to 0} \frac{f(x + \Delta x) - f(x)}{\Delta x}$$

$$= \lim_{\Delta x \to 0} \frac{\frac{1}{4}(x + \Delta x)^3 - \frac{1}{4}x^3}{\Delta x}$$

$$= \lim_{\Delta x \to 0} \frac{\frac{1}{4}[3x^2\Delta x + 3x(\Delta x)^2 + (\Delta x)^3]}{\Delta x}$$

$$= \lim_{\Delta x \to 0} \frac{1}{4}[3x^2 + 3x\Delta x + (\Delta x)^2]$$

$$= \frac{3}{4}x^2.$$

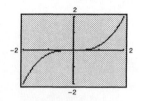

59. $f(x) = -\dfrac{1}{2}x^3$

x	-2	$-\frac{3}{2}$	-1	$-\frac{1}{2}$	0	$\frac{1}{2}$	1	$\frac{3}{2}$	2
$f(x)$	4	1.6875	0.5	0.0625	0	-0.0625	-0.5	-1.6875	-4
$f'(x)$	-6	-3.375	-1.5	-0.375	0	-0.375	-1.5	-3.375	-6

Analytically, the slope of $f(x) = -\frac{1}{2}x^3$ is

$$
\begin{aligned}
m &= \lim_{\Delta x \to 0} \frac{f(x + \Delta x) - f(x)}{\Delta x} \\[2mm]
&= \lim_{\Delta x \to 0} \frac{-\frac{1}{2}(x + \Delta x)^3 + \frac{1}{2}x^3}{\Delta x} \\[2mm]
&= \lim_{\Delta x \to 0} \frac{-\frac{1}{2}[x^3 + 3x^2\Delta x + 3x(\Delta x)^2 + (\Delta x)^3] + \frac{1}{2}x^3}{\Delta x} \\[2mm]
&= \lim_{\Delta x \to 0} \frac{-\frac{1}{2}[3x^2\Delta x + 3x(\Delta x)^2 + (\Delta x)^3]}{\Delta x} \\[2mm]
&= \lim_{\Delta x \to 0} -\frac{1}{2}[3x^2 + 3x(\Delta x) + (\Delta x)^2] \\[2mm]
&= -\frac{3}{2}x^2.
\end{aligned}
$$

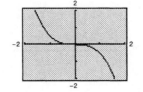

61.
$$
\begin{aligned}
f'(x) &= \lim_{\Delta x \to 0} \frac{f(x + \Delta x) - f(x)}{\Delta x} \\[2mm]
&= \lim_{\Delta x \to 0} \frac{(x + \Delta x)^2 - 4(x + \Delta x) - (x^2 - 4x)}{\Delta x} \\[2mm]
&= \lim_{\Delta x \to 0} \frac{2x\Delta x + (\Delta x)^2 - 4\Delta x}{\Delta x} \\[2mm]
&= \lim_{\Delta x \to 0} (2x + \Delta x - 4) \\[2mm]
&= 2x - 4
\end{aligned}
$$

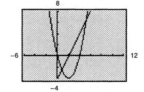

The x-intercept of the derivative indicates a point of horizontal tangency for f.

63.
$$
\begin{aligned}
f'(x) &= \lim_{\Delta x \to 0} \frac{f(x + \Delta x) - f(x)}{\Delta x} \\[2mm]
&= \lim_{\Delta x \to 0} \frac{(x + \Delta x)^3 - 3(x + \Delta x) - (x^3 - 3x)}{\Delta x} \\[2mm]
&= \lim_{\Delta x \to 0} \frac{x^3 + 3x^2\Delta x + 3x(\Delta x)^2 + (\Delta x)^3 - 3x - 3\Delta x - x^3 + 3x}{\Delta x} \\[2mm]
&= \lim_{\Delta x \to 0} \frac{3x^2\Delta x + 3x(\Delta x)^2 + (\Delta x)^3 - 3\Delta x}{\Delta x} \\[2mm]
&= \lim_{\Delta x \to 0} (3x^2 + 3x\Delta x + (\Delta x)^2 - 3) \\[2mm]
&= 3x^2 - 3
\end{aligned}
$$

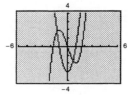

The x-intercepts of the derivative, $x = \pm 1$, indicate points of horizontal tangency for f.

65. $f(x) = -x$

Other answers possible.

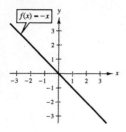

67. The graph of $f(x) = x^2 + 1$ is smooth at $(0, 1)$, but the graph of $g(x) = |x| + 1$ has a sharp point at $(0, 1)$. The function g is not differentiable at $(0, 1)$.

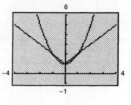

69. False. $f(x) = |x|$ is continuous, but not differentiable at $x = 0$.

71. True

Section 2.2   Some Rules for Differentiation

1. (a) $y = x^2$

$y' = 2x$

At $(1, 1)$, $y' = 2$.

(b) $y = x^{1/2}$

$y' = \frac{1}{2}x^{-1/2} = \frac{1}{2\sqrt{x}}$

At $(1, 1)$, $y' = \frac{1}{2}$.

3. (a) $y = x^{-1}$

$y' = -x^{-2} = -\frac{1}{x^2}$

At $(1, 1)$, $y' = -1$.

(b) $y = x^{-1/3}$

$y' = -\frac{1}{3}x^{-4/3} = -\frac{1}{3x^{4/3}}$

At $(1, 1)$, $y' = -\frac{1}{3}$.

5. $y' = 0$

7. $f(x) = 4x + 1$

$f'(x) = 4$

9. $g(x) = x^2 + 4x - 1$

$g'(x) = 2x + 4$

11. $f'(t) = -6t + 2$

13. $s'(t) = 3t^2 - 2$

15. $y' = 4\left(\frac{4}{3}\right)t^{1/3} = \frac{16}{3}t^{1/3}$

17. $f(x) = 4\sqrt{x} = 4x^{1/2}$

$f'(x) = 4\left(\frac{1}{2}\right)x^{(1/2)-1} = 2x^{-1/2} = \frac{2}{\sqrt{x}}$

19. $y' = 4(-2)x^{-2-1} + 2(2)x^{2-1}$

$= -8x^{-3} + 4x^1 = -\frac{8}{x^3} + 4x$

	Function	Rewrite	Differentiate	Simplify
21.	$y = \dfrac{1}{4x^3}$	$y = \dfrac{1}{4}x^{-3}$	$y' = -\dfrac{3}{4}x^{-4}$	$y' = -\dfrac{3}{4x^4}$
23.	$y = \dfrac{1}{(4x)^3}$	$y = \dfrac{1}{64}x^{-3}$	$y' = -\dfrac{3}{64}x^{-4}$	$y' = -\dfrac{3}{64x^4}$
25.	$y = \dfrac{\sqrt{x}}{x}$	$y = x^{-(1/2)}$	$y' = -\dfrac{1}{2}x^{-(3/2)}$	$y' = -\dfrac{1}{2x^{3/2}}$

27. $f(x) = \dfrac{1}{x} = x^{-1}$

$f'(x) = -x^{-2} = -\dfrac{1}{x^2} \Rightarrow f'(1) = -1$

29. $f(x) = -\frac{1}{2}x(1 + x^2) = -\frac{1}{2}x - \frac{1}{2}x^3$

$f'(x) = -\frac{1}{2} - \frac{3}{2}x^2$

$f'(1) = -\frac{1}{2} - \frac{3}{2} = -2$

31. $y = (2x + 1)^2 = 4x^2 + 4x + 1$

$y' = 8x + 4 = 4(2x + 1)$

At $(0, 1)$, $y' = 4$.

33. $f(x) = x^2 - 4x^{-1} - 3x^{-2}$

$f'(x) = 2x + 4x^{-2} + 6x^{-3} = 2x + \frac{4}{x^2} + \frac{6}{x^3}$

35. $f(x) = x^2 - 2x - \frac{2}{x^4} = x^2 - 2x - 2x^{-4}$

$f'(x) = 2x - 2 + 8x^{-5} = 2x - 2 + \frac{8}{x^5}$

37. $f(x) = x(x^2 + 1) = x^3 + x$

$f'(x) = 3x^2 + 1$

39. $f(x) = (x + 4)(2x^2 - 1) = 2x^3 + 8x^2 - x - 4$

$f'(x) = 6x^2 + 16x - 1$

41. $f(x) = \frac{2x^3 - 4x^2 + 3}{x^2} = 2x - 4 + 3x^{-2}$

$f'(x) = 2 - 6x^{-3} = 2 - \frac{6}{x^3} = \frac{2x^3 - 6}{x^3} = \frac{2(x^3 - 3)}{x^3}$

43. $f(x) = \frac{4x^3 - 3x^2 + 2x + 5}{x^2} = 4x - 3 + 2x^{-1} + 5x^{-2}$

$f'(x) = 4 - 2x^{-2} - 10x^{-3} = 4 - \frac{2}{x^2} - \frac{10}{x^3}$

$= \frac{4x^3 - 2x - 10}{x^3}$

$= \frac{2(2x^3 - x - 5)}{x^3}$

45. $f(x) = x^{4/5} + x$

$f'(x) = \frac{4}{5}x^{-1/5} + 1 = \frac{4}{5x^{1/5}} + 1$

47. $y = -2x^4 + 5x^2 - 3$

$y' = -8x^3 + 10x$

At $(1, 0)$, the slope is $m = y' = -8 + 10 = 2$. The equation of the tangent line is

$y - 0 = 2(x - 1)$

$y = 2x - 2.$

49. $f(x) = x^{1/3} + x^{1/5}$

$f'(x) = \frac{1}{3}x^{-2/3} + \frac{1}{5}x^{-4/5} = \frac{1}{3x^{2/3}} + \frac{1}{5x^{4/5}}$

$f'(1) = \frac{1}{3} + \frac{1}{5} = \frac{8}{15}$

$y - 2 = \frac{8}{15}(x - 1)$

$y = \frac{8}{15}x + \frac{22}{15}$

51. $y' = -4x^3 + 6x = 2x(3 - 2x^2) = 0$

$x = 0, \; x = \pm\sqrt{\frac{3}{2}} = \pm\frac{\sqrt{6}}{2}$

If $x = \pm\frac{\sqrt{6}}{2}$, then

$y = -\left(\frac{\sqrt{6}}{2}\right)^4 + 3\left(\frac{\sqrt{6}}{2}\right)^2 - 1$

$= -\frac{9}{4} + 3\left(\frac{3}{2}\right) - 1$

$= \frac{5}{4}.$

The function has horizontal tangent lines at the points

$(0, -1), \left(-\frac{\sqrt{6}}{2}, \frac{5}{4}\right),$ and $\left(\frac{\sqrt{6}}{2}, \frac{5}{4}\right).$

53. $y' = x + 5 = 0$ when $x = -5$.

The function has a horizontal tangent line at the point $\left(-5, -\frac{25}{2}\right)$.

55. (a)

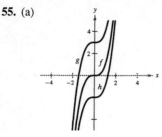

(b) $f'(x) = g'(x) = h'(x) = 3x^2$

$f'(1) = g'(1) = h'(1) = 3$

(c)

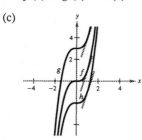

57. (a) $h(x) = f(x) - 2$

$h'(x) = f'(x) - 0 = f'(x)$

$h'(1) = f'(1) = 3$

(b) $h(x) = 2f(x)$

$h'(x) = 2f'(x)$

$h'(1) = 2f'(1) = 2(3) = 6$

(c) $h(x) = -f(x)$

$h'(x) = -f'(x)$

$h'(1) = -f'(1) = -3$

(d) $h(x) = -1 + 2f(x)$

$h'(x) = 0 + 2f'(x) = 2f'(x)$

$h'(1) = 2f'(1) = 2(3) = 6$

59. $S = 8.70947t^4 - 341.0927t^3 + 4885.752t^2 - 30,118.17t + 68,395.3$

$(t = 7$ corresponds to 1997, $7 \leq t \leq 13.)$

(a) $S'(t) = 34.83788t^3 - 1023.2781t^2 + 9771.504t - 30,118.17$

1998: $S'(8) \approx 401.1$

2001: $S'(11) \approx -79.1$

2003: $S'(13) \approx 516.2$

(b) The results are close to the estimates.

(c) The units of S' are millions of dollars per year per year.

61. $C = 0.60x + 250$

$R = 1.00x$

$P = R - C$

$= 1.00x - (0.60x + 250)$

$= 0.40x - 250$

$\dfrac{dP}{dx} = 0.40$

Therefore, the derivative is constant and is equal to the profit on each candy bar sold.

63.

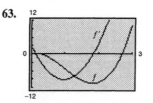

There are horizontal tangents at $(0.11, 0.14)$ and $(1.84, -10.49)$.

65. False. Let $f(x) = x$ and $g(x) = x + 1$.

Then $f'(x) = g'(x) = 1$, but $f(x) \neq g(x)$.

Section 2.3   Rates of Change: Velocity and Marginals

1. (a) 1980–1985: $\dfrac{115 - 60}{5} = \$11$ billion per year

(b) 1985–1990: $\dfrac{150 - 115}{5} = \$7$ billion per year

(c) 1990–1995: $\dfrac{180 - 150}{5} = \$6$ billion per year

(d) 1995–2000: $\dfrac{260 - 180}{5} = \$16$ billion per year

(e) 1980–2002: $\dfrac{270 - 60}{22} = \$9.5$ billion per year

(f) 1990–2002: $\dfrac{270 - 150}{12} = \$10$ billion per year

Answers will vary.

3. $f'(t) = 2$

Average rate of change: $\dfrac{\Delta y}{\Delta t} = \dfrac{f(2) - f(1)}{2 - 1} = \dfrac{11 - 9}{1}$

$= 2$

Instantaneous rates of change: $f'(1) = 2, \quad f'(2) = 2$

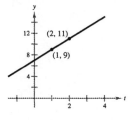

5. $h'(x) = 2x - 4$

Average rate of change: $\dfrac{\Delta y}{\Delta x} = \dfrac{h(2) - h(-2)}{2 - (-2)} = \dfrac{-2 - 14}{4}$

$= -4$

Instantaneous rates of change: $h'(-2) = -8, \ h'(2) = 0$

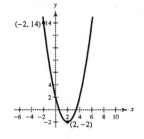

7. $f'(x) = -\dfrac{1}{x^2}$

Average rate of change: $\dfrac{\Delta y}{\Delta x} = \dfrac{f(4) - f(1)}{4 - 1} = \dfrac{(1/4) - 1}{3}$

$= -\dfrac{1}{4}$

Instantaneous rates of change: $f'(1) = -1, \quad f'(4) = -\dfrac{1}{16}$

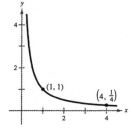

9. $g'(x) = 4x^3 - 2x$

Average rate of change: $\dfrac{g(3) - g(1)}{3 - 1} = \dfrac{74 - 2}{2} = 36$

Instantaneous rates of change: $g'(1) = 2, \quad g'(3) = 102$

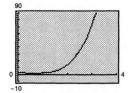

11. (a) $\approx \dfrac{0 - 1400}{3} \approx -467.$

The number of visitors to the park is decreasing at an average rate of 467 people per month from September to December.

(b) At $t = 8$, the instantaneous rate of change is about 0. Answers will vary. For example, $[4, 11]$.

13. $E'(t) = \dfrac{1}{27}(9 + 6t - 3t^2) = \dfrac{1}{9}(3 + 2t - t^2)$

(a) $\dfrac{E(1) - E(0)}{1 - 0} = \dfrac{(11/27) - 0}{1 - 0} = \dfrac{11}{27}$

$E'(0) = \dfrac{1}{3}$

$E'(1) = \dfrac{4}{9}$

(b) $\dfrac{E(2) - E(1)}{2 - 1} = \dfrac{(22/27) - (11/27)}{2 - 1} = \dfrac{11}{27}$

$E'(1) = \dfrac{4}{9}$

$E'(2) = \dfrac{1}{3}$

(c) $\dfrac{E(3) - E(2)}{3 - 2} = \dfrac{1 - (22/27)}{3 - 2} = \dfrac{5}{27}$

$E'(2) = \dfrac{1}{3}$

$E'(3) = 0$

(d) $\dfrac{E(4) - E(3)}{4 - 3} = \dfrac{(20/27) - 1}{4 - 3} = -\dfrac{7}{27}$

$E'(3) = 0$

$E'(4) = -\dfrac{5}{9}$

15. $s = -16t^2 + 555$

(a) Average velocity $= \dfrac{5(3) - 5(2)}{3 - 2} = \dfrac{411 - 491}{1} = -80\,\text{ft/sec}$

(b) $v = s'(t) = -32t$, $v(2) = -64\,\text{ft/sec}$, $v(3) = -96\,\text{ft/sec}$

(c) $\quad s = -16t^2 + 555 = 0$

$16t^2 = 555$

$t^2 = \dfrac{555}{16}$

$t \approx 5.89\,\text{seconds}$

(d) $v(5.89) \approx -188.5\,\text{ft/sec}$

17. $\dfrac{dC}{dx} = 1.47$

19. $\dfrac{dC}{dx} = 470 - 0.5x$

21. $\dfrac{dR}{dx} = 50 - x$

23. $\dfrac{dR}{dx} = -18x^2 + 16x + 200$

25. $\dfrac{dP}{dx} = -4x + 72$

27. $\dfrac{dP}{dx} = -0.0005x + 12.2$

29. $C = 3.6\sqrt{x} + 500$

(a) $C(10) - C(9) \approx \$0.584$

(b) $C'(x) = 1.8/\sqrt{x}$

$C'(9) = \$0.6$ per unit.

(c) The answers are very close.

31. $p = -0.05x^2 + 20x - 1000$

(a) $p(151) - p(150) = 879.95 - 875 = \$4.95$

(b) $p'(x) = -0.1x + 20$

$p'(150) = -0.1(150) + 20 = \$5.00$ per unit.

(c) The results are nearly the same.

33. $T = -0.0375t^2 + 0.3t + 100.4$

(a)

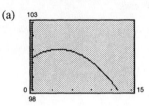

(b) For $t < 4$, the slopes are positve and the fever is going up. For $t > 4$, the slopes are negative and the fever is diminshing.

(c) $T(0) = 100.4°F$

   $T(4) = 101°F$

   $T(8) = 100.4°F$

   $T(12) = 98.6°F$

(d) $\dfrac{dT}{dt} = -0.075t + 0.3$

(e) $T'(0) = 0.3°F$ per hour

   $T'(4) = 0°F$ per hour

   $T'(8) = -0.3°F$ per hour

   $T'(12) = -0.6°F$ per hour

35. $p = 5 - 0.001x, \ C = 35 + 1.5x$

(a) $R = xp = x(5 - 0.001x) = 5x - 0.001x^2$

(b) $P = R - C = (5x - 0.001x^2) - (35 + 1.5x)$

$$= -0.001x^2 + 3.5x - 35$$

(c) $R'(x) = 5 - 0.002x$

$P'(x) = 3.5 - 0.002x$

x	600	1200	1800	2400	3000
dR/dx	3.8	2.6	1.4	0.2	-1.0
dP/dx	2.3	1.1	-0.1	-1.3	-2.5
P	1705	2725	3025	2605	1465

37. (a) The slope is $\dfrac{500 - 400}{0.50 - 0.75} = -400$. Hence

$$x - 400 = -400(p - 0.75)$$

$$x = -400p + 700$$

$$p = \frac{1}{400}(700 - x).$$

(p is price, x is number of glasses sold.)

$$P = R - C = xp - C$$

$$= \frac{x}{400}(700 - x) - (0.05x + 20)$$

$$= -0.0025x^2 + 1.7x - 20$$

(b)

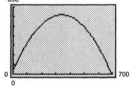

$P'(x) = -0.005x + 1.7$

$P'(200) > 0 \implies$ Profit increasing, slope positive

$P'(400) < 0 \implies$ Profit decreasing, slope negative

(c) $P'(200) = 0.7$

$P'(400) = -0.3$

39. $(36,000, 6), (30,000, 7)$

$$\text{Slope} = \frac{7 - 6}{30,000 - 36,000} = \frac{1}{-6000}$$

$$p - 6 = \frac{-1}{6000}(x - 36,000)$$

$$p = \frac{-1}{6000}x + 12 \qquad \text{(demand function)}$$

(a) $P = R - C = xp - (0.20x + 85,000)$

$$= x\left(\frac{-1}{6000}x + 12\right) - 0.2x - 85,000$$

$$= \frac{-x^2}{6000} + 11.8x - 85,000$$

(b)

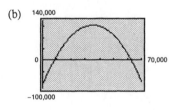

(c) $P'(x) = \dfrac{-x}{3000} + 11.8$

$P'(18,000) = 5.8$ dollars, positive

$P'(36,000) = -0.20$ dollars, negative

41. $p = \dfrac{50}{\sqrt{x}}$, $1 \le x < 8000$, $C = 0.5x + 500$

$$P = R - C = xp - C = x\left(\dfrac{50}{\sqrt{x}}\right) - (0.5x + 500)$$

$$= 50\sqrt{x} - 0.5x - 500$$

$$P'(x) = \dfrac{25}{\sqrt{x}} - 0.5$$

(a) $P'(900) = \$0.33$ per unit

(b) $P'(1600) = \$0.13$ per unit

(c) $P'(2500) = \$0$ per unit

(d) $P'(3600) = \$-0.08$ per unit

$P'(2500) = 0$ indicates that $x = 2500$ is the optimal value of x. Hence, $p = 50/\sqrt{x} = 50/\sqrt{2500} = \$1.00$.

43. (a) 1790 to 1900: $\dfrac{480 \text{ miles}}{110 \text{ years}} \approx 4.4 \text{ mi/yr}$

(b) 1900 to 2000: $\dfrac{350 \text{ miles}}{100 \text{ years}} \approx 3.5 \text{ mi/yr}$

(c) 1790 to 2000: $\dfrac{810 \text{ miles}}{210 \text{ years}} \approx 3.85 \text{ mi/yr}$

45. $N = f(p)$

(a) $f'(1.479)$ is the rate of change of a gallon of gasoline when the price is \$1.479/gallon.

(b) In general, it should be negative. Demand tends to decrease as price increases. Answers will vary.

47. $f(x) = \frac{1}{4}x^3$, $f'(x) = \frac{3}{4}x^2$

f has a horizontal tangent at $x = 0$.

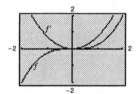

49. The population in each phase is increasing. During the acceleration phase the growth is the greatest. Therefore, the slopes of the tangent lines are greater than the slopes of the tangent lines during the lag phase and the deceleration phase. Possible reasons for the changing rates could be seasonal growth or food supplies.

Section 2.4   The Product and Quotient Rules

1. $f'(x) = x^2(9x^2) + 2x(3x^3 - 1) = 15x^4 - 2x$

$f'(1) = 13$

3. $f'(x) = \frac{1}{3}(6x^2) = 2x^2$

$f'(0) = 0$

5. $g'(x) = (x^2 - 4x + 3)(1) + (2x - 4)(x - 2)$

$$= 3x^2 - 12x + 11$$

$g'(4) = (16 - 16 + 3) + (8 - 4)(4 - 2) = 3 + 8 = 11$

7. $h'(x) = \dfrac{(x - 5)(1) - (x)(1)}{(x - 5)^2} = \dfrac{-5}{(x - 5)^2}$

$h'(6) = -5$

9. $f(t) = \dfrac{2t^2 - 3}{3t + 1}$

$$f'(t) = \dfrac{(3t + 1)(4t) - (2t^2 - 3)3}{(3t + 1)^2}$$

$$= \dfrac{6t^2 + 4t + 9}{(3t + 1)^2}$$

$$f'(3) = \dfrac{6(9) + 4(3) + 9}{(9 + 1)^2} = \dfrac{75}{100} = \dfrac{3}{4}$$

11. $g(x) = \dfrac{2x + 1}{x - 5}$

$$g'(x) = \dfrac{(x - 5)(2) - (2x + 1)(1)}{(x - 5)^2} = \dfrac{-11}{(x - 5)^2}$$

$$g'(6) = \dfrac{-11}{(6 - 5)^2} = -11$$

13. $f(t) = \dfrac{t^2 - 1}{t + 4}$

$f'(t) = \dfrac{(t + 4)(2t) - (t^2 - 1)(1)}{(t + 4)^2} = \dfrac{t^2 + 8t + 1}{(t + 4)^2}$

$f'(1) = \dfrac{1 + 8 + 1}{5^2} = \dfrac{2}{5}$

Function	Rewrite	Differentiate	Simplify
15. $y = \dfrac{x^2 + 2x}{x}$	$y = x + 2, \quad x \neq 0$	$y' = 1, \quad x \neq 0$	$y' = 1, \quad x \neq 0$
17. $y = \dfrac{7}{3x^3}$	$y = \dfrac{7}{3}x^{-3}$	$y' = -7x^{-4}$	$y' = -\dfrac{7}{x^4}$
19. $y = \dfrac{4x^2 - 3x}{8\sqrt{x}}$	$y = \dfrac{1}{2}x^{3/2} - \dfrac{3}{8}x^{1/2}$	$y' = \dfrac{3}{4}x^{1/2} - \dfrac{3}{16}x^{-1/2}$	$y' = \dfrac{3}{4}\sqrt{x} - \dfrac{3}{16\sqrt{x}}$
21. $y = \dfrac{x^2 - 4x. + 3}{x - 1}$	$y = x - 3, \, x \neq 1$	$y' = 1, \, x \neq 1$	$y' = 1, \, x \neq 1$

23. $f'(x) = (x^3 - 3x)(4x + 3) + (3x^2 - 3)(2x^2 + 3x + 5)$

$\quad = 4x^4 + 3x^3 - 12x^2 - 9x + 6x^4 + 9x^3 + 9x^2 - 9x - 15$

$\quad = 10x^4 + 12x^3 - 3x^2 - 18x - 15$

25. $g(t) = (2t^3 - 1)^2 = 4t^6 - 4t^3 + 1$

$\quad g'(t) = 24t^5 - 12t^2 = 12t^2(2t^3 - 1)$

27. $f(x) = \sqrt[3]{x}\left(\sqrt{x} + 3\right) = x^{1/3}(x^{1/2} + 3) = x^{5/6} + 3x^{1/3}$

$\quad f'(x) = \dfrac{5}{6}x^{-1/6} + x^{-2/3} = \dfrac{5}{6x^{1/6}} + \dfrac{1}{x^{2/3}}$

29. $f(x) = \dfrac{3x - 2}{2x - 3}$

$\quad f'(x) = \dfrac{(2x - 3)(3) - (3x - 2)(2)}{(2x - 3)^2} = \dfrac{-5}{(2x - 3)^2}$

31. $f(x) = \dfrac{3 - 2x - x^2}{x^2 - 1}$

$\quad = \dfrac{(3 + x)(1 - x)}{(x + 1)(x - 1)} = \dfrac{-(3 + x)}{x + 1}, \quad x \neq 1$

$\quad f'(x) = \dfrac{(x + 1)(-1) + (3 + x)(1)}{(x + 1)^2} = \dfrac{2}{(x + 1)^2}, \quad x \neq 1$

33. $f(x) = x\left(1 - \dfrac{2}{x + 1}\right) = x - \dfrac{2x}{x + 1}$

$\quad f'(x) = 1 - \dfrac{(x + 1)(2) - (2x)(1)}{(x + 1)^2}$

$\quad = 1 - \dfrac{2}{(x + 1)^2} = \dfrac{x^2 + 2x - 1}{(x + 1)^2}$

35. $g(s) = \dfrac{s^2 - 2s + 5}{\sqrt{s}} = s^{3/2} - 2s^{1/2} + 5s^{-1/2}$

$\quad g'(s) = \dfrac{3}{2}s^{1/2} - s^{-1/2} - \dfrac{5}{2}s^{-3/2} = \dfrac{3s^2 - 2s - 5}{2s^{3/2}}$

37. $g(x) = \dfrac{x - 3}{x + 4}(x^2 + 2x + 1) = \dfrac{x^3 - x^2 - 5x - 3}{x + 4}$

$\quad g'(x) = \dfrac{(x + 4)(3x^2 - 2x - 5) - (x^3 - x^2 - 5x - 3)(1)}{(x + 4)^2}$

$\quad = \dfrac{3x^3 + 10x^2 - 13x - 20 - x^3 + x^2 + 5x + 3}{(x + 4)^2}$

$\quad = \dfrac{2x^3 + 11x^2 - 8x - 17}{(x + 4)^2}$

39. $f(x) = (x - 1)^2(x - 2) = (x^2 - 2x + 1)(x - 2)$

$f'(x) = (x^2 - 2x + 1)(1) + (x - 2)(2x - 2)$

$f'(0) = 1 + 4 = 5$

$y + 2 = 5(x - 0)$

$\quad y = 5x - 2$

41. $f(x) = \dfrac{x - 2}{x + 1}$

$f'(x) = \dfrac{(x + 1) - (x - 2)}{(x + 1)^2} = \dfrac{3}{(x + 1)^2}$

$f'(1) = \dfrac{3}{4}$

$y + \dfrac{1}{2} = \dfrac{3}{4}(x - 1)$

$\quad y = \dfrac{3}{4}x - \dfrac{5}{4}$

43. $f(x) = \dfrac{x + 5}{x - 1}(2x + 1) = \dfrac{2x^2 + 11x + 5}{(x - 1)}$

$f'(x) = \dfrac{(x - 1)(4x + 11) - (2x^2 + 11x + 5)}{(x - 1)^2}$

$\quad\quad = \dfrac{2x^2 - 4x - 16}{(x - 1)^2}$

$f'(0) = -16$

$y + 5 = -16(x - 0)$

$\quad y = -16x - 5$

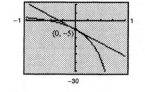

45. $f'(x) = \dfrac{(x - 1)(2x) - x^2(1)}{(x - 1)^2} = \dfrac{x^2 - 2x}{(x - 1)^2}$

$f'(x) = 0$ when $x^2 - 2x = x(x - 2) = 0$, therefore $x = 0$ or $x = 2$. Thus, the horizontal tangent lines occur at $(0, 0)$ and $(2, 4)$.

47. $f'(x) = \dfrac{(x^3 + 1)(4x^3) - x^4(3x^2)}{(x^3 + 1)^2} = \dfrac{x^6 + 4x^3}{(x^3 + 1)^2}$

$f'(x) = 0$ when $x^6 + 4x^3 = x^3(x^3 + 4) = 0$. Thus, the horizontal tangent lines occur at $(0, 0)$ and $\left(\sqrt[3]{-4}, -2.117\right)$.

49. $f(x) = x(x + 1) = x^2 + x$

$f'(x) = 2x + 1$

51. $f(x) = x(x + 1)(x - 1)$

$\quad\quad = x^3 - x$

$f'(x) = 3x^2 - 1$

53. $x = 275\left(1 - \dfrac{3p}{5p + 1}\right)$

$\dfrac{dx}{dp} = -275\left[\dfrac{(5p + 1)3 - (3p)(5)}{(5p + 1)^2}\right] = -275\left[\dfrac{3}{(5p + 1)^2}\right]$

When $p = 4$,

$\dfrac{dx}{dp} = -275\left[\dfrac{3}{(21)^2}\right] \approx \$-1.87$ per unit.

57. $P' = 500\left[\dfrac{(50 + t^2)(4) - (4t)(2t)}{(50 + t^2)^2}\right] = 500\left[\dfrac{200 - 4t^2}{(50 + t^2)^2}\right]$

When $t = 2$, $P' = 500\left[\dfrac{184}{(54)^2}\right] \approx 31.55$ bacteria/hour.

59. (a) $x = \dfrac{k}{p^2}, \quad x \geq 5$

$16 = \dfrac{k}{1000^2}$

$16{,}000{,}000 = k$

$x = \dfrac{16{,}000{,}000}{p^2}$

$p^2 = \dfrac{16{,}000{,}000}{x}$

$p = \dfrac{4000}{\sqrt{x}}$

(b) $C = 250x + 10{,}000$

(c) $P = R - C = xp - C$

$= x\left(\dfrac{4000}{\sqrt{x}}\right) - (250x + 10{,}000)$

$= 4000\sqrt{x} - 250x - 10{,}000$

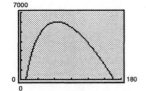

Let $x = 64$ and $p = \$500$.

55. $f'(t) = \dfrac{(t^2 + 1)(2t - 1) - (t^2 - t + 1)(2t)}{(t^2 + 1)^2} = \dfrac{t^2 - 1}{(t^2 + 1)^2}$

(a) $f'(0.5) = -0.48$ per week

(b) $f'(2) = 0.12$ per week

(c) $f'(8) \approx 0.015$ per week

61. (a) $f(x) = x^2 + 1$

$f'(x) = 2x$

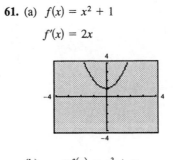

(b) $xf(x) = x^3 + x$

$\dfrac{d}{dx}[xf(x)] = 3x^2 + 1$

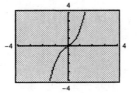

(c) $\dfrac{1}{f(x)} = \dfrac{1}{x^2 + 1}$

$\dfrac{d}{dx}\left[\dfrac{1}{f(x)}\right] = \dfrac{(x^2 + 1)(0) - (1)(2x)}{(x^2 + 1)^2} = -\dfrac{2x}{(x^2 + 1)^2}$

Since (a) and (b) are increasing functions, only (c) could represent a demand function.

63. $$C = 100\left(\frac{200}{x^2} + \frac{x}{x + 30}\right), \quad x \geq 1$$

$$C' = 100\left[-2(200x^{-3}) + \frac{(x + 30) - x}{(x + 30)^2}\right] = 100\left[-\frac{400}{x^3} + \frac{30}{(x + 30)^2}\right]$$

(a) $C'(10) = 100\left(-\frac{400}{10^3} + \frac{30}{40^2}\right) = -38.125$

(b) $C'(15) \approx -10.37$

(c) $C'(20) \approx -3.8$

Increasing the order size reduces the cost per item.

65. $P = \dfrac{1.47 - 0.311t + 0.0173t^2}{1 - 0.206t + 0.0112t^2} \cdot \dfrac{10,000}{10,000}$

$ = \dfrac{14,700 - 3110t + 173t^2}{10,000 - 2060t + 112t^2}$

$P'(t) = \dfrac{(10,000 - 2060t + 112t^2)(-3110 + 346t) - (14,700 - 3110t + 173t^2)(-2060 + 224t)}{(10,000 - 2060t + 112t^2)^2}$

$ = \dfrac{-5(403t^2 - 8360t + 40,900)}{4(28t^2 - 515t + 2500)^2}$

$P'(5) \approx -0.0294$

$P'(7) \approx -0.0373$

$P'(9) \approx 0.1199$

$P'(11) \approx 0.0577$

These derivatives give the rate of change of price at year t.

Section 2.5   The Chain Rule

$y = f(g(x))$	$u = g(x)$	$y = f(u)$		$y = f(g(x))$	$u = g(x)$	$y = f(u)$
1. $y = (6x - 5)^4$	$u = 6x - 5$	$y = u^4$	**3.**	$y = (4 - x^2)^{-1}$	$u = 4 - x^2$	$y = u^{-1}$
5. $y = \sqrt{5x - 2}$	$u = 5x - 2$	$y = \sqrt{u}$	**7.**	$y = \dfrac{1}{3x + 1}$	$u = 3x + 1$	$y = u^{-1}$

9. $f(x) = \dfrac{2}{1 - x^3} = 2(1 - x^3)^{-1}$ is most efficiently done by the General Power Rule (c).

11. $f(x) = \sqrt[3]{8^2}$ is most efficiently done by the Constant Rule (b).

13. (a) First rewrite $f(x) = x + 2x^{-1}$, then use the simple power rule.

15. (c) Rewrite as $f(x) = 2(x - 2)^{-1}$ and use the General Power Rule.

17. $y' = 3(2x - 7)^2(2) = 6(2x - 7)^2$

19. $g'(x) = 3(4 - 2x)^2(-2) = -6(4 - 2x)^2$

21. $h'(x) = 2(6x - x^3)(6 - 3x^2) = 6x(6 - x^2)(2 - x^2)$

23. $f'(x) = \dfrac{2}{3}(x^2 - 9)^{-1/3}(2x) = \dfrac{4x}{3(x^2 - 9)^{1/3}}$

25. $f(t) = \sqrt{t + 1} = (t + 1)^{1/2}$

$f'(t) = \dfrac{1}{2}(t + 1)^{-1/2}(1) = \dfrac{1}{2\sqrt{t + 1}}$

27. $s(t) = \sqrt{2t^2 + 5t + 2} = (2t^2 + 5t + 2)^{1/2}$

$s'(t) = \dfrac{1}{2}(2t^2 + 5t + 2)^{-1/2}(4t + 5) = \dfrac{4t + 5}{2\sqrt{2t^2 + 5t + 2}}$

29. $y = \sqrt[3]{9x^2 + 4} = (9x^2 + 4)^{1/3}$

$y' = \dfrac{1}{3}(9x^2 + 4)^{-2/3}(18x) = \dfrac{6x}{(9x^2 + 4)^{2/3}}$

31. $f(x) = -3(2 - 9x)^{1/4}$

$f'(x) = -3\left(\dfrac{1}{4}\right)(2 - 9x)^{-3/4}(-9) = \dfrac{27}{4(2 - 9x)^{3/4}}$

33. $h'(x) = -\dfrac{4}{3}(4 - x^3)^{-7/3}(-3x^2) = \dfrac{4x^2}{(4 - x^3)^{7/3}}$

35. $f'(x) = 2(3)(x^2 - 1)^2(2x) = 12x(x^2 - 1)^2$

$f'(2) = 24(3^2) = 216$

$f(2) = 54$

$y - 54 = 216(x - 2)$

$y = 216x - 378$

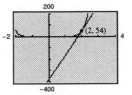

37. $f(x) = \sqrt{4x^2 - 7} = (4x^2 - 7)^{1/2}$

Point: $(2, f(2)) = (2, 3)$

$f'(x) = \dfrac{1}{2}(4x^2 - 7)^{-1/2}(8x) = \dfrac{4x}{\sqrt{4x^2 - 7}}$

When $x = 2$, the slope is $f'(2) = \dfrac{8}{3}$, and the equation of the tangent line is

$y - 3 = \dfrac{8}{3}(x - 2)$

$y = \dfrac{8}{3}x - \dfrac{7}{3}$

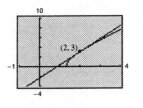

39. $f'(x) = \dfrac{1}{2}(x^2 - 2x + 1)^{-1/2}(2x - 2)$

$= \dfrac{x - 1}{\sqrt{x^2 - 2x + 1}}$

$= \dfrac{x - 1}{|x - 1|}$

$f'(2) = 1$

$f(2) = 1$

$y - 1 = 1(x - 2)$

$y = x - 1$

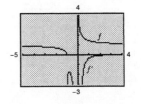

41. $f(x) = \dfrac{\sqrt{x} + 1}{x^2 + 1}$

$f'(x) = \dfrac{1 - 3x^2 - 4x^{3/2}}{2\sqrt{x}(x^2 + 1)^2}$

f has a horizontal tangent when $f' = 0$.

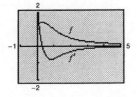

43. $f(x) = \sqrt{\dfrac{x + 1}{x}}$

$f'(x) = -\dfrac{1}{2x^{3/2}\sqrt{x + 1}}$

f' is never 0.

45. $y = (x - 2)^{-1}$

$$y' = (-1)(x - 2)^{-2}(1) = -\frac{1}{(x - 2)^2}$$

47. $y = -4(t + 2)^{-2}$

$$y' = 8(t + 2)^{-3} = \frac{8}{(t + 2)^3}$$

49. $f(x) = (x^2 - 3x)^{-2}$

$$f'(x) = -2(x^2 - 3x)^{-3}(2x - 3)$$

$$= \frac{6 - 4x}{(x^2 - 3x)^3} = \frac{-2(2x - 3)}{x^3(x - 3)}$$

51. $g(t) = (t^2 - 2)^{-1}$

$$g'(t) = -(t^2 - 2)^{-2}(2t) = \frac{-2t}{(t^2 - 2)^2}$$

53. $f'(x) = x(3)(3x - 9)^2(3) + (3x - 9)^3(1)$

$$= (3x - 9)^2[9x + (3x - 9)]$$

$$= 9(x - 3)^2(12x - 9)$$

$$= 27(x - 3)^2(4x - 3)$$

55. $y = x(2x + 3)^{1/2}$

$$y' = x\left[\frac{1}{2}(2x + 3)^{-1/2}(2)\right] + (2x + 3)^{1/2}$$

$$= (2x + 3)^{-1/2}[x + (2x + 3)]$$

$$= \frac{3(x + 1)}{\sqrt{2x + 3}}$$

57. $y = t^2(t - 2)^{1/2}$

$$y' = t^2\left[\frac{1}{2}(t - 2)^{-1/2}(1)\right] + 2t(t - 2)^{1/2}$$

$$= \frac{1}{2}(t - 2)^{-1/2}[t^2 + 4t(t - 2)]$$

$$= \frac{t^2 + 4t(t - 2)}{2\sqrt{t - 2}}$$

$$= \frac{t(5t - 8)}{2\sqrt{t - 2}}$$

59. $f(x) = \sqrt{\dfrac{3 - 2x}{4x}} = \sqrt{\dfrac{3}{4}x^{-1} - \dfrac{1}{2}} = \left(\dfrac{3}{4}x^{-1} - \dfrac{1}{2}\right)^{1/2}$

$$f'(x) = \frac{1}{2}\left(\frac{3}{4}x^{-1} - \frac{1}{2}\right)^{-1/2}\left(-\frac{3}{4}x^{-2}\right)$$

$$= \frac{-3}{8x^2\sqrt{(3/4x) - (1/2)}}$$

$$= -\frac{3}{4x^{3/2}\sqrt{3 - 2x}}$$

61. $f(x) = (x^2 + 1)^{1/2} - (x^2 - 1)^{1/2}$

$$f'(x) = \frac{1}{2}(x^2 + 1)^{-1/2}(2x) - \frac{1}{2}(x^2 - 1)^{-1/2}(2x)$$

$$= \frac{x}{\sqrt{x^2 + 1}} - \frac{x}{\sqrt{x^2 - 1}}$$

63. $y = \left(\dfrac{6 - 5x}{x^2 - 1}\right)^2$

$$y' = 2\left(\frac{6 - 5x}{x^2 - 1}\right)\left[\frac{(x^2 - 1)(-5) - (6 - 5x)(2x)}{(x^2 - 1)^2}\right]$$

$$= \frac{2(6 - 5x)(5x^2 - 12x + 5)}{(x^2 - 1)^3}$$

65. $f(t) = \dfrac{36}{(3 - t)^2} = 36(3 - t)^{-2}$

$$f'(t) = -72(3 - t)^{-3}(-1) = \frac{72}{(3 - t)^3}$$

$$f'(0) = \frac{72}{27} = \frac{8}{3}$$

$$y - 4 = \frac{8}{3}(t - 0)$$

$$y = \frac{8}{3}t + 4$$

67. $f(t) = (t^2 - 9)(t + 2)^{1/2}$

$$f'(t) = (t^2 - 9)\frac{1}{2}(t + 2)^{-1/2} + (t + 2)^{1/2}(2t)$$

$$f'(-1) = (-8)\frac{1}{2} + (-2) = -6$$

$$y - (-8) = -6[x - (-1)]$$

$$y = -6x - 14$$

69. $f(x) = \dfrac{x + 1}{\sqrt{2x - 3}}$

$f'(x) = \dfrac{\sqrt{2x - 3}(1) - (x + 1)(2x - 3)^{-1/2}}{(2x - 3)}$

$f'(2) = \dfrac{1 - 3}{1} = -2$

$y - 3 = -2(x - 2)$

$y = -2x + 7$

71. $A = 1000\left(1 + \dfrac{r}{12}\right)^{60}$

$A' = 1000(60)\left(1 + \dfrac{r}{12}\right)^{59}\left(\dfrac{1}{12}\right) = 5000\left(1 + \dfrac{r}{12}\right)^{59}$

(a) $A'(0.08) = 50\left(1 + \dfrac{0.08}{12}\right)^{59} \approx \$74.00$ per percentage point

(b) $A'(0.10) = 50\left(1 + \dfrac{0.10}{12}\right)^{59} \approx \$81.59$ per percentage point

(c) $A'(0.12) = 50\left(1 + \dfrac{0.12}{12}\right)^{59} \approx \$89.94$ per percentage point

73. $N = 400[1 - 3(t^2 + 2)^{-2}] = 400 - 1200(t^2 + 2)^{-2}$

$\dfrac{dN}{dt} = 2400(t^2 + 2)^{-3}(2t) = \dfrac{4800t}{(t^2 + 2)^3}$

The rate of growth of N is decreasing.

t	0	1	2	3	4
$\dfrac{dN}{dt}$	0	177.78	44.44	10.82	3.29

75. (a) $V = \dfrac{k}{\sqrt[3]{t + 1}}$

When $t = 0$, $V = 10{,}000 \Rightarrow k = 10{,}000$.

Therefore, $V = \dfrac{10{,}000}{\sqrt[3]{t + 1}}$.

(b) $V' = -\dfrac{10{,}000}{3(t + 1)^{4/3}}$

When $t = 1$, $\dfrac{dV}{dt} = -\dfrac{10{,}000}{3(2)^{4/3}} \approx -\$1322.83$ per year.

(c) When $t = 3$, $\dfrac{dV}{dt} = -\dfrac{10{,}000}{3(4)^{4/3}} \approx -\$524.97$ per year.

77. False. By the chain rule, $y' = \frac{1}{2}(1 - x)^{-1/2}(-1) = -\frac{1}{2}(1 - x)^{-1/2}$

Section 2.6   Higher-Order Derivatives

1. $f'(x) = -4$

$f''(x) = 0$

3. $f'(x) = 2x + 7$

$f''(x) = 2$

5. $g'(t) = t^2 - 8t + 2$

$g''(t) = 2t - 8$

7. $f(t) = \dfrac{3}{4}t^{-2}$

$f'(t) = -\dfrac{3}{2}t^{-3}$

$f''(t) = \dfrac{9}{2}t^{-4} = \dfrac{9}{2t^4}$

9. $f(x) = 3(2 - x^2)^3$

$f'(x) = 9(2 - x^2)^2(-2x) = -18x(2 - x^2)^2$

$f''(x) = (-18x)2(2 - x^2)(-2x) + (2 - x^2)^2(-18)$

$= 18(2 - x^2)[4x^2 - (2 - x^2)]$

$= 18(2 - x^2)(5x^2 - 2)$

11. $f'(x) = \dfrac{(x - 1)(1) - (x + 1)(1)}{(x - 1)^2}$

$= -\dfrac{2}{(x - 1)^2} = -2(x - 1)^{-2}$

$f''(x) = 4(x - 1)^{-3}(1) = \dfrac{4}{(x - 1)^3}$

13. $y = x^2(x^2 + 4x + 8) = x^4 + 4x^3 + 8x^2$

$y' = 4x^3 + 12x^2 + 16x$

$y'' = 12x^2 + 24x + 16$

15. $f'(x) = 5x^4 - 12x^3$

$f''(x) = 20x^3 - 36x^2$

$f'''(x) = 60x^2 - 72x$

17. $f(x) = 5x(x + 4)^3$

$= 5x(x^3 + 12x^2 + 48x + 64)$

$= 5x^4 + 60x^3 + 240x^2 + 320x$

$f'(x) = 20x^3 + 180x^2 + 480x + 320$

$f''(x) = 60x^2 + 360x + 480$

$f'''(x) = 120x + 360$

19. $f(x) = \dfrac{3}{16}x^{-2}$

$f'(x) = -\dfrac{3}{8}x^{-3}$

$f''(x) = \dfrac{9}{8}x^{-4}$

$f'''(x) = -\dfrac{9}{2}x^{-5} = -\dfrac{9}{2x^5}$

21. $g'(t) = 20t^3 + 20t$

$g''(t) = 60t^2 + 20$

$g''(2) = 60(4) + 20 = 260$

23. $f(x) = (4 - x)^{1/2}$

$f'(x) = -\dfrac{1}{2}(4 - x)^{-1/2}$

$f''(x) = -\dfrac{1}{4}(4 - x)^{-3/2}$

$f'''(x) = -\dfrac{3}{8}(4 - x)^{-5/2} = \dfrac{-3}{8(4 - x)^{5/2}}$

$f'''(-5) = \dfrac{-3}{8(9)^{5/2}} = \dfrac{-1}{648}$

25. $f(x) = x^2(3x^2 + 3x - 4) = 3x^4 + 3x^3 - 4x^2$

$f'(x) = 12x^3 + 9x^2 - 8x$

$f''(x) = 36x^2 + 18x - 8$

$f'''(x) = 72x + 18$

$f'''(-2) = 72(-2) + 18 = -126$

27. $f''(x) = 4x$

29. $f''(x) = \dfrac{2x - 2}{x} = 2 - \dfrac{2}{x}$

$f'''(x) = 0 - \left(-\dfrac{2}{x^2}\right) = \dfrac{2}{x^2}$

31. $f^{(5)}(x) = 2(x + 1)$

$f^{(6)}(x) = 2$

33. $f'(x) = 3x^2 - 18x + 27$

$f''(x) = 6x - 18 = 0$

$f''(x) = 0$ when $x = 3$.

35. $f(x) = (x + 3)(x - 4)(x + 5)$

$= x^3 + 4x^2 - 17x - 60$

$f'(x) = 3x^2 + 8x - 17$

$f''(x) = 6x + 8$

$f''(x) = 0$ when $6x + 8 = 0$

$x = -\dfrac{4}{3}.$

37. $f(x) = x\sqrt{x^2 - 1}$

$$f'(x) = x\frac{1}{2}(x^2 - 1)^{-1/2}(2x) + (x^2 - 1)^{1/2}$$

$$= \frac{x^2}{\sqrt{x^2 - 1}} + \sqrt{x^2 - 1}$$

$$f''(x) = \frac{(x^2 - 1)^{1/2}(2x) - x^2\frac{1}{2}(x^2 - 1)^{-1/2}(2x)}{x^2 - 1} + \frac{1}{2}(x^2 - 1)^{-1/2}(2x)$$

$$= \frac{(x^2 - 1)(2x) - x^3}{(x^2 - 1)^{3/2}} + \frac{x}{(x^2 - 1)^{1/2}} \cdot \frac{x^2 - 1}{x^2 - 1}$$

$$= \frac{2x^3 - 3x}{(x^2 - 1)^{3/2}}$$

$$f''(x) = 0 \implies 2x^3 - 3x = x(2x^2 - 3) = 0$$

$$\implies x = \pm\sqrt{\frac{3}{2}} = \pm\frac{\sqrt{6}}{2}$$

(**Note:** $x = 0$ is not in the domain of f.)

39. $f'(x) = \dfrac{(x^2 + 3)(1) - (x)(2x)}{(x^2 + 3)^2} = \dfrac{3 - x^2}{(x^2 + 3)^2} = (3 - x^2)(x^2 + 3)^{-2}$

$$f''(x) = (3 - x^2)[-2(x^2 + 3)^{-3}(2x)] + (x^2 + 3)^{-2}(-2x)$$

$$= -2x(x^2 + 3)^{-3}[2(3 - x^2) + (x^2 + 3)]$$

$$= \frac{-2x(9 - x^2)}{(x^2 + 3)^3}$$

$$= \frac{2x(x^2 - 9)}{(x^2 + 3)^3}$$

$$f''(x) = 0 \text{ when } 2x(x^2 - 9) = 0$$

$$x = 0, \pm 3.$$

41. (a) $s(t) = -16t^2 + 144t$

(b) $v(t) = s'(t) = -32t + 144$

$a(t) = v'(t) = -32$

(c) $v(t) = 0 = -32t + 144$ when $t = \frac{144}{32} = 4.5$ sec.

$s(4.5) = 324$ feet

(d) $s(t) = 0$ when $16t^2 = 144t$, or $t = 0, 9$ sec.

$v(9) = -144$ ft/sec, which is the same speed as the initial velocity.

43. $\dfrac{d^2s}{dt^2} = \dfrac{(t + 10)(90) - (90t)(1)}{(t + 10)^2} = \dfrac{900}{(t + 10)^2}$

t	0	10	20	30	40	50	60
$\dfrac{ds}{dt}$	0	45	60	67.5	72	75	77.14
$\dfrac{d^2s}{dt^2}$	9	2.25	1	0.56	0.36	0.25	0.18

As time increases, the acceleration decreases. After 1 minute, the automobile has traveled about 77.14 feet.

45. $f(x) = x^2 - 6x + 6$

$f'(x) = 2x - 6$

$f''(x) = 2$

The degrees of the successive derivatives decrease by 1.

47. The degree of f is 3, and the degrees of the successive derivatives decrease by 1.

49. (a)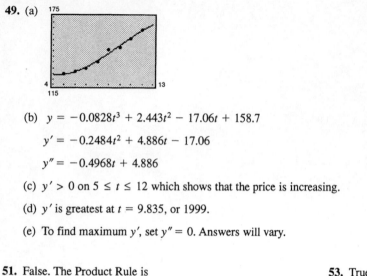

(b) $y = -0.0828t^3 + 2.443t^2 - 17.06t + 158.7$

   $y' = -0.2484t^2 + 4.886t - 17.06$

   $y'' = -0.4968t + 4.886$

(c) $y' > 0$ on $5 \le t \le 12$ which shows that the price is increasing.

(d) y' is greatest at $t = 9.835$, or 1999.

(e) To find maximum y', set $y'' = 0$. Answers will vary.

51. False. The Product Rule is

   $[f(x)g(x)]' = f'(x)g(x) + f(x)g'(x).$

53. True.

   $h'(c) = f'(c)g(c) + f(c)g'(c) = 0$

55. True

57. Let $y = xf(x).$

   Then, $y' = xf'(x) + f(x)$

   $y'' = xf''(x) + f'(x) + f'(x) = xf''(x) + 2f'(x)$

   $y''' = xf'''(x) + f''(x) + 2f''(x) = xf'''(x) + 3f''(x).$

   In general

   $y^{(n)} = [xf(x)]^{(n)} = xf^{(n)}(x) + nf^{(n-1)}(x).$

Section 2.7   Implicit Differentiation

1. $5xy = 1$

   $5xy' + 5y = 0$

   $5xy' = -5y$

   $y' = -\dfrac{y}{x}$

3. $y^2 = 1 - x^2$

   $2yy' = -2x$

   $y' = -\dfrac{x}{y}$

5. $x^2y^2 - 4y = 1$

   $x^2 2yy' + 2xy^2 - 4y' = 0$

   $(2x^2y - 4)y' = -2xy^2$

   $y' = \dfrac{2xy^2}{4 - 2x^2y}$

   $\quad = \dfrac{xy^2}{2 - x^2y}$

7. $4y^2 - xy = 2$

   $8yy' - xy' - y = 0$

   $(8y - x)y' = y$

   $y' = \dfrac{y}{8y - x}$

9. $\dfrac{2y - x}{y^2 - 3} = 5$

$2y - x = 5y^2 - 15$

$2\dfrac{dy}{dx} - 1 = 10y\dfrac{dy}{dx}$

$-1 = (10y - 2)\dfrac{dy}{dx}$

$\dfrac{dy}{dx} = -\dfrac{1}{2(5y - 1)}$

11. $\dfrac{x + y}{2x - y} = 1$

$x + y = 2x - y$

$1 + y' = 2 - y'$

$2y' = 1$

$y' = \dfrac{1}{2}$

Note: $x + y = 2x - y$ simplifies to $2y = x$ and hence $y' = \frac{1}{2}$.

13. $x^2 + y^2 = 49$

$2x + 2yy' = 0$

$y' = -\dfrac{x}{y}$

At $(0, 7)$, $y' = -\dfrac{0}{7} = 0$.

15. $y + xy = 4$

$\dfrac{dy}{dx} + x\dfrac{dy}{dx} + y = 0$

$\dfrac{dy}{dx}(1 + x) = -y$

$\dfrac{dy}{dx} = -\dfrac{y}{x + 1}$

At $(-5, -1)$, $\dfrac{dy}{dx} = -\dfrac{1}{4}$.

17. $x^3 - xy + y^2 = 4$

$3x^2 - x\dfrac{dy}{dx} - y + 2y\dfrac{dy}{dx} = 0$ (Product Rule)

$\dfrac{dy}{dx}(2y - x) = y - 3x^2$

$\dfrac{dy}{dx} = \dfrac{y - 3x^2}{2y - x}$

At $(0, -2)$, $\dfrac{dy}{dx} = \dfrac{1}{2}$.

19. $x^3y^3 - y = x$

$3x^3y^2\dfrac{dy}{dx} + 3x^2y^3 - \dfrac{dy}{dx} = 1$

$\dfrac{dy}{dx}(3x^3y^2 - 1) = 1 - 3x^2y^3$

$\dfrac{dy}{dx} = \dfrac{1 - 3x^2y^3}{3x^3y^2 - 1}$

At $(0, 0)$, $\dfrac{dy}{dx} = -1$.

21. $x^{1/2} + y^{1/2} = 9$

$\dfrac{1}{2}x^{-1/2} + \dfrac{1}{2}y^{-1/2}\dfrac{dy}{dx} = 0$

$x^{-1/2} + y^{-1/2}\dfrac{dy}{dx} = 0$

$\dfrac{dy}{dx} = \dfrac{-x^{-1/2}}{y^{-1/2}} = -\sqrt{\dfrac{y}{x}}$

At $(16, 25)$, $\dfrac{dy}{dx} = -\dfrac{5}{4}$.

23. $x^{2/3} + y^{2/3} = 5$

$\dfrac{2}{3}x^{-1/3} + \dfrac{2}{3}y^{-1/3}\dfrac{dy}{dx} = 0$

$\dfrac{dy}{dx} = \dfrac{-x^{-1/3}}{y^{-1/3}} = -\dfrac{y^{1/3}}{x^{1/3}} = -\sqrt[3]{\dfrac{y}{x}}$

At $(8, 1)$, $\dfrac{dy}{dx} = -\dfrac{1}{2}$.

25. $3x^2 - 2y + 5 = 0$

$6x - 2\dfrac{dy}{dx} = 0$

$\dfrac{dy}{dx} = 3x$

At $(1, 4)$, $\dfrac{dy}{dx} = 3$.

27. $x^2 + y^2 = 4$

$2x + 2yy' = 0$

$y' = -\dfrac{x}{y}$

At $(0, 2)$, $y' = 0$.

29. $4x^2 + 9y^2 = 36$

$8x + 18yy' = 0$

$y' = -\dfrac{4x}{9y}$

At $\left(\sqrt{5}, \dfrac{4}{3}\right)$, $y' = -\dfrac{4\sqrt{5}}{9(4/3)} = -\dfrac{\sqrt{5}}{3}$.

31. Implicitly: $2x + 2y\dfrac{dy}{dx} = 0$

$$\frac{dy}{dx} = -\frac{x}{y}$$

Explicitly: $y = \pm\sqrt{25 - x^2}$

$$\frac{dy}{dx} = \pm\left(\frac{1}{2}\right)(25 - x^2)^{-1/2}(-2x)$$

$$= \pm\frac{-x}{\sqrt{25 - x^2}} = -\frac{x}{\pm\sqrt{25 - x^2}} = -\frac{x}{y}$$

At $(-4, 3)$, $\dfrac{dy}{dx} = \dfrac{4}{3}$.

33. Implicitly: $1 - 2yy' = 0$

$$y' = \frac{1}{2y}$$

Explicitly: $y = \pm\sqrt{x - 1}$

$$= \pm(x - 1)^{1/2}$$

$$y' = \pm\frac{1}{2}(x - 1)^{-1/2}(1)$$

$$= \pm\frac{1}{2\sqrt{x - 1}}$$

$$= \frac{1}{2(\pm\sqrt{x - 1})}$$

$$= \frac{1}{2y}$$

At $(2, -1)$, $y' = -\frac{1}{2}$.

35. $x^2 + y^2 = 169$

$$2x + 2y\frac{dy}{dx} = 0$$

$$\frac{dy}{dx} = -\frac{x}{y}$$

At $(5, 12)$:

$$m = -\frac{5}{12}$$

$$y - 12 = -\frac{5}{12}(x - 5)$$

$$y = -\frac{5}{12}x + \frac{169}{12}$$

At $(-12, 5)$:

$$m = \frac{12}{5}$$

$$y - 5 = \frac{12}{5}(x + 12)$$

$$y = \frac{12}{5}x + \frac{169}{5}$$

37. $y^2 = 5x^3$

$$2yy' = 15x^2$$

$$y' = \frac{15x^2}{2y}$$

At $\left(1, \sqrt{5}\right)$:

$$y' = \frac{15}{2\sqrt{5}}$$

$$y - \sqrt{5} = \frac{15}{2\sqrt{5}}(x - 1)$$

$$2\sqrt{5}y - 10 = 15x - 15$$

$$15x - 2\sqrt{5}y - 5 = 0$$

At $\left(1, -\sqrt{5}\right)$:

$$y' = \frac{-15}{2\sqrt{5}}$$

$$y + \sqrt{5} = \frac{-15}{2\sqrt{5}}(x - 1)$$

$$2\sqrt{5}y + 10 = -15x + 15$$

$$15x + 2\sqrt{5}y - 5 = 0$$

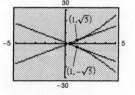

39. $x^3 + y^3 = 8$

$3x^2 + 3y^2 y' = 0$

$3y^2 y' = -3x^2$

$y' = -\dfrac{x^2}{y^2}$

At $(0, 2)$, $y' = 0$ and $y = 2$ (tangent line).

At $(2, 0)$, y' is undefined and $x = 2$ (tangent line).

41. $p = 0.006x^4 + 0.02x^2 + 10, \quad x \geq 0$

$\dfrac{dp}{dp} = 1 = 0.024x^3 \dfrac{dx}{dp} + 0.04x \dfrac{dx}{dp}$

$1 = (0.024x^3 + 0.04x) \dfrac{dx}{dp}$

$\dfrac{dx}{dp} = \dfrac{1}{0.024x^3 + 0.04x}$

43. $p = \sqrt{\dfrac{200 - x}{2x}}, \quad 0 < x \leq 200$

$2xp^2 = 200 - x$

$2x(2p) + 2\dfrac{dx}{dp}p^2 = -\dfrac{dx}{dp}$

$(2p^2 + 1)\dfrac{dx}{dp} = -4xp$

$\dfrac{dx}{dp} = \dfrac{-4xp}{2p^2 + 1}$

45. (a) $100x^{0.75}y^{0.25} = 135{,}540$

$100x^{0.75}\left(0.25y^{-0.75}\dfrac{dy}{dx}\right) + y^{0.25}(75x^{-0.25}) = 0$

$\dfrac{25x^{0.75}}{y^{0.75}}\dfrac{dy}{dx} = -\dfrac{75y^{0.25}}{x^{0.25}}$

$\dfrac{dy}{dx} = -\dfrac{3y}{x}$

When $x = 1500$ and $y = 1000$, $\dfrac{dy}{dx} = -2$.

(b)

If more labor is used, then less capital is available.
If more capital is used, then less labor is available.

Section 2.8   Related Rates

1. $y = x^2 - \sqrt{x}, \quad \dfrac{dy}{dt} = 2x\dfrac{dx}{dt} - \dfrac{1}{2\sqrt{x}}\dfrac{dx}{dt} = \left(2x - \dfrac{1}{2\sqrt{x}}\right)\dfrac{dx}{dt}$

(a) When $x = 4$ and $\dfrac{dx}{dt} = 8$, $\dfrac{dy}{dt} = \left(2(4) - \dfrac{1}{2\sqrt{4}}\right)8 = 62$.

(b) When $x = 16$ and $\dfrac{dy}{dt} = 12$, we have

$12 = \left(2(16) - \dfrac{1}{2\sqrt{16}}\right)\dfrac{dx}{dt} = \left(32 - \dfrac{1}{8}\right)\dfrac{dx}{dt} = \dfrac{255}{8}\dfrac{dx}{dt}, \quad \dfrac{dx}{dt} = \dfrac{812}{255} = \dfrac{32}{85}.$

3. $xy = 4, \quad x\dfrac{dy}{dt} + y\dfrac{dx}{dt} = 0, \quad \dfrac{dy}{dt} = \left(-\dfrac{y}{x}\right)\dfrac{dx}{dt}, \quad \dfrac{dx}{dt} = \left(-\dfrac{x}{y}\right)\dfrac{dy}{dt}$

(a) When $x = 8$, $y = \dfrac{1}{2}$, and $\dfrac{dx}{dt} = 10$, $\dfrac{dy}{dt} = -\dfrac{1/2}{8}(10) = -\dfrac{5}{8}$.

(b) When $x = 1$, $y = 4$, and $\dfrac{dy}{dt} = -6$, $\dfrac{dx}{dt} = -\dfrac{1}{4}(-6) = \dfrac{3}{2}$.

5. $A = \pi r^2$, $\dfrac{dr}{dt} = 2$, $\dfrac{dA}{dt} = 2\pi r \dfrac{dr}{dt}$

(a) When $r = 6$, $\dfrac{dA}{dt} = 2\pi(6)(2) = 24\pi$ in²/min.

(b) When $r = 24$, $\dfrac{dA}{dt} = 2\pi(24)(2) = 96\pi$ in²/min.

7. $A = \pi r^2$, $\dfrac{dA}{dt} = 2\pi r \dfrac{dr}{dt}$

If $\dfrac{dr}{dt}$ is constant, then $\dfrac{dA}{dt}$ is not constant; $\dfrac{dA}{dt}$ is proportional to r.

9. $V = \dfrac{4}{3}\pi r^3$, $\dfrac{dV}{dt} = 20$, $\dfrac{dV}{dt} = 4\pi r^2 \dfrac{dr}{dt}$, $\dfrac{dr}{dt} = \left(\dfrac{1}{4\pi r^2}\right)\dfrac{dV}{dt}$

(a) When $r = 1$, $\dfrac{dr}{dt} = \dfrac{1}{4\pi(1)^2}(20) = \dfrac{5}{\pi}$ ft/min.

(b) When $r = 2$, $\dfrac{dr}{dt} = \dfrac{1}{4\pi(2)^2}(20) = \dfrac{5}{4\pi}$ ft/min.

11. (a) $C = 125{,}000 + 0.75x$

$\dfrac{dC}{dt} = 0.75\dfrac{dx}{dt} = 0.75(150) = 112.5$ dollars per week

(b) $R = 250x - \dfrac{1}{10}x^2$

$\dfrac{dR}{dt} = 250\dfrac{dx}{dt} - \dfrac{1}{5}x\dfrac{dx}{dt} = 250(150) - \dfrac{1}{5}(1000)(150)$

$= 7500$ dollars per week

(c) $P = R - C$

$\dfrac{dP}{dt} = \dfrac{dR}{dt} - \dfrac{dC}{dt} = 7500 - 112.5$

$= 7387.5$ dollars per week

13. $V = x^3$, $\dfrac{dx}{dt} = 3$, $\dfrac{dV}{dt} = 3x^2 \dfrac{dx}{dt}$

(a) When $x = 1$, $\dfrac{dV}{dt} = 3(1)^2(3) = 9$ cm³/sec.

(b) When $x = 10$, $\dfrac{dV}{dt} = 3(10)^2(3) = 900$ cm³/sec.

15. $y = x^2$, $\dfrac{dx}{dt} = 2$, $\dfrac{dy}{dt} = 2x \dfrac{dx}{dt}$

(a) When $x = -3$, $\dfrac{dy}{dt} = 2(-3)(2) = -12$ cm/min.

(b) When $x = 0$, $\dfrac{dy}{dt} = 2(0)(2) = 0$ cm/min.

(c) When $x = 1$, $\dfrac{dy}{dt} = 2(1)(2) = 4$ cm/min.

(d) When $x = 3$, $\dfrac{dy}{dt} = 2(3)(2) = 12$ cm/min.

17. Let y be the distance from the ground to the top of the ladder and let x be the distance from the house to the base of the ladder.

$x^2 + y^2 = 25^2$, $2x\dfrac{dx}{dt} + 2y\dfrac{dy}{dt} = 0$, $\dfrac{dy}{dt} = \dfrac{-x}{y}\dfrac{dx}{dt} = \dfrac{-2x}{y}$ since $\dfrac{dx}{dt} = 2$.

(a) When $x = 7$, $y = \sqrt{576} = 24$, $\dfrac{dy}{dt} = \dfrac{-2(7)}{24} = \dfrac{-7}{12}$ ft/sec.

(b) When $x = 15$, $y = \sqrt{400} = 20$, $\dfrac{dy}{dt} = \dfrac{-2(15)}{20} = \dfrac{-3}{2}$ ft/sec.

(c) When $x = 24$, $y = 7$, $\dfrac{dy}{dt} = \dfrac{-2(24)}{7} = \dfrac{-48}{7}$ ft/sec.

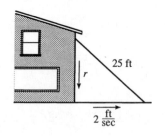

19. (a) $L^2 = x^2 + y^2$, $\dfrac{dx}{dt} = -450$, $\dfrac{dy}{dt} = -600$, and $\dfrac{dL}{dt} = \dfrac{x(dx/dt) + y(dy/dt)}{L}$

When $x = 150$ and $y = 200$, $L = 250$ and $\dfrac{dL}{dt} = \dfrac{150(-450) + 200(-600)}{250} = -750$ mph.

(b) $t = \dfrac{250}{750} = \dfrac{1}{3}$ hr $= 20$ min

21. $s^2 = 90^2 + x^2$, $x = 26$, $\dfrac{dx}{dt} = -30$

$2s\dfrac{ds}{dt} = 2x\dfrac{dx}{dt} \Longrightarrow \dfrac{ds}{dt} = \dfrac{x}{s}\dfrac{dx}{dt}$

When $x = 26$,

$\quad s = \sqrt{90^2 + 26^2} \approx 93.68$

$\quad \dfrac{ds}{dt} = \dfrac{26}{93.68}(-30) \approx -8.33$ ft/sec.

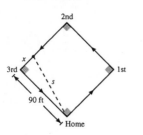

23. $V = \pi r^2 h$, $h = 0.08$, $V = 0.08\pi r^2$, $\dfrac{dV}{dt} = 0.16\pi r\dfrac{dr}{dt}$

When $r = 750$ and $\dfrac{dr}{dt} = \dfrac{1}{2}$,

$\quad \dfrac{dV}{dt} = 0.16\pi(750)\left(\dfrac{1}{2}\right)$

$\quad\quad = 60\pi$

$\quad\quad \approx 188.5$ ft^3/min.

25. $P = R - C = xp - C = x(6000 - 0.4x^2) - (2400x + 5200) = -0.4x^3 + 3600x - 5200$

$\dfrac{dP}{dt} = -1.2x^2\dfrac{dx}{dt} + 3600\dfrac{dx}{dt}$

$\dfrac{dx}{dt} = \dfrac{1}{3600 - 1.2x^2}\dfrac{dP}{dt}$

When $x = 44$ and $\dfrac{dP}{dt} = 6384$, $\dfrac{dx}{dt} = \dfrac{1}{3600 - 1.2(44)^2}(6384) = 5$ units per week.

Review Exercises for Chapter 2

1. Slope $\approx \dfrac{-4}{2} = -2$

3. Slope ≈ 0

5. $t = 7$: slope $\approx \$5500$ million per year per year
(sales are increasing)

$\quad t = 10$: slope $\approx \$7500$ million per year per year
(sales are increasing)

$\quad t = 12$: slope $\approx \$3500$ million per year per year
(sales are increasing)

7. At $t = 0$, slope ≈ 180.

At $t = 4$, slope ≈ -70.

At $t = 6$, slope ≈ -900.

9. $f'(x) = \lim_{\Delta x \to 0} \dfrac{f(x + \Delta x) - f(x)}{\Delta x}$

$= \lim_{\Delta x \to 0} \dfrac{-3(x + \Delta x) - 5 - (-3x - 5)}{\Delta x}$

$= \lim_{\Delta x \to 0} \dfrac{-3\Delta x}{\Delta x} = -3$

$f'(-2) = -3$

11. $f'(x) = \lim_{\Delta x \to 0} \dfrac{f(x + \Delta x) - f(x)}{\Delta x}$

$= \lim_{\Delta x \to 0} \dfrac{(x + \Delta x)^2 - 4(x + \Delta x) - (x^2 - 4x)}{\Delta x}$

$= \lim_{\Delta x \to 0} \dfrac{x^2 + 2x\Delta x + (\Delta x)^2 - 4\Delta x - x^2}{\Delta x}$

$= \lim_{\Delta x \to 0} \dfrac{2x\Delta x + (\Delta x)^2 - 4\Delta x}{\Delta x}$

$= \lim_{\Delta x \to 0} (2x + \Delta x - 4) = 2x - 4$

$f'(1) = 2(1) - 4 = -2$

13. $f'(x) = \lim_{\Delta x \to 0} \dfrac{f(x + \Delta x) - f(x)}{\Delta x}$

$= \lim_{\Delta x \to 0} \dfrac{\sqrt{x + \Delta x + 9} - \sqrt{x + 9}}{\Delta x} \cdot \dfrac{\sqrt{x + \Delta x + 9} + \sqrt{x + 9}}{\sqrt{x + \Delta x + 9} + \sqrt{x + 9}}$

$= \lim_{\Delta x \to 0} \dfrac{(x + \Delta x + 9) - (x + 9)}{\Delta x\left[\sqrt{x + \Delta x + 9} + \sqrt{x + 9}\right]}$

$= \lim_{\Delta x \to 0} \dfrac{1}{\sqrt{x + \Delta x + 9} + \sqrt{x + 9}} = \dfrac{1}{2\sqrt{x + 9}}$

$f'(-5) = \frac{1}{4}$

15. $f'(x) = \lim_{\Delta x \to 0} \dfrac{f(x + \Delta x) - f(x)}{\Delta x}$

$= \lim_{\Delta x \to 0} \dfrac{\dfrac{1}{x + \Delta x - 5} - \dfrac{1}{x - 5}}{\Delta x}$

$= \lim_{\Delta x \to 0} \dfrac{(x - 5) - (x + \Delta x - 5)}{\Delta x(x + \Delta x - 5)(x - 5)}$

$= \lim_{\Delta x \to 0} \dfrac{-1}{(x + \Delta x - 5)(x - 5)} = \dfrac{-1}{(x - 5)^2}$

$f'(6) = -1$

17. $f(x) = 8 - 5x$

$f'(x) = -5$

$f'(3) = -5$

19. $f(x) = -\frac{1}{2}x^2 + 2x$

$f'(x) = -x + 2$

$f'(2) = -2 + 2 = 0$

21. $f(x) = \sqrt{x} + 2 = x^{1/2} + 2$

$f'(x) = \frac{1}{2}x^{-1/2} = \dfrac{1}{2\sqrt{x}}$

$f'(9) = \dfrac{1}{2\sqrt{9}} = \dfrac{1}{6}$

23. $f(x) = \dfrac{5}{x} = 5x^{-1}$

$f'(x) = -5x^2 = -\dfrac{5}{x^2}$

$f'(1) = \dfrac{-5}{1^2} = -5$

25. $y = \dfrac{x + 1}{x - 1}$ is not differentiable at $x = 1$.

27. $y = \begin{cases} -x - 2, & x \le 0 \\ x^3 + 2, & x > 0 \end{cases}$ is not differentiable at $x = 0$.

29. $g(t) = \dfrac{2}{3}t^{-2}$

$g'(t) = -\dfrac{4}{3}t^{-3} = -\dfrac{4}{3t^3}$

$g'(1) = -\dfrac{4}{3}$

$y - \dfrac{2}{3} = -\dfrac{4}{3}(t - 1)$

$y = -\dfrac{4}{3}t + 2$

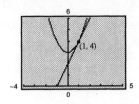

31. $f(x) = x^2 + 3, \quad (1, 4)$

$f'(x) = 2x$

$f'(1) = 2$

$y - 4 = 2(x - 1)$

$y = 2x + 2$

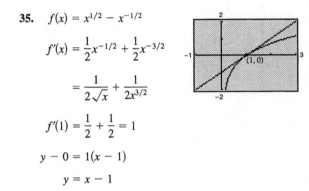

33. $y' = 44x^3 - 10x$

$y'(-1) = -34$

$y - 7 = -34(x + 1)$

$y = -34x - 27$

35. $f(x) = x^{1/2} - x^{-1/2}$

$f'(x) = \dfrac{1}{2}x^{-1/2} + \dfrac{1}{2}x^{-3/2}$

$\qquad = \dfrac{1}{2\sqrt{x}} + \dfrac{1}{2x^{3/2}}$

$f'(1) = \dfrac{1}{2} + \dfrac{1}{2} = 1$

$y - 0 = 1(x - 1)$

$y = x - 1$

37. $f(x) = \dfrac{x^2 + 3}{x} = x + 3x^{-1}$

$f'(x) = 1 - 3x^{-2}$

$f'(1) = -2$

$y - 4 = -2(x - 1)$

$y = -2x + 6$

39. $f(x) = x^2 + 3x - 4, \quad [0, 1]$

Average rate of change $= \dfrac{f(1) - f(0)}{1 - 0} = \dfrac{0 - (-4)}{1} = 4$

$f'(x) = 2x + 3$

$f'(0) = 3$

$f'(1) = 5$

41. (a) 1998-2002: $\dfrac{58{,}360 - 30{,}457}{12 - 8} \approx \$6976$ million per year per year

(b) 1998: $S'(8) \approx \$7317$ million per year per year

2002: $S'(12) \approx \$3876$ million per year per year

(c) Sales were increasing in 1998 thru 2002, and grew at an average rate of \$6976 million over the period 1998−2002.

43. $P = -0.001059t^4 + 0.03015t^3 - 0.2850t^2 + 1.007t + 0.50$

(a) $P'(t) = -0.004236t^3 + 0.09045t^2 - 0.57t + 1.007$

(b) 1997: $P'(7) \approx \$-0.004$ per pound

2000: $P'(10) \approx \$0.116$ per pound

2002: $P'(12) \approx \$-0.128$ per pound

(c)

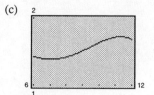

Price is increasing on $(7, 11)$, or 1997 to 2001.
Decreasing on $(6, 7)$ and $(11, 12)$, or 1996-1997 and 201-2002.

(d) Negative slope: $7 < t < 11$

Positive slope: $6 < t < 7$ and $11 < t < 12$

(e) When prices increase, slope is positive.
When prices decrease, slope is negative.

45. (a) $s(t) = -16t^2 + 276$

(b) Average velocity $= \dfrac{s(2) - s(0)}{2 - 0} = \dfrac{-64}{2}$

$= -32$ ft/sec

(c) $v(t) = -32t$

$v(2) = -64$ ft/sec

$v(3) = -96$ ft/sec

(d) $s(t) = -16t^2 + 276 = 0 \Rightarrow t^2 = \dfrac{276}{16} = 17.25$

$t \approx 4.15$ sec

(e) $v(4.15) \approx -32(4.15) = -132.8$ velocity

Speed $= 132.8$ ft/sec at impact

47. $R = 27.50x$

$C = 15x + 2500$

$P = R - C$

$= 27.50x - (15x + 2500)$

$= 12.50x - 2500$

49. $\dfrac{dC}{dx} = 320$

51. $C = 370 + 2.55\sqrt{x}$

$C'(x) = \dfrac{1}{2}(2.55)(x^{-1/2}) = \dfrac{1.275}{\sqrt{x}}$

53. $R = 200x - \dfrac{1}{5}x^2$

$R'(x) = 200 - \dfrac{2}{5}x$

55. $R = \dfrac{35x}{\sqrt{x-2}} = 35x(x-2)^{-1/2}$

$\dfrac{dR}{dx} = 35x\left[-\dfrac{1}{2}(x-2)^{-3/2}\right] + 35(x-2)^{-1/2}$

$= \dfrac{35}{2}(x-2)^{-3/2}[-x + 2(x-2)]$

$= \dfrac{35(x-4)}{2(x-2)^{3/2}}$

57. $\dfrac{dP}{dx} = -0.0006x^2 + 12x - 1$

59. $f(x) = x^3(5 - 3x^2) = 5x^3 - 3x^5$

$f'(x) = 15x^2 - 15x^4 = 15x^2(1 - x^2)$

61. $y = (4x - 3)(x^3 - 2x^2) = 4x^4 - 11x^3 + 6x^2$

$y' = 16x^3 - 33x^2 + 12x$

63. $f(x) = \dfrac{6x - 5}{x^2 + 1}$

$$f'(x) = \dfrac{(x^2 + 1)(6) - (6x - 5)(2x)}{(x^2 + 1)^2}$$

$$= \dfrac{6 + 10x - 6x^2}{(x^2 + 1)^2}$$

$$= \dfrac{2(3 + 5x - 3x^2)}{(x^2 + 1)^2}$$

65. $f(x) = (5x^2 + 2)^3$

$$f'(x) = 3(5x^2 + 2)^2(10x)$$

$$= 30x(5x^2 + 2)^2$$

67. $h(x) = \dfrac{2}{\sqrt{x + 1}} = 2(x + 1)^{-1/2}$

$$h'(x) = 2\left(-\dfrac{1}{2}\right)(x + 1)^{-3/2}(1)$$

$$= -\dfrac{1}{(x + 1)^{3/2}}$$

69. $g(x) = x\sqrt{x^2 + 1} = x(x^2 + 1)^{1/2}$

$$g'(x) = x\left[\dfrac{1}{2}(x^2 + 1)^{-1/2}(2x)\right] + (1)(x^2 + 1)^{1/2}$$

$$= (x^2 + 1)^{-1/2}[x^2 + (x^2 + 1)]$$

$$= \dfrac{2x^2 + 1}{\sqrt{x^2 + 1}}$$

71. $f(x) = x(1 - 4x^2)^2$

$$= x(1 - 8x^2 + 16x^4)$$

$$= x - 8x^3 + 16x^5$$

$$f'(x) = 1 - 24x^2 + 80x^4$$

73. $h(x) = [x^2(2x + 3)]^3 = x^6(2x + 3)^3$

$$h'(x) = x^6[3(2x + 3)^2(2)] + 6x^5(2x + 3)^3$$

$$= 6x^5(2x + 3)^2[x + (2x + 3)]$$

$$= 18x^5(2x + 3)^2(x + 1)$$

75. $f(x) = x^2(x - 1)^5$

$$f'(x) = x^2 5(x - 1)^4 + 2x(x - 1)^5$$

$$= x(x - 1)^4[5x + 2(x - 1)]$$

$$= x(x - 1)^4(7x - 2)$$

77. $h(t) = \dfrac{\sqrt{3t + 1}}{(1 - 3t)^2}$

$$h'(t) = \dfrac{(1 - 3t)^2(1/2)(3t + 1)^{-1/2}(3) - (3t + 1)^{1/2}(2)(1 - 3t)(-3)}{(1 - 3t)^4}$$

$$= \dfrac{(3t + 1)^{-1/2}[(1 - 3t)(3/2) + (3t + 1)6]}{(1 - 3t)^3}$$

$$= \dfrac{3(9t + 5)}{2\sqrt{3t + 1}(1 - 3t)^3}$$

79. $T = \dfrac{1300}{t^2 + 2t + 25} = 1300(t^2 + 2t + 25)^{-1}$

$$T'(t) = -1300(t^2 + 2t + 25)^{-2}(2t + 2) = \dfrac{-2600(t + 1)}{(t^2 + 2t + 25)^2}$$

(a) $T'(1) = \dfrac{-325}{49} \approx -6.63$

$T'(3) = \dfrac{-13}{2} \approx -6.5$

$T'(5) = \dfrac{-13}{3} \approx -4.33$

$T'(10) = \dfrac{-1144}{841} \approx -1.36$

(b)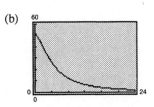

The rate of decrease is approaching zero.

81. $f(x) = 3x^2 + 7x + 1$

$f'(x) = 6x + 7$

$f''(x) = 6$

83. $f'''(x) = -6x^{-4}$

$f^{(4)}(x) = 24x^{-5}$

$f^{(5)}(x) = -120x^{-6} = \dfrac{-120}{x^6}$

85. $f'(x) = 7x^{5/2}$

$f''(x) = \dfrac{35}{2}x^{3/2}$

87. $f''(x) = 6x^{1/3}$

$f'''(x) = 2x^{-2/3} = \dfrac{2}{x^{2/3}}$

89. (a) $s(t) = -16t^2 + 5t + 30$

(b) $s(t) = 0 = -16t^2 + 5t + 30$

Using the Quadratic Formula or a graphing utility, $t \approx 1.534$ seconds.

(c) $v(t) = s'(t) = -32t + 5$

$v(1.534) \approx -44.09$ ft/sec

(d) $a(t) = v'(t) = -32$ ft/sec^2

91. $\qquad x^2 + 3xy + y^3 = 10$

$2x + 3x\dfrac{dy}{dx} + 3y + 3y^2\dfrac{dy}{dx} = 0$

$\dfrac{dy}{dx}(3x + 3y^2) = -2x - 3y$

$\dfrac{dy}{dx} = \dfrac{-2x - 3y}{3x + 3y^2} = -\dfrac{2x + 3y}{3(x + y^2)}$

93. $y^2 - x^2 + 8x - 9y - 1 = 0$

$2yy' - 2x + 8 - 9y' = 0$

$(2y - 9)y' = 2x - 8$

$\dfrac{dy}{dx} = \dfrac{2x - 8}{2y - 9}$

95. $\qquad y^2 = x - y$

$2yy' = 1 - y'$

$2yy' + y' = 1$

$(2y + 1)y' = 1$

$y' = \dfrac{1}{2y + 1}$

At $(2, 1), y' = \frac{1}{3}$.

$y - 1 = \dfrac{1}{3}(x - 2)$

$y = \dfrac{1}{3}x + \dfrac{1}{3}$

97. $\qquad y^2 - 2x = xy$

$2yy' - 2 = xy' + y$

$(2y - x)y' = y + 2$

$y' = \dfrac{y + 2}{2y - x}$

At $(1, 2), y' = \dfrac{4}{3}$.

$y - 2 = \dfrac{4}{3}(x - 1)$

$y = \dfrac{4}{3}x + \dfrac{2}{3}$

99. $b = 8h, \quad 0 \le h \le 5$

$V = \dfrac{1}{2}bh(20) = 10bh = 10(8h)h = 80h^2$

$\dfrac{dV}{dt} = 160h\dfrac{dh}{dt}$

$\dfrac{dh}{dt} = \dfrac{1}{160h}\dfrac{dV}{dt} = \dfrac{1}{16h} \left[\text{since } \dfrac{dV}{dt} = 10 \right]$

When $h = 4, \dfrac{dh}{dt} = \dfrac{1}{16(4)} = \dfrac{1}{64}$ ft/min.

Practice Test for Chapter 2

1. Use the definition of the derivative to find the derivative of $f(x) = 2x^2 + 3x - 5$.

2. Use the definition of the derivative to find the derivative of $f(x) = \dfrac{1}{x - 4}$.

3. Use the definition of the derivative to find the equation of the tangent line to the graph of $f(x) = \sqrt{x - 2}$ at the point $(6, 2)$.

4. Find $f'(x)$ for $f(x) = 5x^3 - 6x^2 + 15x - 9$.

5. Find $f'(x)$ for $f(x) = \dfrac{6x^2 - 4x + 1}{x^2}$.

6. Find $f'(x)$ for $f(x) = \sqrt[3]{x^2} + \sqrt[5]{x^3}$.

7. Find the average rate of change of $f(x) = x^3 - 11$ over the interval $[0, 2]$. Compare this to the instantaneous rate of change at the endpoints of the interval.

8. Given the cost function $C = 6200 + 4.31x - 0.0001x^2$, find the marginal cost of producing x units.

9. Find $f'(x)$ for $f(x) = (x^3 - 4x)(x^2 + 7x - 9)$.

10. Find $f'(x)$ for $f(x) = \dfrac{x + 7}{x^2 - 8}$.

11. Find $f'(x)$ for $f(x) = x^3\left(\dfrac{x - 3}{x + 5}\right)$.

12. Find $f'(x)$ for $f(x) = \dfrac{\sqrt{x}}{x^2 + 4x - 1}$.

13. Find $f'(x)$ for $f(x) = (6x - 5)^{12}$.

14. Find $f'(x)$ for $f(x) = 8\sqrt{4 - 3x}$.

15. Find $f'(x)$ for $f(x) = -\dfrac{3}{(x^2 + 1)^3}$.

16. Find $f'(x)$ for $f(x) = \sqrt{\dfrac{10x}{x + 2}}$.

17. Find $f'''(x)$ for $f(x) = x^4 - 9x^3 + 17x^2 - 4x + 121$.

18. Find $f^{(4)}(x)$ for $f(x) = \sqrt{3 - x}$.

19. Use implicit differentiation to find $\dfrac{dy}{dx}$ for $x^5 + y^5 = 100$.

20. Use implicit differentiation to find $\dfrac{dy}{dx}$ for $x^2y^3 + 2x - 3y + 11 = 0$.

21. Use implicit differentiation to find $\dfrac{dy}{dx}$ for $\sqrt{xy + 4} = 5y - 4x$.

22. Use implicit differentiation to find $\dfrac{dy}{dx}$ for $y^3 = \dfrac{x^3 + 4}{x^3 - 4}$.

23. Let $y = 3x^2$. Find $\dfrac{dx}{dt}$ when $x = 2$ and $\dfrac{dy}{dt} = 5$.

24. The area A of a circle is increasing at a rate of 10 in.^{2}/min. Find the rate of change of the radius r when $r = 4$ inches.

25. The volume of a cone is $V = \left(\dfrac{1}{3}\right)\pi r^2 h$. Find the rate of change of the height when $\dfrac{dV}{dt} = 200$, $h = \dfrac{r}{2}$, and $h = 20$ inches.

Graphing Calculator Required

26. Graph $f(x) = \dfrac{x^2}{x - 2}$ and its derivative on the same set of coordinate axes. From the graph of $f(x)$, determine any points at which the graph has horizontal tangent lines. What is the value of $f'(x)$ at these points?

27. Use a graphing utility to graph $\sqrt[3]{x} + \sqrt[3]{y} = 3$. Then find and sketch the tangent line at the point $(8, 1)$.

CHAPTER 3
Applications of the Derivative

C H A P T E R   3
Applications of the Derivative

Section 3.1   Increasing and Decreasing Functions

Solutions to Odd-Numbered Exercises

1. $f(x) = \dfrac{x^2}{x^2 + 4}$

$f'(x) = \dfrac{(x^2 + 4)(2x) - (x^2)(2x)}{(x^2 + 4)^2} = \dfrac{8x}{(x^2 + 4)^2}$

At $\left(-1, \frac{1}{5}\right)$, f is decreasing since $f'(-1) = -\frac{8}{25}$.

At $(0, 0)$, f has a critical number since $f'(0) = 0$.

At $\left(1, \frac{1}{5}\right)$, f is increasing since $f'(1) = \frac{8}{25}$.

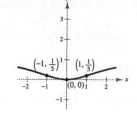

3. $f(x) = (x + 2)^{2/3}$

$f'(x) = \dfrac{2}{3}(x + 2)^{-1/3} = \dfrac{2}{3\sqrt[3]{x + 2}}$

At $(-3, 1)$, f is decreasing since $f'(-3) = -\frac{2}{3}$.

At $(-2, 0)$, f has a critical number since $f'(-2)$ is undefined.

At $(-1, 1)$, f is increasing since $f'(-1) = \frac{2}{3}$.

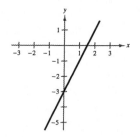

5. $f(x) = -(x + 1)^2$

$f'(x) = -2(x + 1)$

f has a critical number at $x = -1$. Moreover, f is increasing on $(-\infty, -1)$ and decreasing on $(-1, \infty)$.

7. $f(x) = x^4 - 2x^2$

$f'(x) = 4x^3 - 4x = 4x(x^2 - 1)$

f has critical numbers at $x = 0, \pm 1$. Moreover, f is increasing on $(-1, 0)$, $(1, \infty)$ and decreasing on $(-\infty, -1)$, $(0, 1)$.

9. $f(x) = 2x - 3$

$f'(x) = 2$

Since the derivative is positive for all x, the function is increasing for all x.
Thus, there are no critical numbers.
Increasing on $(-\infty, \infty)$.

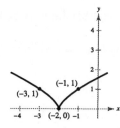

11. $g(x) = -(x - 1)^2$

$g'(x) = -2(x - 1) = 0$

Critical number: $x = 1$

Interval	$-\infty < x < 1$	$1 < x < \infty$
Sign of g'	$g' > 0$	$g' < 0$
Conclusion	Increasing	Decreasing

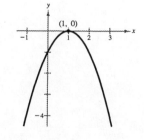

13. $y = x^2 - 5x$

$y' = 2x - 5 = 0$

Critical number: $x = \frac{5}{2}$

Interval	$-\infty < x < \frac{5}{2}$	$\frac{5}{2} < x < \infty$
Sign of y'	$y' < 0$	$y' > 0$
Conclusion	Decreasing	Increasing

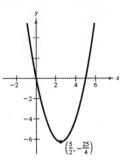

15. $y = x^3 - 6x^2$

$y' = 3x^2 - 12x = 3x(x - 4) = 0$

Critical numbers: $x = 0$ and $x = 4$

Interval	$-\infty < x < 0$	$0 < x < 4$	$4 < x < \infty$
Sign of y'	$y' > 0$	$y' < 0$	$y' > 0$
Conclusion	Increasing	Decreasing	Increasing

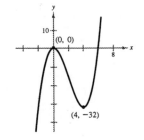

17. $f(x) = \sqrt{x^2 - 1}$. Domain: $(-\infty, -1] \cup [1, \infty)$

$f'(x) = \frac{1}{2}(x^2 - 1)^{-1/2}(2x) = \frac{x}{\sqrt{x^2 - 1}} = 0$

Critical numbers: $x = \pm 1$ ($x = 0$ not in domain)

Interval	$-\infty < x < -1$	$1 < x < \infty$
Sign of f'	$f' < 0$	$f' > 0$
Conclusion	Decreasing	Increasing

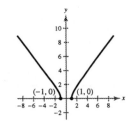

19. $f(x) = -2x^2 + 4x + 3$

$f'(x) = -4x + 4 = 0$

Critical number: $x = 1$

Interval	$-\infty < x < 1$	$1 < x < \infty$
Sign of f'	$f' > 0$	$f' < 0$
Conclusion	Increasing	Decreasing

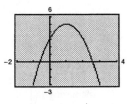

21. $y = 3x^3 + 12x^2 + 15x$

$y' = 9x^2 + 24x + 15 = 3(x + 1)(3x + 5) = 0$

Critical numbers: $x = -1, x = -\frac{5}{3}$

Interval	$-\infty < x < -\frac{5}{3}$	$-\frac{5}{3} < x < -1$	$-1 < x < \infty$
Sign of y'	$y' > 0$	$y' < 0$	$y' > 0$
Conclusion	Increasing	Decreasing	Increasing

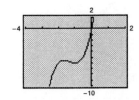

23. $f(x) = x\sqrt{x+1} = x(x+1)^{1/2}$

$$f'(x) = x\left[\frac{1}{2}(x+1)^{-1/2}(1)\right] + (x+1)^{1/2}(1)$$

$$= \frac{1}{2}(x+1)^{-1/2}[x + 2(x+1)]$$

$$= \frac{3x+2}{2\sqrt{x+1}}$$

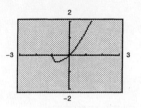

Domain: $[-1, \infty)$

Critical numbers: $x = -1, -\frac{2}{3}$

Interval	$-1 < x < -\frac{2}{3}$	$-\frac{2}{3} < x < \infty$
Sign of f'	$f' < 0$	$f' > 0$
Conclusion	Decreasing	Increasing

25. $f(x) = x^4 - 2x^3$

$$f'(x) = 4x^3 - 6x^2 = 2x^2(2x - 3) = 0$$

Critical numbers: $x = 0$ and $x = \frac{3}{2}$

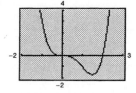

Interval	$-\infty < x < 0$	$0 < x < \frac{3}{2}$	$\frac{3}{2} < x < \infty$
Sign of f'	$f' < 0$	$f' < 0$	$f' > 0$
Conclusion	Decreasing	Decreasing	Increasing

27. $f(x) = \frac{x}{x^2 + 4}$

$$f'(x) = \frac{(x^2 + 4)(1) - (x)(2x)}{(x^2 + 4)^2} = \frac{4 - x^2}{(x^2 + 4)^2}$$

Critical numbers: $x = \pm 2$

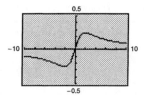

Interval	$-\infty < x < -2$	$-2 < x < 2$	$2 < x < \infty$
Sign of f'	$f' < 0$	$f' > 0$	$f' < 0$
Conclusion	Decreasing	Increasing	Decreasing

29. $f(x) = \frac{2x}{16 - x^2}$

$$f'(x) = \frac{(16 - x^2)2 - 2x(-2x)}{(16 - x^2)^2} = \frac{2x^2 + 32}{(16 - x^2)^2}$$

No critical numbers

Discontinuities: $x = \pm 4$

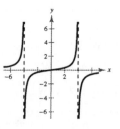

Interval	$-\infty < x < -4$	$-4 < x < 4$	$4 < x < \infty$
Sign of f'	$f' > 0$	$f' > 0$	$f' > 0$
Conclusion	Increasing	Increasing	Increasing

31. $f(x) = \begin{cases} 4 - x^2, & x \le 0 \\ -2x, & x > 0 \end{cases}$

$f'(x) = \begin{cases} -2x, & x < 0 \\ -2, & x > 0 \end{cases}$

$f'(0)$ is undefined.

Discontinuity: $x = 0$

Critical number: $x = 0$

Interval	$-\infty < x < 0$	$0 < x < \infty$
Sign of f'	$f' > 0$	$f' < 0$
Conclusion	Increasing	Decreasing

33. $y = \begin{cases} 3x + 1, & x \le 1 \\ 5 - x^2, & x > 1 \end{cases}$

$y' = \begin{cases} 3, & x < 1 \\ -2x, & x > 1 \end{cases}$

$y'(1)$ is undefined.

Critical number: $x = 1$

($x = 0$ is not a critical number.)

Interval	$-\infty < x < 1$	$1 < x < \infty$
Sign of y'	$y' > 0$	$y' < 0$
Conclusion	Increasing	Decreasing

35. $C = 10\left(\dfrac{1}{x} + \dfrac{x}{x + 3}\right), \quad 1 \le x$

(a) $\dfrac{dC}{dx} = 10\left[-x^{-2} + \dfrac{(x + 3)(1) - (x)(1)}{(x + 3)^2}\right]$

$= 10\left[-\dfrac{1}{x^2} + \dfrac{3}{(x + 3)^2}\right]$

$= 10\left[\dfrac{-(x + 3)^2 + 3x^2}{x^2(x + 3)^2}\right]$

$= 10\left[\dfrac{2x^2 - 6x - 9}{x^2(x + 3)^2}\right]$

By the Quadratic Formula, $2x^2 - 6x - 9 = 0$ when

$x = \dfrac{6 \pm \sqrt{108}}{4} = \dfrac{3 \pm 3\sqrt{3}}{2}.$

The only critical number in the domain is $x \approx 4.10$. Thus, C is decreasing on the interval $[1, 4.10)$ and increasing on $(4.10, \infty)$.

(b)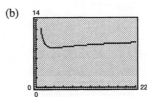

(c) $C = 9$ when $x = 2$ and $x = 15$. Use $x = 4$ to minimize C.

37. Since $s'(t) = 96 - 32t = 0$, the critical number is $t = 3$. Therefore, the ball is moving up on the interval $(0, 3)$ and moving down on $(3, 6)$.

39. $y = 2.743t^3 - 171.55t^2 + 3462.3t + 15,265, \ 0 \le t \le 30$

($t = 0$ corresponds to 1970.)

(a)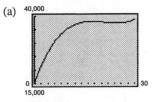

Increasing: $0 < t < 17.1$ (1970 to 1987)

$24.6 < t < 30$ (1994 to 2000)

Decreasing: $17.1 < t < 24.6$ (1987 to 1994)

(b) $y' = 8.229t^2 - 343.1t + 3462.3$

Zeros: $t \approx 17.1, 24.6$

$y' > 0$ for $0 < t < 17.1$ and $24.6 < t < 30$.

$y' < 0$ for $17.1 < t < 24.6$.

Section 3.2   Extrema and the First-Derivative Test

1. $f(x) = -2x^2 + 4x + 3$

$f'(x) = 4 - 4x = 4(1 - x)$

Critical number: $x = 1$

Interval	$(-\infty, 1)$	$(1, \infty)$
Sign of f'	$+$	$-$
f	Increasing	Decreasing

Relative maximum: $(1, 5)$

3. $f(x) = x^2 - 6x$

$f'(x) = 2x - 6 = 2(x - 3)$

Critical number: $x = 3$

Interval	$(-\infty, 3)$	$(3, \infty)$
Sign of f'	$-$	$+$
f	Decreasing	Increasing

Relative minimum: $(3, -9)$

5. $g(x) = 6x^3 - 15x^2 + 12x$

$g'(x) = 18x^2 - 30x + 12$

$\qquad = 6(3x^2 - 5x + 2)$

$\qquad = 6(x - 1)(3x - 2)$

Critical numbers: $x = 1, \frac{2}{3}$

Interval	$\left(-\infty, \frac{2}{3}\right)$	$\left(\frac{2}{3}, 1\right)$	$(1, \infty)$
Sign of g'	$+$	$-$	$+$
g	Increasing	Decreasing	Increasing

Relative maximum: $\left(\frac{2}{3}, \frac{28}{9}\right)$

Relative minimum: $(1, 3)$

7. $h(x) = -(x + 4)^3$

$h'(x) = -3(x + 4)^2$

Critical number: $x = -4$

Interval	$(-\infty, -4)$	$(-4, \infty)$
Sign of h'	$-$	$-$
h	Decreasing	Decreasing

No relative extrema

9. $f(x) = x^3 - 6x^2 + 15$

$f'(x) = 3x^2 - 12x = 3x(x - 4)$

Critical numbers: $x = 0, x = 4$

Interval	$(-\infty, 0)$	$(0, 4)$	$(4, \infty)$
Sign of f'	$+$	$-$	$+$
f	Increasing	Decreasing	Increasing

Relative maximum: $(0, 15)$

Relative minimum: $(4, -17)$

11. $f(x) = x^4 - 2x^3$

$f'(x) = 4x^3 - 6x^2 = 2x^2(2x - 3)$

Critical numbers: $x = 0, x = \frac{3}{2}$

Interval	$(-\infty, 0)$	$\left(0, \frac{3}{2}\right)$	$\left(\frac{3}{2}, \infty\right)$
Sign of f'	$-$	$-$	$+$
f	Decreasing	Decreasing	Increasing

Relative minimum: $\left(\frac{3}{2}, -\frac{27}{16}\right)$

13. $f(x) = (x - 1)^{2/3}$

$f'(x) = \frac{2}{3}(x - 1)^{-1/3}$

$\qquad = \frac{2}{3(x - 1)^{1/3}}$

Critical number: $x = 1$

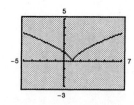

Interval	$(-\infty, 1)$	$(1, \infty)$
Sign of f'	$f' < 0$	$f' > 0$
Conclusion	Decreasing	Increasing

Relative minimum: $(1, 0)$

15. $g(t) = t - \dfrac{1}{2}t^{-2}$

$g'(t) = 1 + t^{-3}$

$\qquad = \dfrac{t^3 + 1}{t^3}$

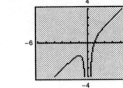

Critical number: $t = -1$

$t = 0$: Discontinuity

Interval	$-\infty < t < -1$	$-1 < t < 0$	$0 < t < \infty$
Sign of g'	$g' > 0$	$g' < 0$	$g' > 0$
Conclusion	Increasing	Decreasing	Increasing

Relative maximum: $\left(-1, -\dfrac{3}{2}\right)$

17. $f(x) = \dfrac{x}{x + 1}$

$f'(x) = \dfrac{(x + 1) - x}{(x + 1)^2} = \dfrac{1}{(x + 1)^2}$

No critical numbers

Discontinuity: $x = -1$

Interval	$(-\infty, -1)$	$(-1, \infty)$
Sign of f'	$+$	$+$
f	Increasing	Increasing

No relative extrema

19. $f(x) = 2(3 - x)$, $[-1, 2]$

$f'(x) = -2$

No critical numbers

x-value	Endpoint $x = -1$	Endpoint $x = 2$
$f(x)$	8	2
Conclusion	Maximum	Minimum

21. $f(x) = 5 - 2x^2$, $[0, 3]$

$f'(x) = -4x$

Critical number: $x = 0$ (also endpoint)

x-value	Endpoint $x = 0$	Endpoint $x = 3$
$f(x)$	5	-13
Conclusion	Maximum	Minimum

23. $f(x) = x^3 - 3x^2$, $[-1, 3]$

$f'(x) = 3x^2 - 6x = 3x(x - 2)$

Critical numbers: $x = 0$ and $x = 2$

x-value	Endpoint $x = -1$	Critical $x = 0$	Critical $x = 2$	Endpoint $x = 3$
$f(x)$	-4	0	-4	0
Conclusion	Minimum	Maximum	Minimum	Maximum

25. $h(s) = \dfrac{1}{3-s} = (3-s)^{-1}, \quad [0, 2]$

$h'(s) = -(3-s)^{-2}(-1) = \dfrac{1}{(3-s)^2}$

No critical numbers

s-value	Endpoint $s = 0$	Endpoint $s = 2$
$h(s)$	$\frac{1}{3}$	1
Conclusion	Minimum	Maximum

27. $f(x) = 3x^{2/3} - 2x, \quad [-1, 2]$

$f'(x) = 2x^{-1/3} - 2 = \dfrac{2}{x^{1/3}} - 2 = \dfrac{2(1 - x^{1/3})}{x^{1/3}}$

Critical numbers: $x = 0, 1$

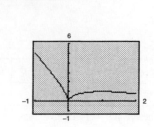

x-value	Endpoint $x = -1$	Critical $x = 0$	Critical $x = 1$	Endpoint $x = 2$
$f(x)$	5	0	1	0.762
Conclusion	Maximum	Minimum		

29. $h(t) = (t - 1)^{2/3}, \quad [-7, 2]$

$h'(t) = \dfrac{2}{3}(t - 1)^{-1/3} = \dfrac{2}{3(t - 1)^{1/3}}$

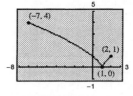

t-value	Endpoint $t = -7$	Critical $t = 1$	Endpoint $t = 2$
$h(t)$	4	0	1
Conclusion	Maximum	Minimum	

31. $f(x) = 0.4x^3 - 1.8x^2 + x - 3, \quad [0, 5]$

Maximum: $(5, 7)$
Minimum: $(2.69, -5.55)$

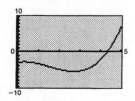

33. $f(x) = \frac{4}{3}x\sqrt{3 - x}, \quad [0, 3]$

Maximum: $(2, 2.\overline{6})$
Minimum: $(0, 0), (3, 0)$

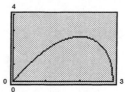

35. $f(x) = \dfrac{4x}{x^2 + 1}$

$f'(x) = \dfrac{(x^2 + 1)(4) - 4x(2x)}{(x^2 + 1)^2} = \dfrac{4(1 - x^2)}{(x^2 + 1)^2}$

Critical number: $x = 1$

x-value	Endpoint $x = 0$	Critical $x = 1$	Interval $(1, \infty)$
$f(x)$	0	2	$0 < f(x) < 2$
Conclusion	Minimum	Maximum	f is decreasing.

37. $f(x) = \dfrac{2x}{x^2 + 4}$

$f'(x) = \dfrac{(x^2 + 4)2 - 2x(2x)}{(x^2 + 4)^2}$

$= \dfrac{8 - 2x^2}{(x^2 + 4)^2}$

$= \dfrac{2(2 - x)(2 + x)}{(x^2 + 4)^2}$

On $[0, \infty)$, $x = 2$ is a critical number.

x-value	Endpoint $x = 0$	Critical $x = 2$
$f(x)$	0	$\frac{1}{2}$
Conclusion	Minimum	Maximum

39. $f(x) = 3x^5 - 10x^3$

$f'(x) = 15x^4 - 30x^2$

$f''(x) = 60x^3 - 60x$

$f'''(x) = 180x^2 - 60 = 60(3x^2 - 1)$

Critical numbers for f'' in $[0, 1]$: $x = \dfrac{1}{\sqrt{3}} = \dfrac{\sqrt{3}}{3}$

x-value	Endpoint $x = 0$	Critical $x = \frac{1}{\sqrt{3}}$	Endpoint $x = 1$
$\lvert f''(x) \rvert$	0	$\dfrac{40}{\sqrt{3}}$	0
Conclusion		Maximum	

41. $f(x) = 15x^4 - \left(\dfrac{2x - 1}{2}\right)^6$, $\quad [0, 1]$

$f'(x) = 60x^3 - 6\left(\dfrac{2x - 1}{2}\right)^5$

$f''(x) = 180x^2 - 30\left(\dfrac{2x - 1}{2}\right)^4$

$f'''(x) = 360x - 120\left(\dfrac{2x - 1}{2}\right)^3$

$f^{(4)}(x) = 360 - 360\left(\dfrac{2x - 1}{2}\right)^2$

$f^{(5)}(x) = -720\left(\dfrac{2x - 1}{2}\right)$

Critical number of $f^{(4)}$: $x = \dfrac{1}{2}$

x-value	Endpoint $x = 0$	Critical $x = \frac{1}{2}$	Endpoint $x = 1$
$\lvert f^{(4)}(x) \rvert$	270	360	270
Conclusion		Maximum	

43. $C = 3x + 20{,}000x^{-1}$, $\quad 0 < x \le 200$

$C' = 3 - 20{,}000x^{-2}$

$= \dfrac{3x^2 - 20{,}000}{x^2}$

Critical numbers: $x = \sqrt{\dfrac{20{,}000}{3}} \approx 81.65 \approx 82$ units

$C(82) \approx 489.90$, which is the minimum by the First-Derivative Test.

45. Demand: $(6000, 0.80), (5600, 1.00)$

$$m = \frac{1 - 0.8}{5600 - 6000} = \frac{0.2}{-400} = -0.0005$$

$$p - 1 = -0.0005(x - 5600)$$

$$p = -0.0005x + 3.80$$

Cost $= C = 5000 + 0.40x$

Profit $= P = R - C$

$$= xp - C$$

$$= x(-0.0005x + 3.80) - (5000 + 0.40x)$$

$$= -0.0005x^2 + 3.40x - 5000$$

$$P' = -0.001x + 3.40 = 0 \Rightarrow x = 3400$$

$$p(3400) = \$2.10 \text{ per can}$$

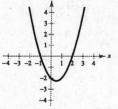

47. (a) $P = 0.00000583t^3 + 0.005003t^2 + 0.13775t + 4.658$,
$-10 \le t \le 200$

($t = 0$ corresponds to 1800.)

Minimum at $t = -10$ (1790) and Maximum at $t = 200$ (2000).

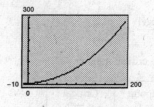

$P(-10) \approx 3.77$ million, $P(200) \approx 278.97$ million

(b) $P' > 0$ on $(-10, 200) \Rightarrow$ minimum at $t = -10$ and maximum at $t = 200$.

(c) The answers are the same.

Section 3.3 Concavity and the Second-Derivative Test

1. $y = x^2 - x - 2$

$y' = 2x - 1$

$y'' = 2$

Concave upward on $(-\infty, \infty)$

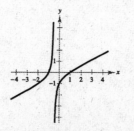

3. $f(x) = \dfrac{x^2 - 1}{2x + 1}$

$$f'(x) = \frac{(2x + 1)(2x) - (x^2 - 1)(2)}{(2x + 1)^2}$$

$$= \frac{2x^2 + 2x + 2}{(2x + 1)^2}$$

$$= (2x^2 + 2x + 2)(2x + 1)^{-2}$$

$$f''(x) = (2x^2 + 2x + 2)[-2(2x + 1)^{-3}(2)] + (2x + 1)^{-2}(4x + 2)$$

$$= -8(x^2 + x + 1)(2x + 1)^{-3} + 2(2x + 1)^{-2}(2x + 1)$$

$$= 2(2x + 1)^{-3}[-4(x^2 + x + 1) + (2x + 1)(2x + 1)]$$

$$= 2(2x + 1)^{-3}[-4x^2 - 4x - 4 + 4x^2 + 4x + 1]$$

$$= \frac{-6}{(2x + 1)^3}$$

$f''(x) \ne 0$ for any value of x.

$x = -\frac{1}{2}$ is a discontinuity.

Concave upward on $\left(-\infty, -\frac{1}{2}\right)$

Concave downward on $\left(-\frac{1}{2}, \infty\right)$

5. $f(x) = 24(x^2 + 12)^{-1}$

$f'(x) = -24(2x)(x^2 + 12)^{-2}$

$f''(x) = -48[x(-2)(2x)(x^2 + 12)^{-3} + (x^2 + 12)^{-2}]$

$\quad = \dfrac{-48(-4x^2 + x^2 + 12)}{(x^2 + 12)^3}$

$\quad = \dfrac{144(x^2 - 4)}{(x^2 + 12)^3}$

$f''(x) = 0$ when $x = \pm 2$.

Concave upward on $(-\infty, -2)$ and $(2, \infty)$

Concave downward on $(-2, 2)$

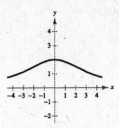

7. $f(x) = -x^3 + 6x^2 - 9x - 1$

$f'(x) = -3x^2 + 12x - 9$

$f''(x) = -6x + 12$

$f''(x) = 0$ at $x = 2$.

Concave upward on $(-\infty, 2)$

Concave downward on $(2, \infty)$

9. $f(x) = 6x - x^2$

$f'(x) = 6 - 2x = 0$

Critical number: $x = 3$

$\quad f''(x) = -2$

$\quad f''(3) = -2 < 0$

Thus, $(3, 9)$ is a relative maximum.

11. $f(x) = x^3 - 5x^2 + 7x$

$f'(x) = 3x^2 - 10x + 7 = (3x - 7)(x - 1)$

Critical numbers: $x = 1$, $x = \frac{7}{3}$

$\quad f''(x) = 6x - 10$

$\quad f''(1) = -4 < 0$

$\quad f''\left(\frac{7}{3}\right) = 4 > 0$

Thus, $(1, 3)$ is a relative maximum and $\left(\frac{7}{3}, \frac{49}{27}\right)$ is a relative minimum.

13. $f(x) = x^{2/3} - 3$

$f'(x) = \dfrac{2}{3}x^{-1/3} = \dfrac{2}{3\sqrt[3]{x}}$

Critical number: $x = 0$

The Second-Derivative Test does not apply, so we use the First-Derivative Test to conclude that $(0, -3)$ is a relative minimum.

15. $f(x) = \sqrt{x^2 + 1}$

$f'(x) = \dfrac{1}{2}(x^2 + 1)^{-1/2}(2x) = \dfrac{x}{\sqrt{x^2 + 1}} = 0$

Critical number: $x = 0$

$\quad f''(x) = x\left[-\dfrac{1}{2}(x^2 + 1)^{-3/2}(2x)\right] + (x^2 + 1)^{-1/2}(1)$

$\quad\quad = (x^2 + 1)^{-3/2}[-x^2 + (x^2 + 1)]$

$\quad\quad = \dfrac{1}{(x^2 + 1)^{3/2}}$

$\quad f''(0) = 1$

Thus, $(0, 1)$ is a relative minimum.

17. $f(x) = \dfrac{x}{x - 1}$

$f'(x) = \dfrac{(x - 1)(1) - (x)(1)}{(x - 1)^2}$

$\quad = -\dfrac{1}{(x - 1)^2}$

No critical numbers

No relative extrema

19. $f(x) = \frac{1}{2}x^4 - \frac{1}{3}x^3 - \frac{1}{2}x^2$

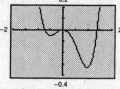

Relative maximum: $(0, 0)$

Relative minimum: $(-0.5, -0.0521)$

21.

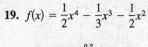

Relative minimum: $(0, 5)$

Relative maximum: $(2, 9)$

23. $f' > 0$ (increasing)

$f'' > 0$ (concave upward)

25. $f' < 0$ (decreasing)

$f'' < 0$ (concave downward)

27. $f(x) = x^3 - 9x^2 + 24x - 18$

$f'(x) = 3x^2 - 18x + 24$

$f''(x) = 6x - 18 = 0 \Rightarrow x = 3$

Interval	$(-\infty, 3)$	$(3, \infty)$
Sign of f''	$-$	$+$
Conclusion	Concave downward	Concave upward

Inflection point: $(3, 0)$

29. $f(x) = (x - 1)^3(x - 5)$

$f'(x) = (x - 1)^3(1) + (x - 5)(3)(x - 1)^2(1)$

$\quad\quad = (x - 1)^2[(x - 1) + 3(x - 5)]$

$\quad\quad = (x - 1)^2(4x - 16)$

$\quad\quad = 4(x - 1)^2(x - 4)$

$f''(x) = 4(x - 1)^2(1) + (x - 4)8(x - 1)(1)$

$\quad\quad = 4(x - 1)[(x - 1) + 2(x - 4)]$

$\quad\quad = 4(x - 1)(3x - 9)$

$\quad\quad = 12(x - 1)(x - 3) = 0 \Rightarrow x = 1 \text{ or } x = 3$

Interval	$(-\infty, 1)$	$(1, 3)$	$(3, \infty)$
Sign of f''	$+$	$-$	$+$
Conclusion	Concave upward	Concave downward	Concave upward

Inflection points: $(1, 0), (3, -16)$

31. $g(x) = 2x^4 - 8x^3 + 12x^2 + 12x$

$g'(x) = 8x^3 - 24x^2 + 24x + 12$

$g''(x) = 24x^2 - 48x + 24 = 24(x - 1)^2$

$g'' > 0$ on $(-\infty, 1)$ and $(1, \infty)$.

No inflection points

33. $h(x) = (x - 2)^3(x - 1)$

$h'(x) = (4x - 5)(x - 2)^2$

$h''(x) = 6(2x - 3)(x - 2)$

Inflection points: $\left(\frac{3}{2}, -\frac{1}{16}\right), (2, 0)$

35. $f(x) = x^3 - 12x$

$f'(x) = 3x^2 - 12 = 3(x^2 - 4)$

Critical numbers: $x = \pm 2$

$\qquad f''(x) = 6x$

$\qquad f''(2) = 12 > 0$

$\qquad f''(-2) = -12 < 0$

Relative maximum: $(-2, 16)$

Relative minimum: $(2, -16)$

$\qquad f''(x) = 0$ when $x = 0$.

$\qquad f''(x) < 0$ on $(-\infty, 0)$.

$\qquad f''(x) > 0$ on $(0, \infty)$.

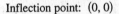

Inflection point: $(0, 0)$

37. $f(x) = x^3 - 6x^2 + 12x$

$f'(x) = 3x^2 - 12x + 12 = 3(x - 2)^2$

Critical number: $x = 2$

$\qquad f''(x) = 6(x - 2)$

$\qquad f''(x) = 0$ when $x = 2$.

Since $f'(x) > 0$ when $x \neq 2$ and the concavity changes at $x = 2$, $(2, 8)$ is an inflection point.

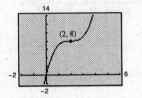

No relative extrema

39. $f(x) = \frac{1}{4}x^4 - 2x^2$

$f'(x) = x^3 - 4x = x(x + 2)(x - 2)$

Critical numbers: $x = \pm 2, \; x = 0$

$\qquad f''(x) = 3x^2 - 4$

$\qquad f''(-2) = 8 > 0$

$\qquad f''(0) = -4 < 0$

$\qquad f''(2) = 8 > 0$

Relative maximum: $(0, 0)$

Relative minima: $(\pm 2, -4)$

$\qquad f''(x) = 3x^2 - 4 = 0$ when $x = \pm\dfrac{2\sqrt{3}}{3}$.

$\qquad f''(x) > 0$ on $\left(-\infty, -\dfrac{2\sqrt{3}}{3}\right)$.

$\qquad f''(x) < 0$ on $\left(-\dfrac{2\sqrt{3}}{3}, \dfrac{2\sqrt{3}}{3}\right)$.

$\qquad f''(x) > 0$ on $\left(\dfrac{2\sqrt{3}}{3}, \infty\right)$.

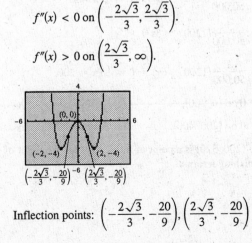

Inflection points: $\left(-\dfrac{2\sqrt{3}}{3}, -\dfrac{20}{9}\right), \left(\dfrac{2\sqrt{3}}{3}, -\dfrac{20}{9}\right)$

41. $g(x) = (x - 2)(x + 1)^2 = x^3 - 3x - 2$

$g'(x) = 3x^2 - 3 = 3(x - 1)(x + 1)$

Critical numbers: $x = \pm 1$

$\qquad f''(x) = 6x$

$\qquad f''(-1) = -6 < 0$

$\qquad f''(1) = 6 > 0$

Relative maximum: $(-1, 0)$

Relative minimum: $(1, -4)$

$\qquad f''(x) = 6x = 0$ when $x = 0$.

$\qquad f''(x) < 0$ on $(-\infty, 0)$.

$\qquad f''(x) > 0$ on $(0, \infty)$.

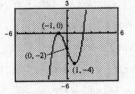

Inflection point: $(0, -2)$

43. $g(x) = x\sqrt{x+3}$

The domain of g is $[-3, \infty)$.

$$g'(x) = x\left[\frac{1}{2}(x+3)^{-1/2}\right] + \sqrt{x+3} = \frac{3x+6}{2\sqrt{x+3}}$$

Critical numbers: $x = -3, \ x = -2$

By the First-Derivative Test, $(-2, -2)$ is a relative minimum.

$$g''(x) = \frac{\left(2\sqrt{x+3}\right)(3) - (3x+6)\left(1/\sqrt{x+3}\right)}{4(x+3)}$$

$$= \frac{3(x+4)}{4(x+3)^{3/2}}$$

$x = -4$ is not in the domain of g. On $[-3, \infty)$, $g''(x) > 0$ and is concave upward.

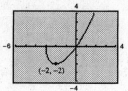

45. $f(x) = \dfrac{4}{1+x^2}$

$$f'(x) = \frac{-8x}{(1+x^2)^2}$$

Critical number: $x = 0$

$$f''(x) = \frac{-8(1-3x^2)}{(1+x^2)^3}$$

$$f''(0) = -8 < 0$$

Thus, $(0, 4)$ is a relative maximum.

$f''(x) = 0$ when $1 - 3x^2 = 0$, $x = \pm\dfrac{\sqrt{3}}{3}$.

$$f''(x) > 0 \text{ on } \left(-\infty, -\frac{\sqrt{3}}{3}\right).$$

$$f''(x) < 0 \text{ on } \left(-\frac{\sqrt{3}}{3}, \frac{\sqrt{3}}{3}\right).$$

$$f''(x) > 0 \text{ on } \left(\frac{\sqrt{3}}{3}, \infty\right).$$

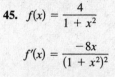

Inflection points: $\left(\dfrac{\sqrt{3}}{3}, 3\right), \left(-\dfrac{\sqrt{3}}{3}, 3\right)$

47. The function has x-intercepts at $(2, 0)$ and $(4, 0)$. On $(-\infty, 3)$, f is decreasing, and on $(3, \infty)$, f is increasing. A relative minimum occurs when $x = 3$. The graph of f is concave upward.

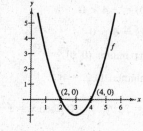

49. (a) $f'(x) > 0$ on $(-\infty, 0)$ where f is increasing.

(b) $f'(x) < 0$ on $(0, \infty)$ where f is decreasing.

(c) f' is not increasing. f is not concave upward.

(d) f' is decreasing on $(-\infty, \infty)$ where f is concave downward.

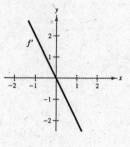

51. $R = \dfrac{1}{50,000}(600x^2 - x^3), \quad 0 \le x \le 400$

$$R' = \frac{1}{50,000}(1200x - 3x^2)$$

$$R'' = \frac{1}{50,000}(1200 - 6x) = 0 \text{ when } x = 200.$$

$R'' > 0$ on $(0, 200)$.

$R'' < 0$ on $(200, 400)$.

Since $(200, 320)$ is a point of inflection, it is the point of diminishing returns.

53. $C = 0.5x^2 + 15x + 5000$

$\overline{C} = 0.5x + 15 + \dfrac{5000}{x}$

$\overline{C}' = 0.5 - \dfrac{5000}{x^2}$

Critical numbers: $x = \pm 100$

$x = 100$ units

55. $N(t) = -0.12t^3 + 0.54t^2 + 8.22t, \quad 0 \le t \le 4$

$N'(t) = -0.36t^2 + 1.08t + 8.22$

$N''(t) = -0.72t + 1.08 = 0$

$t = \dfrac{1.08}{0.72} = 1.5$ hours

Thus, the time is 8:30 P.M.

57. $x = 10,000\left[\dfrac{t^2}{9 + t^2}\right]$

$\dfrac{dx}{dt} = 10,000\left[\dfrac{(9 + t^2)(2t) - (t^2)(2t)}{(9 + t^2)^2}\right]$

$= 10,000\left[\dfrac{18t}{(9 + t^2)^2}\right]$

$= 180,000\left[\dfrac{t}{(9 + t^2)^2}\right]$

$\dfrac{d^2x}{dt^2} = 180,000\left[\dfrac{(9 + t^2)^2(1) - t(2)(9 + t^2)(2t)}{(9 + t^2)^4}\right]$

$= 180,000\left[\dfrac{9 - 3t^2}{(9 + t^2)^3}\right] = 0$

$9 = 3t^2 \Rightarrow t = \sqrt{3} \approx 1.732$ years

59. $f(x) = \dfrac{1}{2}x^3 - x^2 + 3x - 5, \quad [0, 3]$

$f'(x) = \dfrac{3}{2}x^2 - 2x + 3$

$f''(x) = 3x - 2$

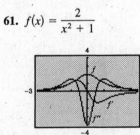

Minimum: $(0, -5)$

Maximum: $(3, 8.5)$

Point of inflection: $\left(\tfrac{2}{3}, -3.2963\right)$

61. $f(x) = \dfrac{2}{x^2 + 1}$

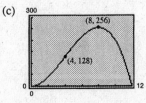

Relative maximum: $(0, 2)$

Inflection points: $(0.58, 1.5), (-0.58, 1.5)$

63. (a) Relative maximum: $(2, 2150)$

Relative minimum: $(1, 2050)$

Absolute maximum: $(0, 2250)$

Absolute minimum: $(7.5, 1740)$

The market opened at the maximum for the day and closed at the minimum. At approximately 9:30 A.M. the market started to recover after falling, and at approximately 10:30 A.M. the market started to fall again.

(b) $(4, 2010)$ is the approximate point of inflection. At approximately 12:30 P.M. the market began to fall at a greater rate.

65. $N(t) = -t^3 + 12t^2, \quad 0 \le t \le 12$

(a) $N'(t) = -3t^2 + 24t = 3t(8 - t) = 0 \Rightarrow t = 8$

At $t = 8$, 256 people will be infected.

(b) $N''(t) = -6t + 24 = 0 \Rightarrow t = 4$.

At $t = 4$, the virus is spreading most rapidly.

(c)

Section 3.4   Optimization Problems

1. Let x be the first number and y be the second number. Then $x + y = 110$ and $y = 110 - x$. Thus, the product of x and y is given by the following.

$$P = xy = x(110 - x)$$

$$P' = 110 - 2x$$

$P' = 0$ when $x = 55$. Since $P''(55) = -2 < 0$, the product is a maximum when $x = 55$ and $y = 110 - 55 = 55$.

3. Let x be the first number and y be the second number. Then $x + 2y = 36$ and $x = 36 - 2y$. The product of x and y is given by the following.

$$P = xy = (36 - 2y)y$$

$$P' = 36 - 4y$$

$P' = 0$ when $y = 9$. Since $P''(9) = -4 < 0$, the product is a maximum when $y = 9$ and $x = 36 - 2(9) = 18$.

5. Let x be the first number and y be the second number. Then $xy = 192$ and $y = 192/x$. The sum of x and y is given by the following.

$$S = x + y = x + \frac{192}{x}$$

$$S' = 1 - \frac{192}{x^2}$$

$S' = 0$ when $x = \sqrt{192}$. Since $S''(x) > 0$ when $x > 0$, S is minimum when $x = \sqrt{192} = 8\sqrt{3}$ and $y = 192/\sqrt{192} = \sqrt{192} = 8\sqrt{3}$.

7. $S = x + \dfrac{1}{x}, \quad x > 0$

$$S' = 1 - \frac{1}{x^2} = \frac{x^2 - 1}{x^2}$$

Critical number: $x = 1$

$$S'' = \frac{2}{x^3}$$

Since $S''(1) = 2 > 0$, $(1, 2)$ is a relative minimum and the sum is a minimum when $x = 1$.

9. Let x be the length and y be the width of the rectangle. Then $2x + 2y = 100$ and $y = 50 - x$. The area is given by the following.

$$A = xy = x(50 - x)$$

$$A' = 50 - 2x$$

$A' = 0$ when $x = 25$. Since $A''(25) = -2 < 0$, A is maximum when $x = 25$ meters and $y = 50 - 25 = 25$ meters.

11. Let x and y be the length and width of the rectangle. Then the area is $xy = 64$ and $y = 64/x$. The perimeter is given by the following.

$$P = 2x + 2y = 2x + 2\left(\frac{64}{x}\right)$$

$$P' = 2 - \frac{128}{x^2} = \frac{2(x^2 - 64)}{x^2}$$

$P' = 0$ when $x = 8$. Since $P'' = 256/x^3$ and $P''(8) > 0$, the perimeter is a minimum when $x = 8$ and $y = 64/8 = 8$ feet.

13. Let x and y be the lengths shown in the figure. Then $4x + 3y = 200$ and $y = (200 - 4x)/3$. The area of the corrals is given by the following.

$$A = 2xy = 2x\left(\frac{200 - 4x}{3}\right) = \frac{8}{3}(50x - x^2)$$

$$A' = \frac{8}{3}(50 - 2x)$$

$A' = 0$ when $x = 25$. Since $A''(25) = -\frac{16}{3} < 0$, A is maximum when $x = 25$ feet and $y = \frac{100}{3}$ feet.

15. (a) $9 + 9 + 4(3)(11) = 150$ in^2

$$6(25) = 150 \text{ in}^2$$

$$36 + 36 + 4(6)(3.25) = 150 \text{ in}^2$$

(b) $V = 3(3)(11) = 99$ in^2

$V = 5(5)(5) = 125$ in^3

$V = 6(6)(3.25) = 117$ in^3

(c) Let the base measure x by x, and the height measure y.

Surface area $= 150 = 2x^2 + 4xy \Rightarrow y = \frac{1}{4x}(150 - 2x^2)$

$$V = x^2y = x^2\left[\frac{1}{4x}(150 - 2x^2)\right] = \frac{75}{2}x - \frac{x^3}{2}$$

$$V'(x) = \frac{75}{2} - \frac{3x^2}{2} = 0 \Rightarrow x^2 = 25 \Rightarrow x = 5 \Rightarrow y = 5$$

$V''(x) = -3x < 0 \Rightarrow x = 5$ is a maximum.

Dimensions: $5 \times 5 \times 5$ inches

17. Perimeter $= 16 = 2y + x + \pi\left(\frac{x}{2}\right)$

$$32 = 4y + 2x + \pi x$$

$$y = \frac{32 - 2x - \pi x}{4}$$

Area $= A = xy + \frac{\pi}{2}\left(\frac{x}{2}\right)^2 = x\left(\frac{32 - 2x - \pi x}{4}\right) + \frac{\pi x^2}{8}$

$$= 8x - \frac{1}{2}x^2 - \frac{\pi}{4}x^2 + \frac{\pi}{8}x^2$$

$A'(x) = 8 - x - \frac{\pi x}{2} + \frac{\pi x}{4} = 8 - x\left(1 + \frac{\pi}{4}\right)$

$A'(x) = 0$ when $x = \frac{8}{1 + (\pi/4)} = \frac{32}{4 + \pi}$

$A''(x) = -(1 + \pi/4) < 0 \Rightarrow$ this x is a maximum.

$x = \frac{32}{4 + \pi} \Rightarrow y = \frac{16}{4 + \pi}$

Dimensions: $x = \frac{32}{4 + \pi}$ feet

$$y = \frac{16}{4 + \pi} \text{ feet}$$

19. Let x be the length of the cut (the height of the box); then the length of the box is $3 - 2x$ and its width is $2 - 2x$. The volume is given by the following.

$$V = x(3 - 2x)(2 - 2x), \quad 0 < x < 1$$

$$= 4x^3 - 10x^2 + 6x$$

$$V' = 12x^2 - 20x + 6$$

$V' = 0$ when $x = \left(5 \pm \sqrt{7}\right)/6$, but we do not consider $\left(5 + \sqrt{7}\right)/6$ since it is not in the domain.

Length: $3 - 2\left(\frac{5 - \sqrt{7}}{6}\right) \approx 2.215$ feet

Width: $2 - 2\left(\frac{5 - \sqrt{7}}{6}\right) \approx 1.215$ feet

Height: $\frac{5 - \sqrt{7}}{6} \approx 0.392$ foot

Volume $= \frac{10 + 7\sqrt{7}}{27}$ ft$^3 \approx 1.056$ ft^3

21. Use the values $(16, 80)$ and $(17, 76)$ to determine the linear equation relating the number of trees to the yield.

$$y - 80 = \frac{80 - 76}{16 - 17}(x - 16)$$

$$y - 80 = -4x + 64$$

$$y = -4x + 144$$

The number of apples is $N(x) = xy = x(-4x + 144)$.

$$N(x) = -4x^2 + 144x$$

$$N'(x) = -8x + 144$$

Critical number: $x = 18$

By the First-Derivative Test, $(18, 1296)$ is a relative maximum. So, she should plant 18 trees to yield 1296 apples.

23. Let x and y be the lengths shown in the figure. Then $xy = 30$ and $y = 30/x$. The area of the page is given by the following.

$$A = (x + 4)(y + 2) = (x + 4)\left(\frac{30}{x} + 2\right) = 38 + 2x + \frac{120}{x}$$

$$A' = 2 - \frac{120}{x^2} = \frac{2(x^2 - 60)}{x^2}$$

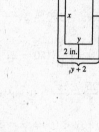

$A' = 0$ when $x = \sqrt{60} = 2\sqrt{15}$. The area of the page is minimum when dimensions are

$$x + 4 = 2\sqrt{15} + 4 \text{ inches}$$

$$y + 2 = \frac{30}{2\sqrt{15}} + 2 = \sqrt{15} + 2 \text{ inches.}$$

25. (a) $\dfrac{y - 2}{0 - 1} = \dfrac{0 - 2}{x - 1}$

$$y - 2 = \frac{2}{x - 1}$$

$$y = 2 + \frac{2}{x - 1}$$

$$L = \sqrt{x^2 + y^2} = \sqrt{x^2 + \left(2 + \frac{2}{x - 1}\right)^2}$$

$$= \sqrt{x^2 + 4 + \frac{8}{x - 1} + \frac{4}{(x - 1)^2}}, \quad x > 1$$

(b)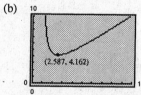

L is a minimum when $x \approx 2.587$ and $L \approx 4.162$.

(c) Area $= A(x) = \dfrac{1}{2}xy = \dfrac{1}{2}x\left(2 + \dfrac{2}{x - 1}\right) = x + \dfrac{x}{x - 1}$

$$A'(x) = 1 + \frac{(x - 1) - x}{(x - 1)^2} = 1 - \frac{1}{(x - 1)^2} = 0$$

$$(x - 1)^2 = 1 \implies x - 1 = \pm 1 \implies x = 2 \implies y = 4, A = 4$$

Vertices: $(0, 0), (2, 0), (0, 4)$

27. The area is given by the following.

$$A = 2xy = 2x\sqrt{r^2 - x^2}$$

$$A' = 2x\left(-\frac{x}{\sqrt{r^2 - x^2}}\right) + 2\sqrt{r^2 - x^2} = \frac{2(r^2 - 2x^2)}{\sqrt{r^2 - x^2}}$$

$A' = 0$ when $x = r/\sqrt{2}$. The length of the rectangle is $2x = \sqrt{2}r$ and the width is $\sqrt{r^2 - x^2} = r/\sqrt{2} = (\sqrt{2}/2)r$.

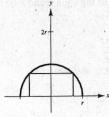

29. The volume is given by the following.

$$V = \pi r^2 h = \pi x^2\left(2\sqrt{r^2 - x^2}\right) = 2\pi x^2\sqrt{r^2 - x^2}$$

$$V' = 2\pi\left[x^2\left(\frac{-x}{\sqrt{r^2 - x^2}}\right) + 2x\sqrt{r^2 - x^2}\right] = 2\pi\left[\frac{2r^2x - 3x^3}{\sqrt{r^2 - x^2}}\right]$$

$V' = 0$ when $x = \sqrt{2/3}\,r$. The maximum volume is

$$V = 2\pi\left(\frac{2}{3}r^2\right)\sqrt{r^2 - \left(\frac{2}{3}\right)r^2} = \frac{4\pi r^3}{3\sqrt{3}}.$$

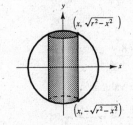

31. The distance between a point (x, y) on the graph and the point $(0, 4)$ is

$$d = \sqrt{(x - 0)^2 + (y - 4)^2}$$

$$= \sqrt{x^2 + (x^2 + 1 - 4)^2}$$

$$= \sqrt{x^2 + (x^2 - 3)^2}.$$

We can minimize d by minimizing its square $L = d^2$.

$$L = x^2 + (x^2 - 3)^2 = x^4 - 5x^2 + 9$$

$$L' = 4x^3 - 10x = 2x(2x^2 - 5) \Longrightarrow x = 0, \pm\sqrt{\tfrac{5}{2}}$$

Hence, the points are $\left(\pm\sqrt{\tfrac{5}{2}}, \tfrac{7}{2}\right)$.

33. The length and girth is $4x + y = 108$. Thus, $y = 108 - 4x$. The volume is

$$V = x^2 y = x^2(108 - 4x) = 108x^2 - 4x^3$$

$$V' = 216x - 12x^2 = 12x(18 - x).$$

$V' = 0$ when $x = 0$ and $x = 18$. Thus, the maximum volume occurs when $x = 18$ inches and $y = 36$ inches. The dimensions are 18 inches by 18 inches by 36 inches.

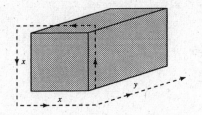

35. Let x be the length of a side of the square and r be the radius of the circle. Then the combined perimeter is $4x + 2\pi r = 16$, which implies that

$$x = \frac{16 - 2\pi r}{4} = 4 - \frac{\pi r}{2}.$$

The combined area of the circle and square is

$$A = x^2 + \pi r^2 = \left(4 - \frac{\pi r}{2}\right)^2 + \pi r^2$$

$$A' = 2\left(4 - \frac{\pi r}{2}\right)\left(-\frac{\pi}{2}\right) + 2\pi r = \frac{1}{2}(\pi^2 r + 4\pi r - 8\pi).$$

$A' = 0$ when

$$r = \frac{8\pi}{\pi^2 + 4\pi} = \frac{8}{\pi + 4}$$

and the corresponding x-value is

$$x = 4 - \frac{\pi[8/(\pi + 4)]}{2} = \frac{16}{\pi + 4}.$$

This is a minimum by the Second-Derivative Test $\left[A'' = \frac{1}{2}(\pi^2 + 4\pi) > 0\right]$.

37. Using the formula Distance = (Rate)(Time), we have $T = D/R$.

$$T = T_{\text{rowed}} + T_{\text{walked}} = \frac{D_{\text{rowed}}}{R_{\text{rowed}}} + \frac{D_{\text{walked}}}{R_{\text{walked}}} = \frac{\sqrt{x^2 + 4}}{2} + \frac{\sqrt{1 + (3 - x)^2}}{4}$$

$$T' = \frac{x}{2\sqrt{x^2 + 4}} - \frac{3 - x}{4\sqrt{1 + (3 - x)^2}}$$

By setting $T' = 0$, we have the following.

$$\frac{x^2}{4(x^2 + 4)} = \frac{(3 - x)^2}{16[1 + (3 - x)^2]}$$

$$\frac{x^2}{x^2 + 4} = \frac{9 - 6x + x^2}{4(10 - 6x + x^2)}$$

$$4(x^4 - 6x^3 + 10x^2) = (x^2 + 4)(9 - 6x + x^2)$$

$$x^4 - 6x^3 + 9x^2 + 8x - 12 = 0$$

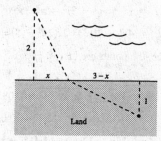

Using a graphing utility, the solution on $[0, 3]$ is $x = 1$ mile.

Possible rational roots: $\pm 1, \pm 2, \pm 3, \pm 4, \pm 6, \pm 12$

By testing, we find that $x = 1$ mile.

39. Use the points $(1, 4)$, $(2, 3.90)$.

$$y - 4 = \frac{3.90 - 4}{2 - 1}(x - 1)$$

$$y = -0.1x + 4.1 \quad (x = \text{week}, y = \text{dollars/bushel})$$

Now use the points $(1, 120)$, $(2, 124)$.

$$z - 120 = \frac{124 - 120}{2 - 1}(x - 1)$$

$$z = 4x + 116 \quad (x = \text{week}, z = \text{bushels})$$

$$\text{Total value} = f(x) = yz = (-0.1x + 4.1)(4x + 116)$$

$$= -0.4x^2 + 4.8x + 475.6$$

$$f'(x) = -0.8x + 4.8 = 0 \Rightarrow x = 6$$

So, harvest strawberries in the sixth week, which yields $4(6) + 116 = 140$ bushels for a value of $f(6) = \$490$.

41. $\text{Area} = l \cdot w = 2x\sqrt{100 - x^2} = A(x)$

$$A'(x) = \frac{2(100 - 2x^2)}{\sqrt{100 - x^2}}$$

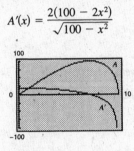

$\text{Length} = l = 10\sqrt{2} \approx 14.14$

$\text{Width} = w = 5\sqrt{2} \approx 7.07$

Section 3.5   Business and Economics Applications

1. $R = 800x - 0.2x^2$

$R' = 800 - 0.4x$

$R' = 0$ when $x = \frac{800}{0.4} = 2000$. R is maximum when $x = 2000$.

3. $R = 400x - x^2$

$R' = 400 - 2x = 0$ when $x = 200$.

Thus, R is maximum when $x = 200$ units.

5. $\overline{C} = 1.25x + 25 + \dfrac{8000}{x}$

$\overline{C}' = 1.25 - \dfrac{8000}{x^2}$

$\overline{C}' = 0$ when

$$1.25x^2 = 8000$$

$$x^2 = 6400$$

$$x = 80 \text{ units.}$$

7. $\overline{C} = 2x + 255 + \dfrac{5000}{x}$

$\overline{C}' = 2 - \dfrac{5000}{x^2}$

$\overline{C}' = 0$ when $2 = 5000/x^2$, which implies that $x^2 = 2500$ and $x = 50$ units.

9. $P = xp - C$

$\quad = (90x - x^2) - (100 + 30x)$

$\quad = -x^2 + 60x - 100$

$P' = -2x + 60$

$P' = 0$ when $x = 30$. Thus, the maximum profit occurs at $x = 30$ units and $p = 90 - 30 = \$60$.

11. $P = xp - C$

$\quad = x(70 - 0.001x) - (8000 + 50x + 0.03x^2)$

$\quad = -0.031x^2 + 20x - 8000$

$P' = -0.062x + 20$

$P' = 0$ when $x \approx 322.58$ and $p \approx \$69.68$.

13. $\overline{C} = 2x + 5 + \dfrac{18}{x}$

$\overline{C}' = 2 - \dfrac{18}{x^2}$

$\overline{C}' = 0$ when $x = 3$. Thus, the average cost is minimum when $x = 3$ units and $\overline{C}(3) = \$17$ per unit.

$\quad C' = \text{Marginal cost} = 4x + 5$

$C'(3) = 17 = \overline{C}(3)$

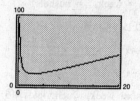

15. (a) $P = xp - C$

$\quad = x\left(100 - \dfrac{1}{2}x^2\right) - (40x + 37.5)$

$\quad = -\dfrac{1}{2}x^3 + 60x - 37.5$

$P' = -\dfrac{3}{2}x^2 + 60$

$P' = 0$ when $x = \sqrt{40} = 2\sqrt{10} \approx 6.32$ units. The price is $p = 100 - \frac{1}{2}(40) = \$80$.

(b) $\overline{C} = 40 + \dfrac{37.5}{x}$

When $x = 2\sqrt{10}$, the average cost per unit is

$$\overline{C}(2\sqrt{10}) = 40 + \dfrac{37.5}{2\sqrt{10}} \approx \$45.93.$$

17. $P' = -6s^2 + 70s - 100$

$\quad = -2(3s^2 - 35s + 50)$

$\quad = -2(3s - 5)(s - 10)$

Critical numbers: $s = \frac{5}{3}$ and $s = 10$

$\quad\quad P'' = -12s + 70$

$P''\left(\frac{5}{3}\right) = 50 > 0 \Longrightarrow \text{Minimum}$

$P''(10) = -50 < 0 \Longrightarrow \text{Maximum}$

$P'' = -12s + 70 = 0$ when $s = \frac{35}{6}$. The maximum profit occurs when $s = 10$ (or $\$10,000$) and the point of diminishing returns occurs at $s = \frac{35}{6}$ (or $\$5833.33$).

19. Let $x = $ number of units purchased, $p = $ price per unit, and $P = $ profit.

$\quad p = 90 - (0.10)(x - 100)$

$\quad = 100 - 0.10x, \quad x \geq 100$

$\quad P = xp - C$

$\quad = x(100 - 0.10x) - 60x$

$\quad = 40x - 0.10x^2$

$P' = 40 - 0.20x$

$P' = 0$ when $x = 200$ radios.

21. $(40, 300), (45, 275)$

$$\text{Slope} = \frac{275 - 300}{45 - 40} = -5$$

$$x - 300 = -5(p - 40)$$

$$x = -5p + 500$$

$$R = xp = (-5p + 500)p = -5p^2 + 500p$$

$$R' = -10p + 500 = 0 \Longrightarrow p = \$50$$

23. Let T be the total cost.

$$T = 8(5280)\sqrt{x^2 + \frac{1}{4}} + 6(5280)(6 - x)$$

$$= 2(5280)\left[4\sqrt{x^2 + \frac{1}{4}} + 18 - 3x\right]$$

$$= 2(5280)\left[2\sqrt{4x^2 + 1} + 18 - 3x\right]$$

$$\frac{dT}{dx} = 2(5280)\left[\frac{2(8x)}{2\sqrt{4x^2 + 1}} - 3\right] = 2(5280)\left(\frac{8x - 3\sqrt{4x^2 + 1}}{\sqrt{4x^2 + 1}}\right)$$

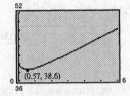

$dT/dx = 0$ when $8x = 3\sqrt{4x^2 + 1}$ which implies that

$$64x^2 = 9(4x^2 + 1)$$

$$28x^2 = 9$$

$$x^2 = \frac{9}{28}$$

$$x = \frac{3}{2\sqrt{7}} \approx 0.57 \text{ mile.}$$

25. $C = \left(\frac{v^2}{600} + 10\right)\left(\frac{110}{v}\right) = \frac{11}{60}v + \frac{1100}{v}$

$$C' = \frac{11}{60} - \frac{1100}{v^2} = \frac{11v^2 - 66,000}{60v^2}$$

$$C' = 0 \text{ when } v = \sqrt{\frac{66,000}{11}} \approx 77.46 \text{ mph.}$$

27. Since $dp/dx = -3$, the price elasticity of demand is

$$\eta = \frac{p/x}{dp/dx} = \frac{(400 - 3x)/x}{-3} = 1 - \frac{400}{3x}.$$

When $x = 20$, we have

$$\eta = 1 - \frac{400}{3(20)} = -\frac{17}{3}.$$

Since $|\eta(20)| = \frac{17}{3} > 1$, the demand is elastic.

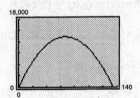

Elastic: $\left(0, \frac{200}{3}\right)$

Inelastic: $\left(\frac{200}{3}, \frac{400}{3}\right)$

29. $p = 20 - 0.0002x, \dfrac{dp}{dx} = -0.0002$

$$\eta = \frac{p/x}{dp/dx} = \frac{(20 - 0.0002x)/x}{-0.0002}$$

$$= \frac{-100,000}{x} + 1$$

When $x = 30$, $p = \dfrac{-100,000}{30} + 1 = -3332.\overline{3}$.

Since $|\eta(30)| > 1$, the demand is elastic.

$R = px = (20 - 0.0002x)x = 20x - 0.0002x$

$|\eta| = 1 = \left| \dfrac{-100,000}{x} + 1 \right| \Rightarrow 1 = \pm\left(1 - \dfrac{100,000}{x}\right) \Rightarrow x = 50,000$

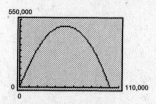

Elastic: $(0, 50,000)$

Inelastic: $(50,000, 100,000)$

31. Since $dp/dx = -200/x^3$, the price elasticity of demand is

$$\eta = \frac{p/x}{dp/dx} = \frac{[(100/x^2) + 2]/x}{-(200/x^3)} = -\frac{1}{2} - \frac{x^2}{100}.$$

When $x = 10$, we have

$$\eta = -\frac{1}{2} - \frac{(10)^2}{100} = -\frac{3}{2}.$$

Since $|\eta(10)| = \frac{3}{2} > 1$, the demand is elastic.

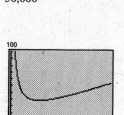

Elastic: $\left(5\sqrt{2}, \infty\right)$

Inelastic: $\left(0, 5\sqrt{2}\right)$

33. (a) If $p = 2$, $x = 64$, and p increases by 5%.

$p = 2 + 2(0.05) = 2.1$

$x = (2.1)^2 - 20(2.1) + 100 = 62.41$

The percentage increase in x is

$$\frac{62.41 - 64}{64} \approx -0.0248 \approx -2.5\%.$$

(b) At $(p, x) = (2, 64)$, the average elasticity of demand is

$$\frac{\% \text{ change in } x}{\% \text{ change in } p} \approx \frac{-0.0248}{0.05} \approx -0.496.$$

(c) The exact elasticity of demand at $(p, x) = (2, 64)$ is

$$\eta = \frac{p/x}{dp/dx} = \frac{p}{x}\frac{dx}{dp} = \frac{p}{x}(2p - 20) = \frac{2}{64}(4 - 20) = -0.5.$$

This is very close to the answer in (b).

(d) $R = xp = (p^2 - 20p + 100)p = p^3 - 20p^2 + 100p$

$\dfrac{dR}{dp} = 3p^2 - 40p + 100 = (p - 10)(3p - 10)$

$R' = 0$ when $p = \dfrac{10}{3}, 10$.

$R'' = 6p - 40$. $R''\left(\dfrac{10}{3}\right) < 0$ and $R''(10) > 0$.

Hence, $p = \frac{10}{3}$ is the maximum, corresponding to $x = 400/9 \approx 44$.

35. (a) $p = 20 - 0.02x, \; 0 < x < 1000$

$x = 560 \implies p = 8.8$

$p'(x) = -0.02$

$\eta = \dfrac{p/x}{dp/dx} = \dfrac{8.8/560}{-0.02} = \dfrac{-11}{14} \approx -0.7857 \approx -0.0079\%$

(b) $R = px = (20 - 0.02x)x = 20x - 0.02x^2$

$R'(x) = 20 - 0.04x = 0 \implies x = 500$

$R''(500) < 0 \implies (x, p) = (500, 10)$ is the maximum.

(c) $|\eta| = \left| \dfrac{10/500}{-0.02} \right| = 1$

37. $x = 600 - 50p$

$p = 12 - \dfrac{x}{50}$

$\dfrac{dp}{dx} = -\dfrac{1}{50}$

When $p = 5, \; x = 350,$ and

$\eta = \dfrac{p/x}{dp/dx} = \dfrac{5/350}{-1/50} = \dfrac{-5}{7}.$

Since $|\eta| = \dfrac{5}{7} < 1,$ the demand is inelastic.

No, lower the price will not increase revenue.

39. $x = \dfrac{a}{p^m}, \quad m > 1$

$1 = -\dfrac{am}{p^{m+1}} \dfrac{dp}{dx}$

$\dfrac{dp}{dx} = -\dfrac{p^{m+1}}{am}$

$\eta = \dfrac{p/x}{dp/dx} = \dfrac{p}{x} \cdot \dfrac{-am}{p^{m+1}}$

$= \dfrac{p}{a/p^m} \cdot \dfrac{-am}{p^{m+1}} = \dfrac{p^{m+1}}{a} \cdot \dfrac{-am}{p^{m+1}} = -m$

41. (a) $R = \dfrac{-18.0 + 24.74t}{1 - 0.16t + 0.008t^2}, \; 4 \le t \le 13$

($t = 4$ corresponds to 1994.)

$R' = \dfrac{-3092.5(t^2 - 1.455t - 110.449)}{(t^2 - 20t + 125)^2}$

$R' = 0$ for $t \approx 11.3$. By the First Derivative Test.

The maximum is $t = 11.3$, or 2001.

The minimum is $t = 4$, or 1994 (endpoint).

(b) The revenue was increasing most rapidly at $t = 8.1$, or 1998.

The revenue was decreasing most rapidly at $t = 13$, or 2003.

(c)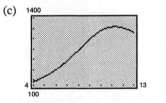

43. Answers will vary.

Section 3.6   Asymptotes

1. A horizontal asymptote occurs at $y = 1$ since

$$\lim_{x \to \infty} \frac{x^2 + 1}{x^2} = 1, \quad \lim_{x \to -\infty} \frac{x^2 + 1}{x^2} = 1.$$

A vertical asymptote occurs at $x = 0$ since

$$\lim_{x \to 0^-} \frac{x^2 + 1}{x^2} = \infty, \quad \lim_{x \to 0^+} \frac{x^2 + 1}{x^2} = \infty.$$

3. A horizontal asymptote occurs at $y = 1$ since

$$\lim_{x \to \infty} \frac{x^2 - 2}{x^2 - x - 2} = 1, \quad \lim_{x \to -\infty} \frac{x^2 - 2}{x^2 - x - 2} = 1.$$

Vertical asymptotes occur at $x = -1$ and $x = 2$ since

$$\lim_{x \to -1^-} \frac{x^2 - 2}{x^2 - x - 2} = -\infty,$$

$$\lim_{x \to -1^+} \frac{x^2 - 2}{x^2 - x - 2} = \infty,$$

$$\lim_{x \to 2^-} \frac{x^2 - 2}{x^2 - x - 2} = -\infty,$$

$$\lim_{x \to 2^+} \frac{x^2 - 2}{x^2 - x - 2} = \infty.$$

5. A horizontal asymptote occurs at $y = \frac{3}{2}$ since

$$\lim_{x \to \infty} \frac{3x^2}{2(x^2 + 1)} = \frac{3}{2} = \lim_{x \to -\infty} \frac{3x^2}{2(x^2 + 1)}.$$

No vertical asymptotes.

7. A horizontal asymptote occurs at $y = \frac{1}{2}$ since

$$\lim_{x \to \infty} \frac{x^2 - 1}{2x^2 - 8} = \lim_{x \to -\infty} \frac{x^2 - 1}{2x^2 - 8} = \frac{1}{2}.$$

Vertical asymptotes occur at $x = \pm 2$ since

$$\lim_{x \to 2^-} \frac{x^2 - 1}{2x^2 - 8} = -\infty, \quad \lim_{x \to 2^+} \frac{x^2 - 1}{2x^2 - 8} = \infty,$$

$$\lim_{x \to -2^-} \frac{x^2 - 1}{2x^2 - 8} = \infty, \quad \lim_{x \to -2^+} \frac{x^2 - 1}{2x^2 - 8} = -\infty.$$

9. The graph of f has a horizontal asymptote at $y = 3$. It matches graph (f).

11. The graph of f has a horizontal asymptote at $y = 0$. It matches graph (c).

13. The graph of f has a horizontal asymptote at $y = 5$. It matches graph (e).

15. $\displaystyle \lim_{x \to -2^-} \frac{1}{(x + 2)^2} = \infty$

17. $\displaystyle \lim_{x \to 3^+} \frac{x - 4}{x - 3} = -\infty$

19. $\displaystyle \lim_{x \to 4^-} \frac{x^2}{x^2 - 16} = -\infty$

21. $\displaystyle \lim_{x \to 0^-} \left(1 + \frac{1}{x}\right) = -\infty$

23. $\displaystyle \lim_{x \to \infty} \frac{2x - 1}{3x + 2} = \frac{2}{3}$

25. $\displaystyle \lim_{x \to \infty} \frac{3x}{4x^2 - 1} = 0$

27. $\displaystyle \lim_{x \to -\infty} \frac{5x^2}{x + 3} = -\infty$

29. $\displaystyle \lim_{x \to \infty} \left(2x - \frac{1}{x^2}\right) = \lim_{x \to \infty} \frac{2x^3 - 1}{x^2} = \infty$

31. $\displaystyle \lim_{x \to -\infty} \left(\frac{2x}{x - 1} + \frac{3x}{x + 1}\right) = \lim_{x \to -\infty} \frac{5x^2 - x}{x^2 - 1} = 5$

33.

x	10^0	10^1	10^2	10^3	10^4	10^5	10^6
$f(x)$	2.000	0.348	0.101	0.032	0.010	0.003	0.001

$$\lim_{x \to \infty} \frac{x+1}{x\sqrt{x}} = 0$$

35.

x	10^0	10^1	10^2	10^3	10^4	10^5	10^6
$f(x)$	0	49.5	49.995	49.99995	50.0	50.0	50.0

$$\lim_{x \to \infty} \frac{x^2 - 1}{0.02x^2} = 50$$

37.

x	-10^6	-10^4	-10^2	10^0	10^2	10^4	10^6
$f(x)$	-2	-2	-1.9996	0.8944	1.9996	2	2

$$\lim_{x \to \infty} \frac{2x}{\sqrt{x^2 + 4}} = 2, \quad \lim_{x \to -\infty} \frac{2x}{\sqrt{x^2 + 4}} = -2$$

39. $y = \dfrac{2 + x}{1 - x}$

x-intercept: $(-2, 0)$

y-intercept: $(0, 2)$

Horizontal asymptote: $y = -1$

Vertical asymptote: $x = 1$

$$y' = \frac{3}{(1 - x)^2}$$

No relative extrema

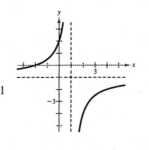

41. $f(x) = \dfrac{x^2}{x^2 + 9}$

Intercept: $(0, 0)$

Horizontal asymptote: $y = 1$

$$f'(x) = \frac{18x}{(x^2 + 9)^2}$$

Relative minimum: $(0, 0)$

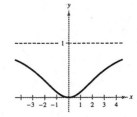

43. $g(x) = \dfrac{x^2}{x^2 - 16}$

Intercept: $(0, 0)$

Horizontal asymptote: $y = 1$

Vertical asymptotes: $x = \pm 4$

$$g'(x) = \frac{-32x}{(x^2 - 16)^2}$$

Relative maximum: $(0, 0)$

45. $x = \dfrac{4}{y^2}$

No intercepts

Horizontal asymptote: $y = 0$

Vertical asymptote: $x = 0$

No relative extrema

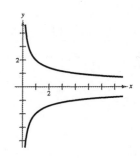

47. $y = \dfrac{2x}{1 - x}$

Intercept: $(0, 0)$

Horizontal asymptote: $y = -2$

Vertical asymptote: $x = 1$

$$y' = \dfrac{2}{(1 - x)^2}$$

No relative extrema

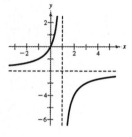

49. $y = 3(1 - x^{-2})$

x-intercepts: $(\pm 1, 0)$

Horizontal asymptote: $y = 3$

$$y = \dfrac{3(x^2 - 1)}{x^2}$$

Vertical asymptote: $x = 0$

$$y' = \dfrac{6}{x^3}$$

No relative extrema

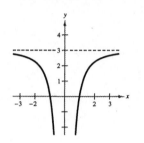

51. $f(x) = \dfrac{1}{x^2 - x - 2}$

y-intercept: $\left(0, -\dfrac{1}{2}\right)$

Horizontal asymptote: $y = 0$

$$f(x) = \dfrac{1}{(x + 1)(x - 2)}$$

Vertical asymptotes: $x = -1, x = 2$

$$f'(x) = -\dfrac{2x - 1}{(x^2 - x - 2)^2}$$

Relative maximum: $\left(\dfrac{1}{2}, -\dfrac{4}{9}\right)$

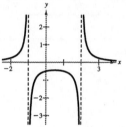

53. $g(x) = \dfrac{x^2 - x - 2}{x - 2}$

$$= x + 1 \text{ for } x \neq 2$$

x-intercept: $(-1, 0)$

y-intercept: $(0, 1)$

No asymptotes

No relative extrema

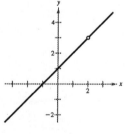

55. $y = \dfrac{2(x^2 - 3)}{x^2 - 2x + 1}$

x-intercepts: $\left(\pm\sqrt{3}, 0\right)$

y-intercept: $(0, -6)$

Vertical asymptote: $x = 1$

Horizontal asymptote: $y = 2$

$$y' = \dfrac{4(3 - x)}{(x - 1)^3}$$

Relative maximum: $(3, 3)$

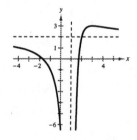

57. (a) $C = 1.35x + 4570$

$$\overline{C} = \dfrac{C}{x} = 1.35 + \dfrac{4570}{x}$$

(b) $\overline{C} = 1.35 + \dfrac{4570}{x}$

When $x = 100$, $\overline{C} = \$47.05$.
When $x = 1000$, $\overline{C} = \$5.92$.

(c) $\displaystyle\lim_{x \to \infty} \left(1.35 + \dfrac{4570}{x}\right) = 1.35 + 0$

$$= \$1.35$$

59. $C = \dfrac{528}{(100 - p)}, \quad 0 \leq p < 100$

(a) $C(25) = \dfrac{528(25)}{100 - 25} = \$176$ million

$C(50) = \dfrac{528(50)}{100 - 50} = \$528$ million

$C(75) = \dfrac{528(75)}{100 - 75} = \$1584$ million

(b) $\displaystyle\lim_{p \to 100^-} \dfrac{528p}{100 - p} = \infty$

(c)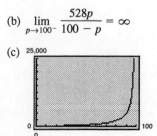

61. $\lim\limits_{n\to\infty} \dfrac{\theta an - \theta a + b}{\theta n - \theta + 1} = \dfrac{\theta a}{\theta} = a$

63. $N = \dfrac{10(3 + 4t)}{1 + 0.1t} = \dfrac{40t + 30}{0.1t + 1} = \dfrac{400t + 300}{t + 10}$

    (a) $N(5) = 153.3\overline{3} \approx 153$ elk

        $N(10) = 215$ elk

        $N(25) \approx 294.29 \approx 294$ elk

    (b) $\lim\limits_{t\to\infty} N(t) = 400$ elk

65. $C = 25.5x + 1000, \; R = 75.5x$

    (a) $\overline{P} = \dfrac{R - C}{x} = \dfrac{75.5x - 25.5x - 1000}{x} = 50 - \dfrac{1000}{x}$

    (b) $\overline{P}(100) = 50 - \dfrac{1000}{100} = 40$

        $\overline{P}(500) = 50 - \dfrac{1000}{500} = 48$

        $\overline{P}(1000) = 50 - \dfrac{1000}{1000} = 49$

    (c) $\lim\limits_{x\to\infty} \overline{P}(x) = 50$

Section 3.7   Curve Sketching: A Summary

1. $y = -x^2 - 2x + 3 = -(x + 3)(x - 1)$

     $y' = -2x - 2 = -2(x + 1)$

     $y'' = -2$

     Intercepts: $(0, 3), (1, 0), (-3, 0)$

     Relative maximum: $(-1, 4)$

     Concave downward

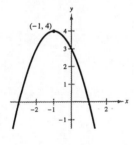

3. $y = x^3 - 4x^2 + 6$

     $y' = 3x^2 - 8x = x(3x - 8)$

     $y'' = 6x - 8 = 2(3x - 4)$

     Relative maximum: $(0, 6)$

     Relative minimum: $\left(\frac{8}{3}, -3.\overline{481}\right)$

     Point of inflection: $\left(\frac{4}{3}, 1.\overline{259}\right)$

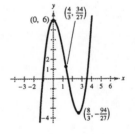

5. $y = 2 - x - x^3$

     $y' = -1 - 3x^2$

     $y'' = -6x$

     No relative extrema

     Point of inflection: $(0, 2)$

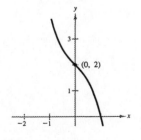

7. $y = 3x^3 - 9x + 1$

     $y' = 9x^2 - 9$

       $= 9(x - 1)(x + 1)$

     $y'' = 18x$

     Relative maximum: $(-, 7)$

     Relative minimum: $(1, -5)$

     Point of inflection: $(0, 1)$

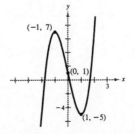

9. $y = 3x^4 + 4x^3 = x^3(3x + 4)$

$y' = 12x^3 + 12x^2 = 12x^2(x + 1)$

$y'' = 36x^2 + 24x = 12x(3x + 2)$

Intercepts: $(0, 0), \left(-\frac{4}{3}, 0\right)$

Relative minimum: $(-1, -1)$

Points of inflection: $(0, 0), \left(-\frac{2}{3}, -\frac{16}{27}\right)$

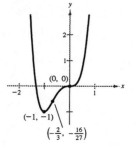

11. $y = x^3 - 6x^2 + 3x + 10$

$\quad = (x + 1)(x - 2)(x - 5)$

$y' = 3x^2 - 12x + 3 = 3(x^2 - 4x + 1)$

$y'' = 6x - 12 = 6(x - 2)$

Intercepts: $(0, 10), (-1, 0), (2, 0), (5, 0)$

Relative maximum: $\left(2 - \sqrt{3}, 10.392\right)$

Relative minimum: $\left(2 + \sqrt{3}, -10.392\right)$

Point of inflection: $(2, 0)$

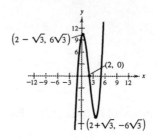

13. $y = x^4 - 8x^3 + 18x^2 - 16x + 5$

$y' = 4x^3 - 24x^2 + 36x - 16 = 4(x^3 - 6x^2 + 9x - 4)$

$\qquad = 4(x - 4)(x - 1)^2$

$y'' = 12x^2 - 48x + 36 = 12(x^2 - 4x + 3) = 12(x - 1)(x - 3)$

Intercepts: $(0, 5), (1, 0), (5, 0)$

Relative minimum: $(4, -27)$

Points of inflection: $(1, 0), (3, -16)$

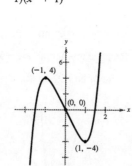

15. $y = x^4 - 4x^3 + 16x$

$y' = 4x^3 - 12x^2 + 16 = 4(x + 1)(x - 2)^2$

$y'' = 12x^2 - 24x = 12x(x - 2)$

Intercepts: $(0, 0), (-1.68, 0)$

Relative minimum: $(-1, -11)$

Points of inflection: $(0, 0), (2, 16)$

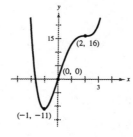

17. $y = x^5 - 5x$

$y' = 5x^4 - 5 = 5(x + 1)(x - 1)(x^2 + 1)$

$y'' = 20x^3$

Intercepts: $(0, 0), \left(\pm \sqrt[4]{5}, 0\right)$

Relative maximum: $(-1, 4)$

Relative minimum: $(1, -4)$

Point of inflection: $(0, 0)$

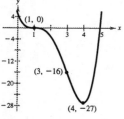

19. $y = \begin{cases} x^2 + 1, & x \le 0 \\ 1 - 2x, & x > 0 \end{cases}$

$y' = \begin{cases} 2x, & x < 0 \\ -2, & x > 0 \end{cases}$

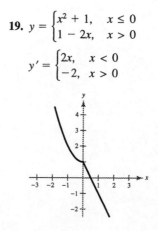

21. $y = \dfrac{x^2 + 2}{x^2 + 1}$

$y' = \dfrac{-2x}{(x^2 + 1)^2}$

$y'' = \dfrac{2(3x^2 - 1)}{x^2 + 1}$

Relative maximum: $(0, 2)$

Points of inflection: $x = \left(\dfrac{\sqrt{3}}{3}, \dfrac{7}{4}\right), \left(-\dfrac{\sqrt{3}}{3}, \dfrac{7}{4}\right)$

Horizontal asymptote: $y = 1$

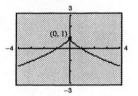

23. $y = 3x^{2/3} - 2x$

$y' = 2x^{-1/3} - 2 = 2(x^{-1/3} - 1)$

$y'' = -\dfrac{2}{3x^{4/3}}$

Intercepts: $(0, 0), \left(\dfrac{27}{8}, 0\right)$

Relative maximum: $(1, 1)$

Relative minimum: $(0, 0)$

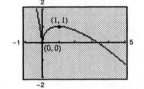

25. $y = 1 - x^{2/3}$

$y' = -\dfrac{2}{3}x^{-1/3} = -\dfrac{2}{3\sqrt[3]{x}}$

$y'' = \dfrac{2}{9}x^{-4/3} = \dfrac{2}{9\sqrt[3]{x^4}}$

Intercepts: $(0, 1), (\pm 1, 0)$

Relative maximum: $(0, 1)$

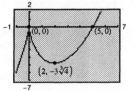

27. $y = x^{1/3} + 1$

$y' = \dfrac{1}{3}x^{-2/3} = \dfrac{1}{3\sqrt[3]{x^2}}$

$y'' = -\dfrac{2}{9}x^{-5/3} = -\dfrac{2}{9\sqrt[3]{x^5}}$

Intercepts: $(0, 1), (-1, 0)$

Point of inflection: $(0, 1)$

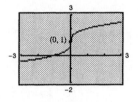

29. $y = x^{5/3} - 5x^{2/3} = x^{2/3}(x - 5)$

$y' = \dfrac{5}{3}x^{2/3} - \dfrac{10}{3}x^{-1/3} = \dfrac{5}{3}x^{-1/3}(x - 2)$

$y'' = \dfrac{10}{9}x^{-1/3} + \dfrac{10}{9}x^{-4/3} = \dfrac{10}{9}x^{-4/3}(x + 1)$

Intercepts: $(0, 0), (5, 0)$

Relative maximum: $(0, 0)$

Relative minimum: $\left(2, -3\sqrt[3]{4}\right)$

Point of inflection: $(-1, -6)$

31. $y = x\sqrt{x^2 - 9}$

Intercepts: $(\pm 3, 0)$

Domain: $|x| \geq 3$

$y' = \dfrac{2x^2 - 9}{\sqrt{x^2 - 9}} \neq 0$

$y'' = \dfrac{x(2x^2 - 27)}{(x^2 - 9)^{3/2}}$

Points of inflection: $x = \pm\sqrt{\dfrac{27}{2}} = \pm\dfrac{3\sqrt{6}}{2} \approx \pm 3.674$

$\left(\pm\dfrac{3\sqrt{6}}{2}, \pm\dfrac{9\sqrt{3}}{2}\right)$

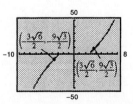

33. $y = \dfrac{5 - 3x}{x - 2}$

$y' = \dfrac{1}{(x - 2)^2} > 0$ No relative extrema

$y'' = \dfrac{-2}{(x - 2)^3}$ No points of inflection

Intercepts: $\left(0, -\frac{5}{2}\right), \left(\frac{5}{3}, 0\right)$

Asymptotes: $x = 2, y = -3$

Domain: all $x \neq 2$

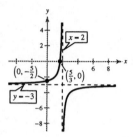

35. $y = \dfrac{2x}{x^2 - 1}$

$y' = \dfrac{-2(x^2 + 1)}{(x^2 - 1)^2}$

$y'' = \dfrac{4x(x^2 + 3)}{(x^2 - 1)^3}$

Point of inflection: $(0, 0)$

Intercept: $(0, 0)$

Horizontal asymptote: $y = 0$

Vertical asymptotes: $x = \pm 1$

Domain: $(-\infty, -1), (-1, 1), (1, \infty)$

Symmetry with respect to the origin

No relative extrema

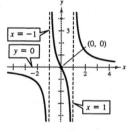

37. $y = x\sqrt{4 - x}$

$y' = \dfrac{8 - 3x}{2\sqrt{4 - x}}$

$y'' = \dfrac{3x - 16}{4(4 - x)^{3/2}}$

Intercepts: $(0, 0), (4, 0)$

Relative maximum: $\left(\dfrac{8}{3}, \dfrac{16}{3\sqrt{3}}\right)$

Domain: $(-\infty, 4]$

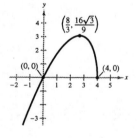

39. $y = \dfrac{x - 3}{x} = 1 - \dfrac{3}{x}$

$y' = \dfrac{3}{x^2}$

$y'' = \dfrac{-6}{x^3}$

Intercept: $(3, 0)$

Horizontal asymptote: $y = 1$

Vertical asymptote: $x = 0$

Domain: $(-\infty, 0), (0, \infty)$

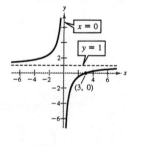

41. $y = \dfrac{x^3}{x^3 - 1}$

$y' = \dfrac{-3x^2}{(x^3 - 1)^2}$

$y'' = \dfrac{6x(2x^3 + 1)}{(x^3 - 1)^3}$

Points of inflection: $(0, 0), \left(-\dfrac{1}{\sqrt[3]{2}}, \dfrac{1}{3}\right)$

Intercept: $(0, 0)$

Horizontal asymptote: $y = 1$

Vertical asymptote: $x = 1$

Domain: $(-\infty, 1), (1, \infty)$

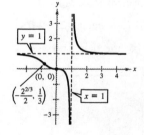

43. Since the graph rises as $x \to -\infty$ and falls as $x \to \infty$, a must be negative.

$$f(x) = -x^3 + x^2 + x + 1$$

(Solution not unique.)

45. Since the graph falls as $x \to -\infty$ and rises as $x \to \infty$, a must be positive.

$$f(x) = x^3 + 1$$

(Solution not unique.)

47. Since $f'(x) = 2$, the graph of f is a line with a slope of 2.

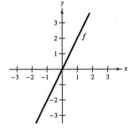

49. Since $f''(x) = 2$, the graph of f' is a line with a slope of 2, and the graph of f is a parabola opening upward.

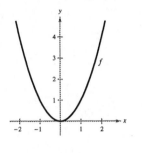

51.

x	$-\infty < x < -1$	$-1 < x < 0$	$0 < x < \infty$
$f'(x)$	$+$	$-$	$+$
$f(x)$	Increasing	Decreasing	Increasing

Relative maximum: $(-1, f(-1))$

Relative minimum: $(0, f(0)) = (0, 0)$

Intercepts: $(-2, 0), (0, 0)$

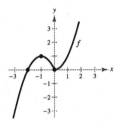

53. (a) $C = 1.80\left(\dfrac{100}{500/s}\right) + 9\left(\dfrac{100}{s}\right)$

$$= 0.36s + \frac{900}{s}, \quad 40 \le s \le 65$$

(b)

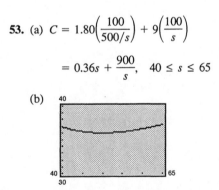

Minimum is $s = 50$, which is the most economical speed.

55. $T = \dfrac{23.011 - 1.0t + 0.048t^2}{1 - 0.204t + 0.014t^2}, \quad 1 \le t \le 12$

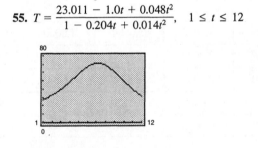

Absolute maximum: $(7.13, 71.23)$

Absolute minimum: $(1.0, 27.23)$

The model predicts the average monthly temperature to be lowest in January $(27.23°)$ and highest in July $(71.23°)$.

57. $g(x) = \dfrac{x^2 + x - 2}{x - 1} = \dfrac{(x-1)(x+2)}{x-1} = x + 2, \quad x \ne 1$

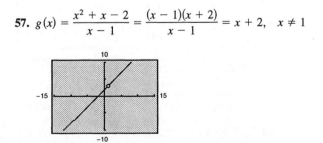

At $x = 1$ there is a hole in the graph, not a vertical asymptote.

Section 3.8   Differentials and Marginal Analysis

1. $dy = 6x\,dx$

3. $dy = 3(4x - 1)^2(4)\,dx$

$\quad = 12(4x - 1)^2\,dx$

5. $dy = \dfrac{1}{2}(x^2 + 1)^{-1/2}(2x)\,dx = \dfrac{x}{\sqrt{x^2 + 1}}\,dx$

7. $f(x) = 5x^2 - 1, \quad x = 1, \quad \Delta x = 0.01$

$\quad \Delta y = f(x + \Delta x) - f(x)$

$\quad\quad = [5(1.01)^2 - 1] - [5(1)^2 - 1]$

$\quad\quad = 0.1005$

9. $f(x) = \dfrac{4}{x^{1/3}}, \quad x = 1, \quad \Delta x = 0.01$

$\quad \Delta y = f(x + \Delta x) - f(x)$

$\quad\quad = \dfrac{4}{(1.01)^{1/3}} - \dfrac{4}{1}$

$\quad\quad \approx -0.013245$

11. $dy = 3x^2\,dx = 0.3$

$\quad \Delta y = (1.1)^3 - 1^3 = 0.331$

13. $dy = 4x^3\,dx = -0.04$

$\quad \Delta y = [(-0.99)^4 + 1] - [(-1)^4 + 1] \approx -0.0394$

15. $dy = 2x\,dx$

$dx = \Delta x$	dy	Δy	$\Delta y - dy$	$\dfrac{dy}{\Delta y}$
1.0000	4.0000	5.0000	1.0000	0.8000
0.5000	2.0000	2.2500	0.2500	0.8889
0.1000	0.4000	0.4100	0.0100	0.9756
0.0100	0.0400	0.0401	0.0001	0.9975
0.0010	0.0040	0.0040	0.0000	0.9998

17. $dy = -\dfrac{2}{x^3}\,dx, \quad \Delta y = \dfrac{1}{(x + \Delta x)^2} - \dfrac{1}{x^2}$

$dx = \Delta x$	dy	Δy	$\Delta y - dy$	$\dfrac{dy}{\Delta y}$
1.000	-0.2500	-0.1389	0.1111	1.7999
0.500	-0.1250	-0.0900	0.0350	1.3889
0.100	-0.0250	-0.0232	0.0018	1.0756
0.010	-0.0025	-0.0025	0.0000	1.0075
0.001	-0.0003	-0.0002	0.0001	1.0007

19. $y = x^{1/4}, dy = \dfrac{1}{4}x^{-3/4}\,dx = \dfrac{1}{4x^{3/4}}\,dx, x = 2$

$\quad \Delta y = (x + \Delta x)^{1/4} - x^{1/4}$

$dx = \Delta x$	dy	Δy	$\Delta y - dy$	$\dfrac{dy}{\Delta y}$
1.0	0.1487	0.1269	-0.0218	1.1717
0.5	0.0743	0.0682	-0.0061	1.0894
0.1	0.0149	0.0146	-0.0003	1.0186
0.01	0.0015	0.0015	-0.000003	1.00187
0.001	0.00015	0.00015	0.0	1.0002

Answers will vary.

21. $f(x) = 2x^3 - x^2 + 1$

   $f'(x) = 6x^2 - 2x$

   $f'(-2) = 24 + 4 = 28$

   $y + 19 = 28(x + 2)$

   $\quad\quad y = 28x + 37 \quad\quad$ Tangent line

   $f(x + \Delta x) = f(-2 + 0.01) = f(-1.99) \approx -18.72$

   $f(x - \Delta x) = f(-2 - 0.01) = f(-2.01) \approx -19.28$

   $y(x + \Delta x) = y(-1.99) = -18.72$

   $y(x - \Delta x) = y(-2.01) = -19.28$

23. $f(x) = \dfrac{x}{x^2 + 1}$

   $f'(x) = \dfrac{(x^2 + 1)(1) - x(2x)}{(x^2 + 1)^2} = \dfrac{1 - x^2}{(x^2 + 1)^2}$

   $f'(0) = 1$

   $y - 0 = 1(x - 0)$

   $\quad\quad y = x \quad\quad\quad\quad$ Tangent line

   $f(x + \Delta x) = f(0 + 0.01) = f(0.01) \approx 0.009999$

   $f(x - \Delta x) = f(0 - 0.01) = f(-0.01) \approx -0.009999$

   $y(x \pm \Delta x) = y(0 \pm 0.01) = y(\pm 0.01) = \pm 0.01$

25. $p = 75 - 0.25x, \quad dp = -0.25 \, dx$

   (a) $x = 7, \ \Delta x = 1$

   $\quad \Delta p = [75 - 0.25(8)] - [75 - 0.25(7)]$

   $\quad\quad = -0.25 = dp$

   (b) $x = 70, \quad \Delta x = 1$

   $\quad \Delta p = [75 - 0.25(71)] - [75 - 0.25(70)]$

   $\quad\quad = -0.25 = dp$

27. $x = 12, \quad dx = \Delta x = 1$

   $\Delta C \approx dC = (0.10x + 4) \, dx$

   $\quad\quad = [0.10(12) + 4](1)$

   $\quad\quad = \$5.20$

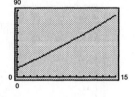

29. $R = 30x - 0.15x^2, \quad x = 75, \quad dx = 1$

   $dR = (30 - 0.30x) \, dx$

   $\quad = [30 - 0.30(75)](1) = \$7.50$

31. $x = 50, \quad dx = \Delta x = 1$

   $\Delta P \approx dP = (-1.5x^2 + 2500) \, dx$

   $\quad\quad = [-1.5(50)^2 + 2500](1)$

   $\quad\quad = -\$1250$

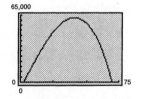

33. $(150, 50), (120, 60)$

   $m = \dfrac{60 - 50}{120 - 150} = -\dfrac{1}{3}$

   $p - 50 = -\dfrac{1}{3}(x - 150)$

   $p = -\dfrac{1}{3}x + 100$

   $R = xp = -\dfrac{1}{3}x^2 + 100x$

When $x = 141$ and $dx = \Delta x = 1$, we have

$$\Delta R \approx dR = \left(-\dfrac{2}{3}x + 100\right) dx = \left[-\dfrac{2}{3}(141) + 100\right](1) = \$6.00.$$

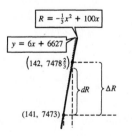

35. $(30,000, 25), (40,000, 20)$

$$m = \frac{20 - 25}{40,000 - 30,000} = \frac{-5}{10,000} = \frac{-1}{2000}$$

$$p - 25 = \frac{-1}{2000}(x - 30,000)$$

$$p = \frac{-1}{2000}x + 40$$

$$C = 275,000 + 17x$$

$$P = R - C = xp - C$$

$$= \left(\frac{-1}{2000}x^2 + 40x\right) - (275,000 + 17x)$$

$$= \frac{-1}{2000}x^2 + 23x - 275,000$$

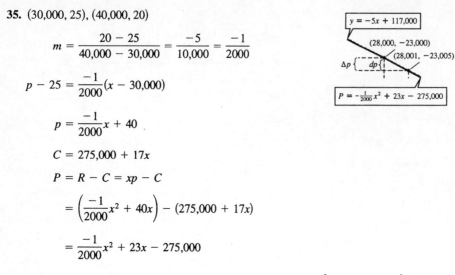

When $x = 28,000$ and $dx = \Delta x = 1$, we have $\Delta P \approx dP = \frac{-1}{1000}x + 23 = \frac{-1}{1000}(28,000) + 23 = -\$5.00.$

37. (a) $dA = 2x\,dx = 2x\Delta x$

$$\Delta A = (x + \Delta x)^2 - x^2$$

$$= 2x\Delta x + (\Delta x)^2$$

(b) See graph.

(c) $\Delta A - dA = (\Delta x)^2$

(See graph.)

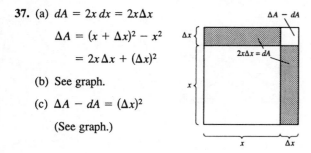

39. $A = \pi r^2, \quad dr = \Delta r = \pm\frac{1}{8}, \quad r = 10$

$$dA = 2\pi r \cdot dr = 2\pi(10)\left(\pm\frac{1}{8}\right) = \pm\frac{5}{2}\pi \text{ in.}^2$$

$$\text{Relative error} = \frac{dA}{A} = \frac{\pm\frac{5}{2}\pi}{100\pi} = \pm\frac{1}{40}$$

41. Let $\Delta r = dr = \pm 0.02$ inch.

$$V = \frac{4}{3}\pi r^3$$

$$dV = 4\pi r^2\,dr = 4\pi(6)^2(\pm 0.02) = \pm 2.88\pi \text{ in.}^3$$

When $r = 6$, the relative error is

$$\frac{dV}{V} = \frac{4\pi r^2\,dr}{(4/3)\pi r^3} = \frac{3}{r}\frac{dr}{} = \frac{3(\pm 0.02)}{6} = \pm 0.01.$$

43. True

Review Exercises for Chapter 3

1. $f(x) = -x^2 + 2x + 4$

$f'(x) = -2x + 2 = 0$ when $x = 1$.

Critical number: $x = 1$

3. $f(x) = x^{3/2} - 3x^{1/2}$ where $x \geq 0$.

$$f'(x) = \frac{3}{2}x^{1/2} - \frac{3}{2}x^{-1/2} = \frac{3}{2}x^{-1/2}(x - 1) = \frac{3(x - 1)}{2\sqrt{x}}$$

Critical numbers: $x = 0$ and $x = 1$

5. $f(x) = x^2 + x - 2$

$f'(x) = 2x + 1$

Critical number: $x = -\frac{1}{2}$

Increasing on $\left(-\frac{1}{2}, \infty\right)$

Decreasing on $\left(-\infty, -\frac{1}{2}\right)$

7. $h(x) = \dfrac{x^2 - 3x - 4}{(x - 3)}$

$h'(x) = \dfrac{x^2 - 6x + 13}{(x - 3)^2} > 0$ for all $x \neq 3$.

Increasing on $(-\infty, 3), (3, \infty)$

9. (a) $T = 0.0385t^4 - 1.122t^3 + 9.67t^2 - 21.8t = 47$

$1 \leq t \leq 12$, $t = 1$ corresponds to January.

$T'(t) = 0.154t^3 - 3.366t^2 + 19.34 - 21.8$

T increasing on $(1.48, 7.28)$.

(b) T decreasing on $(1, 1.48)$ and $(7.28, 12)$.

(c) Temperature increases from mid-January to early July. Decreases otherwise.

(d)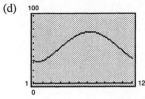

11. $f(x) = 4x^3 - 6x^2 - 2$

$f'(x) = 12x^2 - 12x = 12x(x - 1)$

Critical numbers: $x = 0$, $x = 1$

Relative maximum: $(0, -2)$

Relative minimum: $(1, -4)$

Interval	$(-\infty, 0)$	$(0, 1)$	$(1, \infty)$
Sign of f'	$+$	$-$	$+$
Conclusion	Increasing	Decreasing	Increasing

13. $g(x) = x^2 - 16x + 12$

$g'(x) = 2x - 16 = 2(x - 8)$

Critical number: $x = 8$

Increasing on $(8, \infty)$

Decreasing on $(-\infty, 8)$

Relative minimum: $(8, -52)$

Interval	$(-\infty, 8)$	$(8, \infty)$
Sign of g'	$-$	$+$
Conclusion	Decreasing	Increasing

15. $h(x) = 2x^2 - x^4$

$h'(x) = 4x - 4x^3$

$\qquad = 4x(1 - x^2) = 4x(1 - x)(1 + x)$

Critical numbers: $x = 0, 1, -1$

Relative maxima: $(-1, 1), (1, 1)$

Relative minimum: $(0, 0)$

Interval	$(-\infty, -1)$	$(-1, 0)$	$(0, 1)$	$(1, \infty)$
Sign of h'	$+$	$-$	$+$	$-$
Conclusion	Increasing	Decreasing	Increasing	Decreasing

17. $f(x) = \dfrac{6}{x^2 + 1}$

$f'(x) = \dfrac{-12x}{(x^2 + 1)^2}$

Critical number: $x = 0$

Relative maximum: $(0, 6)$

Interval	$(-\infty, 0)$	$(0, \infty)$
Sign of f'	$+$	$-$
Conclusion	Increasing	Decreasing

19. $h(x) = \dfrac{x^2}{x - 2}$

$h'(x) = \dfrac{x(x - 4)}{(x - 2)^2}$

Critical numbers: $x = 0, 4$

Discontinuity: $x = 2$

Relative maximum: $(0, 0)$

Relative minimum: $(4, 8)$

Interval	$(-\infty, 0)$	$(0, 2)$	$(2, 4)$	$(4, 8)$
Sign of h'	$+$	$-$	$-$	$+$
Conclusion	Increasing	Decreasing	Decreasing	Increasing

21. $f(x) = x^2 + 5x + 6$, $[-3, 0]$

$f'(x) = 2x + 5$

Critical number: $x = -\frac{5}{2}$

x	$f(x)$	
-3	0	
0	6	Maximum
$-\frac{5}{2}$	$-\frac{1}{4}$	Minimum

23. $f(x) = x^3 - 12x + 1$, $[-4, 4]$

$f'(x) = 3x^2 - 12 = 3(x - 2)(x + 2)$

Critical numbers: $x = \pm 2$

x	$f(x)$	
-4	-15	Minimum
4	17	Maximum
-2	17	Maximum
2	-15	Minimum

25. $f(x) = 4\sqrt{x} - x^2$, $[0, 3]$

$f'(x) = \dfrac{2}{\sqrt{x}} - 2x = \dfrac{2 - 2x^{3/2}}{\sqrt{x}}$

Critical number: $x = 1$

x	$f(x)$	
0	0	
1	3	Maximum
3	$4\sqrt{3} - 9 \approx -2.07$	Minimum

Maximum at $(1, 3)$

Minimum at $\left(3, 4\sqrt{3} - 9\right)$

27. $f(x) = 3x^4 - 6x^2 + 2$, $[0, 2]$

$f'(x) = 12x^3 - 12x = 12x(x - 1)(x + 1)$

Critical numbers: $x = 0, 1$

x	$f(x)$	
0	2	
2	26	Maximum
1	-1	Minimum

29. $f(x) = \dfrac{2x}{x^2 + 1}$, $[-1, 2]$

$f'(x) = \dfrac{-2(x^2 - 1)}{(x^2 + 1)^2}$

Critical numbers: $x = 1, -1$

x	$f(x)$	
-1	-1	Minimum
2	$\frac{4}{5}$	
1	1	Maximum

31. $S = 2\pi r^2 + 50r^{-1}$

$S' = 4\pi r - 50r^{-2} = 4\pi r - \dfrac{50}{r^2}$

$S' = 0 \Rightarrow 4\pi r = \dfrac{50}{r^2}$

$r^3 = \dfrac{50}{4\pi}$

$r \approx 1.58$ inches

[*Note:* You can check that $h = 2r$.]

Minimum: $(1.58, 47.33)$

33. $f(x) = (x - 2)^3$

$f'(x) = 3(x - 2)^2$

$f''(x) = 6(x - 2)$

$f''(x) > 0$ for $x > 2$:
concave upward on $(2, \infty)$

$f''(x) < 0$ for $x < 2$:
concave downward on $(-\infty, 2)$

35. $g(x) = \frac{1}{4}(-x^4 + 8x^2 - 12)$

$g'(x) = -x^3 + 4x$

$g''(x) = -3x^2 + 4$

Because g changes concavity at $x \approx \pm 1.1547$, the points of inflection are $(-1.1547, -0.7778)$ and $(1.1547, -0.7778)$. The graph is concave downward on $(-\infty, -1.1547)$ and $(1.1547, \infty)$. The graph is concave upward on $(-1.1547, 1.1547)$.

Note: $\dfrac{2}{\sqrt{3}} \approx 1.1547$

37. $f(x) = \frac{1}{2}x^4 - 4x^3$

$f'(x) = 2x^3 - 12x^2$

$f''(x) = 6x^2 - 24x = 6x(x - 4)$

$f''(x) = 0$ when $x = 0, 4$. Because f changes concavity at $x = 0$ and $x = 4$, the points of inflection are $(0, 0)$ and $(4, -128)$.

39. $f(x) = x^3(x - 3)^2$

$f'(x) = x^2(5x - 9)(x - 3)$

$f''(x) = 2x(10x^2 - 36x + 27)$

Because f changes concavity at $x = 0$, $x = 1.0652$, and $x = 2.5348$, the points of inflection are $(0, 0)$, $(1.0652, 4.5241)$, and $(2.5348, 3.5241)$.

41. $f(x) = x^5 - 5x^3$

$f'(x) = 5x^4 - 15x^2 = 5x^2(x^2 - 3)$

$f''(x) = 20x^3 - 30x$

Critical numbers: $x = 0, \pm\sqrt{3}$

$f''(\sqrt{3}) > 0 \Rightarrow (\sqrt{3}, -6\sqrt{3})$ is a relative minimum.

$f''(-\sqrt{3}) < 0 \Rightarrow (-\sqrt{3}, 6\sqrt{3})$ is a relative maximum.

$f''(0) = 0 \Rightarrow$ test fails.

By the First-Derivative Test, $(0, 0)$ is not a relative extremum.

43. $f(x) = (x - 1)^3(x + 4)^2$

$f'(x) = 5(x + 4)(x + 2)(x - 1)^2$

$f''(x) = 10(x - 1)(2x^2 + 8x + 5)$

Critical numbers: $x = -4, -2, 1$

$f''(-4) < 0 \Rightarrow (-4, 0)$ is a relative maximum.

$f''(-2) > 0 \Rightarrow (-2, -108)$ is a relative minimum.

$f''(1) = 0 \Rightarrow$ test fails.

By the First-Derivative Test, $(1, 0)$ is not a relative extremum.

45. $R = \frac{1}{1500}(150x^2 - x^3), \quad 0 \le x \le 100$

$R' = \frac{1}{1500}(300x - 3x^2)$

$R'' = \frac{1}{1500}(300 - 6x)$

$R'' = 0$ when $x = 50$. The point of diminishing returns is $(50, 166\frac{2}{3})$.

47. $xy = 169$ (product is 169)

Sum $= S = x + y = x + \dfrac{169}{x}$

$S' = 1 - \dfrac{169}{x^2} = 0 \Rightarrow x = 13, \ y = 13$

49. (a) $N = 0.020t^3 - 1.19t^2 + 9.0t + 1746$,

$0 \le t \le 30$, $t = 0$ corresponds to 1970.

$N' = 0.06t^2 - 2.38t + 9$

Maximum: $(4.23, 1764.3)$

Minimum: $(30, 1485)$

(b) $|N'|$ is greatest at $t = 19.8$, or the end of 1989.

(c) The maximum number of daily newspapers was 1764.3 million in 1974, and the minimum was 1485 in 2000. the year 1989 was when the circulation was changing at the greatest rate.

51. $y' = -0.009x^2 + 0.274x + 0.458$

$y'' = -0.018x + 0.274 = 0 \Rightarrow x = \frac{137}{9} \approx 15.2$ years

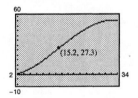

53. $s(r) = c(R^2 - r^2), \ c > 0$

$s'(r) = -2cr = 0 \Rightarrow r = 0$

$s''(r) = -2c < 0$ for all r.

By the Second-Derivative Test, $r = 0$ yields a maximum.

55. $p = 15 - 0.1(n - 20) = 17 - 0.1n, \quad 20 \le n \le N$

Let x be the number of people over 20 in the group.

Revenue $= R = (x + 20)[15 - 0.1(n - 20)]$

$= (x + 20)(15 - 0.1x), \quad x + 20 = n$

$R' = 13 - 0.2x = 0 \Rightarrow x = 65$

Thus, $N = 65 + 20 = 85$ people would maximize revenue.

57. $\dfrac{dC}{dx} = -\dfrac{Qs}{x^2} + \dfrac{r}{2}$

$\dfrac{dC}{dx} = 0$ when $x^2 = \dfrac{2Qs}{r}$ or $x = \sqrt{\dfrac{2Qs}{r}}$.

Since $Q = 10{,}000$, $s = 4.5$ and $r = 5.76$, $x = 125$.

59. $p = 30 - 0.2x$, $0 \le x \le 150$, $\dfrac{dp}{dx} = -0.2$

$\eta = \dfrac{p/x}{dp/dx} = \dfrac{(30 - 0.2x)/x}{-0.2} = \dfrac{x - 150}{x}$

$|\eta| = 1 = \left| \dfrac{x - 150}{x} \right| \implies x = |x - 150|$

$\implies x = 75$

For $0 < x < 75$, $|\eta| > 1$ elastic.

For $75 < x < 150$, $|\eta| < 1$ inelastic.

$x = 75$, unit elasticity.

61. $p = \sqrt{300 - x}$, $0 \le x \le 300$

$\dfrac{dp}{dx} = \dfrac{-1}{2\sqrt{300 - x}}$

$\eta = \dfrac{p/x}{dp/dx} = \dfrac{\sqrt{300 - x}/x}{-1/2\sqrt{300 - x}} = \dfrac{2(x - 300)}{x}$

$|\eta| = 1 \implies x = |2(x - 300)|$

$x = \pm(2x - 600)$

$x = 2x - 600$ or $x = -2x + 600$

$x = 600$ or $x = 200$

Since $0 \le x \le 300$, select $x = 200$.

Elastic: $(0, 200)$

Inelastic: $(200, 300)$

Unit elasticity: $x = 200$

63. Since $x = 4$ is an infinite discontinuity, it is a vertical asymptote. Since

$$\lim_{x \to \infty} \dfrac{2x + 3}{x - 4} = \dfrac{2}{1} = 2$$

$y = 2$ is a horizontal asymptote.

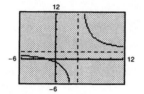

65. $f(x) = \dfrac{\sqrt{9x^2 + 1}}{x}$

$\lim_{x \to \infty} f(x) = 3$, $\lim_{x \to -\infty} f(x) = -3$

Horizontal asymptotes: $y = \pm 3$

Vertical asymptote: $x = 0$

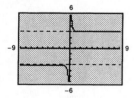

67. $f(x) = \dfrac{3}{x^2 - 5x + 4} = \dfrac{3}{(x - 4)(x - 1)}$

$x = 1$ and $x = 4$ are vertical asymptotes. $y = 0$ is the horizontal asymptote.

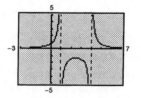

69. $\lim_{x \to 0^+} \left(x - \dfrac{1}{x^3} \right) = -\infty$

71. $\lim_{x \to -1^+} \dfrac{x^2 - 2x + 1}{x + 1} = \infty$

73. $\lim_{x \to \infty} \dfrac{5x^2 + 3}{2x^2 - x + 1} = \lim_{x \to \infty} \dfrac{5 + (3/x^2)}{2 - (1/x) + (1/x^2)} = \dfrac{5}{2}$

75. $\lim_{x \to -\infty} \dfrac{3x^2}{x + 2} = -\infty$

77. $T = \dfrac{0.37s + 23.8}{s}, \quad 0 < s \le 120$

(a)

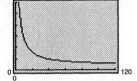

(b) $\displaystyle\lim_{s \to \infty} T = \lim_{s \to \infty} \dfrac{0.37s + 23.8}{s} = 0.37$

79. $f(x) = 4x - x^2 = x(4 - x)$

$f'(x) = 4 - 2x = -2(x - 2)$

$f''(x) = -2$

Intercepts: $(0, 0), (4, 0)$

Domain: $(-\infty, \infty)$

Range: $(-\infty, 4]$

Relative maximum: $(2, 4)$

Concave downward on $(-\infty, \infty)$

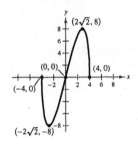

81. $f(x) = x\sqrt{16 - x^2}$

$f'(x) = \dfrac{16 - 2x^2}{\sqrt{16 - x^2}}$

$f''(x) = \dfrac{2x(x^2 - 24)}{(16 - x^2)^{3/2}}$

Domain: $[-4, 4]$

Range: $[-8, 8]$

Intercepts: $(0, 0), (4, 0), (-4, 0)$

Relative maximum: $\left(2\sqrt{2}, 8\right)$

Relative minimum: $\left(-2\sqrt{2}, -8\right)$

Point of inflection: $(0, 0)$

Concave upward on $(-4, 0)$

Concave downward on $(0, 4)$

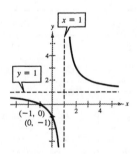

83. $f(x) = \dfrac{x + 1}{x - 1}$

$f'(x) = \dfrac{-2}{(x - 1)^2}$

$f''(x) = \dfrac{4}{(x - 1)^3}$

Domain: all real numbers except 1

Range: all real numbers except 1

Intercepts: $(-1, 0), (0, -1)$

Horizontal asymptote: $y = 1$

Vertical asymptote: $x = 1$

Concave upward on $(1, \infty)$

Concave downward on $(-\infty, 1)$

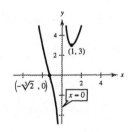

85. $f(x) = x^2 + \dfrac{2}{x}$

$f'(x) = 2x - \dfrac{2}{x^2} = \dfrac{2(x^3 - 1)}{x^2}$

$f''(x) = 2 + \dfrac{4}{x^3} = \dfrac{2(x^3 + 2)}{x^3}$

Domain: all real numbers except 0

Range: all real numbers

Intercept: $\left(-\sqrt[3]{2}, 0\right)$

Vertical asymptote: $x = 0$

Relative minimum: $(1, 3)$

Point of inflection: $\left(-\sqrt[3]{2}, 0\right)$

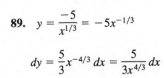

87. $y = 6x^2 - 5$

$dy = 12x \, dx$

89. $y = \dfrac{-5}{x^{1/3}} = -5x^{-1/3}$

$dy = \dfrac{5}{3}x^{-4/3} \, dx = \dfrac{5}{3x^{4/3}} \, dx$

91. $C = 40x^2 + 1225, \quad x = 10$

$dC = 80x \cdot dx = 80(10)(1) = \$800$

93. $R = 6.25x + 0.4x^{3/2}$

$$dR = \left[6.25 + 0.6\sqrt{x}\right] dx$$
$$= \left[6.25 + 0.6\sqrt{225}\right](1)$$
$$= \$15.25$$

95. (a)

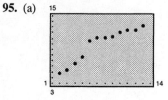

(b) The revenue is increasing.

(c)

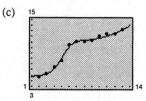

(d) According to the model, the revenue per share is increasing on the entire interval $2 < t < 13$.

(e) R' is increasing on $(2, 5)$ and $(8, 13)$.

   R' is decreasing on $(5, 8)$.

(f) Answers will vary.

97. The radius is 9 inches with a possible error of 0.025 inch.

$$S = 4\pi r^2$$
$$dS = 8\pi r \, dr$$
$$V = \tfrac{4}{3}\pi r^3$$
$$dV = 4\pi r^2 \, dr$$

When $r = 9$ and $dr = \pm 0.025$, we have the following.

$$dS = 8\pi(9)(\pm 0.025) = \pm 1.8\pi \text{ in.}^2$$
$$dV = 4\pi(9)^2(\pm 0.025) = \pm 8.1\pi \text{ in.}^3$$

99.

Quantity	Price	Total revenue	Marginal revenue
1	14.00	14.00	10.00
2	12.00	24.00	6.00
3	10.00	30.00	4.00
4	8.50	34.00	1.00
5	7.00	35.00	-2.00
6	5.50	33.00	

(a) $R = -1.43x^2 + 13.77x + 1.8$

(b) $\dfrac{dR}{dx} = -2.86x + 13.77$

x	1	2	3	4	5	6
$R'(x)$	10.91	8.05	5.19	2.33	-0.53	-3.39

(c) The revenue is maximized for 5 units of output. The quadratic model is maximized at $x \approx 4.8$.

Practice Test for Chapter 3

1. Find the critical numbers and the intervals on which f is increasing or decreasing for $f(x) = x^3 - 6x^2 + 5$.

2. Find the critical numbers and the intervals on which f is increasing or decreasing for $f(x) = 2x\sqrt{1 - x}$.

3. Find the relative extrema of $f(x) = x^4 - 32x + 3$.

4. Find the relative extrema of $f(x) = (x + 3)^{4/3}$.

5. Find the extrema of $f(x) = x^2 - 4x - 5$ on $[0, 5]$.

6. Find the points of inflection of $f(x) = 3x^4 - 24x + 2$.

7. Find the points of inflection of $f(x) = \dfrac{x^2}{1 + x^2}$.

8. Find two positive numbers whose product is 200 such that the sum of the first plus three times the second is a minimum.

9. Three rectangular fields are to be enclosed by 3000 feet of fencing, as shown in the accompanying figure. What dimensions should be used so that the enclosed area will be a maximum?

10. Find the number of units that produces a maximum revenue for $R = 400x^2 - 0.02x^3$.

11. Find the price per unit p that produces a maximum profit P given the cost function $C = 300x + 45{,}000$ and the demand function $p = 21{,}000 - 0.03x^2$.

12. Given the demand function $p = 600 - 0.02x^2$, find η (the price elasticity of demand) when $x = 100$.

13. Find $\lim\limits_{x \to 3^-} \dfrac{x + 4}{x - 3}$.

14. Find $\lim\limits_{x \to \infty} \dfrac{4x^3 - 9x^2 + 1}{1 - 2x^3}$.

15. Sketch the graph of $f(x) = \dfrac{x^2}{x^2 - 9}$.

16. Sketch the graph of $f(x) = \dfrac{x + 2}{x^2 + 5}$.

17. Sketch the graph of $f(x) = x^3 + 3x^2 + 3x - 1$.

18. Sketch the graph of $f(x) = |4 - 2x|$.

19. Sketch the graph of $f(x) = (2 - x)^{2/3}$.

20. Use differentials to approximate $\sqrt[3]{65}$.

Graphing Calculator Required

21. Graph $y = \dfrac{5x}{\sqrt{x^2 + 4}}$ on a graphing calculator and find any asymptotes that exist. Are there any relative extrema?

22. Graph $y = \dfrac{2x^4 - 5x + 1}{x^4 + 1}$ on a graphing calculator. Is it possible for a graph to cross its horizontal asymptote?

C H A P T E R   4
Exponential and Logarithmic Functions

C H A P T E R   4
Exponential and Logarithmic Functions

Section 4.1   Exponential Functions

Solutions to Odd-Numbered Exercises

1. (a) $5(5^3) = 5^4 = 625$

(b) $27^{2/3} = \left(\sqrt[3]{27}\right)^2 = 3^2 = 9$

(c) $64^{3/4} = (2^6)^{3/4} = 2^{9/2} = \sqrt{2^9} = 2^4\sqrt{2} = 16\sqrt{2}$

(d) $81^{1/2} = \sqrt{81} = 9$

(e) $25^{3/2} = \left(\sqrt{25}\right)^3 = 5^3 = 125$

(f) $32^{2/5} = \left(\sqrt[5]{32}\right)^2 = 2^2 = 4$

3. (a) $(5^2)(5^3) = 5^5 = 3125$

(b) $(5^2)(5^{-3}) = 5^{-1} = \dfrac{1}{5}$

(c) $(5^2)^2 = 5^4 = 625$

(d) $5^{-3} = \dfrac{1}{5^3} = \dfrac{1}{125}$

5. (a) $\dfrac{5^3}{25^2} = \dfrac{5^3}{(5^2)^2} = \dfrac{5^3}{5^4} = \dfrac{1}{5}$

(b) $(9^{2/3})(3)(3^{2/3}) = (3^2)^{2/3}(3)(3^{2/3})$

$\qquad\qquad = (3^{4/3})(3^{5/3})$

$\qquad\qquad = 3^{9/3} = 3^3 = 27$

(c) $[(25^{1/2})5^2]\dfrac{1}{3} = [5 \cdot 5^2]^{1/3} = [5^3]^{1/3} = 5$

(d) $(8^2)(4^3) = (64)(64) = 4096$

7. $f(x) = 2^{x-1}$

(a) $f(3) = 2^{3-1} = 2^2 = 4$

(b) $f\left(\dfrac{1}{2}\right) = 2^{1/2-1} = 2^{-1/2} = \dfrac{1}{\sqrt{2}} = \dfrac{\sqrt{2}}{2}$

(c) $f(-2) = 2^{-2-1} = 2^{-3} = \dfrac{1}{8}$

(d) $f\left(-\dfrac{3}{2}\right) = 2^{-3/2-1} = 2^{-5/2} = \dfrac{1}{2^{5/2}} = \dfrac{1}{4\sqrt{2}} = \dfrac{\sqrt{2}}{8}$

9. $g(x) = 1.05^x$

(a) $g(-2) = 1.05^{-2} = \dfrac{1}{1.05^2} = \dfrac{1}{1.1025} \approx 0.907$

(b) $g(120) = 1.05^{120} \approx 348.912$

(c) $g(12) = 1.05^{12} \approx 1.796$

(d) $g(5.5) = 1.05^{5.5} \approx 1.308$

11. $3^x = 81 = 3^4$

$x = 4$

13. $\left(\dfrac{1}{3}\right)^{x-1} = 27$

$3^{-(x-1)} = 3^3$

$-(x - 1) = 3$

$-x + 1 = 3$

$x = -2$

15. $4^3 = (x + 2)^3$

$4 = x + 2$

$x = 2$

17. $x^{3/4} = 8$

$x = 8^{4/3}$

$= \left(\sqrt[3]{8}\right)^4$

$= 2^4$

$= 16$

19. The graph of $f(x) = 3^x$ is an exponential curve with the following characteristics.

Passes through $(0, 1), (1, 3), \left(-1, \frac{1}{3}\right)$

Horizontal asymptote: x-axis

Therefore, it matches graph (e).

21. The graph of $f(x) = -3^x = (-1)(3^x)$ is an exponential curve with the following characteristics.

Passes through $(0, -1), (1, -3), \left(-1, -\frac{1}{3}\right)$

Horizontal asymptote: x-axis

Therefore, it matches graph (a).

23. The graph of $f(x) = 3^{-x} - 1 = \left(\frac{1}{3}\right)^x - 1$ is an exponential curve with the following characteristics.

Passes through $(0, 0), \left(1, -\frac{2}{3}\right), (-1, 2)$

Horizontal asymptote: $y = -1$

Therefore, it matches graph (d).

25. $f(x) = 6^x$

x	-2	-1	0	1	2
$f(x)$	$\frac{1}{36}$	$\frac{1}{6}$	1	6	36

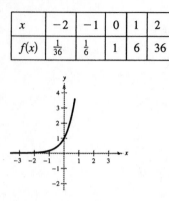

27. $f(x) = 5^{-x}$

x	-2	-1	0	1	2
$f(x)$	25	5	1	$\frac{1}{5}$	$\frac{1}{25}$

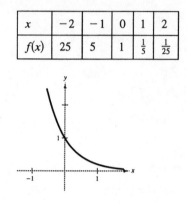

29. $y = 3^{-x^2}$

x	-2	-1	0	1	2
y	$\frac{1}{81}$	$\frac{1}{3}$	1	$\frac{1}{3}$	$\frac{1}{81}$

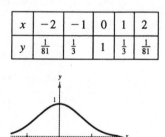

31. $y = 3^{-|x|}$

x	-2	-1	0	1	2
y	$\frac{1}{9}$	$\frac{1}{3}$	1	$\frac{1}{3}$	$\frac{1}{9}$

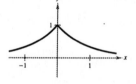

33. $s(t) = \dfrac{3^{-t}}{4} = \dfrac{1}{4(3^t)}$

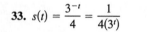

x	-2	-1	0	1	2
$s(t)$	$\frac{9}{4}$	$\frac{3}{4}$	$\frac{1}{4}$	$\frac{1}{12}$	$\frac{1}{36}$

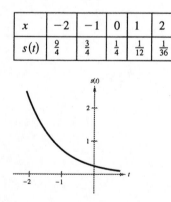

35. $P(t) = 251.27(1.0118)^t$ ($t = 2$ corresponds to 1992)

(a) 2006: $P(16) = 251.27(1.0118)^{16} \approx 303.1$ million

(b) 2012: $P(22) = 251.27(1.0118)^{22} \approx 325.25$ million

37. $V(t) = 64{,}000(2)^{t/15}$

    (a) $V(5) = 64{,}000(2)^{5/15} \approx \$80{,}634.95$ (b) $V(20) = 64{,}000(2)^{20/15} \approx \$161{,}269.89$

39. $V(t) = 16{,}000\left(\frac{3}{4}\right)^{t}$

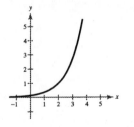

$V(4) = 16{,}000\left(\frac{3}{4}\right)^{4} \approx \$5062.50$

41. $y = 23\left(\frac{1}{2}\right)^{t/45}, \quad t \geq 0$

    (a)

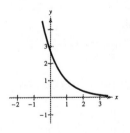

    (b) $y(75) \approx 7.24$ grams

    (c) $y = 1$ for $t \approx 203.56$ years

Section 4.2   Natural Exponential Functions

1. (a) $(e^3)(e^4) = e^{3+4} = e^7$

    (b) $(e^3)^4 = e^{3(4)} = e^{12}$

    (c) $(e^3)^2 = e^6 = \dfrac{1}{e^6}$

    (d) $e^0 = 1$

3. (a) $(e^2)^{5/2} = e^5$

    (b) $(e^2)(e^{1/2}) = e^{2+1/2} = e^{5/2}$

    (c) $(e^{-2})^{-3} = e^6$

    (d) $\dfrac{e^5}{e^{-2}} = e^{5+2} = e^7$

5. $e^{-3x} = e^1$

    $-3x = 1$

    $x = -\dfrac{1}{3}$

7. $e^{\sqrt{x}} = e^3$

    $\sqrt{x} = 3$

    $x = 9$

9. $x^{2/3} = \sqrt[3]{e^2} = e^{2/3}$

    $x = \pm e$

11. $3x^3 = 9e^3$

    $x^3 = 3e^3$

    $x = \sqrt[3]{3e}$

13. $f(x) = e^{2x+1}$. Increasing exponential passing through $(0, e)$. Matches (f)

15. $f(x) = e^{x^2}$. Symmetric with respect to y-axis. Matches (d)

17. $f(x) = e^{\sqrt{x}}$. Domain: $x \geq 0$. Matches (c)

19. $h(x) = e^{x-2}$

Increasing exponential passing through $(0, e^{-2})$.

21. $g(x) = e^{1-x}$

Decreasing exponential passing through $(0, e)$.

23.

x	-5	0	5	10	20
$N(t)$	1359.1	500	183.9	67.7	9.16

$N(t) = 500e^{-0.2t}$

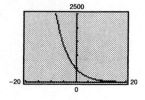

25.

x	-2	-1	0	1	2
$g(x)$	0.036	0.538	1	0.538	0.036

$g(x) = \dfrac{2}{1 + e^{x^2}}$

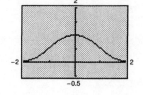

27. $f(x) = (e^x + e^{-x})/2$

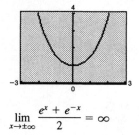

$\displaystyle \lim_{x \to \pm\infty} \frac{e^x + e^{-x}}{2} = \infty$

No horizontal asymptotes

Continuous on the entire real line

29. $f(x) = \dfrac{2}{1 + e^{1/x}}$

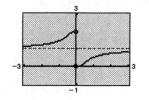

$\displaystyle \lim_{x \to \pm\infty} \frac{2}{1 + e^{1/x}} = \frac{2}{1 + 1} = 1$

Horizontal asymptote: $y = 1$

Discontinuous at $x = 0$

31. $P = 1000, \quad r = 0.03, \quad t = 10$

$$A = P\left(1 + \frac{r}{n}\right)^{nt} = 1000\left(1 + \frac{0.03}{n}\right)^{10n}$$

Continuous compounding: $A = Pe^{rt} = 1000e^{(0.03)(10)}$

n	1	2	4	12	365	Continuous
A	1343.92	1346.86	1348.35	1349.35	1349.84	1349.86

33. $P = 1000, \quad r = 0.03, \quad t = 40$

$$A = P\left(1 + \frac{r}{n}\right)^{nt} = 1000\left(1 + \frac{0.03}{n}\right)^{40n}$$

Continuous compounding: $A = Pe^{rt} = 1000e^{(0.03)(40)}$

n	1	2	4	12	365	Continuous
A	3262.04	3290.66	3305.28	3315.15	3319.95	3320.12

35. $A = Pe^{rt}, \quad A = 100{,}000, \quad r = 0.04 \Rightarrow P = 100{,}000e^{-0.04t}$

t	1	10	20	30	40	50
P	96,078.94	67,032.00	44,932.90	30,119.42	20,189.65	13,533.53

37. $A = P\left(1 + \dfrac{r}{n}\right)^{nt}$, $A = 100{,}000$, $r = 0.05$, $n = 12 \Rightarrow P = \dfrac{100{,}000}{\left(1 + \dfrac{0.05}{12}\right)^{12t}}$

t	1	10	20	30	40	50
P	95,132.82	60,716.10	36,864.45	22,382.66	13,589.88	8251.24

39. $\sqrt{eff} = \left(1 + \dfrac{r}{n}\right)^{n} - 1$, $r = 0.09$

(a) $n = 1$: $\sqrt{eff} = 0.09 = 9\%$

(b) $n = 2$: $\sqrt{eff} = \left(1 + \dfrac{0.09}{2}\right)^{2} - 1 \approx 9.20\%$

(c) $n = 4$: $\sqrt{eff} = \left(1 + \dfrac{0.09}{4}\right)^{4} - 1 \approx 9.31\%$

(d) $n = 12$: $\sqrt{eff} = \left(1 + \dfrac{0.09}{12}\right)^{12} - 1 \approx 9.38\%$

41. $P = \dfrac{A}{\left(1 + \frac{r}{n}\right)^{nt}} = \dfrac{15{,}503.77}{\left(1 + \frac{0.072}{12}\right)^{12(3)}} = \$12{,}500.00$

43. $A = P\left(1 + \dfrac{r}{n}\right)^{nt} = 8000\left(1 + \dfrac{0.045}{12}\right)^{12(2)}$

$\approx \$8751.92$

45. (a) $p(100) = 5000\left(1 - \dfrac{4}{4 + e^{-0.002(100)}}\right) \approx \$849.53$

(b) $p(500) = 5000\left(1 - \dfrac{4}{4 - e^{-0.002(500)}}\right) \approx \$421.12$

$\lim\limits_{x\to\infty} p = 0$

47. (a) $P\left(\dfrac{1}{2}\right) = 1 - e^{-1/6} \approx 0.1535 = 15.35\%$

(b) $P(2) = 1 - e^{-2/3} \approx 0.4866 = 48.66\%$

(c) $P(5) = 1 - e^{-5/3} \approx 0.8111 = 81.11\%$

49. $S = 3557.12e^{0.0475t}$ ($t = 4$ corresponds to 1994)

(a) 1995: $S(5) = 3557.12e^{0.0475(5)} \approx \$4510.7$ million

2000: $S(10) \approx \$5719.9$ million

2003: $S(13) \approx \$6595.9$ million

(b) Using the three points from (a), you obtain $y = 258.73t + 3194.01$, which fits the data well.

(c) $S = 10{,}000 = 3557.12e^{0.0475t}$

Using a graphing utility, $t \approx 21.8$, or 2011.

51. (a)

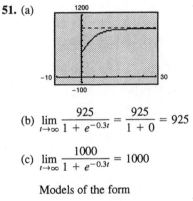

(b) $\lim\limits_{t\to\infty} \dfrac{925}{1 + e^{-0.3t}} = \dfrac{925}{1 + 0} = 925$

(c) $\lim\limits_{t\to\infty} \dfrac{1000}{1 + e^{-0.3t}} = 1000$

Models of the form

$$y = \dfrac{a}{1 + e^{-ct}},\quad c > 0,$$

have a limit of a as $t \to \infty$.

53. $P = \dfrac{0.83}{1 + e^{-0.2n}}$

(a) $n = 10 \Rightarrow P = 0.731$

(b) $P = 0.75 \Rightarrow n \approx 11.19$ or 11 trials

(c) $\lim\limits_{n\to\infty} \dfrac{0.83}{1 + e^{-0.2n}} = \dfrac{0.83}{1 + 0} = 0.83$

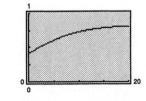

Section 4.3   Derivatives of Exponential Functions

1. $y' = 3e^{3x}$

$y'(0) = 3$

3. $y' = -e^{-x}$

$y'(0) = -1$

5. $y' = 4e^{4x}$

7. $y' = (-2x)e^{-x^2}$

$= -2xe^{-x^2}$

9. $f'(x) = e^{-1/x^2} \cdot \dfrac{d}{dx}(-x^{-2})$

$= 2x^{-3}e^{-1/x^2} = \dfrac{2}{x^3}e^{-1/x^2}$

11. $f'(x) = (x^2 + 1)4e^{4x} + (2x)e^{4x}$

$= e^{4x}(4x^2 + 2x + 4)$

13. $f(x) = \dfrac{2}{(e^x + e^{-x})^3} = 2(e^x + e^{-x})^{-3}$

$f'(x) = -6(e^x + e^{-x})^{-4}(e^x - e^{-x}) = \dfrac{-6(e^x - e^{-x})}{(e^x + e^{-x})^4}$

15. $y' = xe^x + e^x + 4e^{-x}$

17. $y' = (-2 + 2x)e^{-2x+x^2}$

$y'(0) = -2$

$y - 1 = -2(x - 0)$

$y = -2x + 1$

19. $y' = x^2(-e^{-x}) + 2xe^{-x} = xe^{-x}(-x + 2)$

$y'(2) = 0$

$y - \dfrac{4}{e^2} = 0(x - 2)$

$y = \dfrac{4}{e^2} = 4e^{-2}$ (horizontal line)

21. $y = (e^{2x} + 1)^3,\quad (0, 8)$

$y' = 3(e^{2x} + 1)^2 (2e^{2x})$

$y'(0) = 3(4)(2) = 24$

$y - 8 = 24(x - 0)$

$y = 24x + 8$

23. $xe^x + 2ye^x = 0$

$xe^x + e^x + 2ye^x + 2\dfrac{dy}{dx}e^x = 0$

$x + 1 + 2y + 2\dfrac{dy}{dx} = 0$

$\dfrac{dy}{dx} = \dfrac{1}{2}(-x - 1 - 2y)$

Alternatively,

$x + 2y = 0$

$y = -\dfrac{x}{2}$

$\dfrac{dy}{dx} = -\dfrac{1}{2}.$

These answers are the same since $2ye^x = -xe^x$ implies
$2y = -x.$

25. $x^2e^{-x} + 2y^2 - xy = 0$

$2xe^{-x} - x^2e^{-x} + 4y\dfrac{dy}{dx} - x\dfrac{dy}{dx} - y = 0$

$(4y - x)\dfrac{dy}{dx} = y + x^2e^{-x} - 2xe^{-x}$

$\dfrac{dy}{dx} = \dfrac{y + x^2e^{-x} - 2xe^{-x}}{4y - x}$

27. $f'(x) = 6e^{3x} - 6e^{-2x}$

$f''(x) = 18e^{3x} + 12e^{-2x} = 6(3e^{3x} + 2e^{-2x})$

29. $f'(x) = -5e^{-x} + 10e^{-5x}$

$f''(x) = 5e^{-x} - 50e^{-5x}$

31. $f(x) = \dfrac{1}{2 - e^{-x}}$

$f'(x) = \dfrac{-e^{-x}}{(2 - e^{-x})^2}$

$f''(x) = \dfrac{e^{-x}(2 + e^{-x})}{(2 - e^{-x})^3}$

Horizontal asymptote to the right: $y = \dfrac{1}{2}$

Horizontal asymptote to the left: $y = 0$

Vertical asymptote when $2 = e^{-x} \Longrightarrow x \approx -0.693$

No relative extrema or inflection points

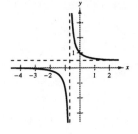

33. $f(x) = x^2 e^{-x}$

$f'(x) = -x^2 e^{-x} + 2xe^{-x} = xe^{-x}(2 - x)$

$f'(x) = 0$ when $x = 0$ and $x = 2$. Since $f''(0) > 0$ and $f''(2) < 0$, we have the following.

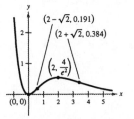

Relative minimum: $(0, 0)$

Relative maximum: $\left(2, \dfrac{4}{e^2}\right)$

Since $f''(x) = 0$ when $x = 2 \pm \sqrt{2}$, the inflection points occur at $\left(2 - \sqrt{2}, 0.191\right)$ and $\left(2 + \sqrt{2}, 0.384\right)$.

x	-2	-1	0	1	2	3
$f(x)$	29.556	2.718	0	0.368	0.541	0.448

35. $f(x) = \dfrac{8}{1 + e^{-0.5x}}$

Horizontal asymptotes:

$y = 8 \quad (\text{as } x \to \infty)$

$y = 0 \ (\text{as } x \to -\infty)$

No vertical asymptotes

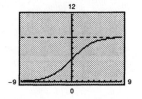

37. (a) $V = 15{,}000e^{-0.6286t}$

$V' = -9429e^{-0.6286t}$

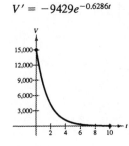

(b) $V'(1) = -9429e^{-0.6286(1)} \approx \$-5028.84 \text{ per year}$

(c) $V'(5) = -9429e^{-0.6286(5)} \approx \$-406.89 \text{ per year}$

(d) $V(0) = 15{,}000, \quad V(10) \approx 29.93$

$\dfrac{V(10) - V(0)}{10 - 0} = \dfrac{29.93 - 15{,}000}{10} \approx -1497.2$

Linear model: $V - 15{,}000 = -1497.2(t - 0)$

$V = -1497.2t + 15{,}000$

(e) Answers will vary.

39. $p = \dfrac{300}{3 + 17e^{-1.57x}}$

(a)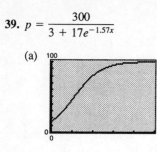

(b) When $x = 2$, $p = 80.3\%$.

(c) p' is increasing most rapidly when $x \approx 1.1$ (point of inflection).

41. $A = 5000e^{0.08t}$

$A' = 400e^{0.08t}$

(a) $A'(1) \approx \$433.31$ per year

(b) $A'(10) \approx \$890.22$ per year

(c) $A'(50) \approx \$21,839.26$ per year

43. $V = 7955e^{-0.0458/t}$

$V' = \dfrac{364.36648e^{-0.0458/t}}{t^2}$

When $t = 5$, $V' = 14.44$.

When $t = 10$, $V' = 3.63$.

When $t = 25$, $V' = 0.58$.

45. Mean $= \mu = 650$

Standard deviation $= \sigma = 12.5$

(a) $f(x) = \dfrac{1}{\sigma\sqrt{2\pi}}\, e^{-(x-)^2/2\sigma^2}$

$= \dfrac{1}{12.5\sqrt{2\pi}}\, e^{-(x-650)^2/312.5}$

(b)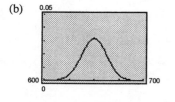

(c) $f'(x) = \dfrac{-4\sqrt{2}(x - 650)e^{-2(x-650)/625}}{15,625\sqrt{\pi}}$

47. $f(x) = \dfrac{1}{\sigma\sqrt{\pi}}e^{-(x-\mu)^2/2\sigma^2} = \dfrac{1}{\sigma\sqrt{2\pi}}e^{-x^2/2\sigma^2}$

For larger σ, the graph becomes flatter.

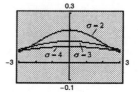

49. (a) $\dfrac{dh}{dt} = 50(-1.6)e^{-1.6t} - 20 = -80e^{-1.6t} - 20$

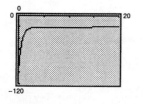

(b)

x	0	1	5	10	20
h'	-100	-36.15	-20.03	-20.00	-20.00

(c) The values in (b) are rates of descent in feet per second. As time increases, the rate is approximately constant at -20 ft/sec.

Section 4.4   Logarithmic Functions

1. $e^{0.6931\cdots} = 2$

3. $e^{-1.6094\cdots} = 0.2$

5. $\ln 1 = 0$

7. $\ln 0.0498 = -3$

9. The graph is a logarithmic curve that passes through the point $(1, 2)$ with a vertical asymptote at $x = 0$. Therefore, it matches graph (c).

11. The graph is a logarithmic curve that passes through the point $(-1, 0)$ with a vertical asymptote at $x = -2$. Therefore, it matches graph (b).

13. $y = \ln(x - 1)$

x	1.5	2	3	4	5
y	-0.69	0	0.69	1.10	1.39

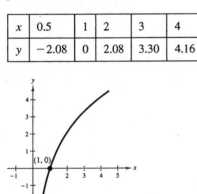

15. $y = \ln 2x$

x	0.25	0.5	1	3	5
y	-0.69	0	0.69	1.79	2.30

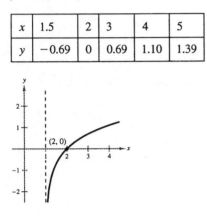

17. $y = 3 \ln x$

x	0.5	1	2	3	4
y	-2.08	0	2.08	3.30	4.16

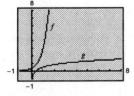

19. $g(x) = \ln\sqrt{x} = \frac{1}{2}\ln x$

$f(g(x)) = f\left(\frac{1}{2}\ln x\right) = e^{2(1/2 \ln x)} = e^{\ln x} = x$

$g(f(x)) = g(e^{2x}) = \frac{1}{2}\ln e^{2x} = \frac{1}{2}(2x)\ln e = x$

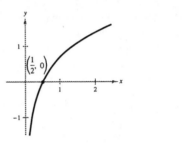

21. $f(g(x)) = f\left(\frac{1}{2} + \ln\sqrt{x}\right) = e^{2[(1/2) + \ln\sqrt{x}] - 1} = e^{2\ln x^{1/2}} = e^{\ln x} = x$

$g(f(x)) = g(e^{2x-1}) = \frac{1}{2} + \ln\sqrt{e^{2x-1}} = \frac{1}{2} + \frac{1}{2}\ln e^{2x-1} = \frac{1}{2} + \frac{1}{2}(2x - 1) = x$

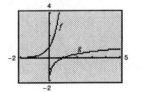

23. $\ln e^{x^2} = x^2$

25. $e^{\ln(5x+2)} = 5x + 2$

27. $-1 + \ln e^{2x} = -1 + 2x = 2x - 1$

29. (a) $\ln 6 = \ln(2 \cdot 3) = \ln 2 + \ln 3 = 0.6931 + 1.0986 = 1.7917$ (b) $\ln \frac{3}{2} = \ln 3 - \ln 2 = 1.0986 - 0.6931 = 0.4055$

(c) $\ln 81 = \ln 3^4 = 4 \ln 3 = 4(1.0986) = 4.3944$ (d) $\ln \sqrt{3} = \left(\frac{1}{2}\right)\ln 3 = \left(\frac{1}{2}\right)(1.0986) = 0.5493$

31. $\ln \frac{2}{3} = \ln 2 - \ln 3$ **33.** $\ln xyz = \ln x + \ln y + \ln z$ **35.** $\ln \sqrt{x^2 + 1} = \ln(x^2 + 1)^{1/2}$
$= \frac{1}{2}\ln(x^2 + 1)$

37. $\ln z(z - 1)^2 = \ln z + \ln(z - 1)^2 = \ln z + 2\ln(z - 1)$

39. $\ln \dfrac{3x(x + 1)}{(2x + 1)^2} = \ln[3x(x + 1)] - \ln(2x + 1)^2 = \ln 3 + \ln x + \ln(x + 1) - 2\ln(2x + 1)$

41. $\ln(x - 2) - \ln(x + 2) = \ln \dfrac{x - 2}{x + 2}$

43. $3\ln x + 2\ln y - 4\ln z = \ln x^3 + \ln y^2 - \ln z^4$ **45.** $3[\ln x + \ln(x + 3) - \ln(x + 4)] = 3\ln \dfrac{x(x + 3)}{x + 4}$
$= \ln\left(\dfrac{x^3 y^2}{z^4}\right)$ $= \ln\left[\dfrac{x(x + 3)}{x + 4}\right]^3$

47. $\dfrac{3}{2}[\ln x(x^2 + 1) - \ln(x + 1)] = \dfrac{3}{2}\ln \dfrac{x(x^2 + 1)}{x + 1}$ **49.** $\dfrac{1}{3}\ln(x + 1) - \dfrac{2}{3}\ln(x - 1) = \ln(x + 1)^{1/3} - \ln(x - 1)^{2/3}$
$= \ln\left[\dfrac{x(x^2 + 1)}{x + 1}\right]^{3/2}$ $= \ln\left[\dfrac{(x + 1)^{1/3}}{(x - 1)^{2/3}}\right]$
$= \ln\left[\dfrac{x + 1}{(x - 1)^2}\right]^{1/3}$

51. $x = 4$ **53.** $x = e^0 = 1$ **55.** $x + 1 = \ln 4$
$x = (\ln 4) - 1 \approx 0.3863$

57. $300e^{-0.2t} = 700$ **59.** $4e^{2x-1} - 1 = 5$
$e^{-0.2t} = \dfrac{7}{3}$ $e^{2x-1} = \dfrac{3}{2}$
$-0.2t = \ln 7 - \ln 3$ $2x - 1 = \ln\left(\dfrac{3}{2}\right)$
$t = \dfrac{\ln 7 - \ln 3}{-0.2} \approx -4.2365$ $x = \dfrac{1 + \ln\left(\frac{3}{2}\right)}{2} \approx 0.7027$

61. $\dfrac{10}{1 + 4e^{-0.01x}} = 2.5$ **63.** $2x \ln 5 = \ln 15$
$1 + 4e^{-0.01x} = \dfrac{10}{2.5} = 4$ $x = \dfrac{\ln 15}{2 \ln 5} \approx 0.8413$
$e^{-0.01x} = \dfrac{3}{4}$
$-0.01x = \ln\left(\dfrac{3}{4}\right)$
$x = -100 \ln\left(\dfrac{3}{4}\right) = 100 \ln\left(\dfrac{4}{3}\right) \approx 28.7682$

65. $500(1.07)^t = 1000$

$t \ln 1.07 = \ln 2$

$t = \dfrac{\ln 2}{\ln 1.07} \approx 10.2448$

67. $1000\left(1 + \dfrac{0.07}{12}\right)^{12t} = 3000$

$\left(1 + \dfrac{0.07}{12}\right)^{12t} = 3$

$12t \ln\left(1 + \dfrac{0.07}{12}\right) = \ln 3$

$t = \dfrac{\ln 3}{12 \ln[1 + (0.07/12)]} \approx 15.7402$

69. $P = 1000, \quad r = 0.05, \quad A = 2000$

(a) $2000 = 1000(1 + 0.05)^t$

$t = \dfrac{\ln 2}{\ln 1.05} \approx 14.2$ years

(b) $2000 = 1000\left(1 + \dfrac{0.05}{12}\right)^{12t}$

$t \approx \dfrac{\ln 2}{12 \ln 1.00417} \approx 13.88$ years

(c) $2000 = 1000\left(1 + \dfrac{0.05}{365}\right)^{365t}$

$t \approx \dfrac{\ln 2}{365 \ln 1.000137} \approx 13.87$ years

(d) $2000 = 1000e^{0.05t}$

$t = \dfrac{\ln 2}{0.05} \approx 13.86$ years

71. $\ln\left(\dfrac{N}{N_0}\right) = -kt$

$\ln\left(\dfrac{3.1}{13.6}\right) = -\left(\dfrac{0.693}{5715}\right)t$

$-1.4787 = -1.2126t$

$t \approx 12{,}194$ years

73. $P = 821.95e^{0.0358t}$ ($t = 0$ corresponds to 1980)

(a) 2000: $P(20) = 821.95e^{0.0358(20)} \approx 1681.9$

Population in 2000: 1,681,900

(b) $821.95e^{0.0358t} = 2500$

$e^{0.0358t} = \dfrac{2500}{821.95} \approx 3.042$

$0.0358t = \ln(3.042)$

$t = \dfrac{\ln(3.042)}{0.0358} \approx 31$

Population will be 2,500,000 in 2011.

75. $0.32 \times 10^{-12} = 10^{-12}\left(\dfrac{1}{2}\right)^{t/5715}$

$0.32 = \left(\dfrac{1}{2}\right)^{t/5715}$

$\ln 0.32 = \dfrac{t}{5715} \ln \dfrac{1}{2}$

$t = \dfrac{5715 \ln 0.32}{\ln\left(\frac{1}{2}\right)} \approx 9395$ years

77. $0.22 \times 10^{-12} = 10^{-12}\left(\dfrac{1}{2}\right)^{t/5715}$

$0.22 = \left(\dfrac{1}{2}\right)^{t/5715}$

$\ln 0.22 = \dfrac{t}{5715} \ln\left(\dfrac{1}{2}\right)$

$t = \dfrac{5715 \ln 0.22}{\ln\left(\frac{1}{2}\right)} \approx 12{,}484$ years

79. (a) $S(0) = 80 - 14 \ln 1 = 80$

(b) $S(4) = 80 - 14 \ln 5 \approx 57.5$

(c) $46 = 80 - \ln(t + 1) \quad \ln(t + 1) = \frac{34}{14} \Longrightarrow \approx 10$ months

81.

x	y	$\dfrac{\ln x}{\ln y}$	$\ln \dfrac{x}{y}$	$\ln x - \ln y$
1	2	0.0000	-0.6931	-0.6931
3	4	0.7925	-0.2877	-0.2877
10	5	1.4307	0.6931	0.6931
4	0.5	-2.0000	2.0794	2.0794

83.

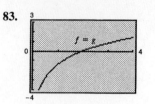

The graphs appear to be identical.

85. False. $\ln(0)$ is undefined.

87. False. $\ln x - \ln 2 = \ln(2x)$

89. False. $\ln u = 2 \ln v = \ln v^2 \implies u = v^2$

Section 4.5   Derivatives of Logarithmic Functions

1. $y = \ln x^3 = 3 \ln x$

$y' = \dfrac{3}{x}$

$y'(1) = 3$

3. $y = \ln x^2 = 2 \ln x$

$y' = \dfrac{2}{x}$

$y'(1) = 2$

5. $y = \ln x^2 = 2 \ln x$

$y' = \dfrac{2}{x}$

7. $y = \ln(x^2 + 3)$

$y' = \dfrac{2x}{x^2 + 3}$

9. $y = \ln \sqrt{x^4 - 4x} = \dfrac{1}{2}\ln(x^4 - 4x)$

$y' = \dfrac{1}{2}\left(\dfrac{4x^3 - 4}{x^4 - 4x}\right) = \dfrac{2(x^3 - 1)}{x(x^3 - 4)}$

11. $y = \dfrac{1}{2}(\ln x)^6$

$y' = \dfrac{1}{2} \cdot 6(\ln x)^5 \cdot \dfrac{1}{x} = \dfrac{3}{x}(\ln x)^5$

13. $y = x^2 \ln x$

$y' = x^2\left(\dfrac{1}{x}\right) + 2x \ln x = x + 2x \ln x = x(1 + \ln x^2)$

15. $y = \ln\left(x\sqrt{x^2 - 1}\right) = \ln x + \dfrac{1}{2}\ln(x^2 - 1)$

$y' = \dfrac{1}{x} + \left(\dfrac{1}{2}\right)\dfrac{1}{x^2 - 1}(2x) = \dfrac{1}{x} + \dfrac{x}{x^2 - 1} = \dfrac{2x^2 - 1}{x(x^2 - 1)}$

17. $y = \ln x - \ln(x + 1)$

$y' = \dfrac{1}{x} - \dfrac{1}{x + 1} = \dfrac{1}{x(x + 1)}$

19. $y = \ln\left(\dfrac{x - 1}{x + 1}\right)^{1/3} = \dfrac{1}{3}[\ln(x - 1) - \ln(x + 1)]$

$y' = \dfrac{1}{3}\left[\dfrac{1}{x - 1} - \dfrac{1}{x + 1}\right] = \dfrac{1}{3}\left[\dfrac{2}{x^2 - 1}\right] = \dfrac{2}{3(x^2 - 1)}$

21. $y = \ln \dfrac{\sqrt{4 + x^2}}{x} = \dfrac{1}{2}\ln(4 + x^2) - \ln x$

$y' = \dfrac{1}{2}\left(\dfrac{2x}{4 + x^2}\right) - \dfrac{1}{x} = -\dfrac{4}{x(4 + x^2)}$

23. $g(x) = e^{-x} \ln x$

$g'(x) = e^{-x}\left(\dfrac{1}{x}\right) + (-e^{-x})\ln x = e^{-x}\left(\dfrac{1}{x} - \ln x\right)$

25. $g(x) = \ln\dfrac{e^x + e^{-x}}{2}$

$= \ln(e^x + e^{-x}) - \ln 2$

$g'(x) = \dfrac{e^x - e^{-x}}{e^x + e^{-x}}$

27. $2^x = e^{x(\ln 2)}$

29. $\log_4 x = \dfrac{1}{\ln 4}\ln x$

31. $\log_2 48 = \dfrac{\ln 48}{\ln 2} \approx 5.585$

33. $\log_3 \dfrac{1}{2} = \dfrac{1}{\ln 3} \ln \dfrac{1}{2} \approx -0.63093$

(calculator)

35. $\log_{1/5} (31) = \dfrac{\ln 31}{\ln (1/5)} \approx -2.134$

37. $y = 3^x$

$y' = (\ln 3)3^x$

39. $f(x) = \log_2 x$

$f'(x) = \dfrac{1}{\ln 2} \cdot \dfrac{1}{x} = \dfrac{1}{x \ln 2}$

41. $h(x) = 4^{2x-3}$

$h'(x) = (\ln 4)4^{2x-3}(2)$

$= 2 \ln 4 \cdot 4^{2x-3}$

43. $y = \log_{10} (x^2 + 6x)$

$y' = \dfrac{1}{\ln 10} \dfrac{1}{x^2 + 6x}(2x + 6) = \dfrac{2x + 6}{(x^2 + 6x) \ln 10}$

45. $y = x2^x$

$y' = x(\ln 2)2^x + 2^x = 2^x(1 + x \ln 2)$

47. $y' = x\left(\dfrac{1}{x}\right) + (1) \ln x = 1 + \ln x$

$y'(1) = 1$

$y - 0 = 1(x - 1)$

$y = x - 1$ Tangent line

49. $y' = \dfrac{1}{\ln 3} \dfrac{1}{3x + 7}(3)$

$y'\left(\dfrac{2}{3}\right) = \dfrac{1}{\ln 3} \dfrac{1}{9}(3) = \dfrac{1}{3 \ln 3}$

$y - 2 = \dfrac{1}{3 \ln 3}\left(x - \dfrac{2}{3}\right)$

$y = \dfrac{1}{3 \ln 3}x + 2 - \dfrac{2}{9 \ln 3}$

51. $x^2 - 3 \ln y + y^2 = 10$

$2x - 3\left(\dfrac{1}{y}\right)\dfrac{dy}{dx} + 2y\dfrac{dy}{dx} = 0$

$2x = \dfrac{dy}{dx}\left(\dfrac{3}{y} - 2y\right)$

$2x = \dfrac{dy}{dx}\left(\dfrac{3 - 2y^2}{y}\right)$

$\dfrac{2xy}{3 - 2y^2} = \dfrac{dy}{dx}$

53. $4x^3 + \ln y^2 + 2y = 2x$

$12x^2 + 2\dfrac{y'}{y} + 2y' = 2$

$\left(\dfrac{2}{y} + 2\right)y' = 2 - 12x^2$

$y' = \dfrac{2 - 12x^2}{(2/y) + 2}$

$= \dfrac{1 - 6x^2}{(1/y) + 1} = \dfrac{(1 - 6x^2)}{1 + y}y$

55. $f(x) = x \ln \sqrt{x} + 2x = \dfrac{1}{2}x \ln x + 2x$

$f'(x) = \dfrac{1}{2}x\left(\dfrac{1}{x}\right) + \dfrac{1}{2} \ln x + 2 = \dfrac{1}{2} \ln x + \dfrac{5}{2}$

$f''(x) = \dfrac{1}{2x}$

57. $f(x) = 5^x$

$f'(x) = (\ln 5)5^x$

$f''(x) = (\ln 5)(\ln 5)5^x = (\ln 5)^2 5^x$

59. $\beta = 10 \log_{10} I - 10 \log_{10}(10^{-16})$

$\dfrac{d\beta}{dI} = \dfrac{10}{(\ln 10)I}$

For $I = 10^{-4}$,

$\dfrac{d\beta}{dI} = \dfrac{10}{(\ln 10)10^{-4}}$

$= \dfrac{10^5}{\ln 10} \approx 43{,}429.4$ decibels per watt per cm^2.

61. $f(x) = 1 + 2x \ln x$

$f'(x) = 2x\left(\dfrac{1}{x}\right) + 2 \ln x = 2 + 2 \ln x$

At $(1, 1)$, the slope of the tangent line is $f'(1) = 2$.

Tangent line:

$y - 1 = 2(x - 1)$

$y = 2x - 1$

63. $f(x) = \ln \dfrac{5(x+2)}{x} = \ln 5 + \ln(x+2) - \ln x$

$f'(x) = \dfrac{1}{x+2} - \dfrac{1}{x}$

At $(-2.5, 0)$, the slope of the tangent line is

$f'(-2.5) = \dfrac{1}{-2.5+2} - \dfrac{1}{-2.5} = -2 + \dfrac{2}{5} = -\dfrac{8}{5}.$

Tangent line: $y - 0 = -\dfrac{8}{5}\left(x + \dfrac{5}{2}\right)$

$y = -\dfrac{8}{5}x - 4$

65. $f(x) = x \log_2 x$

$f'(x) = \log_2 x + x\left(\dfrac{1}{x}\dfrac{1}{\ln 2}\right) = \log_2 x + \dfrac{1}{\ln 2}$

$f'(1) = \dfrac{1}{\ln 2}$

$y - 0 = \dfrac{1}{\ln 2}(x - 1)$

$y = \dfrac{1}{\ln 2}x - \dfrac{1}{\ln 2}$

67. $y = x - \ln x$

$y' = 1 - \dfrac{1}{x} = \dfrac{x-1}{x}$

$y' = 0$ when $x = 1$.

$y'' = \dfrac{1}{x^2}$

Since $y''(1) = 1 > 0$, there is a relative minimum at $(1, 1)$. Moreover, since $y'' > 0$ on $(0, \infty)$, it follows that the graph is concave upward on its domain and there are no inflection points.

69. The domain of the function

$y = \dfrac{\ln x}{x}$

is $(0, \infty)$.

$y' = \dfrac{1 - \ln x}{x^2}$

$y' = 0$ when $x = e$.

$y'' = \dfrac{2 \ln x - 3}{x^3}$

Since $y''(e) < 0$, it follows that $(e, 1/e)$ is a relative maximum. Since $y'' = 0$ when $2 \ln x - 3 = 0$ and $x = e^{3/2}$, there is an inflection point at $(e^{3/2}, 3/(2e^{3/2}))$.

71. $y = x^2 \ln x$

$y' = x(1 + 2 \ln x)$

$y' = 0$ when $x = e^{-1/2}$.

$y'' = 3 + 2 \ln x$

($x = 0$ is not in the domain.)

Since $y''(e^{-1/2}) > 0$, it follows that there is a relative minimum at $(1/\sqrt{e}, -1/(2e))$.
Since $y'' = 0$ when $x = e^{-3/2}$, it follows that there is an inflection point at $(1/e^{3/2}, -3/(2e^3))$.

73. $x = \ln \dfrac{1000}{p} = \ln 1000 - \ln p$

$\dfrac{dx}{dp} = 0 - \dfrac{1}{p} = -\dfrac{1}{p}$

When $p = 10$,

$\dfrac{dy}{dp} = -\dfrac{1}{10}.$

75. $x = \dfrac{500}{\ln(p^2 + 1)} = 500[\ln(p^2 + 1)]^{-1}$

$\dfrac{dx}{dp} = -500[\ln(p^2 + 1)]^{-2}\dfrac{1}{p^2 + 1}(2p)$

$= \dfrac{-1000p}{(p^2 + 1)[\ln(p^2 + 1)]^2}$

If $p = 10$,

$\dfrac{dx}{dp} = \dfrac{-10,000}{101[\ln(101)]^2} \approx -4.65.$

77. $x = \ln \dfrac{1000}{p}$

$e^x = \dfrac{1000}{p}$

$p = 1000e^{-x}$

$\dfrac{dp}{dx} = -1000e^{-x}$

When $p = 10$, $x = \ln \dfrac{1000}{10} = \ln 100$ and $\dfrac{dp}{dx} = -1000e^{-\ln 100} = \dfrac{-1000}{100} = -10$.

Note that $\dfrac{dp}{dx}$ and $\dfrac{dx}{dp}$ are reciprocals of each other.

79. $C = 500 + 300x - 300 \ln x$, $\qquad x \geq 1$

(a) $\overline{C} = \dfrac{C}{x} = \dfrac{500}{x} + 300 - 300\dfrac{\ln x}{x}$

(b) $\overline{C}' = \dfrac{-500}{x^2} - 300\left[\dfrac{x(1/x) - \ln x}{x^2}\right]$

$\qquad = \dfrac{-500}{x^2} - \dfrac{300}{x^2}(1 - \ln x)$

Setting $\overline{C}' = 0$,

$\qquad 500 = -300(1 - \ln x)$

$\qquad \dfrac{5}{3} = \ln x - 1$

$\qquad \dfrac{8}{3} = \ln x$

$\qquad x = e^{8/3} \approx 14.39, \quad \overline{C}(14.39) \approx 279.15$

By the First Derivative Test, this is a minimum.

(c)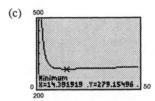

$\overline{C} \approx 279.15$ when $x \approx 14.39$ is the minimum.

81. $S = -210.3 + 103.30 \ln t$

($t = 8$ corresponds to 1998)

(a)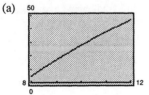

(b) 2000: $S(10) \approx \$27.6$ billion

(c) $S'(t) = \dfrac{103.30}{t}$

$\qquad S'(10) = \dfrac{103.30}{10} = \$10.33$ billion per year per year

83. Answer will vary.

Section 4.6   Exponential Growth and Decay

1. Since $y = 2$ when $t = 0$, it follows that $C = 2$. Moreover, since $y = 3$ when $t = 4$, we have $3 = 2e^{4k}$ and

$$k = \frac{\ln(3/2)}{4} \approx 0.1014.$$

Thus, $y = 2e^{0.1014t}$.

3. Since $y = 4$ when $t = 0$, it follows that $C = 4$. Moreover, since $y = \frac{1}{2}$ when $t = 5$, we have $\frac{1}{2} = 4e^{5k}$ and

$$k = \frac{\ln(1/8)}{5} \approx -0.4159.$$

Thus, $y = 4e^{-0.4159t}$.

5. Using the fact that $y = 1$ when $t = 1$ and $y = 5$ when $t = 5$, we have $1 = Ce^{k}$ and $5 = Ce^{5k}$. From these two equations, we have $Ce^{k} = \left(\frac{1}{5}\right)Ce^{5k}$. Thus,

$$k = \frac{\ln 5}{4} \approx 0.4024$$

and we have $y = Ce^{0.4024t}$. Since $1 = Ce^{0.4024}$, it follows that $C \approx 0.6687$ and $y = 0.6687e^{0.4024t}$.

7. $\dfrac{dy}{dt} = 2y,\ y = 10$ when $t = 0$.

$$y = 10e^{2t}$$

$$\frac{dy}{dt} = 10(2)e^{2t} = 2(10e^{2t}) = 2y$$

Exponential growth

9. $\dfrac{dy}{dt} = -4y,\ y = 30$ when $t = 0$.

$$y = 30e^{-4t}$$

$$\frac{dy}{dt} = 30(-4)e^{-4t} = -4(30e^{-4t}) = -4y$$

Exponential decay

11. From Example 1 we have

$$y = Ce^{kt} = 10e^{[\ln(1/2)/1599]t}.$$

When $t = 1000$,

$$y = 10e^{[\ln(1/2)/1599]1000} \approx 6.48 \text{ grams.}$$

When $t = 10,000$,

$$y = 10e^{[\ln(1/2)/1599]1000} \approx 0.13 \text{ gram.}$$

13. Since $y = Ce^{kt} = Ce^{[\ln(1/2)/5715]t}$, we have

$$2 = Ce^{[\ln(1/2)/5715]10,000} \implies C \approx 6.726.$$

The initial quantity is 6.726 grams.

When $t = 1000$,

$$y = 6.726e^{[\ln(1/2)/5715]1000} \approx 5.96 \text{ grams.}$$

15. Since $y = Ce^{[\ln(1/2)/24,100]t}$, we have

$$2.1 = Ce^{[\ln(1/2)/24,100]1000} \implies C \approx 2.161.$$

Initial quantity is 2.161 grams.

When $t = 10,000$,

$$y = 2.161e^{[\ln(1/2)/24,000]10,000} \approx 1.62 \text{ grams.}$$

17. $y = Ce^{[\ln(1/2)/1620t]}$ (See Example 1.)

When $t = 900$, $y = Ce^{[\ln(1/2)/1620 \cdot 900]} \approx 0.68C$. After 900 years, approximately 68% of the radioactive radium will remain.

19. $0.15C = Ce^{[\ln(1/2)/5715]t}$

$$\ln 0.15 = \frac{\ln(1/2)}{5715}t$$

$$t = \frac{5715 \ln 0.15}{\ln(1/2)} \approx 15{,}642 \text{ years}$$

21. $(0, 5)$: $y_1 = 5e^{k_1 t}$ and $y_2 = 5(2)^{k_2 t}$

$(12, 20)$: $20 = 5e^{12k_1} \implies 12k_1 = \ln 4 \implies k_1 \approx 0.1155$

$$y_1 = 5e^{0.1155t}$$

$(12, 20)$: $20 = 5(2)^{12k_2} \implies \log_2 4 = 12k_2$

$$k_2 = \frac{\log_2 4}{12} = \frac{1}{6}$$

$$y_2 = 5(2)^{t/6}$$

$$k_1 = \ln(2)\, k_2$$

23. The model is $y = Ce^{kt}$. Since $y = 150$ when $t = 0$, we have $C = 150$. Furthermore,

$$450 = 150e^{k5}$$

$$3 = e^{5k}$$

$$k = \frac{\ln 3}{5}.$$

Therefore, $y = 150e^{[(\ln 3)/5]t} \approx 150e^{0.2197t}$.

(a) When $t = 10$, $y = 150^{[(\ln 3)/5]10} = 1350$ bacteria.

(b) To find the time required for the population to double, solve for t.

$$300 = 150e^{[(\ln 3)/5]t}$$

$$2 = e^{[(\ln 3)/5]t}$$

$$\ln 2 = \frac{\ln 3}{5}t$$

$$t = \frac{5 \ln 2}{\ln 3} \approx 3.15 \text{ hours}$$

(c) No, the doubling time is always 3.15 hours.

25. Since $A = 1000e^{0.12t}$, the time to double is given by $2000 = 1000e^{0.12t}$ and we have

$$t = \frac{\ln 2}{0.12} \approx 5.776 \text{ years.}$$

Amount after 10 years: $A = 1000e^{1.2} \approx \$3320.12$

Amount after 25 years: $A = 1000e^{0.12(25)} \approx \$20,085.54$

27. Since $A = 750e^{rt}$ and $A = 1500$ when $t = 7.75$, we have

$$1500 = 750e^{7.75r}$$

$$r = \frac{\ln 2}{7.75} \approx 0.0894 = 8.94\%.$$

Amount after 10 years: $A = 750e^{0.0894(10)} \approx \$1833.67$

Amount after 25 years: $A = 750e^{0.0894(25)} \approx \$7009.86$

29. Since $A = 500e^{rt}$ and $A = 1292.85$ when $t = 10$, we have

$$1292.85 = 500e^{10r}$$

$$r = \frac{\ln(1292.85/500)}{10} \approx 0.095 = 9.5\%.$$

The time to double is given by

$$1000 = 500e^{0.095t}$$

$$t = \frac{\ln 2}{0.095} \approx 7.296 \text{ years.}$$

Amount after 25 years: $A = 500e^{0.095(25)} \approx \$5375.51$

31. (a) $P(1 + i)^t = P\left(1 + \dfrac{r}{n}\right)^{nt}$

$$\sqrt[t]{(1 + i)^t} = \sqrt[t]{\left(1 + \frac{r}{n}\right)^{nt}}$$

$$1 + i = \left(1 + \frac{r}{n}\right)^n$$

$$i = \left(1 + \frac{r}{n}\right)^n - 1$$

(b) If $r = 0.06$ and $n = 12$, then

$$i = \left(1 + \frac{0.06}{12}\right)^{12} - 1 \approx 0.0617 \text{ or } 6.17\%.$$

33.

Number of compoundings per year	4	12	365	Continuous
Effective yield	5.095%	5.116%	5.127%	5.127%

$n = 4$: $i = \left(1 + \dfrac{0.05}{4}\right)^4 - 1 \approx 0.05095 \approx 5.095\%$

$n = 12$: $i = \left(1 + \dfrac{0.05}{12}\right)^{12} - 1 \approx 0.05116 \approx 5.116\%$

$n = 365$: $i = \left(1 + \dfrac{0.05}{365}\right)^{365} - 1 \approx 0.05127 \approx 5.127\%$

Continuous: $i = e^{0.05} - 1 \approx 0.05127 \approx 5.127\%$

35. $2P = Pe^{rt}$

$2 = e^{rt}$

$\ln 2 = rt$

$t = \dfrac{\ln 2}{r} \approx \dfrac{0.6931}{r} \approx \dfrac{0.70}{r}$

If r is entered as a percentage and not as a decimal, then

$t \approx 100\left(\dfrac{0.70}{r}\right) = \dfrac{70}{r}.$

37. (a) Let $t = 0$ represent 1993.

$y = Ce^{kt} = 83.3e^{kt}$ (using $y(0) = 83.3$)

$446.6 = 83.3e^{k(10)}$ (using $y(10) = 446.6$)

$k = \dfrac{1}{10}\ln\left(\dfrac{446.6}{83.3}\right) \approx 0.1679$

$y = 83.3e^{0.1679t}$

In 2008, $t = 15$ and $y \approx \$1033.8$ million.

(b) $y - 83.3 = \dfrac{446.6 - 83.3}{10 - 0}(t - 0)$

$y = 83.3 + 36.33t$ linear model

In 2008, $t = 15$ and $y = \$628.25$ million.

(c)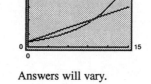

Answers will vary.

39. $S = Ce^{k/t}$

(a) Since $S = 5$ when $t = 1$, we have $5 = Ce^{k}$ and

$\lim_{t \to \infty} Ce^{k/t} = C = 30.$

Therefore, $5 = 30e^{k}$, $k = \ln(1/6) \approx -1.7918$, and $S = 30e^{-1.7918/t}$.

(b) $S(5) = 30e^{-1.7918/5} \approx 20.9646 \approx 20{,}965$ units

(c)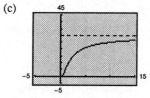

41. $N = 30(1 - e^{kt})$

Since $19 = 30(1 - e^{20k})$, it follows that

$30e^{20k} = 11$

$k = \dfrac{\ln(11/30)}{20} \approx -0.0502$

$N = 30(1 - e^{-0.0502t})$

$25 = 30(1 - e^{-0.0502t})$

$e^{-0.0502t} = \dfrac{1}{6}$

$t = \dfrac{\ln 6}{0.0502} \approx 36$ days.

43. $(0, 742000)$, $(2, 632000)$

$y = 742000e^{kt}$

$632000 = 742000e^{2k}$

$k = \dfrac{1}{2}\ln\left(\dfrac{632}{742}\right) \approx -0.0802$

$y = 742000e^{-0.0802t}$

For 2003, $t = 5$ and $y \approx \$496{,}880.$

45. (a) Since $p = Ce^{kx}$ and $p = 5$ when $x = 300$, and $p = 4$ when $x = 400$, we have the following.

$$5 = Ce^{300k} \text{ and } 4 = Ce^{400k}$$

$$C = \frac{5}{e^{300k}} \text{ and } C = \frac{4}{e^{400k}}$$

$$\frac{5}{e^{300k}} = \frac{4}{e^{400k}}$$

$$5e^{400k} = 4e^{300k}$$

$$e^{100k} = \frac{4}{5}$$

$$k = \frac{1}{100} \ln \frac{4}{5}$$

Therefore, $C = \dfrac{5}{e^{300[(1/100)\ln(4/5)]}} = \dfrac{5}{e^{3\ln(4/5)}} = \dfrac{625}{64}$ and $p = \dfrac{625}{64}e^{[\ln(4/5)/100]x} \approx 9.7656e^{-0.0022x}$.

(b) $R = xp = \dfrac{625}{64}xe^{[\ln(4/5)/100]x}$

$$R' = \frac{625}{64}\left\{ x\left[\frac{\ln(4/5)}{100}e^{[\ln(4/5)/100]x}\right] + e^{[\ln(4/5)/100]x}\right\} = \frac{625}{64}e^{[\ln(4/5)/100]x}\left[\frac{\ln(4/5)}{100}x + 1\right]$$

$R' = 0$ when $x = \dfrac{-100}{\ln(4/5)} \approx 448$ units.

$p(448) \approx \$3.59$

47. $A = Ve^{-0.04t} = 100{,}000e^{0.6\sqrt{t}}e^{-0.04t} = 100{,}000e^{(0.6\sqrt{t}-0.04t)}$

$$A'(t) = 100{,}000\left(\frac{0.6}{2\sqrt{t}} - 0.04\right)e^{(0.6\sqrt{t}-0.04t)} = 0$$

$$\frac{0.6}{2\sqrt{t}} = 0.04$$

$$\sqrt{t} = \frac{0.6}{(0.04)2} = 7.5$$

$$t = 56.25 \approx 56$$

The timber should be harvested in 2046.

Review Exercises for Chapter 4

1. $32^{3/5} = [(32)^{1/5}]^3 = 2^3 = 8$

3. $\left(\frac{1}{25}\right)^{-3/2} = (25)^{3/2} = 5^3 = 125$

5. $\left(\frac{25}{4}\right)^0 = 1$

7. $\dfrac{6^3}{36^2} = \dfrac{6 \cdot 6^2}{36 \cdot 6^2} = \dfrac{6}{36} = \dfrac{1}{6}$

9. $5^x = 625$
$x = 4$

11. $x^{3/4} = 8$
$x^{1/4} = 2$
$x = 2^4 = 16$

13. $e^{-1/2} = e^{x-1}$
$-\frac{1}{2} = x - 1$
$x = \frac{1}{2}$

15. $\dfrac{x^3}{3} = e^3$
$x^3 = 3e^3$
$x = e\sqrt[3]{3}$

17. $R = 378.914(1.067)^t$

    (a) 1995: $R(5) = 378.914(1.067)^5 \approx \$524.0$ million   (b) Answers will vary.

            1998: $R(8) \approx \$636.6$ million

            2001: $R(11) \approx \$773.3$ million

19. $f(x) = 9^{x/2}$ **21.** $f(t) = \left(\dfrac{1}{6}\right)^t$ **23.** $f(x) = \left(\dfrac{1}{2}\right)^{2x} + 4$

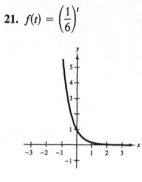

25. $f(x) = e^{-x} + 1$. Decaying exponential passing through $(0, 2)$.   **27.** $f(x) = 1 - e^x$

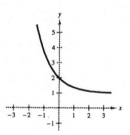

29. $p = 12{,}500 - \dfrac{10{,}000}{2 + e^{-0.001x}}$

       $\displaystyle\lim_{x \to \infty} p = 12{,}500 - \dfrac{10{,}000}{2 + 0} = 7500$

31. $f(x) = 2e^{x-1}$

    (a) $f(2) = 2e^{2-1} = 2e \approx 5.4366$

    (b) $f\left(\dfrac{1}{2}\right) = 2e^{1/2-1} = \dfrac{2}{\sqrt{e}} \approx 1.2131$

    (c) $f(10) = 2e^{10-1} = 2e^9 \approx 16206.1679$

33. $g(t) = 12e^{-0.2t}$

    (a) $g(17) = 12e^{-3.4} \approx 0.4005$

    (b) $g(50) = 12e^{-10} \approx 0.0005448$

    (c) $g(100) = 12e^{-20} \approx 2.4734 \times 10^{-8} \approx 0$

35. (a) $P = \dfrac{10{,}000}{1 + 19e^{-t/5}}, \quad t \geq 0$

    (b) When $t = 4$, $P \approx 1049$ fish.

    (c) Yes, P approaches 10,000 fish as $t \to \infty$.

    (d) The population is increasing most rapidly at the inflection point, around $t = 15$ months ($P = 5000$).

37.

n	1	2	4	12	365	Continuous
A	1216.65	1218.99	1220.19	1221.00	1221.39	1221.40

$$A = P\left(1 + \frac{r}{n}\right)^{nt} = 1000\left(1 + \frac{0.04}{n}\right)^{5n}$$

$$A = Pe^{rt} = 1000e^{(0.04)5} \approx 1221.50 \ \text{(Continuous)}$$

39. (a) $A = Pe^{rt} = 2000e^{0.05(10)} \approx \$3297.44$

    (b) $A = P\left(1 + \frac{r}{n}\right)^{nt} = 2000\left(1 + \frac{0.06}{4}\right)^{4(10)} \approx \$3628.04$

    Account (b) will be greater.

41. (a) $r_{eff} = \left(1 + \frac{r}{n}\right)^{n} - 1$

$$= \left(1 + \frac{0.06}{4}\right)^{4} - 1$$

$$\approx 0.0614 \quad \text{or} \quad 6.14\%$$

    (b) $r_{eff} = \left(1 + \frac{r}{n}\right)^{7} - 1$

$$= \left(1 + \frac{0.06}{12}\right)^{12} - 1$$

$$\approx 0.0617 \quad \text{or} \quad 6.17\%$$

43. $P = \dfrac{A}{\left(1 + \frac{r}{n}\right)^{nt}}$

$$= \dfrac{12000}{\left(1 + \frac{0.065}{4}\right)^{4(3)}}$$

$$\approx \$9889.50$$

45. $P = 22.9 + 0.87t - 0.019t^2 + 1.2960e^{-t}, \quad 0 \le t \le 20$

    ($t = 0$ corresponds to 1980)

    1980: $P(0) = 22.9 + 1.296 = 24.196$ million

    1995: $P(15) \approx 31.7$ million

    2000: $P(20) \approx 32.7$ million

47. $y = 4e^{x^2}$

$$y' = 4e^{x^2}(2x) = 8xe^{x^2}$$

49. $y = \dfrac{x}{e^{2x}}$

$$y' = \dfrac{e^{2x}(1) - x2e^{2x}}{(e^{2x})^2} = \dfrac{1 - 2x}{e^{2x}}$$

51. $y = \sqrt{4e^{4x}} = (4e^{4x})^{1/2} = 2e^{2x}$

$$y' = 4e^{2x}$$

53. $y = \dfrac{5}{1 + e^{2x}} = 5(1 + e^{2x})^{-1}$

$$y' = -5(1 + e^{2x})^{-2}(2e^{2x}) = \dfrac{-10e^{2x}}{(1 + e^{2x})^2}$$

55. $f(x) = 4e^{-x}$

No relative extrema

No inflection point

Horizontal asymptote: $y = 0$

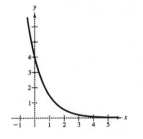

57. $f(x) = x^3 e^x$

$f'(x) = x^2 e^x (3 + x)$

$f''(x) = xe^x(x^2 + 6x + 6)$

$(-3, -1.344)$ is a relative minimum.

There are inflection points at $(0, 0)$, $\left(-3 + \sqrt{3}, -0.574\right)$, and $\left(-3 - \sqrt{3}, -0.933\right)$.

$y = 0$ is a horizontal asymptote.

59. $f(x) = \dfrac{1}{xe^x}$

$f'(x) = \dfrac{-x - 1}{x^2 e^x}$

$f''(x) = \dfrac{x^2 + 2x + 2}{x^3 e^x}$

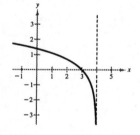

$(-1, -2.718)$ is a relative maximum.

$y = 0$ is a horizontal asymptote.

$x = 0$ is a vertical asymptote.

61. $f(x) = xe^{2x}$

$f'(x) = (2x + 1)e^{2x}$

$f''(x) = (4x + 4)e^{2x}$

Critical number $x = -\dfrac{1}{2}$

Inflection point: $\left(-1, \dfrac{-1}{e^2}\right)$

Relative minimum: $\left(-\dfrac{1}{2}, \dfrac{-1}{2e}\right)$

Horizontal asymptote: $y = 0$

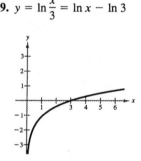

63. $\ln 12 \approx 2.4849$

$e^{2.4849} \approx 12$

65. $e^{1.5} \approx 4.4817$

$\ln 4.4817 \approx 1.5$

67. $y = \ln(4 - x)$

69. $y = \ln \dfrac{x}{3} = \ln x - \ln 3$

71. $\ln \sqrt{x^2(x - 1)} = \dfrac{1}{2} \ln[x^2(x - 1)]$

$\qquad = \dfrac{1}{2}[\ln x^2 + \ln(x - 1)]$

$\qquad = \ln x + \dfrac{1}{2} \ln(x - 1)$

73. $\ln \dfrac{x^2}{(x + 1)^3} = \ln x^2 - \ln(x + 1)^3$

$\qquad = 2 \ln x - 3 \ln(x + 1)$

75. $\ln \left(\dfrac{1 - x}{3x}\right)^3 = 3 \ln \left(\dfrac{1 - x}{3x}\right)$

$\qquad = 3[\ln(1 - x) - \ln 3x]$

$\qquad = 3[\ln(1 - x) - \ln 3 - \ln x]$

77. $e^{\ln x} = 3$

$x = 3$

79. $\ln x = 3e^{-1} = \dfrac{3}{e}$

$e^{3/e} = x \approx 3.015$

81. $\ln 2x - \ln(3x - 1) = 0$

$\ln 2x = \ln(3x - 1)$

$2x = 3x - 1$

$x = 1$

83. $e^{2x-1} - 6 = 0$

$e^{2x-1} = 6$

$2x - 1 = \ln 6$

$x = \dfrac{1 + \ln 6}{2} \approx 1.396$

85. $\ln x + \ln(x - 3) = 0$

$\ln[x(x - 3)] = 0$

$x(x - 3) = 1$

$x^2 - 3x - 1 = 0$

$x = \dfrac{3 \pm \sqrt{13}}{2}$

$x = \dfrac{3 + \sqrt{13}}{2} \approx 3.3028$ is the only solution in the domain.

87. $e^{-1.386x} = 0.25$

$-1.386x = \ln 0.25$

$x = \dfrac{\ln 0.25}{-1.386} \approx 1.0002$

89. $100(1.21)^x = 110$

$1.21^x = \dfrac{110}{100} = 1.1$

$x \ln 1.21 = \ln 1.1$

$x = \dfrac{\ln 1.1}{\ln 1.21} \approx 0.5$

91. $\dfrac{40}{1 - 5e^{-0.01x}} = 200$

$1 - 5e^{-0.01x} = \dfrac{1}{5}$

$5e^{-0.01x} = \dfrac{4}{5}$

$e^{-0.01x} = \dfrac{4}{25}$

$-0.01x = \ln\!\left(\dfrac{4}{25}\right)$

$x = -100 \ln\!\left(\dfrac{4}{25}\right) = 100 \ln\!\left(\dfrac{25}{4}\right) \approx 183.26$

93. (a) $M = 100{,}000\!\left(\dfrac{\dfrac{0.08}{12}}{1 - \left(\dfrac{1}{(0.08/12) + 1}\right)^{12t}}\right)$

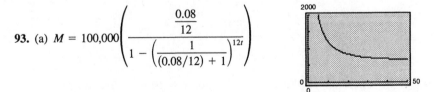

(b) A 30-year term has a smaller monthly payment, but takes more time to pay off than a 20-year term.

95. $f(x) = \ln 3x^2 = \ln 3 + 2 \ln x$

$f'(x) = \dfrac{2}{x}$

97. $y = \ln\dfrac{x(x - 1)}{x - 2} = \ln x + \ln(x - 1) - \ln(x - 2)$

$y' = \dfrac{1}{x} + \dfrac{1}{x - 1} - \dfrac{1}{x - 2}$

99. $f(x) = \ln e^{2x+1} = 2x + 1$

$f'(x) = 2$

101. $y = \dfrac{\ln x}{x^3}$

$y' = \dfrac{x^3(1/x) - 3x^2 \cdot \ln x}{x^6} = \dfrac{1 - 3\ln x}{x^4}$

103. $y = \ln(x^2 - 2)^{2/3} = \dfrac{2}{3}\ln(x^2 - 2)$

$y' = \dfrac{2}{3} \cdot \dfrac{2x}{x^2 - 2} = \dfrac{4x}{3(x^2 - 2)}$

105. $f(x) = \ln\left[x^2\sqrt{x+1}\right] = 2\ln x + \dfrac{1}{2}\ln(x+1)$

$f'(x) = \dfrac{2}{x} + \dfrac{1}{2(x+1)}$

107. $y = \ln\dfrac{e^x}{1 + e^x} = \ln e^x - \ln(1 + e^x) = x - \ln(1 + e^x)$

$y' = 1 - \dfrac{e^x}{1 + e^x} = \dfrac{1}{1 + e^x}$

109. $y = \ln(x + 3)$

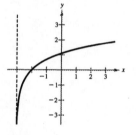

111. $y = \ln\left(\dfrac{10}{x+2}\right) = \ln 10 - \ln(x + 2)$

No relative extrema or inflection points

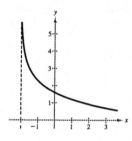

113. $\log_7 49 = \log_7 y^2 = 2\log_7 7 = 2$

115. $\log_{10} 1 = 0$

117. $\log_5 10 = \dfrac{\ln 10}{\ln 5} \approx 1.4307$

119. $\log_{16} 64 = \dfrac{\ln 64}{\ln 16} = 1.5$

121. $y = \log_3(2x - 1)$

$y' = \dfrac{1}{\ln 3} \cdot \dfrac{2}{2x - 1} = \dfrac{2}{(2x - 1)\ln 3}$

123. $y = \log_2 \dfrac{1}{x^2} = \log_2 1 - \log_2 x^2 = -2\log_2 x$

$y' = -2\dfrac{1}{\ln 2} \cdot \dfrac{1}{x} = \dfrac{-2}{x\ln 2}$

125. $V = 20{,}000(0.75)^t$

(a) $V(2) = 11{,}250$

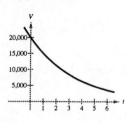

(b) $V'(t) = 20{,}000 \cdot \ln\left(\dfrac{3}{4}\right)(0.75)^t$

$V'(1) = -4315.23$ dollars/year

$V'(4) = -1820.49$ dollars/year

(c) $V = 5000 = 20{,}000(0.75)^t$

$\dfrac{1}{4} = (0.75)^t$

$t = \dfrac{\ln(1/4)}{\ln(3/4)} \approx 4.8$ years

127. $A = Ce^{kt} = 500e^{kt}$

$A = 500$ when $t = 0$.

$A = 300$ after 40 days.

$300 = 500e^{40k}$

$\dfrac{3}{5} = e^{40k}$

$40k = \ln\left(\dfrac{3}{5}\right)$

$k = \dfrac{\ln(3/5)}{40} \approx -0.01277$

$y = 500e^{-0.01277t}$

129. $y = 50e^{kt}$

$42.031 = 50e^{7k}$

$k = \dfrac{1}{7}\ln\left(\dfrac{42.031}{50}\right) \approx -0.02480$

$25 = 50e^{kt}$

$t = \dfrac{1}{k}\ln\left(\dfrac{1}{2}\right) \approx 27.95$ half-life

131. Using the data points $(5, 17.6)$ and $(13, 306.8)$ you obtain the model

$P = 2.949(1.4294)^t = 2.949e^{0.3573t}$.

For 2006, $P(16) \approx \$896.3$ million.

Practice Test for Chapter 4

1. Evaluate each of the following expressions.

   (a) $27^{4/3}$ (b) $4^{-5/2}$ (c) $(8^{2/3})(64^{-1/3})$

2. Solve for x.

   (a) $4^{x+1} = 64$ (b) $x^{6/5} = 64$ (c) $(2x + 3)^{10} = 13^{10}$

3. Sketch the graph of (a) $f(x) = 3^x$, and (b) $g(x) = \left(\frac{4}{9}\right)^x$.

4. Find the amount in an account in which \$2000 is invested for 7 years at 8.5% if the interest is compounded (a) annually, (b) monthly, and (c) continuously.

5. Differentiate $y = e^{3x^2}$.

6. Differentiate $y = e^{\sqrt[3]{x}}$.

7. Differentiate $y = \sqrt{e^x + e^{-x}}$.

8. Differentiate $y = x^3 e^{2x}$.

9. Differentiate $y = \dfrac{e^x + 3}{4x}$.

10. Write $\ln 5 = 1.6094\ldots$ as an exponential equation.

11. Sketch the graph of (a) $y = \ln(x + 2)$, and (b) $y = \ln x + 2$.

12. Write the given expression as a single logarithm.

   (a) $\ln(3x + 1) - \ln(2x - 5)$ (b) $4 \ln x - 3 \ln y - \frac{1}{2} \ln z$

13. Solve for x.

   (a) $\ln x = 17$ (b) $5^{3x} = 2$

14. Differentiate $y = \ln(6x - 7)$.

15. Differentiate $y = \ln\left(\dfrac{x^3}{4x + 10}\right)$.

16. Differentiate $y = \ln \sqrt[3]{\dfrac{x}{x + 3}}$.

17. Differentiate $y = x^4 \ln x$.

18. Differentiate $y = \sqrt{\ln x + 1}$.

19. Find the exponential function $y = Ce^{kt}$ that passes through the following points.

   (a) $(0, 7), \left(4, \frac{1}{3}\right)$ (b) $\left(3, \frac{2}{3}\right), (8, 8)$

20. If \$5000 is invested in an account in which the interest rate of 12% is compounded continuously, find the time required for the investment to double.

Graphing Calculator Required

21. Use a graphing calculator to graph both $y = \ln\left[x^3 \sqrt{x + 3}\right]$ and $y = 3 \ln x + \frac{1}{2} \ln(x + 3)$ on the same set of axes. What do you notice about the graphs?

22. Graph the function $f(t) = \dfrac{4200}{7 + e^{-0.9t}}$ and use the graph to find $\lim\limits_{t \to \infty} f(t)$ and $\lim\limits_{t \to -\infty} f(t)$.

C H A P T E R 5
Integration and Its Applications

CHAPTER 5
Integration and Its Applications

Section 5.1   Antiderivatives and Indefinite Integrals

Solutions to Odd-Numbered Exercises

1. $\dfrac{d}{dx}\left(\dfrac{3}{x^3} + C\right) = \dfrac{d}{dx}(3x^{-3} + C) = -9x^{-4} = \dfrac{-9}{x^4}$

3. $\dfrac{d}{dx}\left(x^4 + \dfrac{1}{x} + C\right) = 4x^3 - \dfrac{1}{x^2}$

5. $\dfrac{d}{dx}\left[\dfrac{4x^{3/2}(x-5)}{5} + C\right] = \dfrac{d}{dx}\left[\dfrac{4}{5}x^{5/2} - 4x^{3/2} + C\right]$

$\qquad = 2x^{3/2} - 6x^{1/2}$

$\qquad = 2\sqrt{x}(x-3)$

7. $\dfrac{d}{dx}\left(\dfrac{2(x^2+3)}{3\sqrt{x}} + C\right) = \dfrac{d}{dx}\left(\dfrac{2}{3}x^{2/3} + 2x^{-1/2} + C\right)$

$\qquad = x^{1/2} - x^{-3/2} = \dfrac{x^2 - 1}{x^{3/2}}$

9. $\displaystyle\int 6\,dx = 6x + C$

$\dfrac{d}{dx}[6x + C] = 6$

11. $\displaystyle\int 5t^2\,dt = \dfrac{5t^3}{3} + C$

$\dfrac{d}{dt}\left[\dfrac{5t^3}{3} + C\right] = \dfrac{3(5t^2)}{3} = 5t^2$

13. $\displaystyle\int 5x^{-3}\,dx = \dfrac{5x^{-2}}{-2} + C = \dfrac{-5}{2x^2} + C$

$\dfrac{d}{dx}\left[-\dfrac{5}{2}x^{-2} + C\right] = 5x^{-3}$

15. $\displaystyle\int du = u + C$

$\dfrac{d}{du}[u + C] = 1$

17. $\displaystyle\int e\,dt = et + C$

$\dfrac{d}{dt}[et + C] = e$

19. $\displaystyle\int y^{3/2}\,dy = \dfrac{y^{5/2}}{\frac{5}{2}} + C = \dfrac{2}{5}y^{5/2} + C$

$\dfrac{d}{dy}\left[\dfrac{2}{5}y^{5/2} + C\right] = y^{3/2}$

	Given	*Rewrite*	*Integrate*	*Simplify*
21.	$\displaystyle\int \sqrt[3]{x}\,dx$	$\displaystyle\int x^{1/3}\,dx$	$\dfrac{x^{4/3}}{4/3} + C$	$\dfrac{3}{4}x^{4/3} + C$
23.	$\displaystyle\int \dfrac{1}{x\sqrt{x}}\,dx$	$\displaystyle\int x^{-3/2}\,dx$	$\dfrac{x^{-1/2}}{-1/2} + C$	$-\dfrac{2}{\sqrt{x}} + C$
25.	$\displaystyle\int x(x^2 + 3)\,dx$	$\displaystyle\int (x^3 + 3x)\,dx$	$\dfrac{x^4}{4} + \dfrac{3x^2}{2} + C$	$\dfrac{x^2}{4}(x^2 + 6) + C$
27.	$\displaystyle\int \dfrac{1}{2x^3}\,dx$	$\dfrac{1}{2}\displaystyle\int x^{-3}\,dx$	$\dfrac{1}{2}\left(\dfrac{x^{-2}}{-2}\right) + C$	$-\dfrac{1}{4x^2} + C$

29. If $f'(x) = 2$, then $f(x) = 2x + C$. For example, $f(x) = 2x$ or $f(x) = 2x + 1$.

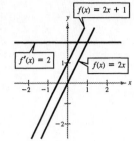

31. If $f'(x) = x$, then $f(x) = \dfrac{x^2}{2} + C$. For example, $f(x) = \dfrac{x^2}{2}$ or $f(x) = \dfrac{x^2}{2} + 2$.

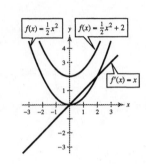

33. $\displaystyle\int (x^3 + 2)\, dx = \dfrac{x^4}{4} + 2x + C$

$\dfrac{d}{dx}\left[\dfrac{x^4}{4} + 2x + C\right] = x^3 + 2$

35. $\displaystyle\int \left(x^{1/3} - \dfrac{1}{2}x^{-1/3}\right) dx = \dfrac{3}{4}x^{4/3} - \dfrac{3}{4}x^{2/3} + C$

$\dfrac{d}{dx}\left[\dfrac{3}{4}x^{4/3} - \dfrac{3}{4}x^{2/3} + C\right] = x^{1/3} - \dfrac{1}{2x^{1/3}}$

37. $\displaystyle\int (x^{2/3} + 1)\, dx = \dfrac{3}{5}x^{5/3} + x + C$

$\dfrac{d}{dx}\left[\dfrac{3}{5}x^{5/3} + x + C\right] = x^{2/3} + 1$

39. $\displaystyle\int \dfrac{1}{3}x^{-4}\, dx = -\dfrac{1}{9}x^{-3} + C = \dfrac{-1}{9x^3} + C$

$\dfrac{d}{dx}\left[\dfrac{-1}{9x^3} + C\right] = \dfrac{1}{3x^4}$

41. $\displaystyle\int \dfrac{2x^3 + 1}{x^3}\, dx = \int (2 + x^{-3})\, dx = 2x - \dfrac{1}{2}x^{-2} + C$

$\qquad = 2x - \dfrac{1}{2x^2} + C$

$\dfrac{d}{dx}\left[2x - \dfrac{1}{2x^2} + C\right] = 2 + \dfrac{1}{x^3}$

43. $\displaystyle\int u(3u^2 + 1)\, du = \int (3u^3 + u)\, du$

$\qquad = \dfrac{3}{4}u^4 + \dfrac{1}{2}u^2 + C$

$\dfrac{d}{du}\left[\dfrac{3}{4}u^4 + \dfrac{1}{2}u^2 + C\right] = 3u^3 + u = u(3u^2 + 1)$

45. $\displaystyle\int (x - 1)(6x - 5)\, dx = \int (6x^2 - 11x + 5)\, dx$

$\qquad = 2x^3 - \dfrac{11}{2}x^2 + 5x + C$

$\dfrac{d}{dx}\left[2x^3 - \dfrac{11}{2}x^2 + 5x + C\right] = 6x^2 - 11x + 5$

47. $\displaystyle\int y^2\sqrt{y}\, dy = \int y^{5/2}\, dy = \dfrac{2}{7}y^{7/2} + C$

$\dfrac{d}{dy}\left[\dfrac{2}{7}y^{7/2} + C\right] = y^{5/2} = y^2\sqrt{y}$

49. $f(x) = \displaystyle\int (3x^{1/2} + 3)\, dx = 2x^{3/2} + 3x + C$

$f(1) = 4 = 2(1) + 3(1) + C = 5 + C \Rightarrow C = -1$

$f(x) = 2x^{3/2} + 3x - 1$

51. $f(x) = \displaystyle\int 6x(x - 1)\, dx = \int (6x^2 - 6x)\, dx$

$\qquad = 2x^3 - 3x^2 + C$

$f(1) = -1 = 2 - 3 + C = -1 + C \Rightarrow C = 0$

$f(x) = 2x^3 - 3x^2$

53. $f(x) = \displaystyle\int \frac{2-x}{x^3}\,dx = \int (2x^{-3} - x^{-2})\,dx = -x^{-2} + x^{-1} + C = \frac{-1}{x^2} + \frac{1}{x} + C$

$f(2) = \dfrac{3}{4} = -\dfrac{1}{4} + \dfrac{1}{2} + C = \dfrac{1}{4} + C \Rightarrow C = \dfrac{1}{2}$

$f(x) = -\dfrac{1}{x^2} + \dfrac{1}{x} + \dfrac{1}{2}$

55. $y = \displaystyle\int (-5x - 2)\,dx = -\frac{5}{2}x^2 - 2x + C$

At $(0, 2)$, $2 = C$. Thus, $y = -\frac{5}{2}x^2 - 2x + 2$.

57. $f(x) = \displaystyle\int \left(6\sqrt{x} - 10\right)dx = \int (6x^{1/2} - 10)\,dx = 4x^{3/2} - 10x + C = 4x\sqrt{x} - 10x + C$

At $(4, 2)$, $2 = 4(4)\sqrt{4} - 10(4) + C$ which implies that $C = 10$. Thus, $f(x) = 4x\sqrt{x} - 10x + 10$.

59. $f'(x) = \displaystyle\int 2\,dx = 2x + C_1$

Since $f'(2) = 4 + C_1 = 5$, we know that $C_1 = 1$. Thus, $f'(x) = 2x + 1$.

$f(x) = \displaystyle\int (2x + 1)\,dx = x^2 + x + C_2$

Since $f(2) = 4 + 2 + C_2 = 10$, we know that $C_2 = 4$. Thus, $f(x) = x^2 + x + 4$.

61. $f'(x) = \displaystyle\int x^{-2/3}\,dx = 3x^{1/3} + C_1$

Since $f'(8) = 3(2) + C_1 = 6$, we have $C_1 = 0$. Thus, $f'(x) = 3x^{1/3}$.

$f(x) = \displaystyle\int 3x^{1/3}\,dx = \frac{9}{4}x^{4/3} + C_2$

Since $f(0) = 0 = C_2$, $f(x) = \frac{9}{4}x^{4/3}$.

63. $C = \displaystyle\int 85\,dx = 85x + k$

When $x = 0$, $C = 5500 = k$. Thus, $C = 85x + 5500$.

65. $C = \displaystyle\int \left(\frac{1}{20}x^{-1/2} + 4\right)dx = \frac{1}{10}x^{1/2} + 4x + k$

When $x = 0$,

$$C = \frac{1}{10}\left(\sqrt{0}\right) + 4(0) + k = 750 \Rightarrow k = 750.$$

Thus,

$$C = \frac{\sqrt{x}}{10} + 4x + 750.$$

67. $R = \displaystyle\int (225 - 3x)\,dx = 225x - \frac{3}{2}x^2 + C$

Since $R = 0$ when $x = 0$, it follows that $C = 0$. Thus, $R = 225x - \frac{3}{2}x^2$ and the demand function is

$$p = \frac{R}{x} = 225 - \frac{3}{2}x.$$

69. $R = \displaystyle\int (225 + 2x - x^2)\,dx = 225x + x^2 - \frac{x^3}{3} + C$

Since $R = 0$ when $x = 0$, it follows that $C = 0$. Thus, $R = 225x + x^2 - x^3/3$, and the demand function is

$$p = \frac{R}{x} = 225 + x - \frac{x^2}{3}.$$

71. $P = \displaystyle\int (-18x + 1650)\,dx = -9x^2 + 1650x + C$

$P(15) = 22{,}725 = -9(15)^2 + 1650(15) + C \Rightarrow C = 0$

$P = -9x^2 + 1650x$

73. $P = \displaystyle\int (-24x + 805)\,dx = -12x^2 + 805x + C$

$P(12) = 8000 = -12(12)^2 + 805(12) + C \Rightarrow C = 68$

$P = -12x^2 + 805x + 68$

75. $v(t) = \displaystyle\int -32\, dt = -32t + C_1$

Since $v(0) = 60$, it follows that $C_1 = 60$. Thus, we have $v(t) = -32t + 60$.

$$s(t) = \int (-32t + 60)\, dt = -16t^2 + 60t + C_2$$

Since $s(0) = 0$, it follows that $C_2 = 0$. Therefore, the position function is $s(t) = -16t^2 + 60t$. Now, since $v(t) = 0$ when $t = \frac{60}{32} = 1.875$ seconds, the maximum height of the ball is $s(1.875) = 56.25$ feet.

77. $v(t) = \displaystyle\int -32\, dt = -32t + C_1$

Letting v_0 be the initial velocity, we have $C_1 = v_0$.

$$s(t) = \int (-32t + v_0)\, dt = -16t^2 + v_0 t + C_2$$

Since $s(0) = 0$, we have $C_2 = 0$. Therefore, the position function is $s(t) = 16t^2 + v_0 t$. At the highest point, the velocity is zero. Therefore, we have $v(t) = -32t + v_0 = 0$, and $t = v_0/32$ seconds. Finally, substituting this value into the position function, we have

$$s\!\left(\frac{v_0}{32}\right) = -16\!\left(\frac{v_0}{32}\right)^2 + v_0\!\left(\frac{v_0}{32}\right) = 550$$

which implies that $v_0{}^2 = 35{,}200$ and the initial velocity should be $v_0 = 40\sqrt{22} \approx 187.617$ ft/sec.

79. (a) $C(x) = \displaystyle\int (2x - 12)\, dx = x^2 - 12x + C_1$

Since $C(0) = 125$, it follows that $C_1 = 125$. Thus, $C(x) = x^2 - 12x + 125$ and the average cost is $C/x = x - 12 + (125/x)$.

(b) $C(50) = 50^2 - 12(50) + 125 = \$2025$

(c) \$125 is fixed, \$1900 is variable.

81. (a) $M = \displaystyle\int (0.636t^2 - 28.48t + 632.7)\,dt = 0.212t^3 - 14.24t^2 + 632.7t + C$

($t = 0$ corresponds to 1970)

When $t = 30$,

$$M = 56{,}497 \implies 56{,}497 = 0.212(30)^3 - 14.24(30)^2 + 632.7(30) + C \implies C = 44{,}608$$

$M = 0.212t^3 - 14.24t^2 + 632.7t + 44{,}608.$

(b) For 2010, $t = 40$ and $M = 60{,}700$, or 60,700,000 married couples. This answer seems reasonable.

83. Answers will vary.

Section 5.2   The General Power Rule

$\displaystyle\int u^n \dfrac{du}{dx}\, dx$	u	$\dfrac{du}{dx}$	$\displaystyle\int u^n \dfrac{du}{dx}\, dx$	u	$\dfrac{du}{dx}$
1. $\displaystyle\int (5x^2 + 1)^2(10x)\, dx$	$5x^2 + 1$	$10x$	**5.** $\displaystyle\int \left(4 + \frac{1}{x^2}\right)^5\!\left(\frac{-2}{x^3}\right) dx$	$4 + \dfrac{1}{x^2}$	$-\dfrac{2}{x^3}$
3. $\displaystyle\int \sqrt{1 - x^2}\,(-2x)\, dx$	$1 - x^2$	$-2x$	**7.** $\displaystyle\int (1 + \sqrt{x})^3 \frac{1}{2\sqrt{x}}\, dx$	$(1 + \sqrt{x})$	$\dfrac{1}{2\sqrt{x}}$

9. $\int (1 + 2x)^4(2)\,dx = \dfrac{(1 + 2x)^5}{5} + C$

11. $\int \sqrt{5x^2 - 4}\,(10x)\,dx = \int (5x^2 - 4)^{1/2}(10x)\,dx$

$$= \dfrac{2}{3}(5x^2 - 4)^{3/2} + C$$

13. $\int (x - 1)^4\,dx = \int (x - 1)^4(1)\,dx$

$$= \dfrac{(x - 1)^5}{5} + C$$

15. $\int x(x^2 - 1)^7\,dx = \dfrac{1}{2} \int (x^2 - 1)^7(2x)\,dx$

$$= \dfrac{1}{2}\dfrac{(x^2 - 1)^8}{8} + C$$

$$= \dfrac{(x^2 - 1)^8}{16} + C$$

17. $\int \dfrac{x^2}{(1 + x^3)^2}\,dx = \dfrac{1}{3} \int (1 + x^3)^{-2}(3x^2)\,dx$

$$= \dfrac{1}{3}\dfrac{(1 + x^3)^{-1}}{-1} + C$$

$$= -\dfrac{1}{3(1 + x^3)} + C$$

19. $\int \dfrac{x + 1}{(x^2 + 2x - 3)^2}\,dx = \dfrac{1}{2} \int (x^2 + 2x - 3)^{-2}2(x + 1)\,dx$

$$= \dfrac{1}{2}\dfrac{(x^2 + 2x - 3)^{-1}}{-1} + C$$

$$= -\dfrac{1}{2(x^2 + 2x - 3)} + C$$

21. $\int \dfrac{x - 2}{\sqrt{x^2 - 4x + 3}}\,dx = \dfrac{1}{2} \int (x^2 - 4x + 3)^{-1/2}(2x - 4)\,dx$

$$= \dfrac{1}{2}(2)(x^2 - 4x + 3)^{1/2} + C$$

$$= \sqrt{x^2 - 4x + 3} + C$$

23. $\int 5u \sqrt[3]{1 - u^2}\,du = 5\left(-\dfrac{1}{2}\right) \int (1 - u^2)^{1/3}(-2u)\,du$

$$= \dfrac{-5}{2}\left(\dfrac{3}{4}\right)(1 - u^2)^{4/3} + C$$

$$= \dfrac{-15(1 - u^2)^{4/3}}{8} + C$$

25. $\int \dfrac{4y}{\sqrt{1 + y^2}}\,dy = 4\left(\dfrac{1}{2}\right) \int (1 + y^2)^{-1/2}(2y)\,dy$

$$= 2(2)(1 + y^2)^{1/2} + C$$

$$= 4\sqrt{1 + y^2} + C$$

27. $\int \dfrac{-3}{\sqrt{2t + 3}}\,dt = -\dfrac{3}{2} \int (2t + 3)^{-1/2}(2)\,dt$

$$= -\dfrac{3}{2}(2)(2t + 3)^{1/2} + C$$

$$= -3\sqrt{2t + 3} + C$$

29. $\int \dfrac{x^3}{\sqrt{1 - x^4}}\,dx = -\dfrac{1}{4} \int (1 - x^4)^{-1/2}(-4x^3)\,dx$

$$= -\dfrac{1}{4}(2)(1 - x^4)^{1/2} + C$$

$$= -\dfrac{\sqrt{1 - x^4}}{2} + C$$

31. $\int \left(1 + \dfrac{4}{t^2}\right)^2\left(\dfrac{1}{t^3}\right)\,dt = -\dfrac{1}{8} \int (1 + 4t^{-2})^2(-8t^{-3})\,dt$

$$= -\dfrac{1}{8}\dfrac{1}{3}(1 + 4t^{-2})^3 + C$$

$$= -\dfrac{1}{24}\left(1 + \dfrac{4}{t^2}\right)^3 + C$$

33. $\int (x^3 + 3x)(x^2 + 1)\,dx = \dfrac{1}{3} \int (x^3 + 3x)^1(3x^2 + 3)\,dx$

$$= \dfrac{1}{3}\dfrac{(x^3 + 3x)^2}{2} + C$$

$$= \dfrac{1}{6}(x^3 + 3x)^2 + C$$

35. Let $u = 6x^2 - 1$. Then $du = 12x \, dx$ and $x \, dx = \frac{1}{12} du$.

$$\int x(6x^2 - 1)^3 \, dx = \int (6x^2 - 1)^3 (x \, dx)$$

$$= \int u^3 \frac{1}{12} du$$

$$= \frac{1}{12} \frac{u^4}{4} + C$$

$$= \frac{1}{48}(6x^2 - 1)^4 + C$$

37. Let $u = 2 - 3x^3$, then $du = -9x^2 \, dx$ which implies that $x^2 \, dx = -\frac{1}{9} du$.

$$\int x^2(2 - 3x^3)^{3/2} \, dx = \int (2 - 3x^3)^{3/2}(x^2) \, dx$$

$$= \int u^{3/2}\left(-\frac{1}{9}\right) du$$

$$= -\frac{1}{9}\left(\frac{2}{5}\right)u^{5/2} + C$$

$$= -\frac{2}{45}(2 - 3x^3)^{5/2} + C$$

39. Let $u = x^2 + 25$, then $du = 2x \, dx$ which implies that $x \, dx = \frac{1}{2} du$.

$$\int \frac{x}{\sqrt{x^2 + 25}} \, dx = \int (x^2 + 25)^{-1/2}(x) \, dx$$

$$= \int u^{-1/2}\left(\frac{1}{2}\right) du$$

$$= \frac{1}{2}(2u^{1/2}) + C$$

$$= \sqrt{u} + C$$

$$= \sqrt{x^2 + 25} + C$$

41. Let $u = x^3 + 3x + 4$, then $du = (3x^2 + 3) \, dx = 3(x^2 + 1) \, dx$ and $(x^2 + 1) \, dx = \frac{1}{3} du$.

$$\int \frac{x^2 + 1}{\sqrt{x^3 + 3x + 4}} \, dx = \int (x^3 + 3x + 4)^{-1/2}(x^2 + 1) \, dx$$

$$= \int u^{-1/2}\left(\frac{1}{3}\right) du$$

$$= \left(\frac{1}{3}\right)2u^{1/2} + C$$

$$= \frac{2}{3}\sqrt{u} + C$$

$$= \frac{2}{3}\sqrt{x^3 + 3x + 4} + C$$

43. (a) $\displaystyle\int (2x - 1)^2 \, dx = \frac{1}{2}\int (2x - 1)^2(2) \, dx$

$$= \frac{1}{2}\frac{(2x - 1)^3}{3} + C_1$$

$$= \frac{1}{6}(2x - 1)^3 + C_1$$

$$= \frac{1}{6}(8x^3 - 12x^2 + 6x - 1) + C_1$$

$$= \frac{4}{3}x^3 - 2x^2 + x - \frac{1}{6} + C_1$$

$$\int (2x - 1)^2 \, dx = \int (4x^2 - 4x + 1) \, dx$$

$$= \frac{4}{3}x^3 - 2x^2 + x + C_2$$

(b) The two answers differ by a constant.

(c) Answers will vary.

45. (a) $\displaystyle\int x(x^2 - 1)^2 \, dx = \frac{1}{2}\int (x^2 - 1)^2(2x) \, dx$

$$= \frac{1}{2}\frac{(x^2 - 1)^3}{3} + C_1$$

$$= \frac{1}{6}(x^6 - 3x^4 + 3x^2 - 1) + C_1$$

$$= \frac{1}{6}x^6 - \frac{1}{2}x^4 + \frac{1}{2}x^2 - \frac{1}{6} + C_1$$

$$\int x(x^2 - 1)^2 \, dx = \int (x^5 - 2x^3 + x) \, dx$$

$$= \frac{1}{6}x^6 - \frac{1}{2}x^4 + \frac{1}{2}x^2 + C_2$$

(b) The two answers differ by a constant.

(c) Answers will vary.

47. $f(x) = \displaystyle\int x\sqrt{1-x^2}\, dx$

$$= -\frac{1}{2}\int (1-x^2)^{1/2}(-2x)\, dx$$

$$= -\frac{1}{2}\left(\frac{2}{3}\right)(1-x^2)^{3/2} + C$$

$$= -\frac{1}{3}(1-x^2)^{3/2} + C$$

Since $f(0) = \frac{4}{3}$, it follows that $C = \frac{5}{3}$ and we have

$$f(x) = -\frac{1}{3}(1-x^2)^{3/2} + \frac{5}{3} = \frac{1}{3}[5 - (1-x^2)^{3/2}].$$

49. (a) $C = \displaystyle\int \frac{4}{\sqrt{x+1}}\, dx$

$$= 4\int (x+1)^{-1/2}\, dx$$

$$= 4(2)(x+1)^{1/2} + K$$

$$= 8\sqrt{x+1} + K$$

Since $C(15) = 50$, it follows that $K = 18$, and we have $C = 8\sqrt{x+1} + 18$.

(b)

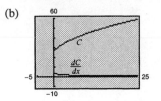

51. $x = \displaystyle\int p\sqrt{p^2 - 25}\, dp$

$$= \frac{1}{2}\int (p^2 - 25)^{1/2}(2p)\, dp$$

$$= \frac{1}{2}\cdot\frac{2}{3}(p^2 - 25)^{3/2} + C$$

$$= \frac{1}{3}(p^2 - 25)^{3/2} + C$$

Since $x = 600$ when $p = 13$, it follows that

$$x = 600 = \frac{1}{3}(13^2 - 25)^{3/2} + C = 576 + C$$

and $C = 24$. Therefore, $x = \frac{1}{3}(p^2 - 25)^{3/2} + 24$.

53. $x = \displaystyle\int -\frac{6000p}{(p^2 - 16)^{3/2}}\, dp$

$$= -\frac{6000}{2}\int (p^2 - 16)^{-3/2}(2p)\, dp$$

$$= -3000(-2)(p^2 - 16)^{-1/2} + C$$

$$= \frac{6000}{\sqrt{p^2 - 16}} + C$$

Since $x = 5000$ when $p = 5$, it follows that $C = 3000$, and we have

$$x = \frac{6000}{\sqrt{p^2 - 16}} + 3000.$$

55. (a) $h = \displaystyle\int \frac{17.6t}{\sqrt{17.6t^2 + 1}}\, dt$

$$= \frac{1}{2}\int (17.6t^2 + 1)^{-1/2}(35.2t)\, dt$$

$$= (17.6t^2 + 1)^{1/2} + C$$

$$h(0) = 6 \Rightarrow 6 = 1 + C \Rightarrow C = 5$$

$$h(t) = \sqrt{17.6t^2 + 1} + 5$$

(b) $h(5) = 26$ inches

57. (a) $Q = \displaystyle\int \frac{0.95}{(x - 19,999)^{0.05}}\, dx = (x - 19,999)^{0.95} + C$

Since $Q = 20,000$ when $x = 20,000$,

$$20,000 = (20,000 - 19,999)^{0.95} + C = 1 + C$$

$$C = 19,999.$$

Thus, $Q = (x - 19,999)^{0.95} + 19,999.$

(b)

x	20,000	50,000	100,000	150,000
Q	20,000	37,916.56	65,491.59	92,151.16
$X - Q$	0	12,083.44	34,508.41	57,848.84

(c)

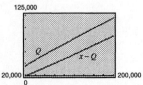

59. $\displaystyle\int \frac{1}{\sqrt{x} + \sqrt{x+1}}\, dx = -\frac{2}{3}x^{3/2} + \frac{2}{3}(x+1)^{3/2} + C$

Section 5.3   Exponential and Logarithmic Integrals

1. $\displaystyle\int e^{2x}(2)\, dx = e^{2x} + C$

3. $\displaystyle\int e^{4x}\, dx = \frac{1}{4}\int e^{4x}(4)\, dx = \frac{1}{4}e^{4x} + C$

5. $\displaystyle\int 9xe^{-x^2}\, dx = -\frac{9}{2}\int e^{-x^2}(-2x)\, dx = -\frac{9}{2}e^{-x^2} + C$

7. $\displaystyle\int 5x^2 e^{x^3}\, dx = \frac{5}{3}\int e^{x^3}(3x^2)\, dx = \frac{5}{3}e^{x^3} + C$

9. $\displaystyle\frac{1}{3}\int e^{x^3 + 3x^2 - 1}(3x^2 + 6x)\, dx = \frac{1}{3}e^{x^3 + 3x^2 - 1} + C$

11. $\displaystyle\int 5e^{2-x}\, dx = -5\int e^{2-x}(-1)\, dx = -5e^{2-x} + C$

13. $\displaystyle\int \frac{1}{x+1}\, dx = \ln|x+1| + C$

15. $\displaystyle -\frac{1}{2}\int \frac{1(-2)}{3 - 2x}\, dx = -\frac{1}{2}\ln|3 - 2x| + C$

17. $\displaystyle\int \frac{2}{3x+5}\, dx = \frac{2}{3}\int \frac{3}{3x+5}\, dx = \frac{2}{3}\ln|3x+5| + C$

19. $\displaystyle\frac{1}{2}\int \frac{x(2)}{x^2 + 1}\, dx = \frac{1}{2}\ln(x^2 + 1) + C = \ln\sqrt{x^2 + 1} + C$

21. $\displaystyle\int \frac{x^2}{x^3 + 1}\, dx = \frac{1}{3}\int \frac{3x^2}{x^3 + 1}\, dx = \frac{1}{3}\ln|x^3 + 1| + C$

23. $\displaystyle\int \frac{x+3}{x^2 + 6x + 7}\, dx = \frac{1}{2}\int \frac{2(x+3)}{x^2 + 6x + 7}\, dx = \frac{1}{2}\ln|x^2 + 6x + 7| + C$

25. $\displaystyle\int \frac{1}{x \ln x}\, dx = \int \frac{1}{\ln x}\left(\frac{1}{x}\right)dx = \ln|\ln x| + C$

27. $\displaystyle\int \frac{e^{-x}}{1 - e^{-x}}\, dx = \int \frac{1}{1 - e^{-x}}(e^{-x}\, dx) = \ln|1 - e^{-x}| + C$

29. $\displaystyle -\frac{1}{2}\int e^{2/x}(-2x^{-2})\, dx = -\frac{1}{2}e^{2/x} + C$

31. $\displaystyle 2\int e^{\sqrt{x}}\frac{1}{2\sqrt{x}}\, dx = 2e^{\sqrt{x}} + C$

33. $\displaystyle\int (e^x - 2)^2\, dx = \int (e^{2x} - 4e^x + 4)\, dx = \frac{1}{2}e^{2x} - 4e^x + 4x + C$

35. $\displaystyle -\int \frac{1}{1 + e^{-x}}(-e^{-x})\, dx = -\ln(1 + e^{-x}) + C$

37. $\displaystyle -2\int \frac{1}{5 - e^{2x}} - 2e^{2x}\, dx = -2\ln|5 - e^{2x}| + C$

39. $\displaystyle\int \frac{e^{2x} + 2e^x + 1}{e^x}\, dx = \int (e^x + 2 + e^{-x})\, dx$

$$= e^x + 2x - e^{-x} + C$$

41. $\displaystyle\int e^x\sqrt{1 - e^x}\, dx = -\int (1 - e^x)^{1/2}(-e^x)\, dx$

$$= -\frac{2}{3}(1 - e^x)^{3/2} + C$$

43. $\displaystyle\int (x-1)^{-2}\, dx = \frac{(x-1)^{-1}}{-1} = \frac{-1}{x-1} + C$

45. $\displaystyle\int 4e^{2x-1}\, dx = 2\int e^{2x-1}(2)\, dx = 2e^{2x-1} + C$

47. $\int \dfrac{x^3 - 8x}{2x^2}\,dx = \int \left(\dfrac{x}{2} - \dfrac{4}{x}\right) dx = \dfrac{x^2}{4} - 4\ln|x| + C$

49. $\int \dfrac{2}{1 + e^{-x}}\,dx = 2\int \dfrac{1}{e^x + 1}\,e^x\,dx = 2\ln(e^x + 1) + C$

51. $\int \dfrac{x^2 + 2x + 5}{x - 1}\,dx = \int \left(x + 3 + \dfrac{8}{x - 1}\right) dx$

$$= \dfrac{1}{2}x^2 + 3x + 8\ln|x - 1| + C$$

53. $\int \dfrac{1 + e^{-x}}{1 + xe^{-x}}\,dx = \int \dfrac{e^x + 1}{e^x + x}\,dx = \ln|e^x + x| + C$

55. Dividing, you obtain

$$\dfrac{x^2 + 4x + 3}{x - 1} = x + 5 + \dfrac{8}{x - 1}.$$

Hence,

$$f(x) = \int \left(x + 5 + \dfrac{8}{x - 1}\right) dx$$

$$= \dfrac{x^2}{2} + 5x + 8\ln|x - 1| + C.$$

$f(2) = 4 \Rightarrow$

$$4 = \dfrac{2^2}{2} + 5(2) + 8\ln|2 - 1| + C$$

$$-8 = C$$

Hence,

$$f(x) = \dfrac{x^2}{2} + 5x + 8\ln|x - 1| - 8.$$

57. (a) $P = \displaystyle\int \dfrac{3000}{1 + 0.25t}\,dt$

$$= \dfrac{3000}{0.25} \int \dfrac{0.25}{1 + 0.25t}\,dt$$

$$= 12{,}000\ln|1 + 0.25t| + C$$

Since $P(0) = 12{,}000\ln 1 + C = 1000$, it follows that $C = 1000$. Therefore,

$$P(t) = 12{,}000\ln|1 + 0.25t| + 1000$$

$$= 1000[12\ln|1 + 0.25t| + 1]$$

$$= 1000[1 + \ln(1 + 0.25t)^{12}].$$

(b) The population when $t = 3$ is

$$P(3) = 1000[1 + \ln(1 + 0.75)^{12}]$$

$$\approx 7715 \text{ bacteria.}$$

(c) Using a graphing utility,

$$12{,}000 = 1000[1 + \ln(1 + 0.25t)^{12}]$$

$$\Rightarrow t \approx 6 \text{ days.}$$

59. (a) $p = \displaystyle\int 0.1e^{-x/500}\,dx = -50e^{-x/500} + C$

Since $x = 600$ when $p = 30$, we have:

$$30 = -50e^{600/500} + C \Rightarrow C \approx 45.06$$

$$p = -50e^{-x/500} + 45.06$$

(b)

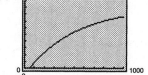

Price increases as demand increases.

(c) When $p = 22$, $x \approx 387$.

61. (a) $S = \displaystyle\int 2621.7e^{0.07t}\,dt$

$$= \dfrac{2621.7}{0.07}e^{0.07t} + C$$

$$= 37{,}452.86e^{0.07t} + C$$

At $t = 11$,

$$118{,}496 = 37{,}452.86e^{0.07(11)} + C$$

$$C \approx 37{,}606.58$$

$$S = 37{,}452.86e^{0.07t} + 37{,}606.58 \quad \text{(in dollars).}$$

(b) For 1999, $t = 9$ and

$$S = 37{,}452.86e^{0.07(9)} + 37{,}606.58$$

$$\approx \$107{,}928.47.$$

63. False. $(\ln 5)^{1/2} \approx 1.269 \neq \dfrac{1}{2}\ln 5 \approx 0.8047.$

$$\ln x^{1/2} = \dfrac{1}{2}\ln x$$

Section 5.4   Area and the Fundamental Theorem of Calculus

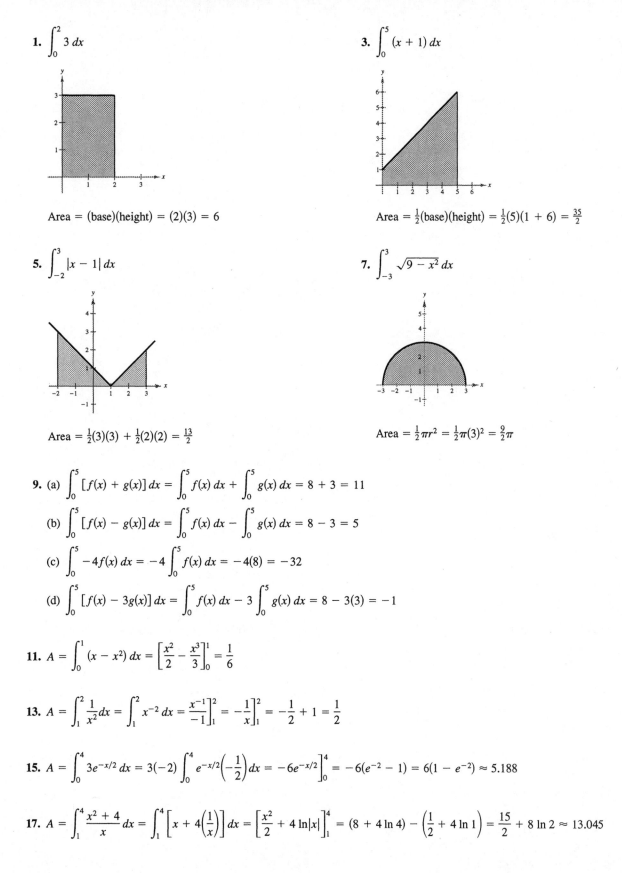

1. $\displaystyle\int_0^2 3\,dx$

Area = (base)(height) = (2)(3) = 6

3. $\displaystyle\int_0^5 (x+1)\,dx$

Area = $\frac{1}{2}$(base)(height) = $\frac{1}{2}$(5)(1 + 6) = $\frac{35}{2}$

5. $\displaystyle\int_{-2}^3 |x-1|\,dx$

Area = $\frac{1}{2}$(3)(3) + $\frac{1}{2}$(2)(2) = $\frac{13}{2}$

7. $\displaystyle\int_{-3}^3 \sqrt{9-x^2}\,dx$

Area = $\frac{1}{2}\pi r^2 = \frac{1}{2}\pi(3)^2 = \frac{9}{2}\pi$

9. (a) $\displaystyle\int_0^5 [f(x)+g(x)]\,dx = \int_0^5 f(x)\,dx + \int_0^5 g(x)\,dx = 8 + 3 = 11$

(b) $\displaystyle\int_0^5 [f(x)-g(x)]\,dx = \int_0^5 f(x)\,dx - \int_0^5 g(x)\,dx = 8 - 3 = 5$

(c) $\displaystyle\int_0^5 -4f(x)\,dx = -4\int_0^5 f(x)\,dx = -4(8) = -32$

(d) $\displaystyle\int_0^5 [f(x)-3g(x)]\,dx = \int_0^5 f(x)\,dx - 3\int_0^5 g(x)\,dx = 8 - 3(3) = -1$

11. $\displaystyle A = \int_0^1 (x-x^2)\,dx = \left[\frac{x^2}{2} - \frac{x^3}{3}\right]_0^1 = \frac{1}{6}$

13. $\displaystyle A = \int_1^2 \frac{1}{x^2}\,dx = \int_1^2 x^{-2}\,dx = \frac{x^{-1}}{-1}\bigg]_1^2 = -\frac{1}{x}\bigg]_1^2 = -\frac{1}{2} + 1 = \frac{1}{2}$

15. $\displaystyle A = \int_0^4 3e^{-x/2}\,dx = 3(-2)\int_0^4 e^{-x/2}\left(-\frac{1}{2}\right)dx = -6e^{-x/2}\bigg]_0^4 = -6(e^{-2}-1) = 6(1-e^{-2}) \approx 5.188$

17. $\displaystyle A = \int_1^4 \frac{x^2+4}{x}\,dx = \int_1^4 \left[x + 4\left(\frac{1}{x}\right)\right]dx = \left[\frac{x^2}{2} + 4\ln|x|\right]_1^4 = (8 + 4\ln 4) - \left(\frac{1}{2} + 4\ln 1\right) = \frac{15}{2} + 8\ln 2 \approx 13.045$

19. $\displaystyle\int_0^1 2x\,dx = x^2\Big]_0^1 = 1 - 0 = 1$

21. $\displaystyle\int_{-1}^0 (2x+1)\,dx = \Big[x^2 + x\Big]_{-1}^0 = 0 - 0 = 0$

23. $\displaystyle\int_{-1}^1 (2t-1)^2\,dt = \frac{1}{6}(2t-1)^3\Big]_{-1}^1 = \frac{1}{6} - \left(\frac{-27}{6}\right) = \frac{14}{3}$

25. $\displaystyle\int_0^3 (x-2)^3\,dx = \frac{(x-2)^4}{4}\Big]_0^3 = \frac{1}{4} - 4 = -\frac{15}{4}$

27. $\displaystyle\int_{-1}^1 \left(\sqrt[3]{t} - 2\right)dt = \left[\frac{3}{4}t^{4/3} - 2t\right]_{-1}^1$

$\qquad\qquad = -\frac{5}{4} - \frac{11}{4}$

$\qquad\qquad = -4$

29. $\displaystyle\int_1^4 \frac{2u-1}{\sqrt{u}}\,du = \int_1^4 (2u^{1/2} - u^{-1/2})\,du$

$\qquad\qquad = \left[\frac{4}{3}u^{3/2} - 2u^{1/2}\right]_1^4$

$\qquad\qquad = \left(\frac{32}{3} - 4\right) - \left(\frac{4}{3} - 2\right)$

$\qquad\qquad = \frac{22}{3}$

31. $\displaystyle\int_{-1}^0 (t^{1/3} - t^{2/3})\,dt = \left[\frac{3}{4}t^{4/3} - \frac{3}{5}t^{5/3}\right]_{-1}^0$

$\qquad\qquad = 0 - \left(\frac{3}{4} + \frac{3}{5}\right)$

$\qquad\qquad = -\frac{27}{20}$

33. $\displaystyle\int_0^4 \frac{1}{\sqrt{2x+1}}\,dx = \frac{1}{2}\int_0^4 (2x+1)^{-1/2}(2)\,dx$

$\qquad\qquad = \left[\frac{1}{2}(2)(2x+1)^{1/2}\right]_0^4$

$\qquad\qquad = \sqrt{2x+1}\,\Big]_0^4$

$\qquad\qquad = 3 - 1$

$\qquad\qquad = 2$

35. $\displaystyle\int_0^1 e^{-2x}\,dx = -\frac{1}{2}e^{-2x}\Big]_0^1$

$\qquad\qquad = -\frac{e^{-2}}{2} + \frac{1}{2}$

$\qquad\qquad = \frac{1}{2}(1 - e^{-2}) \approx 0.432$

37. $\displaystyle\int_1^3 \frac{e^{3/x}}{x^2}\,dx = -\frac{1}{3}\int_1^3 e^{3/x}\left(-\frac{3}{x^2}\right)dx$

$\qquad\qquad = -\frac{1}{3}e^{3/x}\Big]_1^3$

$\qquad\qquad = -\frac{1}{3}(e - e^3)$

$\qquad\qquad = \frac{e^3 - e}{3} \approx 5.789$

39. $\displaystyle\int_0^1 e^{2x}\sqrt{e^{2x} + 1}\,dx = \frac{1}{2}\int_0^1 (e^{2x}+1)^{1/2}\,2e^{2x}\,dx$

$\qquad\qquad = \frac{1}{3}(e^{2x}+1)^{3/2}\Big]_0^1$

$\qquad\qquad = \frac{1}{3}\left[(e^2+1)^{3/2} - 2^{3/2}\right]^0 \approx 7.157$

41. $\displaystyle\int_0^2 \frac{x}{1+4x^2}\,dx = \frac{1}{8}\int_0^2 \frac{1}{1+4x^2}(8x)\,dx$

$\qquad\qquad = \frac{1}{8}\ln(1 + 4x^2)\Big]_0^2$

$\qquad\qquad = \frac{1}{8}\left[(\ln 17 - 0)\right]$

$\qquad\qquad = \frac{1}{8}\ln 17 \approx 0.354$

43. $\displaystyle\int_{-1}^1 |4x|\,dx = \int_{-1}^0 -4x\,dx + \int_0^1 4x\,dx$

$\qquad\qquad = \Big[-2x^2\Big]_{-1}^0$

$\qquad\qquad = (0 + 2) + (2 - 0)$

$\qquad\qquad = 4$

45. $\int_0^4 (2 - |x - 2|)\, dx = \int_0^2 \{2 - [-(x - 2)]\}\, dx + \int_2^4 [2 - (x - 2)]\, dx$

$$= \int_0^2 x\, dx + \int_2^4 (4 - x)\, dx$$

$$= \frac{x^2}{2}\Big]_0^2 + \left[4x - \frac{x^2}{2} \right]_2^4$$

$$= (2 - 0) + (8 - 6)$$

$$= 4$$

47. $\int_{-1}^2 \frac{x}{x^2 - 9}\, dx = \left[\frac{1}{2} \ln|x^2 - 9| \right]_{-1}^2$

$$= \frac{1}{2} \ln 5 - \frac{1}{2} \ln 8$$

$$\approx -0.235$$

49. $\int_0^3 \frac{2e^x}{2 + e^x}\, dx = \left[2 \ln(2 + e^x) \right]_0^3$

$$= 2 \ln(2 + e^3) - 2 \ln 3$$

$$\approx 3.993$$

51. $\int_1^3 (4x - 3)\, dx = \left[2x^2 - 3x \right]_1^3$

$$= (18 - 9) - (2 - 3)$$

$$= 10$$

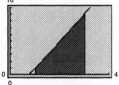

53. $\int_0^1 (x - x^3)\, dx = \left[\frac{x^2}{2} - \frac{x^4}{4} \right]_0^1$

$$= \frac{1}{2} - \frac{1}{4}$$

$$= \frac{1}{4}$$

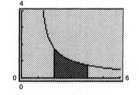

55. $\int_2^4 \frac{3x^2}{x^3 - 1}\, dx = \left[\ln(x^3 - 1) \right]_2^4$

$$= \ln 63 - \ln 7$$

$$= \ln 9$$

57. $A = \int_0^2 (3x^2 + 1)\, dx$

$$= \left[x^3 + x \right]_0^2$$

$$= 10 \text{ square units}$$

59. $A = \int_1^5 \frac{x + 5}{x}\, dx$

$$= \int_1^5 \left[1 + 5\left(\frac{1}{x}\right) \right]\, dx$$

$$= \left[x + 5 \ln|x| \right]_1^5$$

$$= (5 + 5 \ln 5) - (1 + 0)$$

$$= 4 + 5 \ln 5$$

$$\approx 12.047 \text{ square units}$$

61. Average value $= \dfrac{1}{2-(-2)} \displaystyle\int_{-2}^{2} (6-x^2)\, dx$

$$= \frac{1}{4}\left[6x - \frac{x^3}{3}\right]_{-2}^{2}$$

$$= \frac{1}{4}\left(\frac{28}{3} + \frac{28}{3}\right)$$

$$= \frac{14}{3}$$

To find the *x*-values for which $f(x) = \frac{14}{3}$, we let $6 - x^2 = \frac{14}{3}$ and solve for *x*.

$$x^2 = \frac{4}{3}$$

$$x = \pm\frac{2\sqrt{3}}{3} \approx \pm 1.155$$

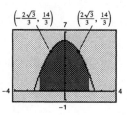

63. Average value $= \dfrac{1}{10-0} \displaystyle\int_{0}^{10} 5e^{0.2(x-10)}\, dx$

$$= \frac{1}{10}(25 - 25e^{-2})$$

$$= \frac{5}{2} - \frac{5}{2}e^{-2}$$

To find the *x*-values for which $f(x) = \frac{5}{2} - \frac{5}{2}e^{-2}$, we let $5e^{0.2(x-10)} = \frac{5}{2} - \frac{5}{2}e^{-2}$ and solve for *x*. Using a graphing utility, we obtain $x \approx 5.807$.

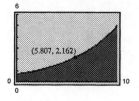

65. Average value $= \dfrac{1}{2-0} \displaystyle\int_{0}^{2} x\sqrt{4 - x^2}\, dx$

$$= \frac{1}{2}\left(-\frac{1}{2}\right) \int_{0}^{2} (4 - x^2)^{1/2}(-2x)\, dx$$

$$= \frac{1}{2}\left(-\frac{1}{2}\right)\left(\frac{2}{3}\right)(4 - x^2)^{3/2}\Big]_{0}^{2}$$

$$= -\frac{1}{6}(4 - x^2)^{3/2}\Big]_{0}^{2}$$

$$= 0 + \frac{4}{3}$$

$$= \frac{4}{3}$$

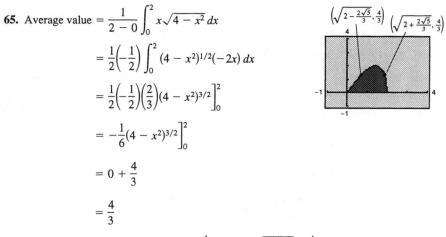

To find the *x*-values for which $f(x) = \frac{4}{3}$, we let $x\sqrt{4 - x^2} = \frac{4}{3}$ and solve for *x* to obtain the following.

$$x^2(4 - x^2) = \frac{16}{9}$$

$$36x^2 - 9x^4 = 16$$

$$9x^4 - 36x^2 + 16 = 0$$

$$x^2 = \frac{36 \pm \sqrt{720}}{18} = 2 \pm \frac{2\sqrt{5}}{3}$$

$$x = \sqrt{2 \pm \frac{2\sqrt{5}}{3}} \text{ in the interval } (0, 2).$$

$$x \approx 1.868 \text{ and } x \approx 0.714$$

67. Average value $= \dfrac{1}{7-0} \displaystyle\int_0^7 \dfrac{5x}{x^2+1}\,dx$

$$= \dfrac{1}{7} \cdot \dfrac{5}{2} \ln(x^2+1)\Big]_0^7$$

$$= \dfrac{5}{14} \ln 50$$

To find the x-values for which $f(x) = \frac{5}{14} \ln 50$, we solve for x as follows:

$$\dfrac{5x}{x^2+1} = \dfrac{5}{14} \ln 50$$

Using a graphing utility, we obtain $x \approx 0.3055$, and $x \approx 3.2732$.

69. Since $f(-x) = 3(-x)^4 = 3x^4 = f(x)$, the function is even.

71. Since $g(-t) \neq g(t)$ nor $g(-t) = -g(t)$, the function is neither even nor odd.

73. Since $f(x) = x^2$ is an **even** function, we have the following.

(a) $\displaystyle\int_{-2}^0 x^2\,dx = \int_0^2 x^2\,dx = \dfrac{8}{3}$ (b) $\displaystyle\int_{-2}^2 x^2\,dx = 2\int_0^2 x^2\,dx = \dfrac{16}{3}$ (c) $\displaystyle\int_0^2 -x^2\,dx = -\int_0^2 x^2\,dx = -\dfrac{8}{3}$

75. $\Delta C = \displaystyle\int_{100}^{103} 2.25\,dx = 2.25x \Big]_{100}^{103} = \$6.75$

77. $\Delta R = \displaystyle\int_{12}^{15} (48-3x)\,dx = \left[48x - \dfrac{3}{2}x^2 \right]_{12}^{15} = \$22.50$

79. $\Delta P = \displaystyle\int_{200}^{203} \dfrac{400-x}{150}\,dx = \dfrac{1}{150}\left[400x - \dfrac{x^2}{2} \right]_{200}^{203} = \$3.97$

81. $A = e^{rt} \displaystyle\int_0^T c(t)e^{-rt}\,dt = e^{0.08(6)} \int_0^6 250e^{-0.08t}\,dt \approx \$1925.23$

83. $A = e^{rt} \displaystyle\int_0^T c(t)e^{-rt}\,dt = e^{0.02(10)} \int_0^{10} 1500e^{-0.02t}\,dt \approx \$16,605.21$

85. $\displaystyle\int_0^5 500\,dt = 500t \Big]_0^5 = \$2500$

87. Capital accumulation $= \displaystyle\int_0^5 500\sqrt{t+1}\,dt = \dfrac{1000}{3}(t+1)^{3/2}\Big]_0^5 = \dfrac{1000}{3}(6)^{3/2} - \dfrac{1000}{3} \approx \$4565.65$

89. $C(x) = 5000\left(25 + 3\displaystyle\int_0^x t^{1/4}\,dt \right)$

$$= 5000\left(25 + \left[\dfrac{12}{5}t^{5/4} \right]_0^x \right)$$

$$= 5000\left(25 + \dfrac{12}{5}t^{5/4} \right)$$

(a) $C(1) = 5000\left[25 + \left(\tfrac{12}{5}\right) \right] = \$137,000.00$

(b) $C(5) = 5000\left[25 + \left(\tfrac{12}{5}\right)(5)^{5/4} \right] \approx \$214,720.93$

(c) $C(10) = 5000\left[25 + \left(\tfrac{12}{5}\right)(10)^{5/4} \right] \approx \$338,393.53$

91. Average balance $= \dfrac{1}{5-0} \displaystyle\int_0^5 2250e^{0.12t}\,dt$

$$= 450 \displaystyle\int_0^5 e^{0.12t}\,dt$$

$$= 450\left(\dfrac{1}{0.12} \right)e^{0.12t}\Big]_0^5$$

$$= 3750(e^{0.6} - 1)$$

$$\approx \$3082.95$$

93. $\dfrac{1}{R-0} \displaystyle\int_0^R k(R^2 - r^2)\, dr = \dfrac{k}{R}\left(R^2 r - \dfrac{r^3}{3}\right)\Big]_0^R = \dfrac{2kR^2}{3}$

95. $\displaystyle\int_3^6 \dfrac{x}{3\sqrt{x^2 - 8}}\, dx \approx 1.4305 = \dfrac{2}{3}\sqrt{7} - \dfrac{1}{3}$

97. $\displaystyle\int_2^5 \left(\dfrac{1}{x^2} - \dfrac{1}{x^3}\right) dx = 0.195 = \dfrac{39}{200}$

Section 5.5   The Area of a Region Bounded by Two Graphs

1. $\displaystyle\int_0^6 [0 - (x^2 - 6x)]\, dx = \left(\dfrac{x^3}{3} - 3x^2\right)\Big]_0^6 = 36$

3. $\displaystyle\int_0^3 [(-x^2 + 2x + 3) - (x^2 - 4x + 3)]\, dx = \int_0^3 (-2x^2 + 6x)\, dx = \left[\dfrac{-2x^3}{3} + 3x^2\right]_0^3 = 9$

5. $A = 2\displaystyle\int_0^1 [0 - 3(x^3 - x)]\, dx = -6\left(\dfrac{x^4}{4} - \dfrac{x^2}{2}\right)\Big]_0^1 = \dfrac{3}{2}$ (by symmetry)

7. $A = \displaystyle\int_0^1 [(e^x - 1) - 0]\, dx = e^x - x\Big]_0^1 = (e - 1) - 1 = e - 2$

9. The region is bounded by the graphs of $y = x + 1$, $y = x/2$, $x = 0$, and $x = 4$, as shown in the figure.

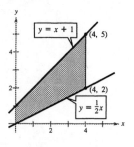

11.

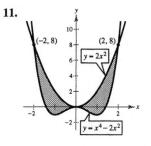

13.

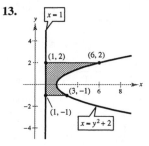

15. The points of intersection are found by evaluating $x = 1$ and $x = 5$ for the function $y = \dfrac{1}{x^2}$.

$$A = \int_1^5 \dfrac{1}{x^2}\, dx = -\dfrac{1}{x}\Big]_1^5 = \dfrac{4}{5}$$

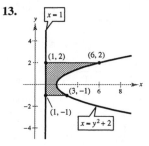

17. $A = 2\displaystyle\int_0^1 \left(\sqrt[3]{x} - x\right) dx = 2\left[\dfrac{3}{4}x^{4/3} - \dfrac{x^2}{2}\right]_0^1 = \dfrac{1}{2}$

19. The points of intersection of the two graphs are found by equating y-values and solving for x.

$$x^2 - 4x + 3 = 3 + 4x - x^2$$

$$2x^2 - 8x = 0$$

$$2x(x - 4) = 0$$

$$x = 0, 4$$

$$A = \int_0^4 [(3 + 4x - x^2) - (x^2 - 4x + 3)]\, dx$$

$$= \int_0^4 (-2x^2 + 8x)\, dx$$

$$= \left[-\frac{2x^3}{3} + 4x^2 \right]_0^4$$

$$= \frac{64}{3}$$

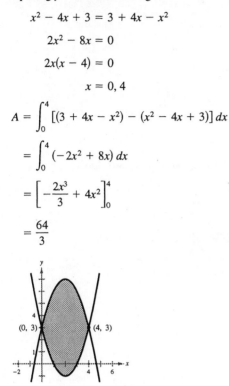

21. $A = \displaystyle\int_0^1 xe^{-x^2}\, dx = -\frac{1}{2}e^{-x^2}\bigg]_0^1$

$$= -\frac{1}{2}e^{-1} + \frac{1}{2} \approx 0.316$$

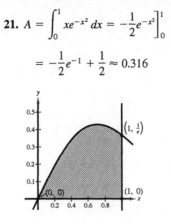

23. $A = \displaystyle\int_1^2 x^2\,dx + \int_2^4 \frac{8}{x}\,dx = \frac{x^3}{3}\bigg]_1^2 + 8\ln x\bigg]_2^4$

$$= \left(\frac{8}{3} - \frac{1}{3} \right) + 8(\ln 4 - \ln 2) = \frac{7}{3} + 8\ln 2$$

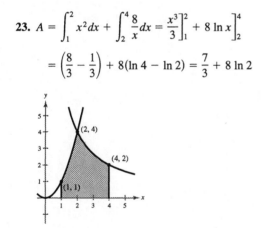

25. $A = \displaystyle\int_1^2 \left[e^{1/2x} - \left(-\frac{1}{x} \right) \right] dx = 2e^{1/2x} + \ln x\bigg]_1^2$

$$= (2e + \ln 2) - 2e^{1/2} \approx 2.832$$

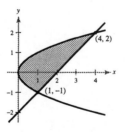

27. The points of intersection are found by setting $f(y)$ equal to $g(y)$ and solving for y.

$$y^2 = y + 2$$

$$y^2 - y - 2 = 0$$

$$(y + 1)(y - 2) = 0$$

$$y = -1, 2$$

$$A = \int_{-1}^2 [(y + 2) - y^2]\,dy = \left[\frac{y^2}{2} + 2y - \frac{y^3}{3} \right]_{-1}^2$$

$$= \left(2 + 4 - \frac{8}{3} \right) - \left(\frac{1}{2} - 2 + \frac{1}{3} \right) = \frac{9}{2}$$

29. $A = \int_0^9 \sqrt{y}\, dy = \frac{2}{3} y^{3/2} \Big]_0^9 = 18$

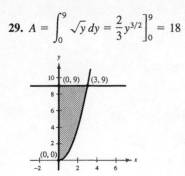

31. To find the points of intersection, we solve:

$$2x = 4 - 2x$$

$$4x = 4$$

$$x = 1$$

$$A = \int_0^1 2x\, dx + \int_1^2 (4 - 2x)\, dx$$

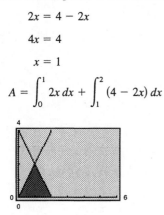

33. $A = \int_1^2 \left(\frac{4}{x} - x\right) dx + \int_2^4 \left(x - \frac{4}{x}\right) dx$

35. The points of intersection of f and g are found by setting $f(x) = g(x)$ and solving for x.

$$x^2 - 4x = 0$$

$$x(x - 4) = 0$$

$$x = 0, 4$$

$$A = \int_0^4 [0 - (x^2 - 4x)]\, dx$$

$$= -\left(\frac{x^3}{3} - 2x^2\right)\Big]_0^4$$

$$= \frac{32}{3}$$

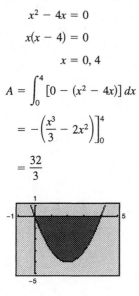

37. The points of intersection are found by setting $f(x) = g(x)$.

$$x^2 + 2x + 1 = x + 1$$

$$x^2 + x = 0$$

$$x(x + 1) = 0$$

$$x = 0, -1$$

$$A = \int_{-1}^0 [(x + 1) - (x^2 - 2x + 1)]\, dx$$

$$= \int_{-1}^0 (-x^2 - x)\, dx$$

$$= \frac{-x^3}{3} - \frac{x^2}{2}\Big]_{-1}^0$$

$$= 0 - \left(\frac{1}{3} - \frac{1}{2}\right)$$

$$= \frac{1}{6}$$

39. The equation of the line passing through $(0, 0)$ and $(4, 4)$ is $y = x$. Therefore, the area is given by the following.

$$A = \int_0^4 x \, dx = \frac{x^2}{2} \bigg]_0^4 = 8$$

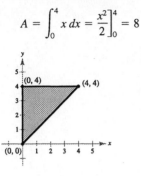

41. The point of equilibrium is found by equating $50 - 0.5x = 0.125x$ to obtain $x = 80$ and $p = 10$.

$$CS = \int_0^{80} [(50 - 0.5x) - 10] \, dx$$

$$= \left[-\frac{0.5x^2}{2} + 40x \right]_0^{80} = 1600$$

$$PS = \int_0^{80} (10 - 0.125x) \, dx$$

$$= \left[10x - \frac{0.125x^2}{2} \right]_0^{80} = 400$$

43. The point of equilibrium is found by equating the supply and demand functions.

$$200 - 0.02x^2 = 100 + x$$

$$0.02x^2 + x - 100 = 0$$

$$x^2 + 50x - 5000 = 0$$

$$(x + 100)(x - 50) = 0$$

$$x = 50 \quad \text{and} \quad p = 150$$

$$CS = \int_0^{50} [(200 - 0.02x^2) - 150] \, dx = \left[50x - \frac{0.02x^3}{3} \right]_0^{50} \approx 1666.67$$

$$PS = \int_0^{50} [150 - (100 + x)] \, dx = \left[50x - \frac{x^2}{2} \right]_0^{50} = \$1250.00$$

45. The point of equilibrium is found by equating the demand and supply functions.

$$\frac{10,000}{\sqrt{x + 100}} = 100\sqrt{0.05x + 10}$$

$$100 = \sqrt{(x + 100)(0.05x + 10)}$$

$$10,000 = 0.05x^2 + 15x + 1000$$

$$5x^2 + 1500x - 900,000 = 0$$

$$5(x^2 + 300x - 180,000) = 0$$

$$5(x + 600)(x - 300) = 0$$

$$x = 300 \quad \text{and} \quad p = 500$$

$$CS = \int_0^{300} \left(\frac{10,000}{\sqrt{x + 100}} - 500 \right) dx$$

$$= \left[20,000\sqrt{x + 100} - 500x \right]_0^{300}$$

$$= 250,000 - 200,000$$

$$= 50,000$$

$$PS = \int_0^{300} \left(500 - 100\sqrt{0.05x + 10} \right) dx$$

$$= \left[500x - \frac{4000}{3}(0.05x + 10)^{3/2} \right]_0^{300}$$

$$= -\frac{50,000}{3} + \frac{40,000\sqrt{10}}{3}$$

$$= \frac{10,000}{3}\left(4\sqrt{10} - 5 \right)$$

$$\approx 25,497$$

47. A typical demand function is decreasing, while a typical supply function is increasing.

49. The model R_1 projects greater revenue than R_2. The difference in total revenue is

$$\int_4^8 (R_1 - R_2)\, dt = \int_4^8 \left[(7.21 + 0.58t) - (7.21 + 0.45t)\right] dt$$

$$= \int_4^8 0.13t\, dt$$

$$= \left[\frac{0.13t^2}{2}\right]_4^8 = \$3.12 \text{ billion.}$$

51. The total savings is given by

$$\int_4^{10} (C_1 - C_2)\, dt = \int_4^{10} \left[(568.5 + 7.15t) - (525.6 + 6.43t)\right] dt$$

$$= \int_4^{10} (42.9 + 0.72t)\, dt$$

$$= \left[42.9t + 0.36t^2\right]_4^{10}$$

$$= \$287.64 \text{ million.}$$

53. (a)

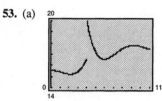

(b) The difference in consumption is given by

$$\int_5^{11} \left[(0.085t^3 - 0.309t^2 + 0.13t + 15.5) - (0.01515t^4 - 0.5348t^3 + 6.864t^2 - 37.68t + 91.4)\right] dt \approx 169.2 \text{ pounds per person.}$$

55. To find the demand function we use the points (6000, 325) and (8000, 300) to obtain

$$p - 325 = \frac{300 - 325}{8000 - 6000}(x - 6000)$$

$$p = -\frac{1}{80}x + 400.$$

The point of equilibrium is obtained by equating $-\frac{1}{80}x + 400 = \frac{11}{400}x$ to obtain $x = 10{,}000$ and $p = 275$.

$$\text{Consumer Surplus} = \int_0^{10{,}000} \left(-\frac{1}{80}x + 400 - 275\right) dx = \$625{,}000$$

$$\text{Producer Surplus} = \int_0^{10{,}000} \left(275 - \frac{11}{400}x\right) dx = \$1{,}375{,}000$$

57. $P = \int_0^{10} (R - C)\, dt$

$= \int_0^{10} \left[(100 + 0.08t) - (60 + 0.2t^2) \right] dt$

$= \int_0^{10} (-0.2t^2 + 0.08t + 40)\, dt$

$= \left[-\dfrac{0.2t^3}{3} + 0.04t^2 + 40t \right]_0^{10}$

$\approx \$337.33$ million

59. $y(20) - y(0) \approx 2.92$ or 2.92%

$y(40) - y(20) \approx 7.09$ or 7.09%

$y(60) - y(40) \approx 14.58$ or 14.58%

$y(80) - y(60) \approx 26.74$ or 26.74%

$y(100) - y(80) \approx 44.89$ or 44.89%

(Note: The sum is approximately 100%)

Quintile	Lowest	2nd	3rd	4th	Highest
Percent	2.92	7.09	14.58	26.74	44.89

Section 5.6   The Definite Integral as the Limit of a Sum

1. The midpoints of the four intervals are $\frac{1}{8}, \frac{3}{8}, \frac{5}{8}$, and $\frac{7}{8}$. The approximate area is

$$A \approx \frac{1 - 0}{4} \left[f\!\left(\frac{1}{8}\right) + f\!\left(\frac{3}{8}\right) + f\!\left(\frac{5}{8}\right) + f\!\left(\frac{7}{8}\right) \right] = \frac{1}{4}\left[\frac{11}{4} + \frac{9}{4} + \frac{7}{4} + \frac{5}{4} \right] = 2.$$

The exact area is $A = \int_0^1 (-2x + 3)\, dx = \left[-x^2 + 3x \right]_0^1 = 2.$

3. The midpoints of the four intervals are $\frac{1}{8}, \frac{3}{8}, \frac{5}{8}$, and $\frac{7}{8}$. The approximate area is

$$A \approx \frac{1 - 0}{4} \left[\sqrt{\frac{1}{8}} + \sqrt{\frac{3}{8}} + \sqrt{\frac{5}{8}} + \sqrt{\frac{7}{8}} \right] = \frac{1}{4}\left[\frac{\sqrt{2}}{4} + \frac{\sqrt{6}}{4} + \frac{\sqrt{10}}{4} + \frac{\sqrt{14}}{4} \right] \approx 0.6730.$$

The exact area is $A = \int_0^1 \sqrt{x}\, dx = \frac{2}{3}x^{3/2} \Big]_0^1 = \frac{2}{3} \approx 0.6667.$

5. The midpoints of the four intervals are $-\frac{3}{4}, -\frac{1}{4}, \frac{1}{4}$, and $\frac{3}{4}$. The approximate area is

$$A \approx \frac{1 - (-1)}{4} \left[\frac{41}{16} + \frac{33}{16} + \frac{33}{16} + \frac{41}{16} \right] = \frac{1}{2}\left[\frac{148}{16} \right] = \frac{37}{8} = 4.625.$$

The exact area is $A = \int_{-1}^1 (x^2 + 2)\, dx = \left[\frac{x^3}{3} + 2x \right]_{-1}^1 = \frac{14}{3} \approx 4.6667.$

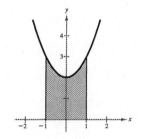

7. The midpoints of the four intervals are $\frac{5}{4}, \frac{7}{4}, \frac{9}{4}$, and $\frac{11}{4}$. The approximate area is

$$A \approx \frac{3 - 1}{4} \left[f\!\left(\frac{5}{4}\right) + f\!\left(\frac{7}{4}\right) + f\!\left(\frac{9}{4}\right) + f\!\left(\frac{11}{4}\right) \right]$$

$$= \frac{1}{2}\left[\frac{25}{8} + \frac{49}{8} + \frac{81}{8} + \frac{121}{8} \right]$$

$$= \frac{69}{4}$$

$$= 17.25.$$

The exact area is $A = \int_1^3 2x^2\, dx = \frac{2x^3}{3} \Big]_1^3 = \frac{52}{3} \approx 17.3333.$

9. The midpoints of the four intervals are $\frac{1}{8}, \frac{3}{8}, \frac{5}{8}$, and $\frac{7}{8}$. The approximate area is

$$A \approx \frac{1-0}{4}\left[\frac{7}{512} + \frac{45}{512} + \frac{75}{512} + \frac{49}{512}\right] \approx 0.0859.$$

The exact area is $A = \int_0^1 (x^2 - x^3)\, dx = \left[\frac{x^3}{3} - \frac{x^4}{4}\right]_0^1 = \frac{1}{12} \approx 0.0833.$

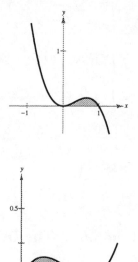

11. The midpoints of the four intervals are $\frac{1}{8}, \frac{3}{8}, \frac{5}{8}$, and $\frac{7}{8}$. The approximate area is

$$A \approx \frac{1-0}{4}\left[\frac{49}{512} + \frac{75}{512} + \frac{45}{512} + \frac{7}{512}\right] = \frac{11}{128} \approx 0.0859.$$

The exact area is

$$A = \int_0^1 x(1-x)^2\, dx = \int_0^1 (x^3 - 2x^2 + x)\, dx = \left[\frac{x^4}{4} - \frac{2x^3}{3} + \frac{x^2}{2}\right]_0^1 = \frac{1}{12} \approx 0.0833.$$

13. $\int_0^4 (2x^2 + 3)\, dx = \frac{164}{3} \approx 54.6667$

Using the Midpoint Rule with $n = 31$, we get 54.66.

15. $\int_1^2 (2x^2 - x + 1)\, dx = \left[\frac{2x^3}{3} - \frac{x^2}{2} + x\right]_1^2 = \left(\frac{16}{3} - 2 + 2\right) - \left(\frac{2}{3} - \frac{1}{2} + 1\right) = \frac{25}{6} \approx 4.16\overline{6}$

Using the Midpoint Rule with $n = 5$, we get 4.16.

17. The midpoints of the four intervals are $\frac{9}{4}, \frac{11}{4}, \frac{13}{4}$, and $\frac{15}{4}$. The approximate area is

$$A \approx \frac{4-2}{4}\left[f\left(\frac{9}{4}\right) + f\left(\frac{11}{4}\right) + f\left(\frac{13}{4}\right) + f\left(\frac{15}{4}\right)\right] = \frac{1}{2}(3) = 1.5.$$

The exact area is $A = \int_2^4 \frac{1}{4}y\, dy = \frac{y^2}{8}\bigg]_2^4 = 2 - \frac{1}{2} = 1.5.$

(The Midpoint Rule is exact for lines.)

19. The midpoints of the four intervals are $\frac{1}{2}, \frac{3}{2}, \frac{5}{2}, \frac{7}{2}$. The approximate area is

$$A \approx \frac{4-0}{4}\left[f\left(\frac{1}{2}\right) + f\left(\frac{3}{2}\right) + f\left(\frac{5}{2}\right) + f\left(\frac{7}{2}\right)\right]$$

$$= \frac{5}{4} + \frac{13}{4} + \frac{29}{4} + \frac{53}{4} = \frac{100}{4} = 25.$$

The exact area is $A = \int_0^4 (y^2 + 1)\, dy = \left[\frac{y^3}{3} + y\right]_0^4 = 25.3\overline{3}.$

21. Exact: $\int_0^2 x^3\, dx = \frac{x^4}{4}\bigg]_0^2 = 4$

Trapezoidal: $\int_0^2 x^3\, dx \approx \frac{2-0}{2(8)}\left[(0)^3 + 2\left(\frac{1}{4}\right)^3 + 2\left(\frac{1}{2}\right)^3 + 2\left(\frac{3}{4}\right)^3 + 2(1)^3 + 2\left(\frac{5}{4}\right)^3 + 2\left(\frac{3}{2}\right)^3 + 2\left(\frac{7}{4}\right)^3 + (2)^3\right] = 4.0625$

Midpoint: 3.9688

The Midpoint Rule is a better approximation in this example.

23. $\int_{0}^{2} \frac{1}{x+1} \, dx \approx \frac{2-0}{2(4)} \left[1 + 2\left(\frac{2}{3}\right) + 2\left(\frac{1}{2}\right) + 2\left(\frac{2}{5}\right) + \frac{1}{3} \right] \approx 1.1167$

(Exact answer is $\ln 3 \approx 1.0986$.)

25. $\int_{-1}^{1} \frac{1}{x^2+1} \, dx \approx \frac{1-(-1)}{2(4)} \left[\frac{1}{(-1)^2} + 1 + \frac{2}{(-1/2)^2+1} + \frac{2}{(0)^2+1} + \frac{2}{(1/2)^2+1} + \frac{1}{(1)^2+1} \right] = 1.55$

(Exact value is $\pi/2$.)

27. $\int_{0}^{4} \sqrt{2+3x^2} \, dx$

n	Midpoint Rule	Trapezoidal Rule
4	15.3965	15.6055
8	15.4480	15.5010
12	15.4578	15.4814
16	15.4613	15.4745
20	15.4628	15.4713

29. $A = \int_{0}^{3} \sqrt{\frac{x^3}{4-x}} \, dx$

$\approx \frac{3-0}{2(10)} \left[0 + 2f\left(\frac{3}{10}\right) + 2f\left(\frac{3}{5}\right) + 2f\left(\frac{9}{10}\right) + 2f\left(\frac{6}{5}\right) + 2f\left(\frac{3}{2}\right) + 2f\left(\frac{9}{5}\right) + 2f\left(\frac{21}{10}\right) + 2f\left(\frac{12}{5}\right) + 2f\left(\frac{27}{10}\right) + f(3) \right]$

≈ 4.8103

31. $s = \int_{0}^{20} v \, dt$

$\approx \frac{20-0}{2(4)} [v(0) + 2v(5) + 2v(10) + 2v(15) + v(20)]$

$\approx \frac{5}{2} [0 + 58.6 + 102.6 + 132 + 73.3]$

$= \frac{5}{2}(366.5)$

$= 916.25$ feet

33. Midpoint Rule: 3.1468

Trapezoidal Rule: 3.1312

Graphing utility: 3.141593

Section 5.7   Volumes of Solids of Revolution

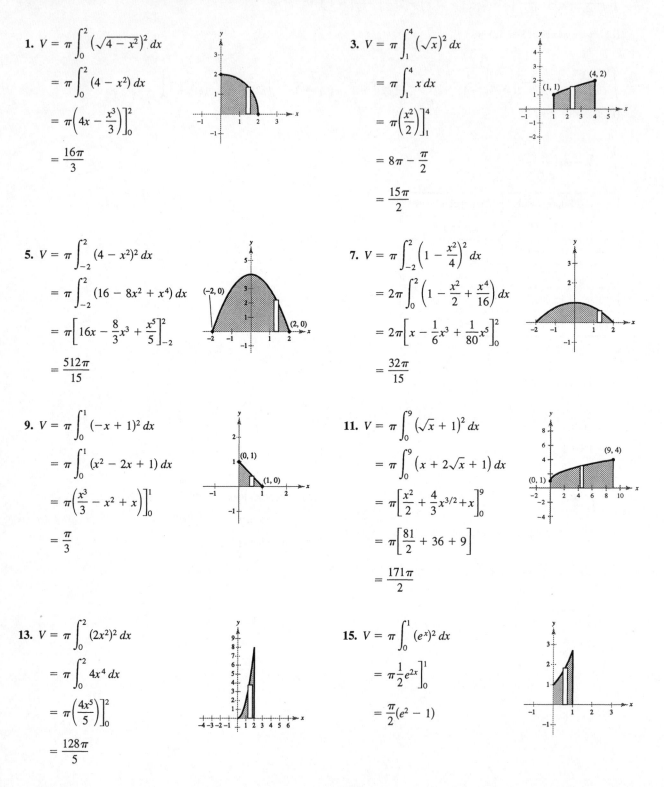

1. $V = \pi \int_0^2 \left(\sqrt{4-x^2}\right)^2 dx$

$= \pi \int_0^2 (4-x^2)\, dx$

$= \pi \left(4x - \dfrac{x^3}{3}\right)\Big]_0^2$

$= \dfrac{16\pi}{3}$

3. $V = \pi \int_1^4 \left(\sqrt{x}\right)^2 dx$

$= \pi \int_1^4 x\, dx$

$= \pi \left(\dfrac{x^2}{2}\right)\Big]_1^4$

$= 8\pi - \dfrac{\pi}{2}$

$= \dfrac{15\pi}{2}$

5. $V = \pi \int_{-2}^2 (4-x^2)^2\, dx$

$= \pi \int_{-2}^2 (16 - 8x^2 + x^4)\, dx$

$= \pi \left[16x - \dfrac{8}{3}x^3 + \dfrac{x^5}{5}\right]_{-2}^2$

$= \dfrac{512\pi}{15}$

7. $V = \pi \int_{-2}^2 \left(1 - \dfrac{x^2}{4}\right)^2 dx$

$= 2\pi \int_0^2 \left(1 - \dfrac{x^2}{2} + \dfrac{x^4}{16}\right) dx$

$= 2\pi \left[x - \dfrac{1}{6}x^3 + \dfrac{1}{80}x^5\right]_0^2$

$= \dfrac{32\pi}{15}$

9. $V = \pi \int_0^1 (-x+1)^2\, dx$

$= \pi \int_0^1 (x^2 - 2x + 1)\, dx$

$= \pi \left(\dfrac{x^3}{3} - x^2 + x\right)\Big]_0^1$

$= \dfrac{\pi}{3}$

11. $V = \pi \int_0^9 \left(\sqrt{x} + 1\right)^2 dx$

$= \pi \int_0^9 \left(x + 2\sqrt{x} + 1\right) dx$

$= \pi \left[\dfrac{x^2}{2} + \dfrac{4}{3}x^{3/2} + x\right]_0^9$

$= \pi \left[\dfrac{81}{2} + 36 + 9\right]$

$= \dfrac{171\pi}{2}$

13. $V = \pi \int_0^2 (2x^2)^2\, dx$

$= \pi \int_0^2 4x^4\, dx$

$= \pi \left(\dfrac{4x^5}{5}\right)\Big]_0^2$

$= \dfrac{128\pi}{5}$

15. $V = \pi \int_0^1 (e^x)^2\, dx$

$= \pi \dfrac{1}{2}e^{2x}\Big]_0^1$

$= \dfrac{\pi}{2}(e^2 - 1)$

17. The points of intersection of the three graphs occur when $y = 0$ and $y = 4$.

$$V = \pi \int_0^4 \left(\sqrt{y}\right)^2 dy$$

$$= \pi \int_0^4 y\, dy$$

$$= \pi \left(\frac{y^2}{2}\right)\Big]_0^4$$

$$= 8\pi$$

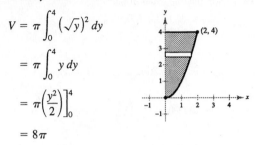

19. $V = \pi \int_0^2 \left(1 - \frac{1}{2}y\right)^2 dy$

$$= \pi \int_0^2 \left[1 - y + \frac{1}{4}y^2\right] dy$$

$$= \pi \left[y - \frac{y^2}{2} + \frac{1}{12}y^3\right]_0^2$$

$$= \pi \left[2 - 2 + \frac{1}{12} \cdot 8\right]$$

$$= \frac{2}{3}\pi$$

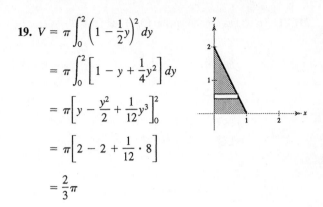

21. The points of intersection of the three graphs occur when $y = 0$ and $y = 1$.

$$V = \pi \int_0^1 (y^{3/2})^2 dy$$

$$= \pi \int_0^1 y^3 \, dy$$

$$= \pi \left(\frac{y^4}{4}\right)\Big]_0^1$$

$$= \frac{\pi}{4}$$

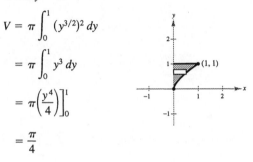

23. The points of intersection of the three graphs occur when $y = 0$ and $y = 2$.

$$V = \pi \int_0^2 (4 - y^2)^2 dy$$

$$= \pi \int_0^2 (16 - 8y^2 + y^4) \, dy$$

$$= \pi \left[16y - \frac{8y^3}{3} + \frac{y^5}{5}\right]_0^2$$

$$= \pi \left[32 - \frac{64}{3} + \frac{32}{5}\right]$$

$$= \frac{256\pi}{15}$$

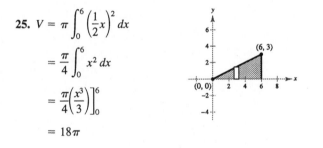

25. $V = \pi \int_0^6 \left(\frac{1}{2}x\right)^2 dx$

$$= \frac{\pi}{4} \int_0^6 x^2 \, dx$$

$$= \frac{\pi}{4}\left(\frac{x^3}{3}\right)\Big]_0^6$$

$$= 18\pi$$

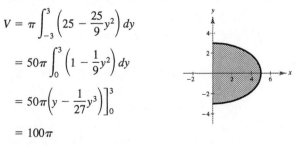

27. A right circular cone can be formed be revolving the region bounded by $y = (r/h)x$, $x = h$, and $y = 0$ about the x-axis.

$$V = \pi \int_0^h \left(\frac{r}{h}x\right)^2 dx$$

$$= \pi \left(\frac{r^2}{h^2} \cdot \frac{x^3}{3}\right)\Big]_0^h$$

$$= \pi \frac{r^2}{h^2} \cdot \frac{h^3}{3}$$

$$= \frac{1}{3}\pi r^2 h$$

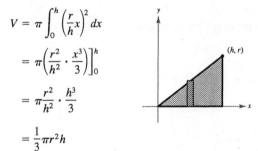

29. The right half of the ellipse is given by $x = \sqrt{25 - (25/9)y^2}$.

$$V = \pi \int_{-3}^3 \left(25 - \frac{25}{9}y^2\right) dy$$

$$= 50\pi \int_0^3 \left(1 - \frac{1}{9}y^2\right) dy$$

$$= 50\pi\left(y - \frac{1}{27}y^3\right)\Big]_0^3$$

$$= 100\pi$$

31. The graph has *x*-intercepts at $(0, 0)$ and $(2, 0)$.

$$V = \pi \int_0^2 \left[\frac{1}{8}x^2\sqrt{2-x}\right]^2 dx$$

$$= \frac{\pi}{64}\int_0^2 (2x^4 - x^5)\,dx$$

$$= \frac{\pi}{64}\left[\frac{2x^5}{5} - \frac{x^6}{6}\right]_0^2$$

$$= \frac{\pi}{30}$$

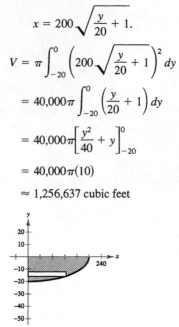

33. (a) $y = 20[(0.005x)^2 - 1]$

Solving for *x*, we obtain

$$x = 200\sqrt{\frac{y}{20} + 1}.$$

$$V = \pi \int_{-20}^0 \left(200\sqrt{\frac{y}{20} + 1}\right)^2 dy$$

$$= 40,000\pi \int_{-20}^0 \left(\frac{y}{20} + 1\right) dy$$

$$= 40,000\pi \left[\frac{y^2}{40} + y\right]_{-20}^0$$

$$= 40,000\pi(10)$$

$$\approx 1,256,637 \text{ cubic feet}$$

(b) The maximum number of fish that can be supported is

$$\frac{1,256,637}{500} \approx 2513 \text{ fish.}$$

35. Using the program on page 366 in the text, and $n = 100$,

$$V = \pi \int_0^7 \left(\sqrt[3]{x+1}\right)^2 dx = \pi \int_0^7 (x+1)^{2/3}\,dx \approx 58.434 \text{ cubic units.}$$

Review Exercises for Chapter 5

1. $\displaystyle\int 16\,dx = 16x + C$

3. $\displaystyle\int (2x^2 + 5x)\,dx = \frac{2}{3}x^3 + \frac{5}{2}x^2 + C$

5. $\displaystyle\int \frac{2}{3\sqrt[3]{x}}\,dx = \int \frac{2}{3}x^{-1/3}\,dx = x^{2/3} + C$

7. $\displaystyle\int \left(\sqrt[3]{x^4} + 3x\right)dx = \int (x^{4/3} + 3x)\,dx$

$$= \frac{3}{7}x^{7/3} + \frac{3}{2}x^2 + C$$

9. $\displaystyle\int \frac{2x^4 - 1}{\sqrt{x}}\,dx = \int (2x^{7/2} - x^{-1/2})\,dx$

$$= 2\frac{x^{9/2}}{\frac{9}{2}} - \frac{x^{1/2}}{\frac{1}{2}} + C$$

$$= \frac{4}{9}x^{9/2} - 2x^{1/2} + C$$

11. $\displaystyle f(x) = \int (3x + 1)\,dx = \frac{3}{2}x^2 + x + C$

$$f(2) = 6 = \frac{3}{2}(2)^2 + 2C = 8 + C \Rightarrow C = -2$$

Hence, $f(x) = \frac{3}{2}x^2 + x - 2$.

13. $f'(x) = \int 2x^2 \, dx = \frac{2}{3}x^3 + C$

$f'(3) = 10 = 18 + C_1 \Rightarrow C_1 = -8$

$f'(x) = \frac{2}{3}x^3 - 8$

$f(x) = \int \left(\frac{2}{3}x^3 - 8\right) dx = \frac{1}{6}x^4 - 8x + C_2$

$f(3) = 6 = \frac{81}{6} - 24 + C_2 \Rightarrow C_2 = 30 - \frac{81}{6} = \frac{33}{2}$

$f(x) = \frac{1}{6}x^4 - 8x + \frac{33}{2}$

15. (a) $s(t) = -16t^2 + 80t$

$v(t) = -32t + 80 = 0$

$32t = 80$

$t = 2.5$ seconds

(b) $s(2.5) = 100$ feet

(c) $v(t) = -32t + 80 = 40$

$32t = 40$

$t = 1.25$ seconds

(d) $s(1.25) = 75$ feet

17. $\int (1 + 5x)^2 \, dx = \int (1 + 10x + 25x^2) \, dx$

$= x + 5x^2 + \frac{25}{3}x^3 + C$

or

$\int (1 + 5x)^2 \, dx = \frac{1}{5}\frac{(1 + 5x)^3}{3} + C_1$

$= \frac{1}{15}(1 + 5x)^3 + C_1$

19. $\int \frac{1}{\sqrt{5x - 1}} \, dx = \frac{1}{5}\int (5x - 1)^{-1/2}(5) \, dx$

$= \frac{1}{5}(2)(5x - 1)^{1/2} + C$

$= \frac{2}{5}\sqrt{5x - 1} + C$

21. $\int x(1 - 4x^2) \, dx = \int (x - 4x^3) \, dx$

$= \frac{x^2}{2} - x^4 + C$

23. $\int (x^4 - 2x)(2x^3 - 1) \, dx = \frac{1}{2}\int (x^4 - 2x)(4x^3 - 2) \, dx$

$= \frac{1}{2} \cdot \frac{1}{2}(x^4 - 2x)^2 + C$

$= \frac{1}{4}(x^4 - 2x)^2 + C$

25. $P = \int 2t(0.001t^2 + 0.5)^{1/4} \, dt = 800(0.001t^2 + 0.5)^{5/4} + C$

$P(0) = 0 = 800(0.5)^{5/4} + C \Rightarrow C \approx -336.36$

$P = 800(0.001t^2 + 0.5)^{5/4} - 336.36$

(a) $P(6) \approx 30.5$ board feet

(b) $P(12) \approx 125.2$ board feet

27. $\int 3e^{-3x} \, dx = -e^{-3x} + C$

29. $\int (x - 1)e^{x^2 - 2x} \, dx = \frac{1}{2}e^{x^2 - 2x} + C$

31. $\int \frac{x^2}{1 - x^3} \, dx = -\frac{1}{3}\int \frac{1}{1 - x^3}(-3x^2) \, dx$

$= -\frac{1}{3}\ln|1 - x^3| + C$

33. $\int \frac{\left(\sqrt{x} + 1\right)^2}{\sqrt{x}} \, dx = \int \frac{x + 2\sqrt{x} + 1}{\sqrt{x}} \, dx$

$= \int (x^{1/2} + 2 + x^{-1/2}) \, dx$

$= \frac{2}{3}x^{3/2} + 2x + 2x^{1/2} + C$

35. $A = \int_0^2 (4 - 2x)\, dx$

$= \left[4x - x^2 \right]_0^2$

$= 4$

37. $A = \int_0^2 (y - 2)^2\, dy$

$= \left[\frac{y^3}{3} - 2y^2 + 4y - \frac{8}{3} \right]_0^2$

$= \frac{8}{3}$

39. $A = \int_0^1 \frac{2}{x + 1}\, dx$

$= 2 \ln(x + 1) \Big]_0^1$

$= 2 \ln 2$

41. $\int_0^4 (2 + x)\, dx = \left[2x + \frac{x^2}{2} \right]_0^4$

$= 16$

43. $\int_{-1}^1 (4t^3 - 2t)\, dt = \left[t^4 - t^2 \right]_{-1}^1$

$= 0$

45. $\int_0^3 \frac{1}{\sqrt{1 + x}}\, dx = \int_0^3 (1 + x)^{-1/2}\, dx$

$= 2\sqrt{1 + x} \Big]_0^3$

$= 2$

47. $\int_1^2 \left(\frac{1}{x^2} - \frac{1}{x^3} \right) dx = \int_1^2 (x^{-2} - x^{-3})\, dx$

$= \left[\frac{x^{-1}}{-1} - \frac{x^{-2}}{-2} \right]_1^2$

$= \left[-\frac{1}{x} + \frac{1}{2x^2} \right]_1^2$

$= -\frac{3}{8} - \left(-\frac{1}{2} \right)$

$= \frac{1}{8}$

49. $\int_1^3 \frac{3 + \ln x}{x}\, dx = \frac{1}{2}(3 + \ln x)^2 \Big]_1^3$

$= \frac{1}{2}[(3 + \ln 3)^2 - 3^2]$

$= \frac{1}{2}[6 \ln 3 + (\ln 3)^2]$

≈ 3.899

51. $\int_{-1}^1 3xe^{x^2 - 1}\, dx = \frac{3}{2} e^{x^2 - 1} \Big]_{-1}^1$

$= \frac{3}{2}(1 - 1)$

$= 0$

53. $C = \int (675 + 0.5x)\, dx = 675x + \frac{1}{4}x^2 + C_1$

$C(51) - C(50) = \$700.25$

55. Average value $= \frac{1}{10 - 5} \int_5^{10} \frac{4}{\sqrt{x - 1}}\, dx$

$= \frac{4}{5} \int_5^{10} (x - 1)^{-1/2}\, dx$

$= \frac{8}{5}(x - 1)^{1/2} \Big]_5^{10}$

$= \frac{8}{5}(3 - 2) = \frac{8}{5}$

To find the values for which $f(x) = \frac{8}{5}$, solve for x in the equation.

$\frac{4}{\sqrt{x - 1}} = \frac{8}{5}$

$\frac{5}{2} = \sqrt{x - 1}$

$x - 1 = \frac{25}{4}$

$x = \frac{29}{4}$

57. Average value $= \dfrac{1}{5-2} \displaystyle\int_2^5 e^{5-x}\, dx$

$\qquad = \dfrac{1}{3}\left[-e^{5-x}\right]_2^5$

$\qquad = \dfrac{1}{3}(-1 + e^3) \approx 6.362$

To find the value of x for which $f(x) = \frac{1}{3}(-1 + e^3)$, we solve $e^{5-x} = \frac{1}{3}(-1 + e^3)$. Using a graphing utility, we obtain $x \approx 3.150$.

59. Average value $= \dfrac{1}{b-a} \displaystyle\int_a^b f(t)\, dt$

$\qquad = \dfrac{1}{2-0} \displaystyle\int_0^2 500e^{0.04t}\, dt$

$\qquad = 250\left[\dfrac{e^{0.04t}}{0.04}\right]_0^2$

$\qquad = 6250[e^{0.08} - 1]$

$\qquad \approx \$520.54$

61. (a) $B = \displaystyle\int (-0.0486t + 0.564)\, dt$

$\qquad = -0.0243t^2 + 0.564t + C$

For $t = 10$, $1.63 = -0.0243(10)^2 + 0.564(10) + C$

$\qquad C = -1.58$

$\qquad B = -0.0243t^2 + 0.564t - 1.58$

(b) $2.00 = -0.0243t^2 + 0.564t - 1.58$

Using a graphing utility, you can see that the price never exceeds \$2.00. The maximum price is \$1.69 at $t = 11$ (2001).

63. Amount $= e^{rt} \displaystyle\int_0^T c(t)e^{-rt}\, dt$

$\qquad = e^{0.06(5)} \displaystyle\int_0^5 3000e^{-0.06t}\, dt$

$\qquad = e^{0.3}(-50{,}000e^{-0.06t})\Big]_0^5$

$\qquad \approx \$17{,}492.94$

65. $\displaystyle\int_{-2}^2 6x^5\, dx = 0$ (odd function)

67. $\displaystyle\int_{-2}^{-1} \dfrac{4}{x^2}\, dx = \displaystyle\int_1^2 \dfrac{4}{x^2}\, dx = 2$ (symmetric about y-axis)

69. Area $= \displaystyle\int_{1/2}^3 \left(4 - \dfrac{1}{x^2}\right) dx$

$\qquad = \left[4x + \dfrac{1}{x}\right]_{1/2}^3$

$\qquad = \left(12 + \dfrac{1}{3}\right) - (2 + 2)$

$\qquad = \dfrac{25}{3}$

71. $A = \displaystyle\int_0^8 \dfrac{4}{\sqrt{x+1}}\, dx$

$\qquad = 4 \displaystyle\int_0^8 (x+1)^{-1/2}\, dx$

$\qquad = \left[4(2)(x+1)^{1/2}\right]_0^8$

$\qquad = \left[8\sqrt{x+1}\right]_0^8$

$\qquad = 8(3-1)$

$\qquad = 16$

73. $(x-3)^2 = 8 - (x-3)^2$

$2(x-3)^2 = 8$

$(x-3)^2 = 4$

$x - 3 = \pm 2$

$x = 1, 5$

$A = \int_1^5 \{[8 - (x-3)^2] - (x-3)^2\}\, dx$

$= \int_1^5 [8 - 2(x-3)^2]\, dx$

$= \left[8x - \dfrac{2}{3}(x-3)^3\right]_1^5$

$= \left(40 - \dfrac{16}{3}\right) - \left(8 + \dfrac{16}{3}\right)$

$= \dfrac{64}{3}$

75. $x = 2 - x^2$

$x^2 + x - 2 = 0$

$(x+2)(x-1) = 0$

$x = -2, 1$

$A = \int_{-2}^1 (2 - x^2 - x)\, dx$

$= \left[2x - \dfrac{x^3}{3} - \dfrac{x^2}{2}\right]_{-2}^1$

$= \dfrac{9}{2}$

77. Demand function $=$ Supply function

$500 - x = 1.25x + 162.5$

$2.25x = 337.5$

$x = 150 \Longrightarrow \text{price} = 350$

Consumer surplus $= \displaystyle\int_0^{150} (\text{demand function} - \text{price})\, dx$

$= \displaystyle\int_0^{150} [(500 - x) - 350]\, dx$

$= \left[150x - \dfrac{x^2}{2}\right]_0^{150}$

$= 11{,}250$

Producer surplus $= \displaystyle\int_0^{150} (\text{price} - \text{supply function})\, dx$

$= \displaystyle\int_0^{150} [350 - (1.25x + 162.5)]\, dx$

$= \left[187.5x - 1.25\dfrac{x^2}{2}\right]_0^{150}$

$= 14{,}062.5$

79. Notice that the models approximately agree at $t = 9$.

$\displaystyle\int_9^{13} (S_1 - S_2)\, dt \approx -5511$

\$5511 million less sales.

81. (a)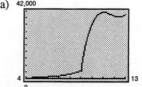

(b) The revenue would have decreased.

(c) $\displaystyle\int_9^{13} [R_2(t) - R_1(t)]\, dt \approx \$85{,}834$ million

83. $\int_0^2 (x^2 + 1)^2 \, dx$

$n = 4$: 13.3203

$n = 20$: 13.7167 (exact 13.7$\overline{3}$)

85. $\int_0^1 \dfrac{1}{x^2 + 1} \, dx$

$n = 4$: 0.7867

$n = 20$: 0.7855

87. $V = \pi \int_1^4 \left(\dfrac{1}{\sqrt{x}}\right)^2 dx$

$= \pi \int_1^4 \dfrac{1}{x} \, dx$

$= \pi \ln x \Big]_1^4$

$= \pi \ln 4$

≈ 4.355

89. $V = \pi \int_0^2 (e^{1-x})^2 \, dx$

$= \pi \int_0^2 e^{2-2x} \, dx$

$= -\dfrac{\pi}{2}\left[e^{2-2x}\right]_0^2$

$= -\dfrac{\pi}{2}[e^{-2} - e^2]$

$= \dfrac{\pi}{2}(e^2 - e^{-2})$

≈ 11.394

91. $V = \pi \int_0^2 \left[(2x + 1)^2 - 1^2\right] dx$

$= \pi \int_0^2 (4x^2 + 4x) \, dx$

$= \pi \left[\dfrac{4}{3}x^3 + 2x^2\right]_0^2$

$= \pi \left(\dfrac{32}{3} + 8\right)$

$= \dfrac{56}{3}\pi$

93. $V = \pi \int_0^1 \left[(x^2)^2 - (x^3)^2\right] dx$

$= \pi \left[\dfrac{x^5}{5} - \dfrac{x^7}{7}\right]_0^1$

$= \dfrac{2}{35}\pi$

95. $x^2 + y^2 = 1$, $y = \dfrac{1}{4}$

$x = \sqrt{1 - \left(\dfrac{1}{4}\right)^2} = \dfrac{\sqrt{15}}{4}$

$V = \pi \int_{-\sqrt{15}/4}^{\sqrt{15}/4} \left[\left(\sqrt{1 - x^2}\right)^2 - \left(\dfrac{1}{4}\right)^2\right] dx$

$= 2\pi \int_0^{\sqrt{15}/4} \left(1 - x^2 - \dfrac{1}{16}\right) dx$

$= \dfrac{5}{16}\pi\sqrt{15}$

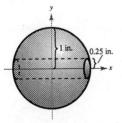

Practice Test for Chapter 5

1. Evaluate $\int (3x^2 - 8x + 5)\, dx$.

2. Evaluate $\int (x + 7)(x^2 - 4)\, dx$.

3. Evaluate $\int \dfrac{x^3 - 9x^2 + 1}{x^2}\, dx$.

4. Evaluate $\int x^3 \sqrt[4]{1 - x^4}\, dx$.

5. Evaluate $\int \dfrac{3}{\sqrt[3]{7x}}\, dx$.

6. Evaluate $\int \sqrt{6 - 11x}\, dx$.

7. Evaluate $\int \left(\sqrt[4]{x} + \sqrt[6]{x} \right) dx$.

8. Evaluate $\int \left(\dfrac{1}{x^4} - \dfrac{1}{x^5} \right) dx$.

9. Evaluate $\int (1 - x^2)^3\, dx$.

10. Evaluate $\int \dfrac{5x}{(1 + 3x^2)^3}\, dx$.

11. Evaluate $\int e^{7x}\, dx$.

12. Evaluate $\int x e^{4x^2}\, dx$.

13. Evaluate $\int e^x (1 + 4e^x)^3\, dx$.

14. Evaluate $\int (e^x + 2)^2\, dx$.

15. Evaluate $\int \dfrac{e^{3x} - 4e^x + 1}{e^x}\, dx$.

16. Evaluate $\int \dfrac{1}{x + 6}\, dx$.

17. Evaluate $\int \dfrac{x^2}{8 - x^3}\, dx$.

18. Evaluate $\int \dfrac{e^x}{1 + 3e^x}\, dx$.

19. Evaluate $\int \dfrac{(\ln x)^6}{x}\, dx$.

20. Evaluate $\int \dfrac{x^2 + 5}{x - 1}\, dx$.

21. Evaluate $\int_0^3 (x^2 - 4x + 2)\, dx$.

22. Evaluate $\int_1^8 x \sqrt[3]{x}\, dx$.

23. Evaluate $\int_{\sqrt{5}}^{\sqrt{13}} \dfrac{x}{\sqrt{x^2 - 4}}\, dx$.

24. Sketch the region bounded by the graphs of $f(x) = x^2 - 6x$ and $g(x) = 0$ and find the area of the region.

25. Sketch the region bounded by the graphs of $f(x) = x^3 + 1$ and $g(x) = x + 1$ and find the area of the region.

26. Sketch the region bounded by the graphs of $f(y) = 1/y^2, x = 0, y = 1$, and $y = 3$ and find the area of the region.

27. Approximate the definite integral by the Midpoint Rule using $n = 4$.

$$\int_0^1 \sqrt{x^3 + 2}\, dx$$

28. Approximate the definite integral by the Midpoint Rule using $n = 4$.

$$\int_3^4 \frac{1}{x^2 - 5}\, dx$$

29. Find the volume of the solid generated by revolving the region bounded by the graphs of $f(x) = 1/\sqrt[3]{x}$, $x = 1$, $x = 8$, and $y = 0$ about the x-axis.

30. Find the volume of the solid generated by revolving the region bounded by the graphs of $y = \sqrt{25 - x}$, $y = 0$, and $x = 0$ about the y-axis.

Graphing Calculator Required

31. Use a program similar to that on page 366 of the textbook to approximate the following integral for $n = 50$ and $n = 100$.

$$\int_0^4 \sqrt{1 + x^4}\, dx$$

32. Use a graphing calculator to sketch the region bounded by $f(x) = 3 - \sqrt{x}$ and $g(x) = 3 - \frac{1}{3}x$. Based on the graph alone (do no calculations), determine which value best approximates the bounded area.

(a) 13   (b) 3   (c) 5   (d) 6

CHAPTER 6
Techniques of Integration

CHAPTER 6
Techniques of Integration

Section 6.1   Integration by Substitution

Solutions to Odd-Numbered Exercises

1. $\displaystyle \int (x-2)^4 \, dx = \frac{(x-2)^5}{5} + C$

3. $\displaystyle \int \frac{2}{(t-9)^2} \, dt = 2 \int (t-9)^{-2} \, dt$

$$= (2)\frac{(t-9)^{-1}}{-1} + C$$

$$= -\frac{2}{t-9} + C$$

$$= \frac{2}{9-t} + C$$

5. $\displaystyle \int \frac{2t-1}{t^2-t+2} \, dt = \ln|t^2-t+2| + C$

7. $\displaystyle \int \sqrt{1+x} \, dx = \int (1+x)^{1/2} \, dx$

$$= \frac{2}{3}(1+x)^{3/2} + C$$

9. $\displaystyle \int \frac{12x+2}{3x^2+x} \, dx = 2 \int \frac{6x+1}{3x^2+x} \, dx$

$$= 2 \ln|3x^2+x| + C$$

$$= \ln(3x^2+x)^2 + C$$

11. $\displaystyle \int \frac{1}{(5x+1)^3} \, dx = \frac{1}{5} \int (5x+1)^{-3}(5) \, dx$

$$= \left(\frac{1}{5}\right)\frac{(5x+1)^{-2}}{-2} + C$$

$$= -\frac{1}{10(5x+1)^2} + C$$

13. $\displaystyle \int \frac{1}{\sqrt{x+1}} \, dx = \int (x+1)^{-1/2} \, dx$

$$= 2(x+1)^{1/2} + C$$

15. $\displaystyle \int \frac{e^{3x}}{1-e^{3x}} \, dx = -\frac{1}{3} \int \frac{-3e^{3x}}{1-e^{3x}} \, dx$

$$= -\frac{1}{3} \ln|1-e^{3x}| + C$$

17. $\displaystyle \int \frac{2x}{e^{3x^2}} \, dx = \int 2xe^{-3x^2} \, dx$

$$= -\frac{1}{3} \int e^{-3x^2} \, (-6x) \, dx$$

$$= -\frac{1}{3} e^{-3x^2} + C$$

$$= \frac{-1}{3e^{3x^2}} + C$$

19. $\displaystyle \int \frac{x^2}{x-1} \, dx = \int \left(x+1+\frac{1}{x-1}\right) dx$

$$= \frac{x^2}{2} + x + \ln|x-1| + C$$

21. $\displaystyle \int x\sqrt{x^2+4} \, dx = \frac{1}{2} \int (x^2+4)^{1/2} 2x \, dx$

$$= \frac{1}{3}(x^2+4)^{3/2} + C$$

23. $\displaystyle \int e^{5x} \, dx = \frac{1}{5} \int e^{5x}(5) \, dx = \frac{1}{5}e^{5x} + C$

25. $\displaystyle\int \frac{e^{-x}}{e^{-x} + 2} = -\int \frac{1}{e^{-x} + 2} (-e^{-x})\, dx$

$\qquad\qquad = -\ln(e^{-x} + 2) + C$

27. Let $u = x + 1, du = dx, x = u - 1.$

$\displaystyle\int \frac{x}{(x + 1)^4}\, dx = \int \frac{u - 1}{u^4}\, du$

$\qquad\qquad = \int u^{-3}\, du - \int u^{-4}\, du$

$\qquad\qquad = \dfrac{u^{-2}}{-2} - \dfrac{u^{-3}}{-3} + C$

$\qquad\qquad = \dfrac{-1}{2u^2} + \dfrac{1}{3u^3} + C$

$\qquad\qquad = \dfrac{-1}{2(x + 1)^2} + \dfrac{1}{3(x + 1)^3} + C$

29. Let $u = 3x - 1$, then $x = (u + 1)/3$ and $dx = (1/3)\, du.$

$\displaystyle\int \frac{x}{(3x - 1)^2}\, dx = \int \frac{(u + 1)/3}{u^2}\left(\frac{1}{3}\right) du$

$\qquad\qquad = \dfrac{1}{9} \int \left(\dfrac{1}{u} + \dfrac{1}{u^2}\right) du$

$\qquad\qquad = \dfrac{1}{9}\left[\ln|u| - \dfrac{1}{u}\right] + C$

$\qquad\qquad = \dfrac{1}{9}\left[\ln|3x - 1| - \dfrac{1}{3x - 1}\right] + C$

31. Let $u = \sqrt{t} - 1, t = (u + 1)^2,$ and $dt = 2(u + 1)\, du.$

$\displaystyle\int \frac{1}{\sqrt{t} - 1}\, dt = \int \frac{2(u + 1)}{u}\, du$

$\qquad\qquad = \int \left(2 + \dfrac{2}{u}\right) du$

$\qquad\qquad = 2u + 2\ln|u| + C$

$\qquad\qquad = 2\left(\sqrt{t} - 1\right) + 2\ln\left|\sqrt{t} - 1\right| + C$

33. $\displaystyle\int \frac{2\sqrt{t} + 1}{t}\, dt = \int \left(2t^{-1/2} + \frac{1}{t}\right) dt$

$\qquad\qquad = 4t^{1/2} + \ln|t| + C$

$\qquad\qquad = 4\sqrt{t} + \ln|t| + C$

35. Let $u = 2x + 1$ which produces $du = 2\, dx$ and $x = \frac{1}{2}(u - 1).$

$\displaystyle\int \frac{x}{\sqrt{2x + 1}}\, dx = \int \frac{\frac{1}{2}(u - 1)}{\sqrt{u}} \frac{du}{2}$

$\qquad\qquad = \dfrac{1}{4} \int (u^{1/2} - u^{-1/2})\, du$

$\qquad\qquad = \dfrac{1}{4}\left[\dfrac{2}{3} u^{3/2} - 2u^{1/2}\right] + C$

$\qquad\qquad = \dfrac{1}{6}(2x + 1)^{3/2} - \dfrac{1}{2}(2x + 1)^{1/2} + C$

$\qquad\qquad = \dfrac{1}{3}\sqrt{2x + 1}\,(x - 1) + C$

37. Let $u = 1 - t$ which produces $du = -dt$ and $t = 1 - u.$

$\displaystyle\int t^2\sqrt{1 - t}\, dt = \int (1 - u)^2 \sqrt{u}(-du)$

$\qquad\qquad = \int (-u^{1/2} + 2u^{3/2} - u^{5/2})\, du$

$\qquad\qquad = -\dfrac{2}{3} u^{3/2} + \dfrac{4}{5} u^{5/2} - \dfrac{2}{7} u^{7/2} + C$

$\qquad\qquad = -\dfrac{2}{3}(1 - t)^{3/2} + \dfrac{4}{5}(1 - t)^{5/2} - \dfrac{2}{7}(1 - t)^{7/2} + C$

$\qquad\qquad = \dfrac{-2}{105}(15t^2 + 12t + 8)(1 - t)^{3/2} + C$

39. $\displaystyle\int_0^4 \sqrt{2x + 1}\, dx = \frac{1}{2} \int_0^4 (2x + 1)^{1/2}(2)\, dx$

$\qquad\qquad = \dfrac{1}{2}\left(\dfrac{2}{3}\right)(2x + 1)^{3/2} \Big]_0^4$

$\qquad\qquad = \dfrac{1}{3}(9)^{3/2} - \dfrac{1}{3}(1)^{3/2}$

$\qquad\qquad = \dfrac{26}{3}$

41. $\displaystyle\int_0^1 3xe^{x^2}\, dx = \frac{3}{2} \int_0^1 2xe^{x^2}\, dx$

$\qquad\qquad = \dfrac{3}{2} e^{x^2} \Big]_0^1$

$\qquad\qquad = \dfrac{3}{2}(e - 1)$

$\qquad\qquad \approx 2.577$

43. Let $u = x + 4$, $x = u - 4$, and $du = dx$. $u = 4$ when $x = 0$, $u = 8$ when $x = 4$.

$$\int_0^4 \frac{x}{(x+4)^2}\, dx = \int_4^8 \frac{u - 4}{u^2}\, du$$

$$= \int_4^8 \left(\frac{1}{u} - 4u^{-2} \right) du$$

$$= \left[\ln u + 4\frac{1}{u} \right]_4^8$$

$$= \left(\ln 8 + \frac{1}{2} \right) - (\ln 4 + 1)$$

$$= \ln 2 - \frac{1}{2}$$

$$\approx 0.193$$

45. Let $u = 1 - x$, then $x = 1 - u$ and $dx = -du$. $u = 1$ when $x = 0$, and $u = 0.5$ when $x = 0.5$.

$$\int_0^{0.5} x(1 - x)^3\, dx = \int_1^{0.5} (1 - u)u^3(-du)$$

$$= \int_1^{0.5} (u^4 - u^3)\, du$$

$$= \left(\frac{u^5}{5} - \frac{u^4}{4} \right) \Big]_1^{0.5}$$

$$= \left(\frac{1}{160} - \frac{1}{64} \right) - \left(\frac{1}{5} - \frac{1}{4} \right)$$

$$= \frac{13}{320}$$

47. Let $u = \sqrt{x - 3}$, then $x = u^2 + 3$ and $dx = 2u\, du$. $u = 0$ when $x = 3$, and $u = 2$ when $x = 7$.

$$\int_3^7 x\sqrt{x - 3}\, dx = \int_0^2 (u^2 + 3)u(2u\, du)$$

$$= 2\int_0^2 (u^4 + 3u^2)\, du$$

$$= 2\left(\frac{u^5}{5} + u^3 \right) \Big]_0^2$$

$$= 2\left(\frac{32}{5} + 8 \right)$$

$$= \frac{144}{5}$$

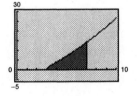

49. Let $u = \sqrt{1 - x}$, then $x = 1 - u^2$ and $dx = -2u\, du$. $u = 2$ when $x = -3$, and $u = 1$ when $x = 0$.

$$\text{Area} = \int_{-3}^0 x^2\sqrt{1 - x}\, dx$$

$$= \int_2^1 (1 - u^2)^2 u(-2u)\, du$$

$$= -2\int_2^1 (u^6 - 2u^4 + u^2)\, du$$

$$= -2\left[\frac{u^7}{7} - 2\frac{u^5}{5} + \frac{u^3}{3} \right]_2^1$$

$$= -2\left[\left(\frac{1}{7} - \frac{2}{5} + \frac{1}{3} \right) - \left(\frac{128}{7} - \frac{64}{5} + \frac{8}{3} \right) \right]$$

$$= -2\left(\frac{8}{105} - \frac{856}{105} \right)$$

$$= \frac{1696}{105}$$

$$\approx 16.1524$$

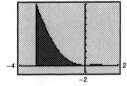

51. Let $u = \sqrt{2x - 1}$, then $x = \frac{1}{2}(u^2 + 1)$, and $dx = u\,du$. $u = 1$ when $x = 1$, and $u = 3$ when $x = 5$.

$$\text{Area} = \int_1^5 \frac{x^2 - 1}{\sqrt{2x - 1}}\,dx$$

$$= \int_1^3 \frac{\left(\dfrac{u^2 + 1}{2}\right)^2 - 1}{u}\,u\,du$$

$$= \frac{1}{4}\int_1^3 (u^4 + 2u^2 - 3)\,du$$

$$= \frac{1}{4}\left[\frac{u^5}{5} + \frac{2u^3}{3} - 3u\right]_1^3$$

$$= \frac{224}{15}$$

$$= 14.9\overline{3}$$

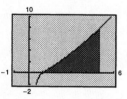

53. Let $u = \sqrt[3]{x + 1}$, then $x = u^3 - 1$ and $dx = 3u^2\,du$. $u = 1$ when $x = 0$, and $u = 2$ when $x = 7$.

$$\int_0^7 x\sqrt[3]{x + 1}\,dx = 3\int_1^2 (u^3 - 1)u^3\,du$$

$$= 3\int_1^2 (u^6 - u^3)\,du$$

$$= 3\left(\frac{u^7}{7} - \frac{u^4}{4}\right)\Big]_1^2$$

$$= 3\left(\frac{128}{7} - 4\right) - 3\left(\frac{1}{7} - \frac{1}{4}\right)$$

$$= \frac{1209}{28}$$

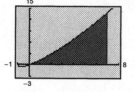

55. Let $u = x + 2$, $x = u - 2$, and $du = dx$. $u = 0$ when $x = -2$, and $u = 2$ when $x = 0$.

$$\int_{-2}^0 -x\sqrt{x + 2}\,dx = \int_0^2 -(u - 2)\sqrt{u}\,du$$

$$= \int_0^2 (-u^{3/2} + 2u^{1/2})\,du$$

$$= \left[-\frac{2}{5}u^{5/2} + \frac{4}{3}u^{3/2}\right]_0^2$$

$$= -\frac{2}{5}\cdot 2^{3/2} + \frac{4}{3}\cdot 2^{3/2}$$

$$= \sqrt{2}\left(\frac{8}{3} - \frac{8}{5}\right)$$

$$= \frac{16\sqrt{2}}{15}$$

57. Use $y = x\sqrt{1 - x^2}$ and $y = 0$ and multiply by 4.

$$A = 4\int_0^1 x\sqrt{1 - x^2}\,dx$$

$$= \frac{4}{-2}\int_0^1 (1 - x^2)^{1/2}(-2x)\,dx$$

$$= -2\left(\frac{2}{3}\right)(1 - x^2)^{3/2}\Big]_0^1$$

$$= -\frac{4}{3}(0 - 1)$$

$$= \frac{4}{3} \text{ square units}$$

59. $V = 2\pi \int_0^1 \left(x\sqrt{1-x^2}\right)^2 dx$

$= 2\pi \int_0^1 x^2(1-x^2)\, dx$

$= 2\pi \left[\dfrac{x^3}{3} - \dfrac{x^5}{5}\right]_0^1$

$= 2\pi \left(\dfrac{1}{3} - \dfrac{1}{5}\right)$

$= \dfrac{4\pi}{15}$ cubic units

61. $\dfrac{1}{1-0} \int_0^1 [f(x) - g(x)]\, dx = \int_0^1 \left[\dfrac{1}{x+1} - \dfrac{x}{(x+1)^2}\right] dx$

$= \int_0^1 \dfrac{1}{(x+1)^2}\, dx$

$= -\dfrac{1}{x+1}\Big]_0^1$

$= -\dfrac{1}{2} + 1$

$= \dfrac{1}{2}$

63. Let $u = \sqrt{1-x}$, then $x = 1 - u^2$ and $dx = -2u\, du$.

$\displaystyle \int \dfrac{15}{4} x\sqrt{1-x}\, dx = \dfrac{1}{2}(1-x)^{3/2}(-3x-2) + C$

(a) $P(0.40 \le x \le 0.80) = \dfrac{1}{2}(1-x)^{3/2}(-3x-2)\Big]_{0.40}^{0.80} = 0.547$

(b) $P(0 \le x \le b) = \dfrac{1}{2}(1-x)^{3/2}(-3x-2)\Big]_0^b = \dfrac{1}{2}\left[(1-b)^{3/2}(-3b-2) + 2\right] = 0.5$

Solving this equation for b produces

$(1-b)^{3/2}(-3b-2) + 2 = 1$

$(1-b)^{3/2}(-3b-2) = -1$

$(1-b)^{3/2}(3b+2) = 1$

$(1-b)^3(3b+2)^2 = 1$

$b \approx 0.586.$

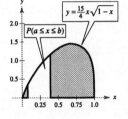

65. (a) $S(t) = t\sqrt{14-t}, \quad 0 \le t \le 14$

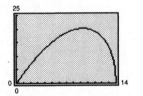

(c) Total snow fall is $14(13.97) \approx 195.6$ inches.

(b) Average $= \dfrac{1}{14-0} \int_0^{14} t\sqrt{14-t}\, dt$

Let $u = 14 - t$, $du = -dt$, $t = 14 - u$. When $t = 0$, $u = 14$ and when $t = 14$, $u = 0$.

Average $= \dfrac{1}{14} \int_{14}^0 (14-u)\sqrt{u}\,(-du)$

$= \dfrac{1}{14} \int_0^{14} (14u^{1/2} - u^{3/2})\, du$

$= \dfrac{1}{14}\left[\dfrac{28}{3}u^{3/2} - \dfrac{2}{5}u^{5/2}\right]_0^{14}$

$= \dfrac{1}{14}\left[\dfrac{28}{3}(14)^{3/2} - \dfrac{2}{5}(14)^{5/2}\right] \approx 13.97$ inches

67. $A = \displaystyle\int_0^4 \sqrt[3]{x}\sqrt{4-x}\, dx \approx \dfrac{4-0}{10}\left[f\!\left(\dfrac{1}{5}\right) + f\!\left(\dfrac{3}{5}\right) + f(1) + f\!\left(\dfrac{7}{5}\right) + f\!\left(\dfrac{9}{5}\right) + f\!\left(\dfrac{11}{5}\right) + f\!\left(\dfrac{13}{5}\right) + f(3) + f\!\left(\dfrac{17}{5}\right) + f\!\left(\dfrac{19}{5}\right)\right] \approx 5.885$

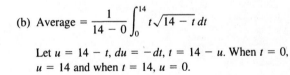

Section 6.2   Integration by Parts and Present Value

1. Let $u = x$ and $dv = e^{3x}\, dx$, then $du = dx$ and $v = \frac{1}{3}e^{3x}$.

$$\int xe^{3x}\, dx = \frac{1}{3}xe^{3x} - \int \frac{1}{3}e^{3x}\, dx$$

$$= \frac{1}{3}xe^{3x} - \frac{1}{9}e^{3x} + C$$

$$= \frac{1}{9}e^{3x}(3x - 1) + C$$

3. Let $u = x^2$ and $dv = e^{-x}\, dx$, then $du = 2x\, dx$ and $v = -e^{-x}$.

$$\int x^2 e^{-x}\, dx = -x^2 e^{-x} + 2\int xe^{-x}\, dx$$

Let $u = x$ and $dv = e^{-x}\, dx$, then $du = dx$ and $v = -e^{-x}$.

$$\int x^2 e^{-x}\, dx = -x^2 e^{-x} + 2\left[-xe^{-x} + \int e^{-x}\, dx \right]$$

$$= -x^2 e^{-x} - 2xe^{-x} - 2e^{-x} + C$$

$$= -e^{-x}(x^2 + 2x + 2) + C$$

5. Let $u = \ln 2x$ and $dv = dx$, then $du = (1/x)\, dx$ and $v = x$.

$$\int \ln 2x\, dx = x \ln 2x - \int x\left(\frac{1}{x}\right) dx$$

$$= x \ln 2x - \int dx$$

$$= x \ln 2x - x + C$$

$$= x(\ln 2x - 1) + C$$

7. $\displaystyle \int e^{4x}\, dx = \frac{1}{4}\int e^{4x}(4)\, dx = \frac{1}{4}e^{4x} + C$

9. Let $u = x$ and $dv = e^{4x}\, dx$, then $du = dx$ and $v = \frac{1}{4}e^{4x}$.

$$\int xe^{4x}\, dx = \frac{1}{4}xe^{4x} - \frac{1}{4}\int e^{4x}\, dx$$

$$= \frac{1}{4}xe^{4x} - \frac{1}{16}e^{4x} + C$$

$$= \frac{e^{4x}}{16}(4x - 1) + C$$

11. $\displaystyle \int xe^{x^2}\, dx = \frac{1}{2}e^{x^2} + C$

13. Let $u = x^2$ and $dv = e^x\, dx$, then $du = 2x\, dx$ and $v = e^x$.

$$\int x^2 e^x\, dx = x^2 e^x - \int 2xe^x\, dx$$

Let $u = 2x$ and $dv = e^x\, dx$, then $du = 2\, dx$ and $v = e^x$.

$$\int x^2 e^x\, dx = x^2 e^x - \left[2xe^x - \int e^x(2\, dx) \right]$$

$$= (x^2 - 2x + 2)e^x + C$$

15. Let $u = \ln(t + 1)$ and $dv = t\, dt$, then $du = [1/(t + 1)]\, dt$ and $v = t^2/2$.

$$\int t \ln(t + 1)\, dt = \frac{t^2}{2}\ln(t + 1) - \frac{1}{2}\int \frac{t^2}{t + 1}\, dt$$

$$= \frac{t^2}{2}\ln(t + 1) - \frac{1}{2}\int \left(t - 1 + \frac{1}{t + 1} \right) dt$$

$$= \frac{t^2}{2}\ln(t + 1) - \frac{1}{2}\left[\frac{t^2}{2} - t + \ln(t + 1) \right] + C$$

$$= \frac{1}{4}[2(t^2 - 1)\ln(t + 1) + t(2 - t)] + C$$

17. Let $u = 1/t$, $du = (-1/t^2)\, dt$.

$$\int \frac{e^{1/t}}{t^2}\, dt = -\int e^u\, du = -e^u + C = -e^{1/t} + C$$

19. Let $u = (\ln x)^2$ and $dv = x\, dx$, then $du = [(2 \ln x)/x]\, dx$ and $v = x^2/2$.

$$\int x(\ln x)^2\, dx = \frac{x^2}{2}(\ln x)^2 - \int x \ln x\, dx$$

Let $u = \ln x$ and $v = x\, dx$, then $du = (1/x)\, dx$ and $v = x^2/2$.

$$\int x(\ln x)^2\, dx = \frac{x^2}{2}(\ln x)^2 - \left[\frac{x^2}{2} \ln x - \int \frac{x}{2}\, dx\right] = \frac{x^2}{2}(\ln x)^2 - \frac{x^2}{2} \ln x + \frac{x^2}{4} + C$$

21. $\displaystyle\int \frac{(\ln x)^2}{x}\, dx = \int (\ln x)^2\left(\frac{1}{x}\right) dx = \frac{(\ln x)^3}{3} + C$

23. Let $u = x - 1$, then $x = u + 1$ and $du = dx$.

$$\int x\sqrt{x - 1}\, dx = \int (u + 1)u^{1/2}\, du$$

$$= \int (u^{3/2} + u^{1/2})\, du$$

$$= \frac{2}{5}u^{5/2} + \frac{2}{3}u^{3/2} + C$$

$$= \frac{2}{5}(x - 1)^{5/2} + \frac{2}{3}(x - 1)^{3/2} + C$$

25. $\displaystyle\int x(x + 1)^2\, dx = \int x(x^2 + 2x + 1)\, dx$

$$= \int (x^3 + 2x^2 + x)\, dx$$

$$= \frac{x^4}{4} + \frac{2x^3}{3} + \frac{x^2}{2} + C$$

27. Let $u = xe^{2x}$ and $dv = (2x + 1)^{-2}\, dx$, then $du = e^{2x}(2x + 1)\, dx$ and $v = -1/[2(2x + 1)]$.

$$\int \frac{xe^{2x}}{(2x + 1)^2}\, dx = -\frac{xe^{2x}}{2(2x + 1)} + \frac{1}{2}\int e^{2x}\, dx$$

$$= -\frac{xe^{2x}}{2(2x + 1)} + \frac{1}{4}e^{2x} + C$$

$$= \frac{e^{2x}}{4(2x + 1)} + C$$

29. Let $u = x^2$ and $dv = e^x\, dx$, then $du = 2x\, dx$ and $v = e^x$.

$$\int_0^1 x^2 e^x\, dx = x^2 e^x \Big]_0^1 - 2\int_0^1 xe^x\, dx$$

Let $u = x$ and $dv = e^x\, dx$, then $du = dx$ and $v = e^x$.

$$\int_0^1 x^2 e^x\, dx = x^2 e^x \Big]_0^1 - 2\left(xe^x \Big]_0^1 - \int_0^1 e^x\, dx\right)$$

$$= e - 2\left(e - \left[e^x\right]_0^1\right)$$

$$= e - 2e + 2(e - 1)$$

$$= e - 2$$

$$\approx 0.718$$

31. Let $u = \ln x$ and $dv = x^5\, dx$, then $du = (1/x)\, dx$ and $v = x^6/6$.

$$\int_1^e x^5 \ln x\, dx = \frac{x^6}{6} \ln x \Big]_1^e - \int_1^e \frac{x^5}{6}\, dx$$

$$= \frac{e^6}{6} - \left[\frac{x^6}{36}\right]_1^e$$

$$= \frac{e^6}{6} - \frac{e^6}{36} + \frac{1}{36}$$

$$= \frac{5}{36}e^6 + \frac{1}{36}$$

$$\approx 56.060$$

33. Use integration by parts to evaluate the indefinite integral.

Let $u = \ln(x + 2)$, $du = \dfrac{1}{x + 2}\, dx$, $dv = dx$, $v = x$.

$$\int \ln(x + 2)\, dx = x \ln(x + 2) - \int x \left(\frac{1}{x + 2} \right) dx = x \ln(x + 2) - \int \left(1 - \frac{2}{x + 2} \right) dx$$

$$= x \ln(x + 2) - x + 2 \ln(x + 2) + C = (x + 2) \ln(x + 2) - x + C$$

$$\int_{-1}^{0} \ln(x + 2)\, dx = \left[(x + 2) \ln(x + 2) - x \right]_{-1}^{0} = (2 \ln 2 - 0) - (\ln(1) - (-1)) = 2 \ln 2 - 1 \approx 0.386$$

35. Area $= \displaystyle\int_0^2 x^3 e^x\, dx$

First use integration by parts three times to evaluate the indefinite integral.
Let $u = x^3$, $dv = e^x\, dx$, $du = 3x^2\, dx$, $v = e^x$.

$$\int x^3 e^x\, dx = x^3 e^x - \int 3x^2 e^x\, dx = x^3 e^x - 3 \int x^2 e^x\, dx$$

Let $u = x^2$, $dv = e^x\, dx$, $du = 2x\, dx$, $v = e^x$.

$$\int x^3 e^x\, dx = x^3 e^x - 3 \left[x^2 e^x - \int 2x e^x\, dx \right] = x^3 e^x - 3x^2 e^x + 6 \int x e^x\, dx$$

Let $u = x$, $dv = e^x\, dx$, $du = dx$, $v = e^x$.

$$\int x^3 e^x\, dx = x^3 e^x - 3x^2 e^x + 6 \left[x e^x - \int e^x\, dx \right] = (x^3 - 3x^2 + 6x - 6) e^x + C$$

$$\text{Area} = \left[(x^3 - 3x^2 + 6x - 6) e^x \right]_0^2 = 2e^2 + 6 \approx 20.778$$

37. $A = \displaystyle\int_1^e x^2 \ln x\, dx$

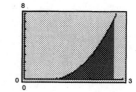

Use integration by parts: $u = \ln x$, $dv = x^2$, $du = (1/x)\, dx$, $v = x^3/3$.

$$\int x^2 \ln x\, dx = \frac{x^3}{3} \ln x - \int \frac{x^3}{3} \left(\frac{1}{x} \right) dx = \frac{x^3}{3} \ln x - \frac{x^3}{9} + C$$

$$A = \left[\frac{x^3}{3} \ln x - \frac{x^3}{9} \right]_1^e = \left(\frac{e^3}{3} - \frac{e^3}{9} \right) - \left(-\frac{1}{9} \right) = \frac{1}{9} + \frac{2e^3}{9} = \frac{1}{9}(1 + 2e^3)$$

39. (a) Let $u = 2x$ and $dv = \sqrt{2x - 3}\, dx$, then $du = 2\, dx$ and $v = \frac{1}{3}(2x - 3)^{3/2}$.

$$\int 2x \sqrt{2x - 3}\, dx = \frac{2}{3} x (2x - 3)^{3/2} - \frac{2}{3} \int (2x - 3)^{3/2}\, dx$$

$$= \frac{2}{3} x (2x - 3)^{3/2} - \frac{2}{15} (2x - 3)^{5/2} + C$$

$$= \frac{2}{15} (2x - 3)^{3/2} (3x + 3) + C$$

$$= \frac{2}{5} (2x - 3)^{3/2} (x + 1) + C$$

(b) Let $u = \sqrt{2x - 3}$, then $x = (u^2 + 3)/2$ and $dx = u\, du$.

$$\int 2x \sqrt{2x - 3}\, dx = \int (u^2 + 3) u^2\, du$$

$$= \frac{u^5}{5} + u^3 + C$$

$$= \frac{u^3}{5} (u^2 + 5) + C$$

$$= \frac{2}{5} (2x - 3)^{3/2} (x + 1) + C$$

41. (a) Let $u = x$ and $dv = \left(1/\sqrt{4 + 5x}\right) dx$, then $du = dx$ and $v = (2/5)\sqrt{4 + 5x}$.

$$\int \frac{x}{\sqrt{4 + 5x}}\, dx = \frac{2}{5}x\sqrt{4 + 5x} - \frac{2}{5}\int \sqrt{4 + 5x}\, dx$$

$$= \frac{2}{5}x\sqrt{4 + 5x} - \frac{4}{75}(4 + 5x)^{3/2} + C$$

$$= \frac{2}{75}\sqrt{4 + 5x}(5x - 8) + C$$

(b) Let $u = \sqrt{4 + 5x}$, then $x = (u^2 - 4)/5$ and $dx = (2u/5)\, du$.

$$\int \frac{x}{\sqrt{4 + 5x}}\, dx = \int \frac{(u^2 - 4)/5}{u}\left(\frac{2u}{5}\right) du$$

$$= \frac{2}{25}\int (u^2 - 4)\, du$$

$$= \frac{2}{25}\left(\frac{u^3}{3} - 4u\right) + C$$

$$= \frac{2}{75}\sqrt{4 + 5x}(5x - 8) + C$$

43. (a) Let $u = \ln x$ and $dv = x^n\, dx$, then $du = (1/x)\, dx$ and $v = x^{n+1}/(n + 1)$.

$$\int x^n \ln x\, dx = \frac{x^{n+1}}{n + 1}\ln x - \int \frac{1}{x}\cdot\frac{x^{n+1}}{n + 1}\, dx$$

$$= \frac{x^{n+1}}{n + 1}\ln x - \frac{1}{n + 1}\int x^n\, dx$$

$$= \frac{x^{n+1}}{n + 1}\ln x - \frac{1}{n + 1}\cdot\frac{x^{n+1}}{n + 1} + C$$

$$= \frac{x^{n+1}}{(n + 1)^2}[-1 + (n + 1)\ln x] + C$$

45. Using $n = 2$ and $a = 5$, we have

$$\int x^2 e^{5x}\, dx = \frac{x^2 e^{5x}}{5} - \frac{2}{5}\int x e^{5x}\, dx.$$

Now, using $n = 1$ and $a = 5$, we have the following.

$$\int x^2 e^{5x}\, dx = \frac{x^2 e^{5x}}{5} - \frac{2}{5}\left[\frac{x e^{5x}}{5} - \frac{1}{5}\int e^{5x}\, dx\right]$$

$$= \frac{x^2 e^{5x}}{5} - \frac{2x e^{5x}}{25} + \frac{2 e^{5x}}{125} + C$$

$$= \frac{e^{5x}}{125}(25x^2 - 10x + 2) + C$$

47. Using $n = -2$,

$$\int x^{-2} \ln x\, dx = \frac{x^{-2+1}}{(-2 + 1)^2}[-1 + (-2 + 1)\ln x] + C$$

$$= \frac{1}{x}(-1 - \ln x) + C$$

$$= -\frac{1}{x} - \frac{\ln x}{x} + C.$$

49. Letting $u = x$ and $dv = e^{-x}\, dx$, we have $du = dx$ and $v = -e^{-x}$, and the area is

$$A = \int_0^4 x e^{-x}\, dx$$

$$= -x e^{-x}\Big]_0^4 + \int_0^4 e^{-x}\, dx$$

$$= -4e^{-4} - \left[e^{-x}\right]_0^4$$

$$= -4e^{-4} - e^{-4} + 1$$

$$= 1 - 5e^{-4}$$

$$\approx 0.908.$$

51. The area is $\displaystyle\int_1^e x \ln x\, dx$.

Use integration by parts. Let $u = \ln x$, $du = (1/x)\, dx$, $dv = x\, dx$, $v = x^2/2$.

$$\int x \ln x\, dx = \frac{x^2}{2}\ln x - \int \frac{x^2}{2}\frac{1}{x}\, dx = \frac{x^2}{2}\ln x - \frac{x^2}{4} + C$$

$$\int_1^e x \ln x\, dx = \left[\frac{x^2}{2}\ln x - \frac{x^2}{4}\right]_1^e = \left(\frac{e^2}{2}\ln(e) - \frac{e^2}{4}\right) - \left(0 - \frac{1}{4}\right) = \frac{e^2}{4} + \frac{1}{4}$$

53. (a) Letting $u = 2 \ln x$ and $dv = dx$, we have $du = (2/x)\, dx$ and $v = x$.

$$\text{Area} = \int_1^e 2 \ln x \, dx$$

$$= 2x \ln x \Big]_1^e - \int_1^e - 2 \, dx$$

$$= 2e - \Big[2x \Big]_1^e$$

$$= 2e - 2e + 2$$

$$= 2$$

(b) Letting $u = (2 \ln x)^2$ and $dv = dx$, we have $du = 8 \ln x (1/x)\, dx$ and $v = x$.

$$\text{Volume} = \pi \int_1^e (2 \ln x)^2 \, dx$$

$$= \pi \left(\Big[x(2 \ln x)^2 \Big]_1^e - \int_1^e 8 \ln x \, dx \right)$$

$$= \pi \left(4e - \Big[8(x \ln x - x) \Big]_1^e \right)$$

$$= \pi (4e - [8])$$

$$= 4\pi (e - 2)$$

$$\approx 9.026$$

51. $\displaystyle \int_0^2 t^3 e^{-4t} \, dt = \frac{3}{128} - \frac{379}{128} e^{-8} \approx 0.222$

57. $\displaystyle \int_0^5 x^4 (25 - x^2)^{3/2} \, dx = \frac{1{,}171{,}875}{256} \pi \approx 14{,}381{,}070$

59. (a) $x = 500(20 + te^{-0.1t}), \quad 0 \le t \le 10$

$$\frac{dx}{dt} = 500(-0.1te^{-0.1t} + e^{-0.1t})$$

$$= 500e^{-0.1t}(-0.1t + 1)$$

On the interval $(0, 10)$, $dx/dt > 0$. Thus, x is increasing. The corporate officers are forecasting an increase in demand over the next decade.

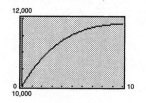

(b) $\displaystyle \int_0^{10} 500(20 + te^{-0.1t}) \, dt = 500 \left[\int_0^{10} 20 \, dt + \int_0^{10} te^{-0.1t} \, dt \right]$

$$= 500(200) + 500 \int te^{-0.1t} \, dt$$

Let $u = t$ and $dv = e^{-0.1t} \, dt$, then $du = dt$ and $v = -10e^{-0.1t}$.

$$= 100{,}000 + 500 \left[-10te^{-0.1t} + 10 \int e^{-0.1t} \, dt \right]_0^{10}$$

$$= 100{,}000 + 500 \left\{ -100e^{-1} - \Big[100e^{-0.1t} \Big]_0^{10} \right\}$$

$$= 100{,}000 + 500(-100e^{-1} - 100e^{-1} + 100)$$

$$= 100{,}000 + 50{,}000(1 - 2e^{-1})$$

$$\approx 113{,}212 \text{ units}$$

(c) $\text{Average} = \dfrac{113{,}212}{10} \approx 11{,}321$ per year

61. (a) $\text{Average} = \displaystyle \int_1^2 (1.6t \ln t + 1) \, dt$

$$= (0.8t^2 \ln t - 0.4t^2 + t) \Big]_1^2$$

$$= 3.2(\ln 2) - 0.2$$

$$\approx 2.0181$$

(b) $\text{Average} = \displaystyle \int_3^4 (1.6t \ln t + 1) \, dt$

$$= (0.8t^2 \ln t - 0.4t^2 + t) \Big]_3^4$$

$$= 12.8(\ln 4) - 7.2(\ln 3) - 1.8$$

$$\approx 8.0346$$

63. $V = \int_0^{t_1} c(t)e^{-rt} \, dt = \int_0^4 5000e^{-0.05t} \, dt = \dfrac{5000}{-0.05} e^{-0.05t} \Big]_0^4 \approx \$18,126.92$

65. $V = \int_0^{t_1} c(t)e^{-rt} = \int_0^{10} (150,000 + 2500t)e^{-0.04t} \, dt$

$\approx \$1,332,474.72$

67. $V = \int_0^{t_1} c(t)e^{-rt} \, dt = \int_0^4 (1000 + 50e^{t/2})e^{-0.06t} \, dt$

$\approx \$4103.07$

69. $c = 150,000 + 75,000t$

(a) $\int_0^4 (150,000 + 75,000t) \, dt = 150,000t + 37,500t^2 \Big]_0^4$

$= \$1,200,000$

(b) $V = \int_0^{t_1} c(t)e^{-rt} \, dt = \int_0^4 (150,000 + 75,000t)e^{-0.04t} \, dt$

$\approx \$1,094,142.26$

71. Future value $= e^{rt_1} \int_0^{t_1} f(t)e^{-rt} \, dt$

$= e^{(0.08)10} \int_0^{10} 3000e^{-0.08t} \, dt$

$= e^{0.8} \left[\dfrac{3000}{-0.08} e^{-0.08t} \right]_0^{10}$

$\approx \$45,957.78$

73. (a) Future value $= e^{(0.07)(10)} \int_0^{10} 1200e^{-(0.07)t} \, dt = e^{0.7} \int_0^{10} 1200e^{-0.07t} \, dt \approx \$17,378.62$

(b) Difference $= \left[e^{(0.10)(15)} \int_0^{15} 1200e^{-0.10t} \, dt \right] - \left[e^{(0.09)(15)} \int_0^{15} 1200e^{-0.09t} \, dt \right]$

$= \left[e^{1.5} \int_0^{15} 1200e^{-0.1t} \, dt \right] - \left[e^{1.35} \int_0^{15} 1200e^{-0.09t} \, dt \right]$

$\approx \$41,780.27 - \$38,099.01$

$= \$3681.26$

75. $\int_1^4 \dfrac{4}{\sqrt{x} + \sqrt[3]{x}} \, dx = 4 \int_1^4 \dfrac{4}{\sqrt{x} + \sqrt[3]{x}} \, dx$

$\approx 4 \left(\dfrac{4-1}{10} \right) \left[f\left(\dfrac{23}{20} \right) + f\left(\dfrac{29}{20} \right) + f\left(\dfrac{35}{20} \right) + f\left(\dfrac{41}{20} \right) + f\left(\dfrac{47}{20} \right) + f\left(\dfrac{53}{20} \right) + f\left(\dfrac{59}{20} \right) + f\left(\dfrac{65}{20} \right) + f\left(\dfrac{71}{20} \right) + f\left(\dfrac{77}{20} \right) \right]$

≈ 4.254

Section 6.3   Partial Fractions and Logistic Growth

1. $\dfrac{2x + 40}{(x - 5)(x + 5)} = \dfrac{A}{x - 5} + \dfrac{B}{x + 5}$

Basic equation: $2x + 40 = A(x + 5) + B(x - 5)$

When $x = 5$: $50 = 10A, A = 5$

When $x = -5$: $30 = -10B, B = -3$

$\dfrac{2(x + 20)}{x^2 - 25} = \dfrac{5}{x - 5} - \dfrac{3}{x + 5}$

3. $\dfrac{8x + 3}{x(x - 3)} = \dfrac{A}{x} + \dfrac{B}{x - 3}$

Basic equation: $8x + 3 = A(x - 3) + Bx$

When $x = 0$: $3 = -3A, A = -1$

When $x = 3$: $27 = 3B, B = 9$

$\dfrac{8x + 3}{x^2 - 3x} = \dfrac{9}{x - 3} - \dfrac{1}{x}$

5. $\dfrac{4x - 13}{(x - 5)(x + 2)} = \dfrac{A}{x - 5} + \dfrac{B}{x + 2}$

Basic equation: $4x - 13 = A(x + 2) + B(x - 5)$

When $x = 5$: $7 = 7A, A = 1$

When $x = -2$: $-21 = -7B, B = 3$

$$\dfrac{4x - 13}{x^2 - 3x - 10} = \dfrac{1}{x - 5} + \dfrac{3}{x + 2}$$

7. $\dfrac{3x^2 - 2x - 5}{x^2(x + 1)} = \dfrac{A}{x} + \dfrac{B}{x^2} + \dfrac{C}{x + 1}$

Basic equation:

$$3x^2 - 2x - 5 = Ax(x + 1) + B(x + 1) + Cx^2$$

When $x = 0$: $-5 = B$

When $x = 1$: $0 = C$

When $x = 1$: $-4 = 2A - 10 \Rightarrow A = 3$

$$\dfrac{3x^2 - 2x - 5}{x^2(x + 1)} = \dfrac{3}{x} - \dfrac{5}{x^2}$$

9. $\dfrac{1}{3}\left[\dfrac{x + 1}{(x - 2)^2}\right] = \dfrac{1}{3}\left[\dfrac{A}{x - 2} + \dfrac{B}{(x - 2)^2}\right]$

Basic equation: $x + 1 = A(x - 2) + B$

When $x = 2$: $3 = B$

When $x = 3$: $4 = A + B \Rightarrow A = 1$

$$\dfrac{x + 1}{3(x - 2)^2} = \dfrac{1}{3}\left[\dfrac{1}{x - 2} + \dfrac{3}{(x - 2)^2}\right]$$

$$= \dfrac{1}{3(x - 2)} + \dfrac{1}{(x - 2)^2}$$

11. $\dfrac{8x^2 + 15x + 9}{(x + 1)^3} = \dfrac{A}{x + 1} + \dfrac{B}{(x + 1)^2} + \dfrac{C}{(x + 1)^3}$

Basic equation:

$$8x^2 + 15x + 9 = A(x + 1)^2 + B(x + 1) + C$$

$$= Ax^2 + (2A + B)x + (A + B + C)$$

Therefore, $A = 8, 2A + B = 15$, and $A + B + C = 9$.
Solving these equations yields $A = 8, B = -1$, and $C = 2$.

$$\dfrac{8x^2 + 15x + 9}{(x + 1)^3} = \dfrac{8}{x + 1} - \dfrac{1}{(x + 1)^2} + \dfrac{2}{(x + 1)^3}$$

13. $\dfrac{1}{(x + 1)(x - 1)} = \dfrac{A}{x + 1} + \dfrac{B}{x - 1}$

Basic equation: $1 = A(x - 1) + B(x + 1)$

When $x = -1$: $1 = -2A, A = -\dfrac{1}{2}$

When $x = 1$: $1 = 2B, B = \dfrac{1}{2}$

$$\int \dfrac{1}{x^2 - 1}\, dx = \dfrac{-1}{2}\int \dfrac{1}{x + 1}\, dx + \dfrac{1}{2}\int \dfrac{1}{x - 1}\, dx$$

$$= \dfrac{-1}{2}\ln|x + 1| + \dfrac{1}{2}\ln|x - 1| + C$$

$$= \dfrac{1}{2}\ln\left|\dfrac{x - 1}{x + 1}\right| + C$$

15. $\dfrac{-2}{(x + 4)(x - 4)} = \dfrac{A}{x + 4} + \dfrac{B}{x - 4}$

Basic equation: $-2 = A(x - 4) + B(x + 4)$

When $x = -4$: $-2 = -8A, A = \dfrac{1}{4}$

When $x = 4$: $-2 = 8B, B = -\dfrac{1}{4}$

$$\int \dfrac{-2}{x^2 - 16}\, dx = \dfrac{1}{4}\int \dfrac{1}{x + 4}\, dx - \dfrac{1}{4}\int \dfrac{1}{x - 4}\, dx$$

$$= \dfrac{1}{4}\ln|x + 4| - \dfrac{1}{4}\ln|x - 4| + C$$

$$= \dfrac{1}{4}\ln\left|\dfrac{x + 4}{x - 4}\right| + C$$

17. $\dfrac{1}{3x^2 - x} = \dfrac{1}{x(3x - 1)} = \dfrac{A}{x} + \dfrac{B}{3x - 1}$

Basic equation: $1 = A(3x - 1) + Bx$

When $x = 0$: $1 = -A \Rightarrow A = -1$

When $x = \dfrac{1}{3}$: $1 = B\left(\dfrac{1}{3}\right) \Rightarrow B = 3$

$$\int \dfrac{1}{3x^2 - x}\, dx = \int \dfrac{-1}{x}\, dx + \int \dfrac{3}{3x - 1}\, dx$$

$$= -\ln|x| + \ln|3x - 1| + C$$

$$= \ln\left|\dfrac{3x - 1}{x}\right| + C$$

19. $\dfrac{1}{x(2x + 1)} = \dfrac{A}{x} + \dfrac{B}{2x + 1}$

Basic equation: $1 = A(2x + 1) + Bx$

When $x = 0$: $A = 1$

When $x = -\dfrac{1}{2}$: $B = -2$

$$\int \dfrac{1}{2x^2 + x}\, dx = \int \dfrac{1}{x}\, dx - \int \dfrac{2}{2x + 1}\, dx$$

$$= \ln|x| - \ln|2x + 1| + C$$

$$= \ln\left|\dfrac{x}{2x + 1}\right| + C$$

21. $\dfrac{3}{(x-1)(x+2)} = \dfrac{A}{x-1} + \dfrac{B}{x+2}$

Basic equation: $3 = A(x+2) + B(x-1)$

When $x = 1$: $3 = 3A, A = 1$

When $x = -2$: $3 = -3B, B = -1$

$$\int \frac{3}{x^2+x-2}\,dx = \int \frac{1}{x-1}\,dx - \int \frac{1}{x+2}\,dx$$

$$= \ln|x-1| - \ln|x+2| + C$$

$$= \ln\left|\frac{x-1}{x+2}\right| + C$$

23. $\dfrac{5-x}{(2x-1)(x+1)} = \dfrac{A}{2x-1} + \dfrac{B}{x+1}$

Basic equation: $5 - x = A(x+1) + B(2x-1)$

When $x = \dfrac{1}{2}$: $4.5 = 1.5A, A = 3$

When $x = -1$: $6 = -3B, B = -2$

$$\int \frac{5-x}{2x^2+x-1}\,dx = 3\int \frac{1}{2x-1}\,dx - 2\int \frac{1}{x+1}\,dx$$

$$= \frac{3}{2}\ln|2x-1| - 2\ln|x+1| + C$$

25. $\dfrac{x^2+12x+12}{x(x+2)(x-2)} = \dfrac{A}{x} + \dfrac{B}{x+2} + \dfrac{C}{x-2}$

Basic equation: $x^2 + 12x + 12 = A(x+2)(x-2) + Bx(x-2) + Cx(x+2)$

When $x = 0$: $12 = -4A, A = -3$

When $x = -2$: $-8 = 8B, B = -1$

When $x = 2$: $40 = 8C, C = 5$

$$\int \frac{x^2+12x+12}{x^3-4x}\,dx = 5\int \frac{1}{x-2}\,dx - \int \frac{1}{x+2}\,dx - 3\int \frac{1}{x}\,dx \;=\; 5\ln|x-2| - \ln|x+2| - 3\ln|x| + C$$

27. $\dfrac{x+2}{x(x-4)} = \dfrac{A}{x-4} + \dfrac{B}{x}$

Basic equation: $x + 2 = Ax + B(x-4)$

When $x = 4$: $6 = 4A, A = \dfrac{3}{2}$

When $x = 0$: $2 = -4B, B = -\dfrac{1}{2}$

$$\int \frac{x+2}{x^2-4x}\,dx = \frac{1}{2}\left[3\int \frac{1}{x-4}\,dx - \int \frac{1}{x}\,dx\right]$$

$$= \frac{1}{2}[3\ln|x-4| - \ln|x|] + C$$

29. $\dfrac{4-3x}{(x-1)^2} = \dfrac{A}{x-1} + \dfrac{B}{(x-1)^2}$

Basic equation: $4 - 3x = A(x-1) + B$

When $x = 1$: $1 = B$

When $x = 0$: $4 = -A + 1 \Rightarrow A = -3$

$$\int \frac{4-3x}{(x-1)^2}\,dx = \int \frac{-3}{x-1}\,dx + \int \frac{1}{(x-1)^2}\,dx$$

$$= -3\ln|x-1| - \frac{1}{x-1} + C$$

31. $\dfrac{3x^2+3x+1}{x(x+1)^2} = \dfrac{A}{x} + \dfrac{B}{x+1} + \dfrac{C}{(x+1)^2}$

Basic equation: $A(x+1)^2 + Bx(x+1) + Cx = 3x^2 + 3x + 1$

When $x = 0$: $A = 1$

When $x = -1$: $-C = 3 - 3 + 1 \Rightarrow C = -1$

When $x = 1$: $4A + 2B + C = 7$

$$4 + 2B - 1 = 7$$

$$B = 2$$

$$\int \frac{3x^2+3x+1}{x(x^2+2x+1)}\,dx = \int \left(\frac{1}{x} + \frac{2}{x+1} + \frac{-1}{(x+1)^2}\right)dx = \ln|x| + 2\ln|x+1| + \frac{1}{x+1} + C$$

33. $\dfrac{1}{9-x^2} = \dfrac{-1}{(x-3)(x+3)} = \dfrac{A}{x-3} + \dfrac{B}{x+3}$

Basic equation: $-1 = A(x+3) + B(x-3)$

When $x = 3$: $-1 = 6A \Longrightarrow A = -\dfrac{1}{6}$

When $x = -3$: $-1 = -6B \Longrightarrow B = \dfrac{1}{6}$

$\displaystyle\int_4^5 \dfrac{1}{9-x^2}\,dx = \int_4^5 \dfrac{-1/6}{x-3}\,dx + \int_4^5 \dfrac{1/6}{x+3}\,dx$

$\qquad = \dfrac{1}{6}\Big[-\ln(x-3) + \ln(x+3)\Big]_4^5$

$\qquad = \dfrac{1}{6}[-\ln 2 + \ln 8 - \ln 7]$

$\qquad = \dfrac{1}{6}\ln\dfrac{4}{7}$

$\qquad \approx -0.093$

35. $\dfrac{x-1}{x^2(x+1)} = \dfrac{A}{x} + \dfrac{B}{x^2} + \dfrac{C}{x+1}$

Basic equation: $x - 1 = Ax(x+1) + B(x+1) + Cx^2$

When $x = 0$: $B = -1$

When $x = -1$: $C = -2$

When $x = 1$: $0 = 2A + 2B + C, 0 = 2A - 4, A = 2$

$\displaystyle\int_1^5 \dfrac{x-1}{x^2(x+1)}\,dx = 2\int_1^5 \dfrac{1}{x}\,dx - \int_1^5 \dfrac{1}{x^2}\,dx - 2\int_1^5 \dfrac{1}{x+1}\,dx$

$\qquad = \left[2\ln|x| + \dfrac{1}{x} - 2\ln|x+1|\right]_1^5$

$\qquad = \left[2\ln\left|\dfrac{x}{x+1}\right| + \dfrac{1}{x}\right]_1^5$

$\qquad = 2\ln\left(\dfrac{5}{3}\right) - \dfrac{4}{5}$

$\qquad \approx 0.222$

37. $\dfrac{x^3}{x^2-2} = x + \dfrac{2x}{x^2-2}$

$\displaystyle\int_0^1 \left(x + \dfrac{2x}{x^2-2}\right)dx = \left[\dfrac{x^2}{2} + \ln|x^2-2|\right]_0^1 = \dfrac{1}{2} - \ln 2 \approx -0.193$

39. First divide to obtain $\dfrac{x^3 - 4x^2 - 3x + 3}{x^2 - 3x} = x - 1 - \dfrac{6x-3}{x^2-3x}$.

Use partial fractions for the $\dfrac{6x-3}{x^2-3x}$ term.

$\dfrac{6x-3}{x(x-3)} = \dfrac{A}{x} + \dfrac{B}{x-3}$

Basic equation: $6x - 3 = A(x-3) + Bx$

When $x = 3$: $15 = 3B \Longrightarrow B = 5$

When $x = 0$: $-3 = -3A \Longrightarrow A = 1$

$\displaystyle\int_1^2 \dfrac{x^3 - 4x^2 - 3x + 3}{x^2-3x}\,dx = \int_1^2 \left(x - 1 - \dfrac{1}{x} - \dfrac{5}{x-3}\right)dx$

$\qquad = \left[\dfrac{x^2}{2} - x - \ln|x| - 5\ln|x-3|\right]_1^2$

$\qquad = (-\ln 2) - \left(-\dfrac{1}{2} - 5\ln 2\right)$

$\qquad = \dfrac{1}{2} + 4\ln 2 \approx 3.273$

41. To find the limits of integration, solve:

$$2 = \frac{14}{16 - x^2}$$

$$32 - 2x^2 = 14$$

$$18 = 2x^2$$

$$x = \pm 3$$

$$A = \int_{-3}^{3} \left(2 - \frac{14}{16 - x^2} \right) dx$$

$$= \int_{-3}^{3} \left(2 - \frac{7}{4} \cdot \frac{1}{x + 4} + \frac{7}{4} \cdot \frac{1}{x - 4} \right) dx$$

$$= \left[2x - \frac{7}{4} \ln|x + 4| + \frac{7}{4} \ln|x - 4| \right]_{-3}^{3}$$

$$= \left(6 - \frac{7}{4} \ln 7 \right) - \left(-6 + \frac{7}{4} \ln 7 \right)$$

$$= 12 - \frac{7}{2} \ln 7 \approx 5.1893$$

43. $A = \int_{2}^{5} \frac{x + 1}{x^2 - x} dx$

$$= \int_{2}^{5} \left(-\frac{1}{x} + \frac{2}{x - 1} \right) dx$$

$$= \left[-\ln x + 2 \ln(x - 1) \right]_{2}^{5}$$

$$= (-\ln 5 + 2 \ln 4) - (-\ln 2)$$

$$= 5 \ln 2 - \ln 5 \approx 1.8563$$

45. $\dfrac{1}{a^2 - x^2} = \dfrac{1}{(a + x)(a - x)}$

$$= \frac{A}{a + x} + \frac{B}{a - x}$$

$$= \frac{1}{2a} \left(\frac{1}{a + x} + \frac{1}{a - x} \right)$$

47. $\dfrac{1}{x(a - x)} = \dfrac{A}{x} + \dfrac{B}{a - x}$

Basic equation: $1 = A(a - x) + Bx$

When $x = a$: $1 = Ba \Rightarrow B = \dfrac{1}{a}$

When $x = 0$: $1 = Aa \Rightarrow A = \dfrac{1}{a}$

$$\frac{1}{x(a - x)} = \frac{1/a}{x} + \frac{1/a}{a - x}$$

49. $V = \pi \int_{1}^{5} \left[\dfrac{10}{x(x + 10)} \right]^2 dx = 100\pi \int_{1}^{5} \dfrac{1}{x^2(x + 10)^2} dx$

(a) $V \approx 1.773$ cubic units

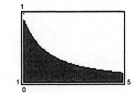

(b) Let $\dfrac{1}{x^2(x + 10)^2} = \dfrac{A}{x} + \dfrac{B}{x^2} + \dfrac{C}{x + 10} + \dfrac{D}{(x + 10)^2}$.

Basic equation: $1 = Ax(x + 10)^2 + B(x + 10)^2 + Cx^2(x + 10) + Dx^2$

When $x = 0$: $1 = 100B \Rightarrow B = \frac{1}{100}$

When $x = -10$: $1 = 100D \Rightarrow D = \frac{1}{100}$

When $x = 1$: $1 = 121A + 121B + 11C + D$ $\left. \begin{array}{l} \\ \\ \end{array} \right\}$ $A = -\frac{1}{500}$

When $x = -1$: $1 = -81A + 81B + 9C + D$ $\left. \begin{array}{l} \\ \\ \end{array} \right\}$ $C = \frac{1}{500}$

$$V = \pi \int_{1}^{5} 100 \left[\frac{-1/500}{x} + \frac{1/100}{x^2} + \frac{1/500}{x + 10} + \frac{1/100}{(x + 10)^2} \right] dx$$

$$= \pi \int_{1}^{5} \left[-\frac{1}{5x} + \frac{1}{x^2} + \frac{1}{5(x + 10)} + \frac{1}{(x + 10)^2} \right] dx$$

$$= \pi \left[-\frac{1}{5} \ln|x| - \frac{1}{x} + \frac{1}{5} \ln|x + 10| - \frac{1}{x + 10} \right]_{1}^{5}$$

$$= \pi \left[\frac{1}{5} \ln \left| \frac{x + 10}{x} \right| - \frac{2x + 10}{x(x + 10)} \right]_{1}^{5} = \pi \left[\left(\frac{1}{5} \ln 3 - \frac{4}{15} \right) - \left(\frac{1}{5} \ln 11 - \frac{12}{11} \right) \right] \approx 1.773 \text{ cubic units}$$

51. (a) $V = \pi \int_{-1}^{1} \left(\dfrac{2}{x^2 - 4}\right)^2 dx \approx 1.9100$

(b) $V = \pi \int_{-1}^{1} \left(\dfrac{2}{x^2 - 4}\right)^2 dx$

$\qquad = \pi \int_{-1}^{2} \left(\dfrac{2}{(x-2)(x+2)}\right)^2 dx$

$\dfrac{4}{(x-2)^2(x+2)^2} = \dfrac{A}{x-2} + \dfrac{B}{(x-2)^2} + \dfrac{C}{x+2} + \dfrac{D}{(x+2)^2}$

Basic equation: $4 = A(x-2)(x+2)^2 + B(x+2)^2 + C(x+2)(x-2)^2 + D(x-2)^2$

When $x = 2$: $4 = 16B \Rightarrow B = \dfrac{1}{4}$

When $x = -2$: $4 = 16D \Rightarrow D = \dfrac{1}{4}$

When $x = 0$: $4 = -8A + 4\left(\dfrac{1}{4}\right) + 8C + 4\left(\dfrac{1}{4}\right) \Rightarrow 1 = -4A + 4C$

When $x = 1$: $4 = -9A + 9\left(\dfrac{1}{4}\right) + 3C + \dfrac{1}{4} \Rightarrow 3 = -18A + 6C$

Solving for A and C, $A = -\frac{1}{8}$, $C = \frac{1}{8}$. Hence,

$V = \pi \int_{-1}^{1} \left(-\dfrac{1}{8} \cdot \dfrac{1}{x-2} + \dfrac{1}{4} \cdot \dfrac{1}{(x-2)^2} + \dfrac{1}{8} \cdot \dfrac{1}{x+2} + \dfrac{1}{4} \cdot \dfrac{1}{(x+2)^2}\right) dx$

$\qquad = \pi \left[-\dfrac{1}{8} \ln|x-2| - \dfrac{1}{4} \cdot \dfrac{1}{x-2} + \dfrac{1}{8} \ln|x+2| - \dfrac{1}{4} \cdot \dfrac{1}{x+2}\right]_{-1}^{1}$

$\qquad = \pi \left[\left(\dfrac{1}{4} + \dfrac{1}{8} \ln 3 - \dfrac{1}{12}\right) - \left(-\dfrac{1}{8} \ln 3 + \dfrac{1}{12} - \dfrac{1}{4}\right)\right]$

$\qquad = \pi \left[\dfrac{1}{3} + \dfrac{1}{4} \ln 3\right] \approx 1.9100.$

53. $\dfrac{1}{y(1000 - y)} = \dfrac{A}{y} + \dfrac{B}{1000 - y}$

Basic equation: $1 = A(1000 - y) + By$

When $y = 0$: $1 = 1000A, A = \dfrac{1}{1000}$

When $y = 1000$: $1 = 1000B, B = \dfrac{1}{1000}$

$\int \dfrac{1}{y(1000 - y)} dy = \int k\, dt$

$\dfrac{1}{1000} \int \left(\dfrac{1}{y} + \dfrac{1}{1000 - y}\right) dy = \int k\, dt$

$\ln|y| - \ln|1000 - y| = 1000(kt + C_1)$

$\ln\left|\dfrac{y}{1000 - y}\right| = 1000(kt + C_1)$

$\dfrac{y}{1000 - y} = e^{1000(kt + C_1)} = e^{1000C_1} e^{1000kt}$

$\dfrac{y}{1000 - y} = Ce^{1000kt}$

—CONTINUED—

53. —CONTINUED—

When $t = 0$, $y = 100$: $\frac{1}{9} = C$, $\frac{y}{1000 - y} = \frac{1}{9}e^{1000kt}$

When $t = 2$, $y = 134$: $\frac{67}{433} = \frac{1}{9}e^{2000k}$, $\ln\frac{603}{433} = 2000k$, $k = \frac{1}{2000}\ln\frac{603}{433}$

Thus, $\frac{y}{1000 - y} = \frac{1}{9}e^{(1/2)\ln(603/433)t}$. Solving for y yields

$$9y \approx e^{0.1656t}(1000 - y)$$

$$y(9 + e^{0.1656t}) = 1000e^{0.1656t}$$

$$y = \frac{1000e^{0.1656t}}{e^{0.1656t} + 9} = \frac{1000}{1 + 9e^{-0.1656t}}.$$

55. $\dfrac{dS}{dt} = \dfrac{2t}{(t + 4)^2}$

$\dfrac{2t}{(t + 4)^2} = \dfrac{A}{t + 4} + \dfrac{B}{(t + 4)^2}$

Basic equation: $2t = A(t + 4) + B$

When $t = -4$: $-8 = B$

When $t = 0$: $0 = 4A - 8 \Rightarrow A = 2$

$$S = \int \frac{2t}{(t + 4)^2}\, dt = \int \left(\frac{2}{t + 4} - \frac{8}{(t + 4)^2}\right) dt = 2\ln(t + 4) + \frac{8}{t + 4} + C$$

When $t = 0$: $S = 0 \Rightarrow 0 = 2\ln 4 + 2 + C \Rightarrow C \approx -4.77259$

When $t = 10$: $S \approx 2\ln(14) + \dfrac{8}{14} - 4.77259 \approx 1.077$ thousand

57. Total revenue $= \displaystyle\int_5^{13} \frac{410t^2 + 28{,}490t + 28{,}080}{-6t^2 + 94t + 100}\, dt$

$= \displaystyle\int_5^{13} \left[\frac{-52{,}370}{9t - 150} - \frac{205}{3}\right] dt$

$= \left[\dfrac{-52{,}370 \ln |3t - 50|}{9} - \dfrac{250t}{3}\right]_5^{13}$

$= \dfrac{52{,}370 \ln (35/11)}{9} - \dfrac{1640}{3}$

$\approx \$6188$ million

Average revenue $= \dfrac{6188}{8} \approx \$773.6$ million per year

59. (a) $N = \displaystyle\int \frac{125e^{-0.125t}}{(1 + 9e^{-0.125t})^2}\, dt$

$= \dfrac{1000}{9} \cdot \dfrac{1}{(1 + 9e^{-0.125t})} + C$

$N(0) = 100 = \dfrac{1000}{9} \cdot \dfrac{1}{(1 + 9)} + C \Rightarrow C = \dfrac{800}{9}$

$N = \dfrac{1000}{9(1 + 9e^{-0.125t})} + \dfrac{800}{9}$

$N(2) \approx 103$

(b) $\displaystyle\lim_{t \to \infty} N = \dfrac{1000}{9} + \dfrac{800}{9} = 200$

61. Answers will vary.

Section 6.4 Integration Tables and Completing the Square

1. Formula 4: $u = x, du = dx, a = 2, b = 3$

$$\int \frac{x}{(2 + 3x)^2} \, dx = \frac{1}{9}\left[\frac{2}{2 + 3x} + \ln|2 + 3x|\right] + C$$

3. Formula 19: $u = x, du = dx, a = 2, b = 3$

$$\int \frac{x}{\sqrt{2 + 3x}} \, dx = \frac{-2(4 - 3x)}{27}\sqrt{2 + 3x} + C$$

$$= \frac{2(3x - 4)}{27}\sqrt{2 + 3x} + C$$

5. Formula 25: $u = x^2, du = 2x \, dx, a = 3$

$$\int \frac{2x}{\sqrt{x^4 - 9}} \, dx = \ln\left|x^2 + \sqrt{x^4 - 9}\right| + C$$

7. Formula 35: $u = x^2, du = 2x \, dx$

$$\int x^3 e^{x^2} \, dx = \frac{1}{2} \int x^2 e^{x^2} 2x \, dx = \frac{1}{2}(x^2 - 1)e^{x^2} + C$$

9. Formula 10: $u = x, du = dx, a = b = 1$

$$\int \frac{1}{x(1 + x)} \, dx = \ln\left|\frac{x}{1 + x}\right| + C$$

11. Formula 26: $u = x, du = dx, a = 3$

$$\int \frac{1}{x\sqrt{x^2 + 9}} \, dx = -\frac{1}{3} \ln\left|\frac{3 + \sqrt{x^2 + 9}}{x}\right| + C$$

13. Formula 32: $u = x, du = dx, a = 2$

$$\int \frac{1}{x\sqrt{4 - x^2}} \, dx = -\frac{1}{2} \ln\left|\frac{2 + \sqrt{4 - x^2}}{x}\right| + C$$

15. Formula 40: $u = x, du = dx$

$$\int x \ln \, x \, dx = \frac{x^2}{4}(-1 + 2 \ln x) + C$$

17. Formula 37: $u = 3x^2, du = 6x \, dx$

$$\int \frac{6x}{1 + e^{3x^2}} \, dx = 3x^2 - \ln(1 + e^{3x^2}) + C$$

19. Formula 21: $u = x^2, du = 2x \, dx, a = 2$

$$\int x\sqrt{x^4 - 4} \, dx = \frac{1}{2} \int \sqrt{(x^2)^2 - 2^2} \; 2x \, dx$$

$$= \frac{1}{4}\left[x^2\sqrt{x^4 - 4} - 4 \ln\left|x^2 + \sqrt{x^4 - 4}\right|\right] + C$$

21. Formula 8: $u = t, du = dt, a = 2, b = 3$

$$\int \frac{t^2}{(2 + 3t)^3} \, dt = \frac{1}{27}\left[\frac{4}{2 + 3t} - \frac{4}{2(2 + 3t)^2} + \ln|2 + 3t|\right] + C$$

23. Formula 15: $u = s, du = ds, a = 3, b = 1$

$$\int \frac{s}{s^2\sqrt{3 + s}} \, ds = \int \frac{1}{s\sqrt{3 + s}} \, ds = \frac{1}{\sqrt{3}} \ln\left|\frac{\sqrt{3 + s} - \sqrt{3}}{\sqrt{3 + s} + \sqrt{3}}\right| + C$$

25. Formula 9: $u = x, du = dx, a = 3, b = 2, n = 5$

$$\int \frac{x^2}{(3 + 2x)^5} \, dx = \frac{1}{8}\left[\frac{-1}{2(3 + 2x)^2} + \frac{6}{3(3 + 2x)^3} - \frac{9}{4(3 + 2x)^4}\right] + C$$

27. Formula 23: $u = x, du = dx, a = 1$

$$\int \frac{1}{x^2\sqrt{1 - x^2}} \, dx = -\frac{\sqrt{1 - x^2}}{x} + C$$

29. Formula 41: $u = x, du = dx, n = 2$

$$\int x^2 \ln x \, dx = \frac{x^3}{9}(-1 + 3 \ln x) + C$$

31. Formula 7: $u = x, du = dx, a = -5, b = 3$

$$\int \frac{x^2}{(3x - 5)^2} \, dx = \frac{1}{27}\left[3x - \frac{25}{3x - 5} + 10 \ln|3x - 5|\right] + C$$

33. Formula 3: $u = \ln x, du = (1/x) \, dx, a = 4, b = 3$

$$\int \frac{\ln x}{x(4 + 3 \ln x)} \, dx = \int \frac{\ln x}{4 + 3 \ln x}\left(\frac{1}{x} \, dx\right) = \frac{1}{9}[3 \ln x - 4 \ln|4 + 3 \ln x|] + C$$

35. Formula 19: $u = x, du = dx, a = b = 1$

$$A = \int_0^8 \frac{x}{\sqrt{x + 1}} \, dx$$

$$= \left[-\frac{2(2 - x)}{3}\sqrt{x + 1}\right]_0^8$$

$$= \frac{40}{3}$$

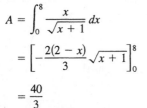

Approximate area: 13.333

37. Formula 37: $u = x^2, du = 2x \, dx$

$$\int_0^2 \frac{x}{1 + e^{x^2}} \, dx = \frac{1}{2}\int_0^4 \frac{1}{1 + e^u} \, du$$

$$= \frac{1}{2}\left[u - \ln(1 + e^u)\right]_0^4$$

$$= \frac{1}{2}[4 - \ln(1 + e^4) + \ln 2]$$

$$= \frac{1}{2}\left[4 + \ln \frac{2}{1 + e^4}\right]$$

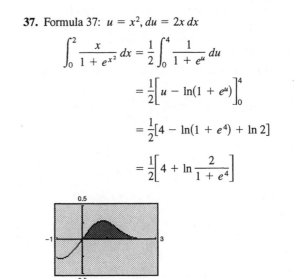

Approximate area: 0.3375

39. Formula 22: $u = x, a = 2, du = dx$

$$\int_0^{\sqrt{5}} x^2\sqrt{x^2 + 4} \, dx = \int_0^{\sqrt{5}} u^2\sqrt{u^2 + 2^2} \, du$$

$$= \frac{1}{8}\left[u(2u^2 + 4)\sqrt{u^2 + 4} - 16 \ln\left|u + \sqrt{u^2 + 4}\right|\right]_0^{\sqrt{5}}$$

$$= \frac{1}{8}\left[\sqrt{5}(14)(3) - 16 \ln\left|\sqrt{5} + 3\right| + 16 \ln 2\right]$$

$$= \frac{1}{4}\left[21\sqrt{5} - 8 \ln\left(\sqrt{5} + 3\right) + 8 \ln 2\right]$$

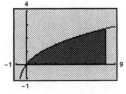

Approximate area: 9.8145

41. Formula 19: $u = x, du = dx, a = 5, b = 2$

$$\int_0^5 \frac{x}{\sqrt{5 + 2x}} \, dx = \frac{-2(10 - 2x)}{12}\sqrt{5 + 2x}\bigg]_0^5$$

$$= 0 + \frac{5}{3}\sqrt{5} = \frac{5\sqrt{5}}{3}$$

43. Formula 38: $u = x, du = dx, n = 0.5$

$$6\int_0^4 \frac{1}{1 + e^{0.5x}} \, dx = 6\left[x - \frac{1}{0.5}\ln(1 + e^{0.5x})\right]_0^4$$

$$= 6\{[4 - 2 \ln(1 + e^2)] - (0 - 2 \ln 2)\}$$

$$\approx 6.795$$

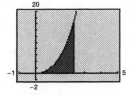

45. (a) Formula 36: $u = x, du = dx, n = 2$

$$\int x^2 e^x \, dx = x^2 e^x - 2\int x e^x \, dx$$

Formula 35: $u = x, du = dx$

$$\int x^2 e^x \, dx = x^2 e^x - 2(x - 1)e^x + C$$

$$= e^x(x^2 - 2x + 2) + C$$

(b) Let $u = x^2$ and $dv = e^x \, dx$, then $du = 2x \, dx$ and $v = e^x$.

$$\int x^2 e^x \, dx = x^2 e^x - 2\int x e^x \, dx$$

Let $u = x$ and $dv = e^x \, dx$, then $du = dx$ and $v = e^x$.

$$\int x^2 e^x \, dx = x^2 e^x - 2\left[x e^x - \int e^x \, dx \right]$$

$$= x^2 e^x - 2x e^x + 2e^x + C$$

$$= e^x(x^2 - 2x + 2) + C$$

47. Formula 12: $u = x, du = dx, a = 1, b = 1$

(a) $\displaystyle\int \frac{1}{x^2(x + 1)} \, dx = -\left[\frac{1}{x} + \ln\left| \frac{x}{1 + x} \right| \right] + C$

(b) $\displaystyle\frac{1}{x^2(x + 1)} = \frac{A}{x} + \frac{B}{x^2} + \frac{C}{x + 1}$

Basic equation: $1 = Ax(x + 1) + B(x + 1) + Cx^2$

When $x = 0$: $1 = B$

When $x = -1$: $1 = C$

When $x = 1$: $1 = 2A + 2B + C = 2A + 3 \Longrightarrow A = -1$

$$\int \frac{1}{x^2(x + 1)} \, dx = \int \left[-\frac{1}{x} + \frac{1}{x^2} + \frac{1}{x + 1} \right] dx$$

$$= -\ln|x| - \frac{1}{x} + \ln|x + 1| + C$$

$$= -\left[\frac{1}{x} + \ln|x| - \ln|x + 1| \right] + C$$

$$= -\left[\frac{1}{x} + \ln\left| \frac{x}{x + 1} \right| \right] + C$$

49. (a) $x^2 + 6x = x^2 + 6x + 9 - 9 = (x + 3)^2 - 9$

(b) $x^2 - 8x + 9 = x^2 - 8x + 16 - 16 + 9$

$$= (x - 4)^2 - 7$$

(c) $x^4 + 2x^2 - 5 = x^4 + 2x^2 + 1 - 1 - 5$

$$= (x^2 + 1)^2 - 6$$

(d) $3 - 2x - x^2 = -(x^2 + 2x - 3)$

$$= -(x^2 + 2x + 1 - 1 - 3)$$

$$= -[(x + 1)^2 - 4]$$

$$= 4 - (x + 1)^2$$

51. (a) $4x^2 + 12x + 15 = 4\left(x^2 + 3x + \frac{9}{4}\right) - 9 + 15$

$$= 4\left(x + \frac{3}{2}\right)^2 + 6$$

(b) $3x^2 - 12x - 9 = 3(x^2 - 4x - 3)$

$$= 3(x^2 - 4x + 4 - 4 - 3)$$

$$= 3[(x - 2)^2 - 7]$$

(c) $x^2 - 2x = x^2 - 2x + 1 - 1 = (x - 1)^2 - 1$

(d) $9 + 8x - x^2 = -(x^2 - 8x - 9)$

$$= -(x^2 - 8x + 16 - 16 - 9)$$

$$= -[(x - 4)^2 - 25]$$

$$= 25 - (x - 4)^2$$

53. Formula 29: $u = x + 3, du = dx, a = \sqrt{17}$

$$\int \frac{1}{x^2 + 6x - 8} \, dx = \int \frac{1}{(x + 3)^2 - 17} \, dx = \frac{1}{2\sqrt{17}} \ln\left| \frac{(x + 3) - \sqrt{17}}{(x + 3) + \sqrt{17}} \right| + C$$

55. Formula 26: $u = x - 1, du = dx, a = 1$

$$\int \frac{1}{(x - 1)\sqrt{x^2 - 2x + 2}} \, dx = \int \frac{1}{(x - 1)\sqrt{(x - 1)^2 + 1}} \, dx = -\ln\left| \frac{1 + \sqrt{x^2 - 2x + 2}}{x - 1} \right| + C$$

57. Formula 29: $u = x - 1, du = dx, a = 2$

$$\int \frac{1}{2x^2 - 4x - 6} \, dx = \frac{1}{2}\int \frac{1}{(x - 1)^2 - 4} \, dx = \frac{1}{8} \ln\left| \frac{x - 3}{x + 1} \right| + C$$

59. Formula 25: $u = x^2 + 1$, $du = 2x \, dx$, $a = 1$

$$\int \frac{x}{\sqrt{x^4 + 2x^2 + 2}} \, dx = \int \frac{x}{\sqrt{(x^2 + 1)^2 + 1}} \, dx$$

$$= \frac{1}{2} \int \frac{2x}{\sqrt{(x^2 + 1)^2 + 1}} \, dx$$

$$= \frac{1}{2} \ln \left| x^2 + 1 + \sqrt{x^4 + 2x^2 + 2} \right| + C$$

61. Formula 38: $u = 4.8 - 1.9t$, $du = -1.9 \, dt$, $n = 1$ (or Formula 37)

$$\text{Average} = \int_3^4 \frac{50}{1 + e^{4.8 - 1.9t}} \, dt$$

$$= -\frac{50}{1.9} \int_3^4 \frac{-1.9}{1 + e^{4.8 - 1.9t}} \, dt$$

$$= -\frac{50}{1.9} \left[4.8 - 1.9t - \ln(1 + e^{4.8 - 1.9t}) \right]_3^4$$

$$\approx 42.58$$

$$\approx 43$$

63. $R = \int_0^2 10{,}000 \left[1 - \frac{1}{(1 + 0.1t^2)^{1/2}} \right] dt$

$$= 10{,}000t \Big]_0^2 - 10{,}000 \int_0^2 \frac{1}{(1 + 0.1t^2)^{1/2}} \, dt$$

Formula 25: $u = \sqrt{0.1}t$, $du = \sqrt{0.1} \, dt$, $a = 1$

$$R = 20{,}000 - \frac{10{,}000}{\sqrt{0.1}} \ln \left| \sqrt{0.1} \, t \right| + \sqrt{0.1t^2 + 1} \Big]_0^2$$

$$\approx \$1138.43$$

65. Average net profit $= \dfrac{1}{13 - 10} \displaystyle\int_{10}^{13} \sqrt{0.04t - 0.3} \, dt$

$$= \frac{1}{3} \left[\frac{\sqrt{2}}{30} (2t - 15)^{3/2} \right]_{10}^{13}$$

$$\approx 0.3976$$

$$\approx \$397.6 \text{ million per year}$$

Section 6.5   Numerical Integration

1. Exact: $\displaystyle\int_0^2 x^2 \, dx = \frac{1}{3} x^3 \Big]_0^2 = \frac{8}{3} \approx 2.6667$

Trapezoidal Rule: $\displaystyle\int_0^2 x^2 \, dx \approx \frac{1}{4} \left[0 + 2\left(\frac{1}{2}\right)^2 + 2(1)^2 + 2\left(\frac{3}{2}\right)^2 + (2)^2 \right] = \frac{11}{4} = 2.7500$

Simpson's Rule: $\displaystyle\int_0^2 x^2 \, dx \approx \frac{1}{6} \left[0 + 4\left(\frac{1}{2}\right)^2 + 2(1)^2 + 4\left(\frac{3}{2}\right)^2 + (2)^2 \right] = \frac{8}{3} \approx 2.6667$

3. Exact: $\displaystyle\int_0^2 (x^4 + 1)\,dx = \frac{x^5}{5} + x\Big]_0^2 = \frac{32}{5} + 2 = \frac{42}{5} = 8.4$

Trapezoidal Rule: $\displaystyle\int_0^2 (x^4 + 1)\,dx \approx \frac{1}{4}\left[1 + 2\left(\frac{1}{16} + 1\right) + 2(1 + 1) + 2\left(\frac{81}{16} + 1\right) + 17\right] = \frac{36.25}{4} = 9.0625$

Simpson's Rule: $\displaystyle\int_0^2 (x^4 + 1)\,dx \approx \frac{1}{6}\left[1 + 4\left(\frac{1}{16} + 1\right) + 2(1 + 1) + 4\left(\frac{81}{16} + 1\right) + 17\right] = \frac{50.5}{6} \approx 8.4167$

5. Exact: $\displaystyle\int_0^2 x^3\,dx = \frac{1}{4}x^4\Big]_0^2 = 4.0000$

Trapezoidal Rule: $\displaystyle\int_0^2 x^3\,dx \approx \frac{1}{8}\left[0 + 2\left(\frac{1}{4}\right)^3 + 2\left(\frac{2}{4}\right)^3 + 2\left(\frac{3}{4}\right)^3 + 2(1)^3 + 2\left(\frac{5}{4}\right)^3 + 2\left(\frac{6}{4}\right)^3 + 2\left(\frac{7}{4}\right)^3 + 8\right] = 4.0625$

Simpson's Rule: $\displaystyle\int_0^2 x^3\,dx \approx \frac{1}{12}\left[0 + 4\left(\frac{1}{4}\right)^3 + 2\left(\frac{2}{4}\right)^3 + 4\left(\frac{3}{4}\right)^3 + 2(1)^3 + 4\left(\frac{5}{4}\right)^3 + 2\left(\frac{6}{4}\right)^3 + 4\left(\frac{7}{4}\right)^3 + 8\right] = 4.0000$

7. Exact: $\displaystyle\int_1^2 \frac{1}{x}\,dx = \ln|x|\Big]_1^2 = \ln 2 \approx 0.6931$

Trapezoidal Rule: $\displaystyle\int_1^2 \frac{1}{x}\,dx \approx \frac{1}{16}\left[1 + 2\left(\frac{8}{9}\right) + 2\left(\frac{4}{5}\right) + 2\left(\frac{8}{11}\right) + 2\left(\frac{2}{3}\right) + 2\left(\frac{8}{13}\right) + 2\left(\frac{4}{7}\right) + 2\left(\frac{8}{15}\right) + \frac{1}{2}\right] \approx 0.6941$

Simpson's Rule: $\displaystyle\int_1^2 \frac{1}{x}\,dx \approx \frac{1}{24}\left[1 + 4\left(\frac{8}{9}\right) + 2\left(\frac{4}{5}\right) + 4\left(\frac{8}{11}\right) + 2\left(\frac{2}{3}\right) + 4\left(\frac{8}{13}\right) + 2\left(\frac{4}{7}\right) + 4\left(\frac{8}{15}\right) + \frac{1}{2}\right] \approx 0.6932$

9. Exact: $\displaystyle\int_0^4 \sqrt{x}\,dx = \frac{2}{3}x^{3/2}\Big]_0^4 = \frac{16}{3} \approx 5.3333$

Trapezoidal Rule: $\displaystyle\int_0^4 \sqrt{x}\,dx \approx \frac{4}{16}\left(0 + 2\sqrt{\tfrac{1}{2}} + 2 + 2\sqrt{\tfrac{3}{2}} + 2\sqrt{2} + 2\sqrt{\tfrac{5}{2}} + 2\sqrt{3} + 2\sqrt{\tfrac{7}{2}} + 2\right) \approx 5.2650$

Simpson's Rule: $\displaystyle\int_0^4 \sqrt{x}\,dx \approx \frac{4}{24}\left(0 + 4\sqrt{\tfrac{1}{2}} + 2 + 4\sqrt{\tfrac{3}{2}} + 2\sqrt{2} + 4\sqrt{\tfrac{5}{2}} + 2\sqrt{3} + 4\sqrt{\tfrac{7}{2}} + 2\right) \approx 5.3046$

11. Exact: $\displaystyle\int_0^1 \frac{1}{1 + x}\,dx = \ln|1 + x|\Big]_0^1 \approx 0.6931$

Trapezoidal Rule: $\displaystyle\int_0^1 \frac{1}{1 + x}\,dx \approx \frac{1}{8}\left[1 + 2\left(\frac{4}{5}\right) + 2\left(\frac{2}{3}\right) + 2\left(\frac{4}{7}\right) + \frac{1}{2}\right] \approx 0.6970$

Simpson's Rule: $\displaystyle\int_0^1 \frac{1}{1 + x}\,dx \approx \frac{1}{12}\left[1 + 4\left(\frac{4}{5}\right) + 2\left(\frac{2}{3}\right) + 4\left(\frac{4}{7}\right) + \frac{1}{2}\right] \approx 0.6933$

13. (a) Trapezoidal Rule: $\displaystyle\frac{1}{8}\left[1 + 2\left(\frac{1}{1 + \frac{1}{16}}\right) + 2\left(\frac{1}{1 + \frac{1}{4}}\right) + 2\left(\frac{1}{1 + \frac{9}{16}}\right) + \frac{1}{2}\right] \approx 0.783$

(b) Simpson's Rule: $\displaystyle\frac{1}{12}\left[1 + 4\left(\frac{1}{1 + \frac{1}{16}}\right) + 2\left(\frac{1}{1 + \frac{1}{4}}\right) + 4\left(\frac{1}{1 + \frac{9}{16}}\right) + \frac{1}{2}\right] \approx 0.785$

15. (a) Trapezoidal Rule: $\displaystyle\frac{1}{8}\left[1 + 2\sqrt{\tfrac{15}{16}} + 2\sqrt{\tfrac{3}{4}} + 2\sqrt{\tfrac{7}{16}} + 0\right] \approx 0.749$

(b) Simpson's Rule: $\displaystyle\frac{1}{12}\left[1 + 4\sqrt{\tfrac{15}{16}} + 2\sqrt{\tfrac{3}{4}} + 4\sqrt{\tfrac{7}{16}} + 0\right] \approx 0.771$

17. (a) Trapezoidal Rule: $\displaystyle\frac{2}{4}\left[e^0 + 2e^{-1} + e^{-4}\right] \approx 0.877$

(b) Simpson's Rule: $\displaystyle\frac{2}{6}\left[e^0 + 4e^{-1} + e^{-4}\right] \approx 0.830$

19. (a) Trapezoidal Rule: $\frac{1}{4}\left[\frac{1}{2} + 2(0.8) + 2(1) + 2(0.8) + 2(0.5) + 2(0.30769) + 0.2\right] \approx 1.88$

    (b) Simpson's Rule: $\frac{1}{6}\left[\frac{1}{2} + 4(0.8) + 2(1) + 4(0.8) + 2(0.5) + 4(0.30769) + 0.2\right] \approx 1.89$

21. $V = \displaystyle\int_0^{t_1} c(t)e^{-rt}\, dt$

$= \displaystyle\int_0^4 \left(6000 + 200\sqrt{t}\right)e^{-0.07t}\, dt$

$\approx \dfrac{4-0}{24}\left[\left(6000 + 200\sqrt{0}\right)e^0 + 4\left(6000 + 200\sqrt{\dfrac{1}{2}}\right)e^{-0.07(1/2)} + 2\left(6000 + 200\sqrt{1}\right)e^{-0.07} + 4\left(6000 + 200\sqrt{\dfrac{3}{2}}\right)e^{-0.07(3/2)}\right.$

$+ 2\left(6000 + 200\sqrt{2}\right)e^{-0.14} + 4\left(6000 + 200\sqrt{\dfrac{5}{2}}\right)e^{-0.07(5/2)} + 2\left(6000 + 200\sqrt{3}\right)e^{-0.21}$

$\left. + 4\left(6000 + 200\sqrt{\dfrac{7}{2}}\right)e^{-0.07(7/2)} + \left(6000 + 200\sqrt{4}\right)e^{-0.28}\right] \approx \$21{,}831.20$

23. $\Delta R = \displaystyle\int_{14}^{16} 5\sqrt{8000 - x^3}\, dx$

$\approx \dfrac{16-14}{12}\left[5\sqrt{8000 - 14^3} + 4(5)\sqrt{8000 - \left(\dfrac{29}{2}\right)^3} + 2(5)\sqrt{8000 - 15^3} + 4(5)\sqrt{8000 - \left(\dfrac{31}{2}\right)^3} + 5\sqrt{8000 - 16^3}\right]$

$\approx \$678.36$

25. $P(a \le x \le b) = \displaystyle\int_a^b \dfrac{1}{\sqrt{2\pi}}e^{-x^2/2}\, dx$

$P(0 \le x \le 1) = \dfrac{1}{\sqrt{2\pi}}\displaystyle\int_0^1 e^{-x^2/2}\, dx$

$\approx \dfrac{1}{\sqrt{2\pi}}\left(\dfrac{1}{18}\right)\left[e^0 + 4e^{-(1/6)^2/2} + 2e^{-(1/3)^2/2} + 4e^{-(1/2)^2/2} + 2e^{-(2/3)^2/2} + 4e^{-(5/6)^2/2} + e^{-1/2}\right]$

$= \dfrac{1}{18\sqrt{2\pi}}\left[1 + 4e^{-1/72} + 2e^{-1/18} + 4e^{-1/8} + 2e^{-2/9} + 4e^{-25/72} + e^{-1/2}\right]$

≈ 0.3413

$= 34.13\%$

27. $P(0 \le x \le 4) = \dfrac{1}{\sqrt{2\pi}}\displaystyle\int_0^4 e^{-x^2/2}\, dx$

$\approx 0.499958 = 49.996\%$

29. $A \approx \dfrac{1000}{3(10)}\left[125 + 4(125) + 2(120) + 4(112) + 2(90) + 4(90) + 2(95) + 4(88) + 2(75) + 4(35) + 0\right]$

$= 89{,}500$ square feet

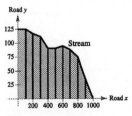

31. $f(x) = x^4$

$f'(x) = 4x^3$

$f''(x) = 12x^2$

$f'''(x) = 24x$

$f^{(4)}(x) = 24$

(a) Trapezoidal Rule: Since $f''(x)$ is maximum in $[0, 2]$ when $x = 2$, we have

$$|\text{Error}| \le \frac{(2 - 0)^3}{12(4^2)}(48) = 2.$$

(b) Simpson's Rule: Since $f^{(4)}(x) = 24$, we have

$$|\text{Error}| \le \frac{(2 - 0)^5}{180(4^4)}(24) = 0.017.$$

33. $f(x) = e^{x^3}$

$f'(x) = 3x^2 e^{x^3}$

$f''(x) = 3(3x^4 + 2x)e^{x^3}$

$f'''(x) = 3(9x^6 + 18x^3 + 2)e^{x^3}$

$f^{(4)}(x) = 3(27x^8 + 108x^5 + 60x^2)e^{x^3}$

(a) Trapezoidal Rule: Since $|f''(x)|$ is maximum in $[0, 1]$ when $x = 1$, we have

$$|\text{Error}| \le \frac{(1 - 0)^3}{12(4^2)}(15e) = \frac{5e}{64} \approx 0.212.$$

(b) Simpson's Rule: Since $|f^{(4)}(x)|$ is maximum in $[0, 1]$ when $x = 1$, we have

$$|\text{Error}| \le \frac{(1 - 0)^5}{180(4^4)}(585e) = \frac{13e}{1024} \approx 0.035.$$

35. $f(x) = x^4$

$f'(x) = 4x^3$

$f''(x) = 12x^2$

$f'''(x) = 24x$

$f^{(4)}(x) = 24$

(a) Trapezoidal Rule: Since $|f''(x)|$ is maximum in $[0, 1]$ when $x = 1$, and $|f''(1)| = 12$, we have:

$$|\text{Error}| \le \frac{(1 - 0)^3}{12n^2}(12) \le 0.0001$$

$$\frac{1}{n^2} < 0.0001$$

$$n^2 > 10,000$$

$$n > 100$$

Let $n = 101$.

(b) Simpson's Rule: Since $|f^{(4)}(x)| = 24$ in $[0, 1]$, we have:

$$|\text{Error}| \le \frac{1}{180n^4}(24) < 0.0001$$

$$n^4 > 1333.33$$

$$n > 6.04$$

Let $n = 8$ (n must be even).

37. $f(x) = e^{2x}$

$f'(x) = 2e^{2x}$

$f''(x) = 4e^{2x}$

$f'''(x) = 8e^{2x}$

$f^{(4)}(x) = 16e^{2x}$

(a) Trapezoidal Rule: Since $|f''(x)|$ is maximum in $[1, 3]$ when $x = 3$, and $|f''(3)| = 4e^6 \approx 1613.715$, we have:

$$|\text{Error}| \le \frac{(3 - 1)^3}{12n^2}(1613.715) < 0.0001$$

$$n^2 > 10,758,100$$

$$n > 3279.95$$

Take $n = 3280$.

(b) Simpson's Rule: Since $|f^{(4)}(x)|$ is maximum in $[1, 3]$ when $x = 3$, and $|f^{(4)}(3)| = 16e^6 \approx 6454.861$, we have:

$$|\text{Error}| \le \frac{(3 - 1)^5}{180n^4}(6454.861) < 0.0001$$

$$n^4 > 11,475,308.44$$

$$n > 58.2$$

Take $n = 60$ (n must be even).

39. $\displaystyle\int_1^4 x\sqrt{x + 4}\, dx \approx 19.5215 \quad (n = 100)$

41. $\displaystyle\int_2^5 10xe^{-x}\, dx \approx 3.6558 \quad (n = 100)$

43. $P_3(x) = ax^3 + bx^2 + cx + d$

$P_3'(x) = 3ax^2 + 2bx + c$

$P_3''(x) = 6ax + 2b$

$P_3'''(x) = 6a$

$P_3^{(4)}(x) = 0$

$$|\text{Error}| \leq \frac{(b-a)^5}{180n^4}[\max|P_3^{(4)}(x)|] = \frac{(b-a)^5}{180n^4}(0) = 0$$

Therefore, Simpson's Rule is exact when used to approximate a cubic polynomial.

$$\int_0^1 x^3\, dx = \frac{1-0}{3(2)}\left[0^3 + 4\left(\frac{1}{2}\right)^3 + (1)^3\right] = \frac{1}{6}\left[0 + \frac{1}{2} + 1\right] = \frac{1}{4}$$

The exact value of this integral is

$$\int_0^1 x^3\, dx = \frac{x^4}{4}\Bigg]_0^1 = \frac{1}{4}$$

which is the same as the Simpson approximation.

45. $y = \dfrac{x^2}{800}$, $y' = \dfrac{x}{400}$, $1 + (y')^2 = 1 + \dfrac{x^2}{160,000}$

$$S = \int_{-200}^{200} \sqrt{\frac{160,000 + x^2}{160,000}}\, dx$$

$$= \int_{-200}^{200} \frac{\sqrt{160,000 + x^2}}{400}\, dx$$

$$= \frac{1}{200}\int_0^{200} \sqrt{160,000 + x^2}\, dx$$

$$\approx \frac{200 - 0}{36(200)}\Bigg[400 + 4\sqrt{160,000 + \left(\frac{50}{3}\right)^2} + 2\sqrt{160,000 + \left(\frac{100}{3}\right)^2} + 4\sqrt{160,000 + (50)^2} + 2\sqrt{160,000 + \left(\frac{200}{3}\right)^2}$$

$$+ 4\sqrt{160,000 + \left(\frac{250}{3}\right)^2} + 2\sqrt{160,000 + (100)^2} + 4\sqrt{160,000 + \left(\frac{350}{3}\right)^2} + 2\sqrt{160,000 + \left(\frac{400}{3}\right)^2}$$

$$+ 4\sqrt{160,000 + (150)^2} + 2\sqrt{160,000 + \left(\frac{500}{3}\right)^2} + 4\sqrt{160,000 + \left(\frac{550}{3}\right)^2} + \sqrt{160,000 + (200)^2}\Bigg]$$

$$\approx 416.1 \text{ feet}$$

Using Formula 21: $u = x$, $du = dx$, $a = 40$

$$\frac{1}{200}\int_0^{200} \sqrt{160,000 + x^2}\, dx = \frac{1}{400}\Bigg[x\sqrt{160,000 + x^2} + 160,000\ln\Big|x + \sqrt{160,000 + x^2}\Big|\Bigg]_0^{200}$$

$$= \frac{1}{400}\Big[200\sqrt{200,000} + 160,000\ln\Big|200 + \sqrt{200,000}\Big| - 160,000\ln\sqrt{160,000}\Big]$$

$$\approx 416.1$$

47. $C = \displaystyle\int_0^{12} \big[8 - \ln(t^2 - 2t + 4)\big]\, dt \approx 58.876$ milligrams

(Simpson's Rule with $n = 100$)

49. $\displaystyle\int_0^6 1000t^2 e^{-t}\, dt \approx 1876$ subscribers

(Simpson's Rule with $n = 100$)

Section 6.6   Improper Integrals

1. This integral converges since

$$\int_0^\infty e^{-x}\,dx = \lim_{b\to\infty}\left[-e^{-x}\right]_0^b = 0 + 1 = 1.$$

3. This integral converges since

$$\int_1^\infty \frac{1}{x^2}\,dx = \lim_{b\to\infty}\left[-\frac{1}{x}\right]_1^b = 0 + 1 = 1.$$

5. This integral diverges since

$$\int_0^\infty e^{x/3}\,dx = \lim_{b\to\infty}\left[3e^{x/3}\right]_0^b = \infty.$$

7. This integral diverges since

$$\int_5^\infty \frac{x}{\sqrt{x^2-16}}\,dx = \lim_{b\to\infty}\left[\sqrt{x^2-16}\right]_5^b = \infty.$$

9. This integral diverges since

$$\int_{-\infty}^0 e^{-x}\,dx = \lim_{a\to-\infty}\left[-e^{-x}\right]_a^0 = -1 + \infty = \infty.$$

11. This integral diverges because

$$\int_1^\infty \frac{e^{\sqrt{x}}}{\sqrt{x}}\,dx = \lim_{b\to\infty}\left[2e^{\sqrt{x}}\right]_1^b = \lim_{b\to\infty}\left[2e^{\sqrt{b}} - 2e\right] = \infty.$$

13. This integral converges since

$$\int_{-\infty}^\infty 2xe^{-3x^2}\,dx = \int_{-\infty}^0 2xe^{-3x^2}\,dx + \int_0^\infty 2xe^{-3x^2}\,dx$$

$$= \lim_{a\to-\infty}\left[-\frac{1}{3}e^{-3x^2}\right]_a^0 + \lim_{b\to\infty}\left[-\frac{1}{3}e^{-3x^2}\right]_0^b$$

$$= \left(-\frac{1}{3} + 0\right) + \left(0 + \frac{1}{3}\right)$$

$$= 0.$$

15. This integral converges since $\displaystyle\int_0^4 \frac{1}{\sqrt{x}}\,dx = \lim_{b\to 0^+}\left[2\sqrt{x}\right]_b^4 = 4.$

17. This integral converges since

$$\int_0^2 \frac{1}{(x-1)^{2/3}}\,dx = \int_0^1 \frac{1}{(x-1)^{2/3}}\,dx + \int_1^2 \frac{1}{(x-1)^{2/3}}\,dx$$

$$= \lim_{b\to 1^-}\left[3(x-1)^{1/3}\right]_0^b + \lim_{a\to 1^+}\left[3(x-1)^{1/3}\right]_a^2$$

$$= 3 + 3$$

$$= 6.$$

19. This integral diverges since

$$\int_0^1 \frac{1}{1-x}\,dx = \lim_{b\to 1^-}\left[-\ln|1-x|\right]_0^b = \infty.$$

21. This integral converges since

$$\int_0^9 \frac{1}{\sqrt{9-x}}\,dx = \lim_{b\to 9^-}\left[-2\sqrt{9-x}\right]_0^b$$

$$= 0 - \left(-2\sqrt{9}\right)$$

$$= 6.$$

23. This integral diverges since $\displaystyle\int_0^1 \frac{1}{x^2}\,dx = \lim_{b\to 0^+}\left[-\frac{1}{x}\right]_b^1 = \infty.$

25. This integral converges since

$$\int_0^2 \frac{1}{\sqrt[3]{x-1}} \, dx = \int_0^1 \frac{1}{\sqrt[3]{x-1}} \, dx + \int_1^2 \frac{1}{\sqrt[3]{x-1}} \, dx$$

$$= \lim_{b \to 1^-} \left[\frac{3}{2}(x-1)^{2/3} \right]_0^b + \lim_{a \to 1^+} \left[\frac{3}{2}(x-1)^{2/3} \right]_a^2$$

$$= -\frac{3}{2} + \frac{3}{2}$$

$$= 0.$$

27. This integral converges since

$$\int_3^4 \frac{1}{\sqrt{x^2-9}} \, dx = \lim_{a \to 3^+} \left[\ln\left(x + \sqrt{x^2-9}\right) \right]_a^4 = \ln\left(4 + \sqrt{7}\right) - \ln 3 \approx 0.7954.$$

29. (a) $A = \int_1^\infty \frac{1}{x^2} \, dx = \lim_{b \to \infty} \left[-\frac{1}{x} \right]_1^b = 1$ (b) $V = \pi \int_1^\infty \left(\frac{1}{x^2} \right)^2 dx = \pi \lim_{b \to \infty} \left[-\frac{1}{3x^3} \right]_1^b = \frac{\pi}{3}$

31. $a = 1, n = 1:$ $\lim_{x \to \infty} xe^{-x}$

x	1	10	25	50
xe^{-x}	0.3679	0.0005	0.0000	0.0000

33. $a = \frac{1}{2}, n = 2:$ $\lim_{x \to \infty} x^2 e^{-(1/2)x}$

x	1	10	25	50
$x^2 e^{-(1/2)x}$	0.6065	0.6738	0.0023	0.0000

35. $\int_0^\infty x^2 e^{-x} \, dx = \lim_{b \to \infty} \left[-x^2 e^{-x} - 2xe^{-x} - 2e^{-x} \right]_0^b = 2$

37. $\int_0^\infty xe^{-2x} \, dx = \lim_{b \to \infty} \left[-\frac{1}{2}xe^{-2x} - \frac{1}{4}e^{-2x} \right]_0^b = (0-0) - \left(0 - \frac{1}{4} \right) = \frac{1}{4}$

39. (a) Present value $= \int_0^{20} 500{,}000 e^{-0.09t} \, dt = \frac{500{,}000}{-0.09} e^{-0.09t} \Big]_0^{20} \approx \$4{,}637{,}228.40$

(b) Present value $= \int_0^\infty 500{,}000 e^{-0.09t} \, dt = \lim_{b \to \infty} \frac{500{,}000}{-0.09} e^{-0.09t} \Big]_0^b \approx \$5{,}555{,}555.56$

41. $C = 650{,}000 + \int_0^n 25{,}000 e^{-0.10t} \, dt = 650{,}000 - \left[250{,}000 e^{-0.10t} \right]_0^n$

(a) For $n = 5$, we have $C = 650{,}000 - \left[250{,}000(e^{-0.50} - 1) \right] \approx \$748{,}367.34$.

(b) For $n = 10$, we have $C = 650{,}000 - \left[250{,}000(e^{-1} - 1) \right] \approx \$808{,}030.14$.

(c) For $n = \infty$, we have

$$C = 650{,}000 - \lim_{n \to \infty} \left[250{,}000 e^{-0.10t} \right]_0^n = 650{,}000 - 250{,}000(0 - 1) = \$900{,}000.$$

43. $\mu = 64.5, \sigma = 2.4$

$$f(x) = \frac{1}{\sigma\sqrt{2\pi}} e^{-(x-\mu)^2/2\sigma^2} = 0.166226 e^{-(x-64.5)^2/11.52}$$

(a) $\int_{60}^{72} f(x) \, dx \approx 0.9687$ (b) $\int_{68}^\infty f(x) \, dx \approx 0.0724$ (c) $\int_{72}^\infty f(x) \, dx \approx 0.0009$

Review Exercises for Chapter 6

1. $\displaystyle\int dt = t + C$

3. $\displaystyle\int (x + 5)^3\, dx = \frac{(x + 5)^4}{4} + C$

5. $\displaystyle\int e^{10x}\, dx = \frac{1}{10}\, e^{10x} + C$

7. $\displaystyle\int \frac{1}{5x}\, dx = \frac{1}{5} \ln|x| + C$

9. $\displaystyle\int x\sqrt{x^2 + 4}\, dx = \frac{1}{2}(x^2 + 4)^{3/2}\left(\frac{2}{3}\right) + C$

$$= \frac{1}{3}(x^2 + 4)^{3/2} + C$$

11. $\displaystyle\int \frac{2e^x}{3 + e^x}\, dx = 2 \ln(3 + e^x) + C$

13. $u = x - 2, x = u + 2, du = dx$

$$\int x(x - 2)^3\, dx = \int (u + 2)u^3\, du$$

$$= \int (u^4 + 2u^3)\, du$$

$$= \frac{u^5}{5} + \frac{u^4}{2} + C$$

$$= \frac{(x - 2)^5}{5} + \frac{(x - 2)^4}{2} + C$$

15. $u = x + 1, x = u - 1, du = dx$

$$\int x\sqrt{x + 1}\, dx = \int (u - 1)u^{1/2}\, du$$

$$= \int (u^{3/2} - u^{1/2})\, du$$

$$= \frac{2}{5}u^{5/2}\left(\frac{-2}{3}\right)u^{3/2} + C$$

$$= \frac{2}{5}(x + 1)^{5/2} - \frac{2}{3}(x + 1)^{3/2} + C$$

$$= \frac{2}{15}(x + 1)^{3/2}[3(x + 1) - 5] + C$$

$$= \frac{2}{15}(x + 1)^{3/2}(3x - 2) + C$$

17. $\displaystyle\int 2x\sqrt{x - 3}\, dx = 2\int (u + 3)\sqrt{u}\, du$

$$= 2\int (u^{3/2} + 3u^{1/2})\, du$$

$$= 2\left(\frac{2}{5}u^{5/2} + 2u^{3/2}\right) + C$$

$$= \frac{4}{5}u^{3/2}(u + 5) + C$$

$$= \frac{4}{5}(x - 3)^{3/2}(x + 2) + C$$

19. Let $u = 1 - x$, then $x = 1 - u$ and $dx = -du$.

$$\int (x + 1)\sqrt{1 - x}\, dx = -\int (2 - u)\sqrt{u}\, du$$

$$= -\int (2u^{1/2} - u^{3/2})\, du$$

$$= -\left(\frac{4}{3}u^{3/2} - \frac{2}{5}u^{5/2}\right) + C$$

$$= -\frac{2}{15}u^{3/2}(10 - 3u) + C$$

$$= -\frac{2}{15}(1 - x)^{3/2}(3x + 7) + C$$

21. $u = x - 2, x = u + 2, du = dx$

$$\int_2^3 x\sqrt{x - 2}\, dx = \int_0^1 (u + 2)u^{1/2}\, du$$

$$= \int_0^1 (u^{3/2} + 2u^{1/2})\, du$$

$$= \left[\frac{2}{5}u^{5/2} + \frac{4}{3}u^{3/2}\right]_0^1$$

$$= \frac{2}{5} + \frac{4}{3} = \frac{26}{15}$$

23. $u = x - 1, x = u + 1, du = dx$

$$\int_1^3 x^2(x - 1)^3 \, dx = \int_0^2 (u + 1)^2 u^3 \, du$$

$$= \int_0^2 (u^5 + 2u^4 + u^3) \, du$$

$$= \left[\frac{u^6}{6} + \frac{2u^5}{5} + \frac{u^4}{4}\right]_0^2$$

$$= \frac{32}{3} + \frac{64}{4} + 4$$

$$= \frac{412}{15} \approx 27.467$$

25. (a) $P(0 \le x \le 0.80) = \int_0^{0.8} \frac{105}{16} x^2 \sqrt{1 - x} \, dx, \quad u = 1 - x, x = 1 - u, du = -dx$

$$\int_1^{0.2} \frac{105}{16}(1 - u)^2 u^{1/2}(-du) = -\frac{105}{16} \int_1^{0.2} (u^{1/2} - 2u^{3/2} + u^{5/2}) \, du$$

$$= -\frac{105}{16}\left[\frac{2}{3}u^{3/2} - \frac{4}{5}u^{5/2} + \frac{2}{7}u^{7/2}\right]_1^{0.2}$$

$$\approx 0.696$$

(b) Solve the equation $\int_0^b \frac{105}{16} x^2 \sqrt{1 - x} \, dx = 0.5$ for b. From part (a):

$$\frac{-105}{16}\left[\frac{2}{3}u^{3/2} - \frac{4}{5}u^{5/2} + \frac{2}{7}u^{7/2}\right]_1^{1 - b} = 0.5$$

$$\left[\frac{2}{3}(1 - b)^{3/2} - \frac{4}{5}(1 - b)^{5/2} + \frac{2}{7}(1 - b)^{7/2}\right] - \left[\frac{2}{3} - \frac{4}{5} + \frac{2}{7}\right] = \frac{-8}{105}$$

$$\left[\frac{2}{3}(1 - b)^{3/2} - \frac{4}{5}(1 - b)^{5/2} + \frac{2}{7}(1 - b)^{7/2}\right] = \frac{8}{105}$$

Using a graphing utility to solve for b, we obtain $b \approx 0.693$.

27. (a) Average net profit $= \frac{1}{13.7} \int_7^{13} (-925.08 + 154.753t \ln t) \, dt$

$$\approx \frac{1}{6}(15{,}970)$$

$$\approx \$2661.67 \text{ million per year}$$

(b) Total net profit $= \int_7^{13} (-925.08 + 154.753t \ln t) \, dt$

$$\approx \$15{,}970 \text{ million}$$

Note: Use integration by parts to integrate $\int t \ln t \, dt$: $u = \ln t, \, du = \frac{1}{t} \, dt, \, dv = t \, dt, \, v = \frac{t^2}{2}$

$$\int t \ln t \, dt = \frac{t^2}{2} \ln t - \int \frac{t^2}{2} \frac{1}{t} \, dt = \frac{t^2}{2} \ln t - \frac{t^2}{4} + C$$

29. Let $u = \ln x$, $dv = \left(1/\sqrt{x}\right) dx$, $du = (1/x) \, dx$, $v = 2\sqrt{x}$.

$$\int \frac{\ln x}{\sqrt{x}} \, dx = 2\sqrt{x} \ln x - \int 2\sqrt{x}\left(\frac{1}{x}\right) dx$$

$$= 2\sqrt{x} \ln x - \int 2x^{-1/2} \, dx$$

$$= 2\sqrt{x} \ln x - 4\sqrt{x} + C$$

31. Use integration by parts and let $u = x - 1$ and $dv = e^x \, dx$, then $du = dx$ and $v = e^x$.

$$\int (x - 1)e^x \, dx = (x - 1)e^x - \int e^x \, dx$$

$$= (x - 1)e^x - e^x + C$$

$$= (x - 2)e^x + C$$

33. Let $u = 2x^2$, $dv = e^{2x}$, $du = 4x \, dx$, $v = \frac{1}{2}e^{2x}$.

$$\int 2x^2 e^{2x} \, dx = (2x^2)\left(\frac{1}{2}e^{2x}\right) - \int \frac{1}{2}e^{2x}(4x \, dx)$$

$$= x^2 e^{2x} - \int 2x e^{2x} \, dx$$

Use integration by parts again for the integral on the right:
$u = 2x$, $dv = e^{2x}$, $du = 2 \, dx$, $v = \frac{1}{2}e^{2x}$.

$$\int 2x^2 e^{2x} \, dx = x^2 e^{2x} - \left[(2x)\left(\frac{1}{2}e^{2x}\right) - \int \frac{1}{2}e^{2x}(2) \, dx\right]$$

$$= x^2 e^{2x} - x e^{2x} + \int e^{2x} \, dx$$

$$= x^2 e^{2x} - x e^{2x} + \frac{1}{2}e^{2x} + C$$

35. Present value $= \displaystyle\int_0^{t_1} c(t)e^{-rt} \, dt$

$$= \int_0^5 10{,}000e^{-0.04t} \, dt$$

$$= -250{,}000e^{-0.04t} \Big]_0^5$$

$$= \$45{,}317.31$$

37. Present value $= \displaystyle\int_0^{t_1} c(t)e^{-rt} \, dt$

$$= \int_0^{10} 12{,}000e^{-0.05t} \, dt$$

$$= (-240{,}000t - 4{,}800{,}000)e^{-0.05t} \Big]_0^{10}$$

$$= \$432{,}979.25$$

39. (a) $\displaystyle\int_0^5 1000e^{-0.05t} \, dt \approx \$4423.98$ (5%)

$$\int_0^5 1000e^{-0.1t} \, dt \approx \$3934.69 \quad (10\%)$$

$$\int_0^5 1000e^{-0.15t} \, dt \approx \$3517.56 \quad (15\%)$$

(b) $\displaystyle\int_0^{20} 100{,}000e^{-0.08t} \, dt \approx \$997{,}629.35$

41. Present value $= \displaystyle\int_0^{10} 6000e^{-0.06t} \, dt = \$45{,}118.84$

43. Use partial fractions.

$$\frac{1}{x(x + 5)} = \frac{A}{x} + \frac{B}{x + 5}$$

Basic equation: $1 = A(x + 5) + Bx$

When $x = 0$: $1 = 5A$, $A = \frac{1}{5}$

When $x = -5$: $1 = -5B$, $B = -\frac{1}{5}$

$$\int \frac{1}{x(x + 5)} \, dx = \frac{1}{5}\int\left[\frac{1}{x} - \frac{1}{x + 5}\right] dx$$

$$= \frac{1}{5}\Big[\ln|x| - \ln|x + 5|\Big] + C$$

$$= \frac{1}{5}\ln\left|\frac{x}{x + 5}\right| + C$$

45. Partial fractions: $\dfrac{4x - 13}{x^2 - 3x - 10} - \dfrac{A}{x - 5} + \dfrac{B}{x + 2}$

Basic equation: $4x - 13 = A(x + 2) + B(x - 5)$

When $x = 2$: $-21 = -7B$, $B = 3$

When $x = 5$: $7 = 7A$, $A = 1$

$$\int \frac{4x - 13}{x^2 - 3x - 10} \, dx = \int\left(\frac{1}{x - 5} + \frac{3}{x + 2}\right) dx$$

$$= \ln|x - 5| + 3\ln|x + 2| + C$$

47. $\dfrac{x^2}{x^2 + 2x - 15} = 1 - \dfrac{2x - 15}{(x + 5)(x - 3)}$

Use partial fractions.

$$\frac{2x - 15}{(x + 5)(x - 3)} = \frac{A}{x + 5} + \frac{B}{x - 3}$$

Basic equation: $2x - 15 = A(x - 3) + B(x + 5)$

When $x = -5$: $-25 = -8A, A = \dfrac{25}{8}$

When $x = 3$: $-9 = 8B, B = -\dfrac{9}{8}$

$$\int \frac{x^2}{x^2 + 2x - 15}\,dx = \int \left[1 - \frac{25}{8}\left(\frac{1}{x + 5}\right) + \frac{9}{8}\left(\frac{1}{x - 3}\right)\right] dx$$

$$= x - \frac{25}{8}\ln|x + 5| + \frac{9}{8}\ln|x - 3| + C$$

49. (a) $y = \dfrac{L}{1 + be^{-kt}} = \dfrac{10{,}000}{1 + be^{-kt}}$

Since $y = 1250$ when $t = 0$,

$$\frac{10{,}000}{1 + b} = 1250 \Longrightarrow 1 + b = 8 \Longrightarrow b = 7.$$

When $t = 24$ (6 months), $y = 6500 = \dfrac{10{,}000}{1 + 7e^{-k(24)}}$.

Solving for k,

$$1 + 7e^{-24k} = \frac{100}{65}$$

$$7e^{-24k} = \frac{35}{65}$$

$$e^{-24k} = \frac{1}{13}$$

$$k = \frac{-1}{24}\ln\left(\frac{1}{13}\right) \approx 0.106873.$$

Thus, $y = \dfrac{10{,}000}{1 + 7e^{-0.106873t}}$.

(b)

t	0	3	6	12	24
y	1250	1645	2134	3400	6500

(c) The sales will be 7500 when $t \approx 28$.

51. Use Formula 23 from the tables and let $u = x$, $a = 5$, and $du = dx$.

$$\int \frac{\sqrt{x^2 + 25}}{x}\,dx = \sqrt{x^2 + 25} - 5\ln\left|\frac{5 + \sqrt{x^2 + 25}}{x}\right| + C$$

53. Use Formula 29 and let $u = x$, $a = 2$.

$$\int \frac{1}{x^2 - 4}\,dx = \frac{1}{4}\ln\left|\frac{x - 2}{x + 2}\right| + C$$

53. Use Formula 19 and let $u = x$, $a = 1 = b$.

$$\int_0^3 \frac{x}{\sqrt{1 + x}}\,dx = \left[-\frac{2(2 - x)}{3}\sqrt{1 + x}\right]_0^3 = \frac{8}{3}$$

57. Use Formula 17 and let $u = x$, $a = 1$, $b = 1$.

$$\int \frac{\sqrt{x + 1}}{x}\,dx = 2\sqrt{1 + x} + \int \frac{1}{x\sqrt{1 + x}}\,dx = 2\sqrt{1 + x} + \ln\left|\frac{\sqrt{1 + x} - 1}{\sqrt{1 + x} + 1}\right| + C \qquad \text{(Formula 15)}$$

59. Use Formula 36 and $u = x - 5, n = 3, du = dx.$

$$\int (x - 5)^3 e^{x-5}\, dx = (x - 5)^3 e^{x-5} - 3 \int (x - 5)^2 e^{x-5}\, dx$$

$$= (x - 5)^3 e^{x-5} - 3\left[\int (x - 5)^2 e^{x-5} - 2 \int (x - 5)e^{x-5}\, dx\right]$$

$$= (x - 5)^3 e^{x-5} - 3(x - 5)^2 e^{x-5} + 6(x - 6)e^{x-5} + C \quad \text{(Formula 35)}$$

61. $\dfrac{1}{x^2 + 4x - 21} = \dfrac{1}{x^2 + 4x + 4 - 25} = \dfrac{1}{(x + 2)^2 - 25}$

$$\int \frac{1}{(x + 1)^2 - 25}\, dx = \frac{1}{10} \ln\left|\frac{(x + 2) - 5}{(x + 2) + 5}\right| + C \quad \text{(Formula 29)}$$

$$= \frac{1}{10} \ln\left|\frac{x - 3}{x + 7}\right| + C$$

63. $x^2 - 10x = x^2 - 10x + 25 - 25 = (x - 5)^2 - 25$

$$\int \sqrt{x^2 - 10x}\, dx = \int \sqrt{(x - 5)^2 - 5^2}\, dx$$

$$= \frac{1}{2}\left[(x - 5)\sqrt{(x - 5)^2 - 5^2} + 25 \ln\left|(x - 5) + \sqrt{(x - 5)^2 - 5^2}\right|\right] + C \quad \text{(Formula 21)}$$

65. $\displaystyle\int_1^3 \frac{1}{x^2}\, dx \approx \frac{3 - 1}{2(4)}\left[1 + 2\left(\frac{4}{9}\right) + 2\left(\frac{1}{4}\right) + 2\left(\frac{4}{25}\right) + \frac{1}{9}\right] = 0.705$

67. $\displaystyle\int_1^2 \frac{1}{1 + \ln x}\, dx = \frac{1}{8}\left(1 + 2\left[\frac{1}{1 + \ln(5/4)}\right] + 2\left[\frac{1}{1 + \ln(3/2)}\right] + 2\left[\frac{1}{1 + \ln(7/4)}\right] + \frac{1}{1 + \ln 2}\right) \approx 0.741$

69. $\displaystyle\int_1^2 \frac{1}{x^3}\, dx \approx \frac{2 - 1}{3(4)}\left[1 + 4\left(\frac{4}{5}\right)^3 + 2\left(\frac{2}{3}\right)^3 + 4\left(\frac{4}{7}\right)^3 + \frac{1}{8}\right] \approx 0.376$

71. $\displaystyle\int_0^1 \frac{x^{3/2}}{2 - x^2}\, dx \approx \frac{1}{12}\left[0 + \frac{8}{31} + \frac{2\sqrt{2}}{7} + \frac{24\sqrt{3}}{23} + 1\right] \approx 0.289$

73. $f(x) = e^{2x}$

$f'(x) = 2e^{2x}$

$f''(x) = 4e^{2x}$

$|f''(x)| \leq 4e^{2(2)} = 4e^4$ on $[0, 2].$

$|E| \leq \dfrac{(2 - 0)^3}{12(4)^2}(4e^4) \leq 9.0997$

75. $f(x) = \dfrac{1}{x - 1} = (x - 1)^{-1}$

$f'(x) = -(x - 1)^{-2}$

$f''(x) = 2(x - 1)^{-3}$

$f'''(x) = -6(x - 1)^{-4}$

$f^{(4)}(x) = 24(x - 1)^{-5} = \dfrac{24}{(x - 1)^5}$

$|f^{(4)}(x)| \leq 24$ on $[2, 4].$

$|E| \leq \dfrac{(4 - 2)^5}{180(4)^4}(24) \approx 0.017$

77. $\displaystyle\int_0^\infty 4xe^{-2x^2}\,dx = \lim_{b\to\infty} -\int_0^b e^{-2x^2}(-4x\,dx)$

$\displaystyle\qquad = \lim_{b\to\infty}\left[-e^{2x^2}\right]_0^b$

$\displaystyle\qquad = 1$

79. $\displaystyle\int_{-\infty}^0 \frac{1}{3x^2}\,dx = \lim_{b\to-\infty}\int_b^{-1}\frac{1}{3}x^{-2}\,dx + \lim_{a\to0^-}\int_{-1}^a \frac{1}{3}x^{-2}\,dx$

$\displaystyle\qquad = \lim_{b\to-\infty}\left[\frac{-1}{3x}\right]_b^{-1} + \lim_{a\to0^-}\left[\frac{-1}{3x}\right]_{-1}^0 \quad \text{(diverges)}$

81. $\displaystyle\int_0^4 \frac{1}{\sqrt{4x}}\,dx = \lim_{a\to0^+}\int_a^4 \frac{1}{2}x^{-1/2}\,dx$

$\displaystyle\qquad = \lim_{a\to0^+}\left[x^{1/2}\right]_a^4$

$\displaystyle\qquad = 2$

83. $\displaystyle\int_2^3 \frac{1}{\sqrt{x-2}}\,dx = \lim_{b\to2^+}\left[2(x-2)^{1/2}\right]_b^3 = 2$

85. (a) Present value $= \displaystyle\int_0^{15} 50{,}000e^{-0.06t}\,dt$

$\displaystyle\qquad = \left[\frac{50{,}000}{-0.06}e^{-0.06t}\right]_0^{15}$

$\displaystyle\qquad \approx \$494{,}525.28$

(b) Present value $= \displaystyle\int_0^\infty 50{,}000e^{-0.06t}\,dt$

$\displaystyle\qquad = \lim_{b\to\infty}\left[\frac{50{,}000}{-0.06}e^{-0.06t}\right]_0^b$

$\displaystyle\qquad \approx \$833{,}333.33$

87. (a) $\displaystyle\int_{500}^{650} y\,dx \approx 0.431 \text{ or } 43.1\%$

(b) $\displaystyle\int_{650}^{800} y\,dx \approx 0.108 \text{ or } 10.8\%$

(c) $\displaystyle\int_{750}^{800} y\,dx \approx 0.013 \text{ or } 1.3\%$

Note: $\displaystyle\int_{200}^{800} y\,dx \approx 0.98 \text{ or } 1.0$

Practice Test for Chapter 6

1. Evaluate $\int x\sqrt{x+3}\,dx.$

2. Evaluate $\int \dfrac{x}{(x-2)^3}\,dx.$

3. Evaluate $\int \dfrac{1}{3x+\sqrt{x}}\,dx.$

4. Evaluate $\int \dfrac{\ln 7x}{x}\,dx.$

5. Evaluate $\int xe^{2x}\,dx.$

6. Evaluate $\int x^3 \ln x\,dx.$

7. Evaluate $\int x^2\sqrt{x-6}\,dx.$

8. Evaluate $\int x^2 e^{4x}\,dx.$

9. Evaluate $\int \dfrac{-5}{x^2+x-6}\,dx.$

10. Evaluate $\int \dfrac{x+12}{x^2+4x}\,dx.$

11. Evaluate $\int \dfrac{5x+3}{(x+2)^2}\,dx.$

12. Evaluate $\int \dfrac{3x^3+9x^2-x+3}{x(x+3)}\,dx.$

13. Evaluate $\int \dfrac{1}{x^2\sqrt{16-x^2}}\,dx.$ (Use tables.)

14. Evaluate $\int (\ln x)^3\,dx.$ (Use tables.)

15. Evaluate $\int \dfrac{1200}{1+e^{0.06x}}\,dx.$ (Use tables.)

16. Approximate the integral using (a) the Trapezoidal Rule and (b) Simpson's Rule.

$$\int_0^4 \sqrt{3+x^3}\,dx, \quad n=8$$

17. Approximate the integral using (a) the Trapezoidal Rule and (b) Simpson's Rule.

$$\int_0^2 e^{-x^2/2}\,dx, \quad n=4$$

18. Determine the divergence or convergence of the integral. Evaluate the integral if it converges.

$$\int_0^9 \frac{1}{\sqrt{x}}\,dx$$

19. Determine the divergence or convergence of the integral. Evaluate the integral if it converges.

$$\int_1^\infty \frac{1}{x-3}\,dx$$

20. Determine the divergence or convergence of the integral. Evaluate the integral if it converges.

$$\int_{-\infty}^0 e^{-3x}\,dx$$

Graphing Calculator Required

21. Use a program similar to that on page 430 of the textbook to approximate

$$\int_0^3 \frac{1}{\sqrt{x^3 + 1}}\, dx$$

when $n = 50$ and $n = 100$.

22. Complete the following chart using Simpson's Rule to determine the convergence or divergence of

$$\int_0^1 \frac{1}{\sqrt{1 - x^2}}\, dx.$$

n	100	1000	10,000
$\displaystyle\int_1^{0.9999999} \frac{1}{\sqrt{1 - x^2}}\, dx$			

Note: The upper limit cannot equal 1 to avoid division by zero. Let $b = 0.9999999$.

C H A P T E R 7
Functions of Several Variables

C H A P T E R 7
Functions of Several Variables

Section 7.1   The Three-Dimensional Coordinate System

Solutions to Odd-Numbered Exercises

1. (a) and (b)

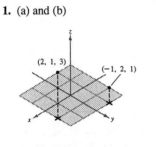

3. (a) and (b)

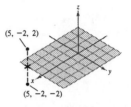

5. $d = \sqrt{(8-4)^2 + (2-1)^2 + (6-5)^2}$

$\quad = \sqrt{18}$

$\quad = 3\sqrt{2}$

7. $d = \sqrt{(-3+1)^2 + (4+5)^2 + (-4-7)^2}$

$\quad = \sqrt{206}$

9. Midpoint $= \left(\dfrac{6+(-2)}{2}, \dfrac{-9+(-1)}{2}, \dfrac{1+5}{2} \right)$

$\qquad\qquad = (2, -5, 3)$

11. Midpoint $= \left(\dfrac{-5+6}{2}, \dfrac{-2+3}{2}, \dfrac{5+(-7)}{2} \right)$

$\qquad\qquad = \left(\dfrac{1}{2}, \dfrac{1}{2}, -1 \right)$

13. $(2, -1, 3) = \left(\dfrac{x+(-2)}{2}, \dfrac{y+1}{2}, \dfrac{z+1}{2} \right)$

$\quad 2 = \dfrac{x-2}{2} \qquad -1 = \dfrac{y+1}{2} \qquad 3 = \dfrac{z+1}{2}$

$\quad 4 = x-2 \qquad\;\; -2 = y+1 \qquad\;\; 6 = z+1$

$\quad x = 6 \qquad\qquad y = -3 \qquad\qquad z = 5(x, y, z)$

$\qquad\qquad\qquad\qquad\qquad\qquad\qquad\qquad = (6, -3, 5)$

15. $\left(\dfrac{3}{2}, 1, 2 \right) = \left(\dfrac{x+2}{2}, \dfrac{y+0}{2}, \dfrac{z+3}{2} \right)$

$\quad \dfrac{3}{2} = \dfrac{x+2}{2} \qquad 1 = \dfrac{y}{2} \qquad 2 = \dfrac{z+3}{2}$

$\quad x = 1 \qquad\qquad y = 2 \qquad z = 1(x, y, z) = (1, 2, 1)$

17. Let $A = (0, 0, 0)$, $B = (2, 2, 1)$, and $C = (2, -4, 4)$. Then we have the following.

$\quad d(AB) = \sqrt{(2-0)^2 + (2-0)^2 + (1-0)^2} = 3$

$\quad d(AC) = \sqrt{(2-0)^2 + (-4-0)^2 + (4-0)^2} = 6$

$\quad d(BC) = \sqrt{(2-2)^2 + (-4-2)^2 + (4-1)^2}$

$\qquad\qquad = 3\sqrt{5}$

The triangle is a right triangle since

$\quad d^2(AB) + d^2(AC) = (3)^2 + (6)^2$

$\qquad\qquad\qquad\qquad = 9 + 36$

$\qquad\qquad\qquad\qquad = 45$

$\qquad\qquad\qquad\qquad = d^2(BC).$

19. Let $A = (-2, 2, 4)$, $B = (-2, 2, 6)$, and $C = (-2, 4, 8)$. Then we have the following.

$\quad d(AB) = \sqrt{[-2-(-2)]^2 + (2-2)^2 + (6-4)^2}$

$\qquad\qquad = 2$

$\quad d(AC) = \sqrt{[-2-(-2)]^2 + (4-2)^2 + (8-4)^2}$

$\qquad\qquad = 2\sqrt{5}$

$\quad d(BC) = \sqrt{[-2-(-2)]^2 + (4-2)^2 + (8-6)^2}$

$\qquad\qquad = 2\sqrt{2}$

The triangle is neither right nor isosceles.

21. $x^2 + (y - 2)^2 + (z - 2)^2 = 4$

23. The midpoint of the diameter is the center.

$$\text{Center} = \left(\frac{2 + 1}{2}, \frac{1 + 3}{2}, \frac{3 - 1}{2} \right) = \left(\frac{3}{2}, 2, 1 \right)$$

The radius is the distance between the center and either endpoint.

$$\text{Radius} = \sqrt{\left(2 - \frac{3}{2} \right)^2 + (1 - 2)^2 + (3 - 1)^2} = \sqrt{\frac{1}{4} + 1 + 4} = \frac{\sqrt{21}}{2}$$

$$\left(x - \frac{3}{2} \right)^2 + (y - 2)^2 + (z - 1)^2 = \frac{21}{4}$$

25. $(x - 1)^2 + (y - 1)^2 + (z - 5)^2 = 9$

27. The midpoint of the diameter is the center.

$$\text{Center} = \left(\frac{2 + 0}{2}, \frac{0 + 6}{2}, \frac{0 + 0}{2} \right) = (1, 3, 0)$$

The radius is the distance between the center and either endpoint.

$$\text{Radius} = \sqrt{(1 - 2)^2 + (3 - 0)^2 + (0 - 0)^2} = \sqrt{10}$$

$$(x - 1)^2 + (y - 3)^2 + (z - 0)^2 = \left(\sqrt{10} \right)^2$$

$$(x - 1)^2 + (y - 3)^2 + z^2 = 10$$

29. The distance from $(-2, 1, 1)$ to the xy-coordinate plane is the radius $r = 1$. Therefore, we have the following.

$$[x - (-2)]^2 + (y - 1)^2 + (z - 1)^2 = 1^2$$

$$(x + 2)^2 + (y - 1)^2 + (z - 1)^2 = 1$$

31.
$$x^2 + y^2 + z^2 - 5x = 0$$
$$\left(x - \tfrac{5}{2} \right)^2 + (y - 0)^2 + (z - 0)^2 = \tfrac{25}{4}$$
Center: $\left(\tfrac{5}{2}, 0, 0 \right)$ Radius: $\tfrac{5}{2}$

33. $(x - 1)^2 + (y + 3)^2 + (z + 4)^2 = 25$

Center: $(1, -3, -4)$ Radius: 5

35. $(x - 1)^2 + (y - 3)^2 + (z - 2)^2 = -\dfrac{3}{2} + 1 + 9 + 4 = \dfrac{25}{2}$

Center: $(1, 3, 2)$

Radius: $\dfrac{5}{\sqrt{2}}$

37. To find the xy-trace, we let $z = 0$.

$$(x - 1)^2 + (y - 3)^2 + (0 - 2)^2 = 25$$

$$(x - 1)^2 + (y - 3)^2 = 21$$

The xy-trace is a circle centered at $(1, 3)$ with radius of $\sqrt{21}$ in the xy-plane.

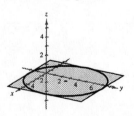

39. To find the xy-trace, we let $z = 0$.

$$x^2 + y^2 + (0)^2 - 6x - 10y + 6(0) + 30 = 0$$

$$(x - 3)^2 + (y - 5)^2 = 4$$

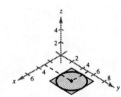

41. $(x - 2)^2 + (y - 2)^2 + (z - 3)^2 = 29$

To find the yz-trace, we let $x = 0$.

$$(0 - 2)^2 + (y - 2)^2 + (z - 3)^2 = 29$$

$$4 + (y - 2)^2 + (z - 3)^2 = 29$$

$$(y - 2)^2 + (z - 3)^2 = 25$$

The yz-trace is a circle centered at $(2, 3)$ with a radius of 5 in the yz-plane.

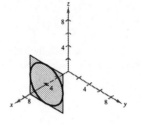

43. (a) To find the trace with $z = 3$, replace z with 3 in the equation of the sphere

$$x^2 + y^2 + 3^2 = 25$$

$$x^2 + y^2 = 16.$$

The trace is a circle centered at $(0, 0, 3)$ with a radius of 4.

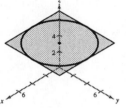

(b) To find the trace with $x = 4$, replace x with 4 in the equation of the sphere

$$4^2 + y^2 + z^2 = 25$$

$$y^2 + z^2 = 9.$$

The trace is a circle centered at $(4, 0, 0)$ with a radius of 3.

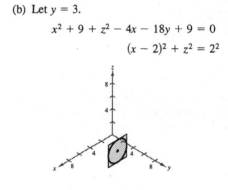

45. (a) Let $x = 2$.

$$4 + y^2 + z^2 - 8 - 6y + 9 = 0$$

$$(y - 3)^2 + z^2 = 2^2$$

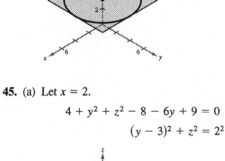

(b) Let $y = 3$.

$$x^2 + 9 + z^2 - 4x - 18y + 9 = 0$$

$$(x - 2)^2 + z^2 = 2^2$$

47. Since the crystal is a cube, $A = (3, 3, 0)$. Thus, $(x, y, z) = (3, 3, 3)$.

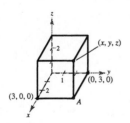

Section 7.2   Surfaces in Space

1. To find the x-intercept, let $y = 0$ and $z = 0$.

$$4x = 12 \Longrightarrow x = 3$$

To find the y-intercept, let $x = 0$ and $z = 0$.

$$2y = 12 \Longrightarrow y = 6$$

To find the z-intercept, let $x = 0$ and $y = 0$.

$$6z = 12 \Longrightarrow z = 2$$

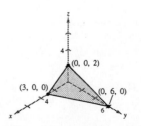

3. To find the *x*-intercept, let $y = 0$ and $z = 0$.

$$3x = 15 \Longrightarrow x = 5$$

To find the *y*-intercept, let $x = 0$ and $z = 0$.

$$3y = 15 \Longrightarrow y = 5$$

To find the *z*-intercept, let $x = 0$ and $y = 0$.

$$5z = 15 \Longrightarrow z = 3$$

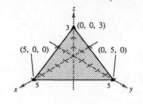

5. To find the *x*-intercept, let $y = 0$ and $z = 0$.

$$2x = 8 \Longrightarrow x = 4$$

To find the *y*-intercept, let $x = 0$ and $z = 0$.

$$-y = 8 \Longrightarrow y = -8$$

To find the *z*-intercept, let $x = 0$ and $y = 0$.

$$3z = 8 \Longrightarrow z = \frac{8}{3}$$

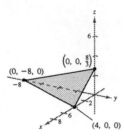

7. Since the coefficients of *x* and *y* are zero, the only intercept is the *z*-intercept of 3. The plane is parallel to the *xy*-plane.

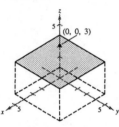

9. The *y*- and *z*-intercepts are both 5, and the plane is parallel to the *x*-axis.

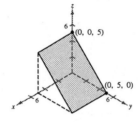

11. The only intercept is the origin. The *xy*-trace is the line $x + y = 0$. The *xz*-trace is the line $x - z = 0$. The *yz*-trace is the line $y - z = 0$.

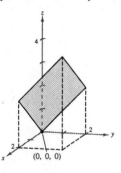

13. For the first plane, $a_1 = 5$, $b_1 = -3$, and $c_1 = 1$, and for the second plane, $a_2 = 1$, $b_2 = 4$, and $c_2 = 7$. Therefore, we have

$$a_1a_2 + b_1b_2 + c_1c_2 = (5)(1) + (-3)(4) + (1)(7)$$
$$= 5 - 12 + 7$$
$$= 0.$$

The planes are perpendicular.

15. For the first plane, $a_1 = 1$, $b_1 = -5$, and $c_1 = -1$, and for the second plane, $a_2 = 5$, $b_2 = -25$, and $c_2 = -5$. Therefore, we have $a_2 = 5a_1$, $b_2 = 5b_1$, $c_2 = 5c_1$.

The planes are parallel.

17. For the first plane, $a_1 = 1$, $b_1 = 2$, and $c_1 = 0$, and for the second plane, $a_2 = 4$, $b_2 = 8$, and $c_2 = 0$. Therefore, we have $a_2 = 4a_1$, $b_2 = 4b_1$, $c_2 = 4c_1$.

The planes are parallel.

19. For the first plane, $a_1 = 2$, $b_1 = 1$, and $c_1 = 0$, and for the second plane, $a_2 = 3$, $b_2 = 0$, and $c_2 = -5$. The planes are not parallel since $3a_1 = 2a_2$ and $3b_1 \neq 2b_2$ and the planes are not perpendicular since

$$a_1a_2 + b_1b_2 + c_1c_2 = (2)(3) + (1)(0) + (0)(-5)$$
$$= 6$$
$$\neq 0.$$

21. For the first plane, $a_1 = 1$, $b_1 = 0$, and $c_1 = 0$, and for the second plane, $a_2 = 0$, $b_2 = 1$, and $c_2 = 0$. Therefore, we have

$$a_1a_2 + b_1b_2 + c_1c_2 = (1)(0) + (0)(1) + (0)(0) = 0.$$

The planes are perpendicular.

23. $D = \dfrac{|ax_0 + by_0 + cz_0 + d|}{\sqrt{a^2 + b^2 + c^2}}$

$= \dfrac{|2(0) + 3(0) + 1(0) - 12|}{\sqrt{(2)^2 + (3)^2 + (1)^2}}$

$= \dfrac{12}{\sqrt{4 + 9 + 1}}$

$= \dfrac{12}{\sqrt{14}}$

$= \dfrac{6\sqrt{14}}{7}$

25. $D = \dfrac{|ax_0 + by_0 + cz_0 + d|}{\sqrt{a^2 + b^2 + c^2}}$

$= \dfrac{|(3)(1) + (-1)(5) + 2(-4) - 6|}{\sqrt{(3)^2 + (-1)^2 + (2)^2}}$

$= \dfrac{16}{\sqrt{14}}$

$= \dfrac{8\sqrt{14}}{7}$

≈ 4.276

27. $D = \dfrac{|ax_0 + by_0 + cz_0 + d|}{\sqrt{a^2 + b^2 + c^2}}$

$= \dfrac{|2(1) - 4(0) + 3(-1) + (-12)|}{\sqrt{4 + 16 + 9}}$

$= \dfrac{13}{\sqrt{29}}$

$= \dfrac{13\sqrt{29}}{29}$

29. $D = \dfrac{|ax_0 + by_0 + cz_0 + d|}{\sqrt{a^2 + b^2 + c^2}}$

$= \dfrac{|2(3) - 3(2) + 4(-1) - 24|}{\sqrt{4 + 9 + 16}}$

$= \dfrac{28}{\sqrt{29}}$

$= \dfrac{28\sqrt{29}}{29}$

31. The graph is an ellipsoid that matches graph (c).

33. The graph of

$$x^2 + y^2 - \frac{z^2}{4} = 1$$

is a hyperboloid of one sheet that matches graph (f).

35. The graph of

$$y = \frac{x^2}{1} + \frac{z^2}{4}$$

is an elliptic paraboloid that matches graph (d).

37. The graph of

$$z = \frac{y^2}{4} - \frac{x^2}{1}$$

is a hyperbolic paraboloid that matches graph (a).

39. $x^2 - y - z^2 = 0$

Trace in xy-plane ($z = 0$):	$y = x^2$	Parabola
Trace in plane $y = 1$:	$x^2 - z^2 = 1$	Hyperbola
Trace in yz-plane ($x = 0$):	$y = -z^2$	Parabola

41. $\dfrac{x^2}{4} + y^2 + z^2 = 1$

Trace in xy-plane ($z = 0$):	$\dfrac{x^2}{4} + y^2 = 1$	Ellipse
Trace in xz-plane ($y = 0$):	$\dfrac{x^2}{4} + z^2 = 1$	Ellipse
Trace in yz-plane ($x = 0$):	$y^2 + z^2 = 1$	Circle

43. The graph is an ellipsoid.

45. $\dfrac{x^2}{1/5} + \dfrac{y^2}{1/5} - \dfrac{z^2}{5} = 1$

The graph is a hyperboloid of one sheet.

47. The graph of $y = x^2 + z^2$ is an elliptic paraboloid.

49. The graph of $z = y^2 - x^2$ is a hyperbolic paraboloid.

51. $\dfrac{y^2}{16} - \dfrac{x^2}{4} - \dfrac{z^2}{4} = 1$

The graph is a hyperboloid of two sheets.

53. The graph of $-z^2 + 9x^2 + y^2 = 0$ is an elliptic cone.

55. The graph of $z = \dfrac{x^2}{3} - \dfrac{y^2}{3}$ is a hyperbolic paraboloid.

57. $z = y^2 - x^2 + 1$

59. $z = \dfrac{x^2}{2} + \dfrac{y^2}{4}$

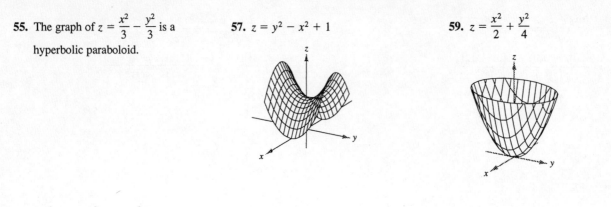

61. $\dfrac{x^2}{3963^2} + \dfrac{y^2}{3963^2} + \dfrac{z^2}{3950^2} = 1$

Section 7.3   Functions of Several Variables

1. (a) $f(3, 2) = \dfrac{3}{2}$ (b) $f(-1, 4) = -\dfrac{1}{4}$

   (c) $f(30, 5) = 6$ (d) $f(5, y) = \dfrac{5}{y}$

   (e) $f(x, 2) = \dfrac{x}{2}$ (f) $f(5, t) = \dfrac{5}{t}$

3. (a) $f(5, 0) = 5$ (b) $f(3, 2) = 3e^2$

   (c) $f(2, -1) = \dfrac{2}{e}$ (d) $f(5, y) = 5e^y$

   (e) $f(x, 2) = xe^2$ (f) $f(t, t) = te^t$

5. (a) $h(2, 3, 9) = \frac{2}{3}$

   (b) $h(1, 0, 1) = 0$

7. (a) $V(3, 10) = \pi(3)^2(10) = 90\pi$

   (b) $V(5, 2) = \pi(5)^2(2) = 50\pi$

9. (a) $A(100, 0.10, 10) = 100\left[\left(1 + \dfrac{0.10}{12}\right)^{120} - 1\right]\left(1 + \dfrac{12}{0.10}\right) \approx \$20{,}655.20$

   (b) $A(275, 0.0925, 40) = 275\left[\left(1 + \dfrac{0.0925}{12}\right)^{480} - 1\right]\left(1 + \dfrac{12}{0.0925}\right) \approx \$1{,}397{,}672.67$

11. (a) $f(1, 2) = \displaystyle\int_1^2 (2t - 3)\,dt = \left[(t^2 - 3t)\right]_1^2 = (-2) - (-2) = 0$

   (b) $f(1, 4) = \displaystyle\int_1^4 (2t - 3)\,dt = \left[(t^2 - 3t)\right]_1^4 = 6$

13. (a) $f(x + \Delta x, y) = (x + \Delta x)^2 - 2y = x^2 + 2x\Delta x + (\Delta x)^2 - 2y$

   (b) $\dfrac{f(x, y + \Delta y) - f(x, y)}{\Delta y} = \dfrac{[x^2 - 2(y + \Delta y)] - (x^2 - 2y)}{\Delta y}$

$$= \dfrac{x^2 - 2y - 2\Delta y - x^2 + 2y}{\Delta y}$$

$$= -\dfrac{2\Delta y}{\Delta y}$$

$$= -2, \quad \Delta y \neq 0$$

15. The domain is the set of all points inside and on the circle $x^2 + y^2 = 16$ since $16 - x^2 - y^2 \geq 0$ and the range is $[0, 4]$.

17. The domain is the set of all points above or below the x-axis since $y \neq 0$. The range is $(0, \infty)$.

19. The domain is the set of all points inside and on the ellipse $9x^2 + y^2 = 9$ since $9 - 9x^2 - y^2 \geq 0$.

21. Since $y \neq 0$, the domain is the set of all points in the xy-plane except those on the x-axis.

23. The domain is the set of all points in the xy-plane except those on the x- and y-axes.

25. The domain is the set of all points in the xy-plane such that $y \geq 0$.

27. The domain is the half plane below the line $y = -x + 4$ since $4 - x - y > 0$.

29. The contour map consists of ellipses

$$x^2 + \frac{y^2}{4} = C.$$

Matches (b).

31. The contour map consists of curves $e^{1-x^2-y^2} = C$, or

$$1 - x^2 - y^2 = \ln C \Longrightarrow x^2 + y^2 = 1 - \ln C, \text{ circles.}$$

Matches (a).

33. $\quad c = -1, \qquad -1 = x + y, \qquad y = -x - 1$

$\quad\quad c = 0, \qquad\quad 0 = x + y, \qquad y = -x$

$\quad\quad c = 2, \qquad\quad 2 = x + y, \qquad y = -x + 2$

$\quad\quad c = 4, \qquad\quad 4 = x + y, \qquad y = -x + 4$

The level curves are parallel lines.

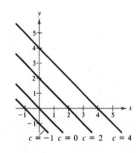

35. $\quad c = 0, \qquad 0 = \sqrt{16 - x^2 - y^2}, \qquad x^2 + y^2 = 16$

$\quad\quad c = 1, \qquad 1 = \sqrt{16 - x^2 - y^2}, \qquad x^2 + y^2 = 15$

$\quad\quad c = 2, \qquad 2 = \sqrt{16 - x^2 - y^2}, \qquad x^2 + y^2 = 12$

$\quad\quad c = 3, \qquad 3 = \sqrt{16 - x^2 - y^2}, \qquad x^2 + y^2 = 7$

$\quad\quad c = 4, \qquad 4 = \sqrt{16 - x^2 - y^2}, \qquad x^2 + y^2 = 0$

The level curves are circles.

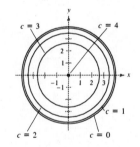

37. $\quad c = 1, \qquad\quad xy = 1$

$\quad\quad c = -1, \qquad xy = -1$

$\quad\quad c = 2, \qquad\quad xy = 2$

$\quad\quad c = -2, \qquad xy = -2$

$\quad\quad c = \pm 3, \qquad xy = \pm 3$

$\quad\quad c = \pm 4, \qquad xy = \pm 4$

$\quad\quad c = \pm 5, \qquad xy = \pm 5$

$\quad\quad c = \pm 6, \qquad xy = \pm 6$

The level curves are hyperbolas.

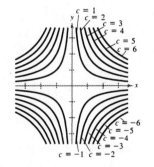

39. $c = \frac{1}{2}$, $\frac{1}{2} = \frac{x}{x^2 + y^2}$, $(x - 1)^2 + y^2 = 1$

$c = -\frac{1}{2}$, $-\frac{1}{2} = \frac{x}{x^2 + y^2}$, $(x + 1)^2 + y^2 = 1$

$c = 1$, $1 = \frac{x}{x^2 + y^2}$, $\left(x - \frac{1}{2}\right)^2 + y^2 = \frac{1}{4}$

$c = -1$, $-1 = \frac{x}{x^2 + y^2}$, $\left(x + \frac{1}{2}\right)^2 + y^2 = \frac{1}{4}$

$c = \frac{3}{2}$, $\frac{3}{2} = \frac{x}{x^2 + y^2}$, $\left(x - \frac{1}{3}\right)^2 + y^2 = \frac{1}{9}$

$c = -\frac{3}{2}$, $-\frac{3}{2} = \frac{x}{x^2 + y^2}$, $\left(x + \frac{1}{3}\right)^2 + y^2 = \frac{1}{9}$

$c = 2$, $2 = \frac{x}{x^2 + y^2}$, $\left(x - \frac{1}{4}\right)^2 + y^2 = \frac{1}{16}$

$c = -2$, $-2 = \frac{x}{x^2 + y^2}$, $\left(x + \frac{1}{4}\right)^2 + y^2 = \frac{1}{16}$

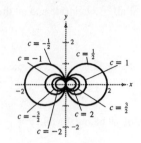

The level curves are circles.

41. $f(1500, 1000) = 100(1500)^{0.75}(1000)^{0.25}$

$\approx 135{,}540$ units

43. $C(80, 20) = 27\sqrt{(80)(20)} + 195(80) + 215(20) + 980$

$\approx \$21{,}960.00$

45. $P(x_1, x_2) = 45(x_1 + x_2) - C_1(x_1) - C_2(x_2)$

$= 45(x_1 + x_2) - [0.02x_1^2 + 4x_1 + 500] - [0.05x_2^2 + 4x_2 + 275]$

(a) $P(250, 150) = 45(250 + 150) - [0.02(250)^2 + 4(250) + 500] - [0.05(150)^2 + 4(150) + 275]$

$= 18{,}000 - 2750 - 2000 = \$13{,}250$

(b) $P(300, 200) = \$15{,}925$

47.

	Inflation Rate, I		
	0.00	0.03	0.05
Tax Rate, R 0.00	\$2593.74	\$1929.99	\$1592.33
0.28	\$2004.23	\$1491.34	\$1230.42
0.35	\$1877.14	\$1396.77	\$1152.40

49. (a) The different colors represent various amplitudes.

(b) No, the level curves are uneven and sporadically spaced.

Section 7.4   Partial Derivatives

1. $f_x(x, y) = 2$

$f_y(x, y) = -3$

3. $f_x(x, y) = \dfrac{5}{2\sqrt{x}}$

$f_y(x, y) = -12y$

5. $f_x(x, y) = \dfrac{1}{y}$

$f_y(x, y) = -xy^{-2} = \dfrac{-x}{y^2}$

7. $f_x(x, y) = \dfrac{1}{2}(x^2 + y^2)^{-1/2}(2x) = \dfrac{x}{\sqrt{x^2 + y^2}}$

$f_y(x, y) = \dfrac{1}{2}(x^2 + y^2)^{-1/2}(2y) = \dfrac{y}{\sqrt{x^2 + y^2}}$

9. $\dfrac{\partial z}{\partial x} = 2xe^{2y}$

$\dfrac{\partial z}{\partial y} = 2x^2e^{2y}$

11. $h_x(x, y) = -2xe^{-(x^2+y^2)}$

$h_y(x, y) = -2ye^{-(x^2+y^2)}$

13. $z = \ln \dfrac{x - y}{(x + y)^2} = \ln(x - y) - 2\ln(x + y)$

$\dfrac{\partial z}{\partial x} = \dfrac{1}{x - y} - \dfrac{2}{x + y} = \dfrac{3y - x}{x^2 - y^2}$

$\dfrac{\partial z}{\partial y} = \dfrac{-1}{x - y} - \dfrac{2}{x + y} = \dfrac{y - 3x}{x^2 - y^2}$

15. $f_x(x, y) = 6xye^{x-y} + 3x^2ye^{x-y}$

$= 3xye^{x-y}(2 + x)$

17. $g_x(x, y) = 3y^2e^{y-x} - 3xy^2e^{y-x}$

$= 3y^2e^{y-x}(1 - x)$

19. Using the solution from Exercise 15, $f_x(1, 1) = 9$.

21. $f_x(x, y) = 6x + y,$ $f_x(2, 1) = 13$

$f_y(x, y) = x - 2y,$ $f_y(2, 1) = 0$

23. $f_x(x, y) = 3ye^{3xy},$ $f_x(0, 4) = 3(4)e^0 = 12$

$f_y(x, y) = 3xe^{3xy},$ $f_y(0, 4) = 0$

25. $f_x(x, y) = \dfrac{(x - y)y - xy}{(x - y)^2} = -\dfrac{y^2}{(x - y)^2},$ $f_x(2, -2) = -\dfrac{4}{16} = -\dfrac{1}{4}$

$f_y(x, y) = \dfrac{(x - y)x - xy(-1)}{(x - y)^2} = \dfrac{x^2}{(x - y)^2},$ $f_y(2, -2) = \dfrac{4}{16} = \dfrac{1}{4}$

27. $f_x(x, y) = \dfrac{1}{x^2 + y^2}(2x) = \dfrac{2x}{x^2 + y^2},$ $f_x(1, 0) = \dfrac{2}{1 + 0} = 2$

$f_y(x, y) = \dfrac{1}{x^2 + y^2}(2y) = \dfrac{2y}{x^2 + y^2},$ $f_y(1, 0) = 0$

29. $w_x = 6xy - 5yz$

$w_y = 3x^2 - 5xz + 10z^2$

$w_z = -5xy + 20yz$

31. $w_x = \dfrac{y(y + z)}{(x + y + z)^2}$

$w_y = \dfrac{x(x + z)}{(x + y + z)^2}$

$w_z = \dfrac{-xy}{(x + y + z)^2}$

33. $w_x = \dfrac{x}{\sqrt{x^2 + y^2 + z^2}},$ at $(2, -1, 2),$ $w_x = \dfrac{2}{3}$

$w_y = \dfrac{y}{\sqrt{x^2 + y^2 + z^2}},$ at $(2, -1, 2),$ $w_y = -\dfrac{1}{3}$

$w_z = \dfrac{z}{\sqrt{x^2 + y^2 + z^2}},$ at $(2, -1, 2),$ $w_z = \dfrac{2}{3}$

35. $w = \ln\sqrt{x^2 + y^2 + z^2} = \dfrac{1}{2}\ln(x^2 + y^2 + z^2)$

$w_x = \dfrac{x}{x^2 + y^2 + z^2},$ $w_x(3, 0, 4) = \dfrac{3}{25}$

$w_y = \dfrac{y}{x^2 + y^2 + z^2},$ $w_y(3, 0, 4) = 0$

$w_z = \dfrac{z}{x^2 + y^2 + z^2},$ $w_z(3, 0, 4) = \dfrac{4}{25}$

37. $w_x = 2z^2 + 3yz,$ $w_x(1, -1, 2) = 8 - 6 = 2$

$w_y = 3xz - 12yz,$ $w_y(1, -1, 2) = 6 + 24 = 30$

$w_z = 4xz + 3xy - 6y^2,$ $w_z(1, -1, 2) = 8 - 3 - 6 = -1$

39. $f_x(x, y) = 2x + 4y - 4 = 0$
$f_y(x, y) = 4x + 2y + 16 = 0$

$$
\begin{aligned}
-4x - 8y &= -8 \\
4x + 2y &= -16 \\
\hline
-6y &= -24 \\
y &= 4 \\
x &= -6
\end{aligned}
$$

Solution: $(-6, 4)$

41. $f_x(x, y) = -\dfrac{1}{x^2} + y = 0 \Rightarrow x^2y = 1$

$f_y(x, y) = -\dfrac{1}{y^2} + x = 0 \Rightarrow y^2x = 1$ $\Bigg\}$ $x = y = 1$

Solution: $(1, 1)$

43. (a) $\dfrac{\partial z}{\partial x} = 2$; at $(2, 1, 6)$, $\dfrac{\partial z}{\partial x} = 2$

(b) $\dfrac{\partial z}{\partial y} = -3$; at $(2, 1, 6)$, $\dfrac{\partial z}{\partial y} = -3$

45. (a) $\dfrac{\partial z}{\partial x} = 2x$; at $(3, 1, 0)$, $\dfrac{\partial z}{\partial x} = 6$

(b) $\dfrac{\partial z}{\partial y} = -18y$; at $(3, 1, 0)$, $\dfrac{\partial z}{\partial y} = -18$

47. (a) $\dfrac{\partial z}{\partial x} = -\dfrac{x}{\sqrt{25 - x^2 - y^2}}$; at $(3, 0, 4)$, $\dfrac{\partial z}{\partial x} = -\dfrac{3}{4}$

(b) $\dfrac{\partial z}{\partial y} = -\dfrac{y}{\sqrt{25 - x^2 - y^2}}$; at $(3, 0, 4)$, $\dfrac{\partial z}{\partial y} = 0$

49. (a) $\dfrac{\partial z}{\partial x} = -2x$; at $(1, 1, 2)$, $\dfrac{\partial z}{\partial x} = -2$

(b) $\dfrac{\partial z}{\partial y} = -2y$; at $(1, 1, 2)$, $\dfrac{\partial z}{\partial y} = -2$

51. $\dfrac{\partial z}{\partial x} = 2x - 2y$, $\dfrac{\partial^2 z}{\partial y \partial x} = -2$

$\dfrac{\partial z}{\partial y} = -2x + 6y$, $\dfrac{\partial^2 z}{\partial x \partial y} = -2$

53. $\dfrac{\partial z}{\partial x} = \dfrac{e^{2xy}(2xy - 1)}{4x^2}$, $\dfrac{\partial^2 z}{\partial y \partial x} = ye^{2xy}$

$\dfrac{\partial z}{\partial y} = \dfrac{1}{2}e^{2xy}$, $\dfrac{\partial^2 z}{\partial x \partial y} = ye^{2xy}$

55. The first partial derivatives are

$$\dfrac{\partial z}{\partial x} = 3x^2 \quad \text{and} \quad \dfrac{\partial z}{\partial y} = -8y$$

and the second partial derivatives are

$$\dfrac{\partial^2 z}{\partial x^2} = 6x, \quad \dfrac{\partial^2 z}{\partial y \partial x} = 0 = \dfrac{\partial^2 z}{\partial x \partial y}, \quad \text{and} \quad \dfrac{\partial^2 z}{\partial y^2} = -8.$$

57. The first partial derivatives are

$$\dfrac{\partial z}{\partial x} = 12x^2 + 3y^2 \quad \text{and} \quad \dfrac{\partial z}{\partial y} = 6xy - 12y^2$$

and the second partial derivatives are

$$\dfrac{\partial^2 z}{\partial x^2} = 24x, \quad \dfrac{\partial^2 z}{\partial y \partial x} = 6y = \dfrac{\partial^2 z}{\partial x \partial y}, \quad \text{and} \quad \dfrac{\partial^2 z}{\partial y^2} = 6x - 24y.$$

59. From Exercise 25, the first partial derivatives are

$$\dfrac{\partial z}{\partial x} = \dfrac{-y^2}{(x - y)^2} \quad \text{and} \quad \dfrac{\partial z}{\partial y} = \dfrac{x^2}{(x - y)^2}$$

and the second partial derivatives are

$$\dfrac{\partial^2 z}{\partial x^2} = \dfrac{2y^2}{(x - y)^3}, \quad \dfrac{\partial^2 z}{\partial y \partial x} = \dfrac{-2xy}{(x - y)^3} = \dfrac{\partial^2 z}{\partial x \partial y},$$

and

$$\dfrac{\partial^2 z}{\partial y^2} = \dfrac{2x^2}{(x - y)^3}.$$

61. The first partial derivatives are

$$\dfrac{\partial z}{\partial x} = e^{-y^2} \quad \text{and} \quad \dfrac{\partial z}{\partial y} = -2xye^{-y^2}$$

and the second partial derivatives are

$$\dfrac{\partial^2 z}{\partial x^2} = 0, \quad \dfrac{\partial^2 z}{\partial y \partial x} = -2ye^{-y^2} = \dfrac{\partial^2 z}{\partial x \partial y},$$

and

$$\dfrac{\partial^2 z}{\partial y^2} = -2xe^{-y^2} + 4xy^2e^{-y^2} = 2xe^{-y^2}(2y^2 - 1).$$

63. $f_x(x, y) = 4x^3 - 6xy^2$ $\qquad$ $f_y(x, y) = -6x^2y + 2y$

$f_{xx}(x, y) = 12x^2 - 6y^2$, $\qquad$ $f_{xx}(1, 0) = 12$

$f_{xy}(x, y) = -12xy$, $\qquad$ $f_{xy}(1, 0) = 0$

$f_{yx}(x, y) = -12xy$, $\qquad$ $f_{yx}(1, 0) = 0$

$f_{yy}(x, y) = -6x^2 + 2$, $\qquad$ $f_{yy}(1, 0) = -4$

65. $f_x(x, y) = \dfrac{1}{x - y}$ $\qquad$ $f_y(x, y) = \dfrac{-1}{x - y}$

$f_{xx}(x, y) = \dfrac{-1}{(x - y)^2}$, $\qquad$ $f_{xx}(2, 1) = -1$

$f_{xy}(x, y) = \dfrac{1}{(x - y)^2}$, $\qquad$ $f_{xy}(2, 1) = 1$

$f_{yx}(x, y) = \dfrac{1}{(x - y)^2}$, $\qquad$ $f_{yx}(2, 1) = 1$

$f_{yy}(x, y) = \dfrac{-1}{(x - y)^2}$, $\qquad$ $f_{yy}(2, 1) = -1$

67. $\dfrac{\partial C}{\partial x} = \dfrac{5y}{\sqrt{xy}} + 149;$ at $(120, 160),$ $\dfrac{\partial C}{\partial x} \approx 154.77$

$\dfrac{\partial C}{\partial y} = \dfrac{5x}{\sqrt{xy}} + 189;$ at $(120, 160),$ $\dfrac{\partial C}{\partial y} \approx 193.33$

69. (a) $\dfrac{\partial f}{\partial x} = 60x^{-0.4}y^{0.4} = 60\left(\dfrac{y}{x}\right)^{0.4};$ at $(1000, 500),$ $\dfrac{\partial f}{\partial x} \approx 45.47$

(b) $\dfrac{\partial f}{\partial y} = 40x^{0.6}y^{-0.6} = 40\left(\dfrac{x}{y}\right)^{0.6};$ at $(1000, 500),$ $\dfrac{\partial f}{\partial y} \approx 60.63$

71. (a) Complementary since $\dfrac{\partial x_1}{\partial p_2} = -\dfrac{5}{2} < 0$ and $\dfrac{\partial x_2}{\partial p_1} = -\dfrac{3}{2} < 0.$

(b) Substitute since $\dfrac{\partial x_1}{\partial p_2} = 1.8 > 0$ and $\dfrac{\partial x_2}{\partial p_1} = 0.75 > 0.$

(c) Complementary since $\dfrac{\partial x_1}{\partial p_2} = \dfrac{-500}{p_2\sqrt{p_1 p_2}} < 0$ and $\dfrac{\partial x_2}{\partial p_1} = \dfrac{-375}{p_1 p_2 \sqrt{p_1}} < 0.$

73. Since both first partials are negative, an increase in the charge for food and housing or tuition will cause a decrease in the number of applicants.

75. (a) $\dfrac{\partial A}{\partial t} = 0.885 + 1.20h;$ at $t = 90°,$ $h = 0.80,$ $\dfrac{\partial A}{\partial t} = 1.845$

$\dfrac{\partial A}{\partial h} = -78.7 + 1.20t;$ at $t = 90°,$ $h = 0.80,$ $\dfrac{\partial A}{\partial h} = 29.3$

(b) The humidity has a greater effect since the coefficient of h is greater.

77. Answers will vary.

Section 7.5   Extrema of Functions of Two Variables

1. The first partial derivatives of f, $f_x(x, y) = 2x + 4$ and $f_y(x, y) = -2y - 8$, are zero at the point $(-2, -4)$. Moreover, since

$f_{xx}(x, y) = 2,$ $f_{yy}(x, y) = -2,$ and $f_{xy}(x, y) = 0,$

it follows that $f_{xx}(2, 4)f_{yy}(2, 4) - [f_{xy}(2, 4)]^2 = -4 < 0.$

Thus, $(-2, -4, 1)$ is a saddle point. There are no relative extrema.

3. The first partial derivatives of f,

$f_x(x, y) = \dfrac{x}{\sqrt{x^2 + y^2 + 1}}$ and $f_y(x, y) = \dfrac{y}{\sqrt{x^2 + y^2 + 1}},$

are zero at the point $(0, 0)$. Moreover, since

$f_{xx}(x, y) = \dfrac{y^2 + 1}{(x^2 + y^2 + 1)^{3/2}},$ $f_{yy}(x, y) = \dfrac{x^2 + 1}{(x^2 + y^2 + 1)^{3/2}},$ and $f_{xy}(x, y) = \dfrac{-xy}{(x^2 + y^2 + 1)^{3/2}},$

it follows that $f_{xx}(0, 0) = 1 > 0$ and $f_{xx}(0, 0)f_{yy}(0, 0) - [f_{xy}(0, 0)]^2 = 1 > 0.$

Thus, $(0, 0, 1)$ is a relative minimum.

5. The first partial derivatives of f, $f_x(x, y) = 2(x - 1)$ and $f_y(x, y) = 2(y - 3)$, are zero at the point $(1, 3)$. Moreover, since

$$f_{xx}(x, y) = 2, \ f_{yy}(x, y) = 2, \ \text{and} \ f_{xy}(x, y) = 0,$$

it follows that $f_{xx}(1, 3) > 0$ and $f_{xx}(1, 3)f_{yy}(1, 3) - [f_{xy}(1, 3)]^2 = 4 > 0$.

Thus, $(1, 3, 0)$ is a relative minimum.

7. The first partial derivatives of f, $f_x(x, y) = 4x + 2y + 2$ and $f_y(x, y) = 2x + 2y$, are zero at the point $(-1, 1)$. Moreover, since

$$f_{xx}(x, y) = 4, \ f_{yy}(x, y) = 2, \ \text{and} \ f_{xy}(x, y) = 2,$$

it follows that $f_{xx}(-1, 1) > 0$ and $f_{xx}(-1, 1)f_{yy}(-1, 1) - [f_{xy}(-1, 0)]^2 = 4 > 0$.

Thus, $(-1, 1, -4)$ is a relative minimum.

9. The first partial derivatives of f, $f_x(x, y) = -10x + 4y + 16$ and $f_y(x, y) = 4x - 2y$, are zero at the point $(8, 16)$. Moreover, since

$$f_{xx}(x, y) = -10, \ f_{yy}(x, y) = -2, \ \text{and} \ f_{xy}(x, y) = 4,$$

it follows that $f_{xx}(8, 16) < 0$ and $f_{xx}(8, 16)f_{yy}(8, 16) - [f_{xy}(8, 16)]^2 = 40$.

Thus, $(8, 16, 74)$ is a relative maximum.

11. The first partial derivatives of f, $f_x(x, y) = 6x - 12$ and $f_y(x, y) = 4y - 4$, are zero at the point $(2, 1)$. Moreover, since

$$f_{xx}(x, y) = 6, \ f_{yy}(x, y) = 4, \ \text{and} \ f_{xy}(x, y) = 0,$$

it follows that $f_{xx}(2, 1) > 0$ and $f_{xx}(2, 1)f_{yy}(2, 1) - [f_{xy}(2, 1)]^2 = 24 > 0$.

Thus, $(2, 1, -7)$ is a relative minimum.

13. The first partial derivatives of f, $f_x(x, y) = 2x + 4$ and $f_y(x, y) = -2y - 4 = -2(y + 2)$, are zero at the point $(-2, -2)$. Moreover, since

$$f_{xx}(x, y) = 2, \ f_{yy}(x, y) = -2, \ \text{and} \ f_{xy}(x, y) = 0,$$

it follows that

$$f_{xx}(1, -2)f_{yy}(1, -2) - [f_{xy}(1, -2)]^2 = -4 < 0.$$

Thus, $(-2, -2, -8)$ is a saddle point.

15. The first partial derivatives of f, $f_x(x, y) = y$ and $f_y(x, y) = x$, are zero at the point $(0, 0)$. Moreover, since

$$f_{xx}(x, y) = 0, \ f_{yy}(x, y) = 0, \ \text{and} \ f_{xy}(x, y) = 1,$$

it follows that

$$f_{xx}(0, 0)f_{yy}(0, 0) - [f_{xy}(0, 0)]^2 = -1 < 0.$$

Thus, $(0, 0, 0)$ is a saddle point.

17. The first partial derivatives of f,

$$f_x(x, y) = 2xe^{1-x^2-y^2}(1 - x^2 - 4y^2) \ \text{and} \ f_y(x, y) = 2ye^{1-x^2-y^2}(4 - x^2 - 4y^2)$$

are zero at $(0, 0)$, $(0, \pm 1)$, and $(\pm 1, 0)$. Moreover, since

$$f_{xx}(x, y) = 2e^{1-x^2-y^2}(1 - 5x^2 + 2x^4 - 4y^2 + 8x^2y^2),$$

$$f_{yy}(x, y) = 2e^{1-x^2-y^2}(4 - x^2 - 20y^2 + 8y^4 + 2x^2y^2), \ \text{and}$$

$$f_{xy}(x, y) = -4xye^{1-x^2-y^2}(5 - x^2 - 4y^2),$$

we can determine that $(0, 0, 0)$ is a relative minimum, $(0, \pm 1, 4)$ are relative maxima, and $(\pm 1, 0, 1)$ are saddle points.

19. The first partial derivatives of f, $f_x(x, y) = ye^{xy}$ and $f_y(x, y) = xe^{xy}$, are zero at the point $(0, 0)$. Moreover, since

$$f_{xx}(x, y) = y^2e^{xy}, \ f_{yy}(x, y) = x^2e^{xy}, \ \text{and} \ f_{xy}(x, y) = e^{xy}(1 + xy),$$

it follows that $f_{xx}(0, 0) = 0$ and $f_{xx}(0, 0)f_{yy}(0, 0) - [f_{xy}(0, 0)]^2 = -1 < 0$.

Thus, $(0, 0, 1)$ is a saddle point.

21. Since

$$d = f_{xx}(x_0, y_0)f_{yy}(x_0, y_0) - [f_{xy}(x_0, y_0)]^2$$

$$= (16)(4) - (8)^2$$

$$= 0,$$

there is insufficient information.

23. Since

$$d = f_{xx}(x_0, y_0)f_{yy}(x_0, y_0) - [f_{xy}(x_0, y_0)]^2$$

$$= (-7)(4) - (9)^2$$

$$< 0,$$

$f(x_0, y_0)$ is a saddle point.

25. The first partial derivatives of f, $f_x(x, y) = 2xy^2$ and $f_y(x, y) = 2x^2y$, are zero at the points $(a, 0)$ and $(0, b)$ where a and b are any real numbers. Since

$$f_{xx}(x, y) = 2y^2, \ f_{yy}(x, y) = 2x^2, \ \text{and} \ f_{xy}(x, y) = 4xy,$$

it follows that $f_{xx}(a, 0)f_{yy}(a, 0) - [f_{xy}(a, 0)]^2 = 0$ and $f_{xx}(0, b)f_{yy}(0, b) - [f_{xy}(0, b)]^2 = 0$ and the Second-Derivative Test fails. We note that $f(x, y) = (xy)^2$ is nonnegative for all $(a, 0, 0)$ and $(0, b, 0)$ where a and b are real numbers. Therefore, $(a, 0, 0)$ and $(0, b, 0)$ are relative minima.

27. The first partial derivatives of f, $f_x(x, y) = 3x^2$ and $f_y(x, y) = 3y^2$, are zero at $(0, 0)$. Moreover, since

$$f_{xx}(x, y) = 6x, \ f_{yy}(x, y) = 6y, \ f_{xy}(x, y) = 0, \ \text{and} \ f_{xx}(0, 0)f_{yy}(0, 0) - [f_{xy}(0, 0)]^2 = 0,$$

the Second-Partials Test fails. By testing "nearby" points, we conclude that $(0, 0, 0)$ is a saddle point.

29. The first partial derivatives of f,

$$f_x(x, y) = \frac{2}{3\sqrt[3]{x}} \ \text{and} \ f_y(x, y) = \frac{2}{3\sqrt[3]{y}},$$

are undefined at the point $(0, 0)$. Since

$$f_{xx}(x, y) = -\frac{2}{9x^{4/3}}, \ f_{yy}(x, y) = -\frac{2}{9y^{4/3}}, \ f_{xy}(x, y) = 0$$

and $f_{xx}(0, 0)$ is undefined, the Second-Derivative Test fails. Since $f(x, y) \geq 0$ for all points in the xy-coordinate plane, $(0, 0, 0)$ is a relative minimum.

31. Critical point: $(x, y, z) = (1, -3, 0)$

    Relative minimum

33. Let x, y, and z be the numbers. The sum is given by $x + y + z = 30$ and $z = 30 - x - y$, and the product is given by $P = xyz = 30xy - x^2y - xy^2$. The first partial derivatives of P are

$$P_x = 30y - 2xy - y^2 = y(30 - 2x - y)$$

$$P_y = 30x - x^2 - 2xy = x(30 - x - 2y).$$

Setting these equal to zero produces the system

$$2x + y = 30$$

$$x + 2y = 30.$$

Solving the system, we have $x = 10, y = 10$, and $z = 10$.

35. The sum is given by

$$x + y + z = 30$$

$$z = 30 - x - y$$

and the sum of the squares is given by $S = x^2 + y^2 + z^2 = x^2 + y^2 + (30 - x - y)^2$. The first partial derivatives of S are

$$S_x = 2x - 2(30 - x - y) = 4x + 2y - 60 \ \text{and}$$

$$S_y = 2y - 2(30 - x - y) = 2x + 4y - 60.$$

Setting these equal to zero produces the system

$$2x + y = 30$$

$$x + 2y = 30.$$

Solving this system yields $x = 10$ and $y = 10$. Thus, the sum of squares is a minimum when $x = y = z = 10$.

37. The first partial derivatives of R are

$$R_{x_1} = -10x_1 - 2x_2 + 42 \text{ and}$$

$$R_{x_2} = -16x_2 - 2x_1 + 102.$$

Setting these equal to zero produces the system

$$5x_1 + x_2 = 21$$

$$x_1 + 8x_2 = 51$$

which yields $x_1 = 3$ and $x_2 = 6$. By the Second-Partials Test, it follows that the revenue is maximized when $x_1 = 3$ and $x_2 = 6$.

39. The revenue function is given by

$$R = x_1p_1 + x_2p_2 = 1000p_1 + 1500p_2 + 3p_1p_2 - 2p_1{}^2 - 1.5p_2{}^2$$

and the first partials of R are $R_{p_1} = 1000 + 3p_2 - 4p_1$ and $R_{p_2} = 1500 + 3p_1 - 3p_2$. Setting these equal to zero produces the system

$$4p_1 - 3p_2 = 1000$$

$$-3p_1 + 3p_2 = 1500.$$

Solving this system yields $p_1 = 2500$ and $p_2 = 3000$, and, by the Second-Partials Test, we conclude that the revenue is maximized when $p_1 = 2500$ and $p_2 = 3000$.

41. The profit is given by

$$P = R - C_1 - C_2$$

$$= [225 - 0.4(x_1 + x_2)](x_1 + x_2) - (0.05x_1{}^2 + 15x_1 + 5400) - (0.03x_2{}^2 + 15x_2 + 6100)$$

$$= -0.45x_1{}^2 - 0.43x_2{}^2 - 0.8x_1x_2 + 210x_1 + 210x_2 - 11,500$$

and the first partial derivatives of P are $P_{x_1} = -0.9x_1 - 0.8x_2 + 210$ and $P_{x_2} = -0.86x_2 - 0.8x_1 + 210$. By setting these equal to zero, we obtain the system

$$0.9x_1 + 0.8x_2 = 210$$

$$0.8x_1 + 0.86x_2 = 210.$$

Solving this system yields $x_1 \approx 94$ and $x_2 \approx 157$, and, by the Second-Partials Test, we conclude that the profit is maximum when $x_1 \approx 94$ and $x_2 \approx 157$.

43. Let $x =$ length, $y =$ width, and $z =$ height. The sum of length and girth is given by

$$x + (2y + 2z) = 144$$

$$x = 144 - 2y - 2z.$$

The volume is given by $V = xyz = 144yz - 2zy^2 - 2yz^2$. Then,

$$V_y = 144z - 4zy - 2z^2 = 2z(72 - 2y - z)$$

$$V_z = 144y - 2y^2 - 4yz = 2y(72 - y - 2z).$$

Setting these equal to zero produces the system

$$2y + z = 72$$

$$2z + y = 72$$

which yields the solution $z = y = 24$. Hence, $x = 48$.

Solution: $x = 48$ inches

$$y = z = 24 \text{ inches}$$

45. $P(p, q, r) = 2pq + 2pr + 2qr$

Since $p + q + r = 1$, $r = 1 - p - q$, and

$$P = 2pq + 2p(1 - p - q) + 2q(1 - p - q)$$

$$= -2p^2 + 2p - 2q^2 + 2q - 2pq$$

$$P_p = -4p + 2 - 2q$$

$$P_q = -4q + 2 - 2p.$$

Solving $P_p = P_q = 0$, we obtain $p = q = \frac{1}{3}$, and hence $r = \frac{1}{3}$. Finally, the maximum proportion is

$$P = 2\left(\tfrac{1}{3}\right)\left(\tfrac{1}{3}\right) + 2\left(\tfrac{1}{3}\right)\left(\tfrac{1}{3}\right) + 2\left(\tfrac{1}{3}\right)\left(\tfrac{1}{3}\right) = \tfrac{6}{9} = \tfrac{2}{3}.$$

47. $D_x(x, y) = 2x - 18 + 2y = 0$

$D_y(x, y) = 4y - 24 + 2x = 0$

To solve $x + y = 9$ and $x + 2y = 12$, subtract the equations $-y = -3$ or $y = 3$. Hence, $x = 6$.

600 mg first drug, 300 mg second drug

49. True

51. False. The origin is a minimum.

Section 7.6   Lagrange Multipliers

1. $F(x, y, \lambda) = xy - \lambda(x + y - 10)$

$\qquad F_x = y - \lambda = 0, \qquad\qquad y = \lambda$

$\qquad F_y = x - \lambda = 0, \qquad\qquad x = \lambda$

$\qquad F_\lambda = -(x + y - 10) = 0, \qquad 2\lambda = 10$

Thus, $\lambda = 5$, $x = 5, y = 5$, and $f(x, y)$ is maximum at $(5, 5)$. The maximum is $f(5, 5) = 25$.

3. $F(x, y, \lambda) = x^2 + y^2 - \lambda(x + y - 4)$

$\qquad F_x = 2x - \lambda = 0, \qquad\qquad x = \frac{1}{2}\lambda$

$\qquad F_y = 2y - \lambda = 0, \qquad\qquad y = \frac{1}{2}\lambda$

$\qquad F_\lambda = -(x + y - 4) = 0, \qquad \lambda = 4$

Thus, $\lambda = 4$, $x = 2, y = 2$, and $f(x, y)$ is minimum at $(2, 2)$. The minimum is $f(2, 2) = 8$.

5. $F(x, y, \lambda) = x^2 - y^2 - \lambda(y - x^2)$

$\qquad F_x = 2x + 2x\lambda = 0, \qquad 2x(1 + \lambda) = 0$

$\qquad F_y = -2y - \lambda = 0, \qquad\qquad y = -\frac{1}{2}\lambda$

$\qquad F_\lambda = -(y - x^2) = 0, \qquad\qquad x = \sqrt{y}$

Thus, $\lambda = -1$, $x = \sqrt{2}/2, y = 1/2$, and $f(x, y)$

is maximum at $\left(\sqrt{2}/2, 1/2\right)$. The maximum is $f\left(\sqrt{2}/2, 1/2\right) = 1/4$.

7. $F(x, y, \lambda) = 3x + xy + 3y - \lambda(x + y - 25)$

$\qquad F_x = 3 + y - \lambda = 0, \qquad\qquad y = \lambda - 3$

$\qquad F_y = 3 + x - \lambda = 0, \qquad\qquad x = \lambda - 3$

$\qquad F_\lambda = -(x + y - 25) = 0, \qquad 2\lambda - 6 = 25$

Thus, $\lambda = \frac{31}{2}$, $x = \frac{25}{2}, y = \frac{25}{2}$, and $f(x, y)$ is maximum at $\left(\frac{25}{2}, \frac{25}{2}\right)$. The maximum is $f\left(\frac{25}{2}, \frac{25}{2}\right) = 231.25$.

9. *Note:* $f(x, y)$ has a maximum value when $g(x, y) = 6 - x^2 - y^2$ is maximum.

$F(x, y, \lambda) = 6 - x^2 - y^2 - \lambda(x + y - 2)$

$\qquad F_x = -2x - \lambda = 0, \qquad\quad \left.\begin{array}{c} -2x = \lambda \\ -2y = \lambda \end{array}\right\} x = y$

$\qquad F_y = -2y - \lambda = 0,$

$\qquad F_\lambda = -(x + y - 2) = 0, \qquad 2x = 2$

Thus, $x = y = 1$ and $f(x, y)$ is maximum at $(1, 1)$. The maximum is $f(1, 1) = 2$.

11. $F(x, y, \lambda) = e^{xy} - \lambda(x^2 + y^2 - 8)$

$\qquad F_x = ye^{xy} - 2x\lambda = 0, \qquad \left.\begin{array}{c} e^{xy} = \dfrac{2x\lambda}{y} \\[2mm] e^{xy} = \dfrac{2y\lambda}{x} \end{array}\right\} x = y$

$\qquad F_y = xe^{xy} - 2y\lambda = 0,$

$\qquad F_\lambda = -(x^2 + y^2 - 8) = 0, \qquad 2x^2 = 8$

Thus, $x = y = 2$ and $f(x, y)$ is maximum at $(2, 2)$. The maximum is $f(2, 2) = e^4$.

13. $F(x, y, z, \lambda) = 2x^2 + 3y^2 + 2z^2 - \lambda(x + y + z - 24)$

$\qquad F_x = 4x - \lambda = 0, \qquad\qquad \lambda = 4x$

$\qquad F_y = 6y - \lambda = 0, \qquad\qquad \lambda = 6y$

$\qquad F_z = 4z - \lambda = 0, \qquad\qquad \lambda = 4z$

$\qquad F_\lambda = -(x + y + z - 24) = 0$

$\qquad \dfrac{\lambda}{4} + \dfrac{\lambda}{6} + \dfrac{\lambda}{4} = 24$

$\qquad\qquad 8\lambda = 24 \cdot 12$

$\qquad\qquad\quad \lambda = 36$

Thus, $x = 9, y = 6, z = 9$, and $f(x, y, z)$ is minimum at $(9, 6, 9)$. The minimum is $f(9, 6, 9) = 432$.

15. $F(x, y, z, \lambda) = x^2 + y^2 + z^2 - \lambda(x + y + z - 1)$

$$\left.\begin{array}{l} F_x = 2x - \lambda = 0 \\ F_y = 2y - \lambda = 0 \\ F_z = 2z - \lambda = 0 \end{array}\right\} x = y = z$$

$$F_\lambda = -(x + y + z - 1) = 0, \; 3x = 1$$

Thus, $x = y = z = \frac{1}{3}$ and $f(x, y, z)$ is minimum at
$f\left(\frac{1}{3}, \frac{1}{3}, \frac{1}{3}\right) = \frac{1}{3}$.

17. $F(x, y, z, \lambda) = x + y + z - \lambda(x^2 + y^2 + z^2 - 1)$

$$\left.\begin{array}{l} F_x = 1 - 2x\lambda = 0 \\ F_y = 1 - 2y\lambda = 0 \\ F_z = 1 - 2z\lambda = 0 \end{array}\right\} x = y = z$$

$$F_\lambda = -(x^2 + y^2 + z^2 - 1) = 0$$

$$3x^2 = 1 \Longrightarrow x = \frac{1}{\sqrt{3}} = y = z$$

$f(x, y, z)$ is maximum at

$$f\left(\frac{1}{\sqrt{3}}, \frac{1}{\sqrt{3}}, \frac{1}{\sqrt{3}}\right) = \frac{3}{\sqrt{3}} = \sqrt{3}.$$

19. $F(x, y, z, w, \lambda) = 2x^2 + y^2 + z^2 + 2x^2 - \lambda(2x + 2y + z + w - z)$

$$F_x = 4x - 2\lambda = 0, \qquad x = \tfrac{1}{2}\lambda$$

$$F_y = 2y - 2\lambda = 0, \qquad y = \lambda = 2x$$

$$F_z = 2z - \lambda = 0, \qquad z = \tfrac{1}{2}\lambda = x$$

$$F_w = 4w - \lambda = 0, \qquad w = \tfrac{1}{4}\lambda = \tfrac{1}{2}x$$

$$F_\lambda = -(2x + 2y + z + w) = 0$$

$$2x + 2(2x) + x + \tfrac{1}{2}x = 2 \Longrightarrow x = \tfrac{4}{15}$$

The maximum is $f\left(\frac{4}{15}, \frac{8}{15}, \frac{4}{15}, \frac{2}{15}\right) = \frac{8}{15}$.

21. $F(x, y, z, \lambda, \eta) = xyz - \lambda(x + y + z - 24) - \eta(x - y + z - 12)$

$$\left.\begin{array}{l} F_x = yz - \lambda - \eta = 0 \\ F_y = xz - \lambda + \eta = 0 \\ F_z = xy - \lambda - \eta = 0 \end{array}\right\} x = z$$

$$F_\lambda = -(x + y + z - 24) = 0, \quad x + (2x - 12) + x = 24$$

$$F_\eta = -(x - y + z - 12) = 0, \quad y = 2x - 12$$

Thus, $x = 9$, $y = 6$, and $z = 9$. The maximum is $f(9, 6, 9) = 486$.

23. $F(x, y, z, \lambda, \eta) = xyz - \lambda(x^2 + z^2 - 5) - \eta(x - 2y)$

$$F_x = yz - 2x\lambda - \eta = 0$$

$$F_y = xz + 2\eta = 0, \qquad \eta = -\frac{xz}{2}$$

$$F_z = xy - 2z\lambda = 0, \qquad \lambda = \frac{xy}{2z}$$

$$F_\lambda = -(x^2 + z^2 - 5) = 0, \qquad z = \sqrt{5 - x^2}$$

$$F_\eta = -(x - 2y) = 0, \qquad y = \frac{x}{2}$$

From F_x, we can write:

$$\frac{x\sqrt{5 - x^2}}{2} - \frac{x^3}{2\sqrt{5 - x^2}} + \frac{x\sqrt{5 - x^2}}{2} = 0$$

$$x\sqrt{5 - x^2} = \frac{x^3}{2\sqrt{5 - x^2}}$$

$$2x(5 - x^2) = x^3$$

$$3x^3 - 10x = 0$$

$$x(3x^2 - 10) = 0$$

Since x, y, and z are positive, we have

$$x = \sqrt{\frac{10}{3}}, y = \frac{1}{2}\sqrt{\frac{10}{3}}, \text{ and } z = \sqrt{\frac{5}{3}}.$$

$$f\left(\sqrt{\frac{10}{3}}, \frac{1}{2}\sqrt{\frac{10}{3}}, \sqrt{\frac{5}{3}}\right) = \frac{5\sqrt{15}}{9}$$

25. $F(x, y, z, \lambda, \mu) = xyz - \lambda(x + 3y - 6) - \mu(x - 2z)$

$\quad F_x = yz - \lambda - \mu \quad = 0$

$\quad F_y = xz - 3\lambda \quad\quad = 0$

$\quad F_z = xy + 2\mu \quad\quad = 0$

$\quad F_\lambda = -x - 3y + 6 = 0 \quad \Rightarrow y = \dfrac{6 - x}{3}$

$\quad F_\mu = -x + 2z \quad\quad = 0 \quad \Rightarrow z = \dfrac{x}{2}$

$\quad yz - \lambda - \mu = yz - \dfrac{xz}{3} + \dfrac{xy}{2} = \dfrac{6 - x}{3}\left(\dfrac{x}{2}\right) - \dfrac{x}{3} \cdot \dfrac{x}{2} + \dfrac{x}{2}\dfrac{(6 - x)}{3}$

$\quad\quad\quad = \dfrac{1}{6}[6x - x^2 - x^2 + 6x - x^2] = \dfrac{1}{6}[12x - 3x^2] = 0$

$\quad\quad\quad \Rightarrow x = 4, \quad y = \dfrac{2}{3}, \quad z = 2$

$\quad f(x, y, z) = 4\left(\dfrac{2}{3}\right)(2) = \dfrac{16}{3}$

27. Maximize $f(x, y, z) = xyz$ subject to the constraint $x + y + z = 120$.

$\quad F(x, y, z, \lambda) = xyz - \lambda(x + y + z - 120)$

$\quad \left.\begin{array}{l} F_x = yz - \lambda = 0 \\ F_y = xz - \lambda = 0 \\ F_z = xy - \lambda = 0 \end{array}\right\} yz = xz = xy \Rightarrow x = y = z$

$\quad F_\lambda = -(x + y + z - 120) = 0, \quad 3x = 120$

Thus, $x = y = z = 40$.

29. Maximize $f(x, y, z) = xyz$ subject to the constraint $x + y + z = S$.

$\quad F(x, y, z, \lambda) = xyz - \lambda(x + y + z - S)$

$\quad \left.\begin{array}{l} F_x = yz - \lambda = 0 \\ F_y = xz - \lambda = 0 \\ F_z = xy - \lambda = 0 \end{array}\right\} yz = xz = xy \Rightarrow x = y = z$

$\quad F_\lambda = -(x + y + z - S) = 0, \quad 3x = S$

Thus, $x = y = z = S/3$.

31. $F(x, y, \lambda) = x^2 + y^2 - \lambda(x + 2y - 5)$

$\quad F_x = 2x - \lambda = 0, \quad\quad x = \dfrac{\lambda}{2}$

$\quad F_y = 2y - 2\lambda = 0, \quad\quad y = \lambda = 2x$

$\quad F_\lambda = -(x + 2y - 5) = 0$

$\quad x + 2(2x) = 5 \Rightarrow x = 1, y = 2$

The minimum distance is $\sqrt{x^2 + y^2} = \sqrt{1 + 4} = \sqrt{5}$.

33. $F(x, y, z, \lambda) = (x - 2)^2 + (y - 1)^2 + (z - 1)^2 - \lambda(x + y + z - 1)$

$\quad \left.\begin{array}{l} F_x = 2(x - 2) - \lambda = 0 \\ F_y = 2(y - 1) - \lambda = 0 \\ F_z = 2(z - 1) - \lambda = 0 \end{array}\right\} \begin{array}{l} x - 2 = y - 1 = z - 1 \\ x - 1 = y = z \end{array}$

$\quad F_\lambda = -(x + y + z - 1) = 0$

Thus, $x = 1, y = z = x - 1 = 0$, and $d = \sqrt{(1 - 2)^2 + (0 - 1)^2 + (0 - 1)^2} = \sqrt{3}$.

35. $F(x, y, z, \lambda) = xyz - \lambda(x + 2y + 2z - 108)$

$\quad \left.\begin{array}{l} F_x = yz - \lambda = 0 \\ F_y = xz - 2\lambda = 0 \end{array}\right\} \quad x = 2y$

$\quad F_z = xy - 2\lambda = 0, \quad y = z$

$\quad F_\lambda = -(x + 2y + 2z - 108) = 0, \quad 6y = 108$

Thus, $x = 36, y = 18$, and $z = 18$. The volume is maximum when the dimensions are $36 \times 18 \times 18$ inches.

37. Minimize $C(x, y, z) = 5xy + 3(xy + 2xz + 2yz) = 8xy + 6xz + 6yz$ subject to the constraint $xyz = 480$.

$F(x, y, z, \lambda) = 8xy + 6xz + 6yz - \lambda(xyz - 480)$

$$\left. \begin{array}{l} F_x = 8y + 6z - \lambda yz = 0 \\ F_y = 8x + 6z - \lambda xz = 0 \\ F_z = 6x + 6y - \lambda xy = 0 \end{array} \right\} \begin{array}{l} x = y \\ 4y = 3z \end{array}$$

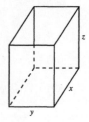

$F_\lambda = -(xyz - 480) = 0, \ y(y)\left(\tfrac{4}{3}y\right) = 480, \ y = 2\sqrt[3]{45}$

Thus, the dimensions are $2\sqrt[3]{45}$ feet $\times\ 2\sqrt[3]{45}$ feet $\times\ \tfrac{8}{3}\sqrt[3]{45}$ feet.

39. $F(x_1, x_2, \lambda) = 0.25x_1{}^2 + 10x_1 + .15x_2{}^2 + 12x_2 - \lambda(x_1 + x_2 - 2000)$

$F_{x_1} = 0.5x_1 + 10 - \lambda = 0, \qquad x_1 = 2\lambda - 20$

$F_{x_2} = 0.3x_2 + 12 - \lambda = 0, \qquad x_2 = \tfrac{10}{3}\lambda - 40$

$F_\lambda = -(x_1 + x_2 - 2000)$

$(2\lambda - 20) + \left(\tfrac{10}{3}\lambda - 40\right) = 2000$

$$\tfrac{16}{3}\lambda = 2060$$

$$\lambda = 386.25$$

Hence, $x_1 = 752.5$ and $x_2 = 1247.5$. To minimize cost, let $x_1 = 753$ units and $x_2 = 1247$ units.

41. (a) $F(x, y, \lambda) = 48x + 36y - \lambda(x^{0.25}y^{0.75} - 200)$

$F_x = 48 - 0.25\lambda x^{-0.75}y^{0.75} = 0$

$F_y = 36 - 0.75\lambda x^{0.25}y^{-0.25} = 0$

$F_\lambda = -(x^{0.25}y^{0.75} - 200) = 0$

This produces $\left(\dfrac{y}{x}\right)^{0.75} = \dfrac{48}{0.25\lambda}$ and $\left(\dfrac{y}{x}\right)^{0.25} = \dfrac{0.75\lambda}{36}$.

Thus,

$$\frac{y}{x} = \left(\frac{48}{0.25\lambda}\right)\left(\frac{0.75\lambda}{36}\right) = 4$$

$$x = \frac{200}{4^{0.75}} = \frac{200}{2\sqrt{2}} = 50\sqrt{2} \approx 71$$

$$y = 4x = 200\sqrt{2} \approx 283.$$

(b) $\dfrac{f_x(x, y)}{f_y(x, y)} = \dfrac{48}{36}$

$$\frac{25x^{-0.75}y^{0.75}}{75x^{0.25}y^{-0.25}} = \frac{48}{36}$$

$$\frac{y}{x} = 4$$

Thus, $y = 4x$ and the conditions of part (a) are met.

43. (a) From Exercise 41, we have $y = 4x$.

$F(x, y, \lambda) = 100x^{0.25}y^{0.75} - \lambda(48x + 36y - 100,000)$

$F_\lambda = -(48x + 36y - 100,000) = 0$

Thus, $x = \tfrac{3125}{6}, \ y = \tfrac{6250}{3}$, and $f\left(\tfrac{3125}{6}, \tfrac{6250}{3}\right) \approx 147,313.91 \approx 147,314$.

(b) $F_x = 25x^{-0.75}y^{0.75} - 48\lambda = 0$

$$\lambda = \frac{25x^{-0.75}y^{0.75}}{48} = \frac{25}{48}\left(\frac{3125}{6}\right)^{-0.75}\left(\frac{6250}{3}\right)^{0.75} \approx 1.4731$$

(c) $147,314 + 25,000\lambda \approx 147,314 + 25,000(1.4731) \approx 184,141.5 \approx 184,142$ units

45. Minimize cost $= x + 2y + 3z.$

Constraint: $12xyz = 0.13$

$F(x, y, z, \lambda) = x + 2y + 3z - \lambda(12xyz - 0.13)$

$$\left.\begin{array}{ll} F_x = 1 - 12\lambda yz = 0, & 12\lambda yz = 1 \\ F_y = 2 - 12\lambda xz = 0, & 12\lambda xz = 2 \\ F_z = 3 - 12\lambda xy = 0, & 12\lambda xy = 3 \end{array}\right\} \begin{array}{l} x = 2y \\ \\ x = 3z \end{array}$$

$F_\lambda = -(12xyz - 0.13) = 0$

$12x\left(\dfrac{x}{2}\right)\left(\dfrac{x}{3}\right) = 0.13, \quad x = \sqrt[3]{0.065} \approx 0.402$

$2x^3 = 0.13, \quad y = \dfrac{1}{2}\sqrt[3]{0.065} \approx 0.201$

$z = \dfrac{1}{3}\sqrt[3]{0.065} \approx 0.134$

47. (a) $f(x, y) = \text{Cost} = 10(2x + 2y) + 4x = 24x + 20y$

Constraint: $g(x, y) = 2xy - 6000 = 0$

$F(x, y, \lambda) = 24x + 20y - \lambda(2xy - 6000)$

$$\left.\begin{array}{l} F_x = 24 - 2\lambda y = 0 \\ F_y = 20 - 2\lambda x = 0 \end{array}\right\} \begin{array}{l} y = 12/\lambda \\ x = 10/\lambda \end{array}$$

$F_\lambda = -2xy + 6000 = 0$

$2xy = 6000$

$2\left(\dfrac{10}{\lambda}\right)\left(\dfrac{12}{\lambda}\right) = 6000$

$\dfrac{1}{\lambda^2} = 25$

$\lambda = \dfrac{1}{5} \Rightarrow x = 50 \text{ and } y = 60$

Dimensions: 50 feet by 120 feet

(b) Cost $= 24(50) + 20(60) = \$2400$

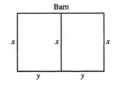

49. Maximize $f(G, P, S) = 0.025G + 0.014P + 0.031S$

Constraints: $G + P + S = 300,000$

$3000G - 3000S + P^2 = 0$

$F(G, P, S, \lambda, \mu) = 0.025G + 0.014P + 0.031S$

$- \lambda(G + P + S - 300,000) - \mu(3000G - 3000S + P^2)$

$\begin{array}{lll} F_G = 0.025 - \lambda - 3000\mu & = 0 & \quad(1) \\ F_P = 0.014 - \lambda - 2P\mu & = 0 & \quad(2) \\ F_S = 0.031 - \lambda + 3000\mu & = 0 & \quad(3) \\ F_\lambda = -G - P - S + 300,000 = 0 & & \quad(4) \\ F_\mu = -3000G + 3000S - P^2 = 0 & & \quad(5) \end{array}$

Adding Equations (1) and (3), we obtain $\lambda = 0.028$ and $\mu = \dfrac{-0.003}{3000}$. From Equation (2), $P = 7000$.

Equations (4) and (5) become

$G + S = 300,000 - 7000 = 293,000$

$-3G + 3S = 49,000.$

Thus, $S = 154,666.67$ and $G = 138,333.33.$

51. Answers will vary.

Section 7.7 Least Squares Regression Analysis

1. (a) $\sum x_i = 0$

$\sum y_i = 4$

$\sum x_i y_i = 6$

$\sum x_i^2 = 8$

$a = \dfrac{3(6) - 0(4)}{3(8) - 0^2} = \dfrac{3}{4}$

$b = \dfrac{1}{3}\left[4 - \dfrac{3}{4}(0)\right] = \dfrac{4}{3}$

The regression line is $y = \frac{3}{4}x + \frac{4}{3}$.

(b) $\left(-\dfrac{3}{2} + \dfrac{4}{3} - 0\right)^2 + \left(\dfrac{4}{3} - 1\right)^2 + \left(\dfrac{3}{2} + \dfrac{4}{3} - 3\right)^2 = \dfrac{1}{6}$

3. (a) $\sum x_i = 4$

$\sum y_i = 8$

$\sum x_i y_i = 4$

$\sum x_i^2 = 6$

$a = \dfrac{4(4) - 4(8)}{4(6) - 4^2} = -2$

$b = \dfrac{1}{4}[8 + 2(4)] = 4$

The regression line is $y = -2x + 4$.

(b) $(4 - 4)^2 + (2 - 3)^2 + (2 - 1)^2 + (0 - 0)^2 = 2$

5. The sum of the squared errors is as follows.

$S = (-2a + b + 1)^2 + (0a + b)^2 + (2a + b - 3)^2$

$\dfrac{\partial S}{\partial a} = 2(-2a + b + 1)(-2) + 2(2a + b - 3)(2) = 16a - 16$

$\dfrac{\partial S}{\partial b} = 2(-2a + b + 1) + 2b + 2(2a + b - 3) = 6b - 4$

Setting these partial derivatives equal to zero produces $a = 1$ and $b = \frac{2}{3}$. Thus, $y = x + \frac{2}{3}$.

7. The sum of the squared errors is as follows.

$S = (-2a + b - 4)^2 + (-a + b - 1)^2 + (b + 1)^2 + (a + b + 3)^2$

$\dfrac{\partial S}{\partial a} = -4(-2a + b - 4) - 2(-a + b - 1) + 2(a + b + 3) = 12a - 4b + 24$

$\dfrac{\partial S}{\partial b} = 2(-2a + b - 4) + 2(-a + b - 1) + 2(b + 1) + 2(a + b + 3) = -4a + 8b - 2$

Setting these partial derivatives equal to zero produces:

$12a - 4b = -24$

$-4a + 8b = 2$

Thus, $a = -2.3$ and $b = -0.9$, and $y = -2.3x - 0.9$.

9. $\sum x_i = 0$

$\sum y_i = 7$

$\sum x_i y_i = 7$

$\sum x_i^2 = 10$

Since $a = \frac{7}{10}$ and $b = \frac{7}{5}$, the regression line is $y = \frac{7}{10}x + \frac{7}{5}$.

11. $\sum x_i = 3$

$\sum y_i = 15$

$\sum x_i y_i = 29$

$\sum x_i^2 = 17$

Since $a = 1$ and $b = 4$, the regression line is $y = x + 4$.

13. $\sum x_i = 0$

$\sum y_i = 7$

$\sum x_i y_i = -13$

$\sum x_i^2 = 20$

Since $a = -\frac{13}{20}$ and $b = \frac{7}{4}$, the regression line is $y = -\frac{13}{20}x + \frac{7}{4}$.

15. $\sum x_i = 13$

$\sum y_i = 12$

$\sum x_i y_i = 46$

$\sum x_i^2 = 51$

Since $a = \frac{37}{43}$ and $b = \frac{7}{43}$, the regression line is $y = \frac{37}{43}x + \frac{7}{43}$.

17. $\sum x_i = 27$

$\sum y_i = 0$

$\sum x_i y_i = -70$

$\sum x_i^2 = 205$

Since $a = -\frac{175}{148}$ and $b = \frac{945}{148}$, the regression line is $y = -\frac{175}{148}x + \frac{945}{148}$.

19. $\sum x_i = 0$

$\sum y_i = 8$

$\sum x_i^2 = 10$

$\sum x_i^3 = 0$

$\sum x_i^4 = 34$

$\sum x_i y_i = 12$

$\sum x_i^2 y_i = 22$

This produces the system

$$34a \qquad + 10c = 22$$
$$10b \qquad = 12$$
$$10a \qquad + 5c = 8$$

which yields $a = \frac{3}{7}$, $b = \frac{6}{5}$, and $c = \frac{26}{35}$, and we have $y = \frac{3}{7}x^2 + \frac{6}{5}x + \frac{26}{35}$.

21. $\sum x_i = 10$

$\sum y_i = 21$

$\sum x_i^2 = 30$

$\sum x_i^3 = 100$

$\sum x_i^4 = 354$

$\sum x_i y_i = 75$

$\sum x_i^2 y_i = 275$

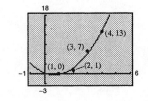

This produces the system

$$354a + 100b + 30c = 275$$
$$100a + 30b + 10c = 75$$
$$30a + 10b + 4c = 21$$

which yields $a = 1.25$, $b = -1.75$, and $c = 0.25$, and we have $y = 1.25x^2 - 1.75x + 0.25$.

23. Linear: $y = 1.4286x + 6$

Quadratic: $y = 0.1190x^2 + 1.6667x + 5.6429$

Quadratic is better.

25. Linear: $y = -68.9143x + 753.9524$

Quadratic: $y = 2.8214x^2 - 83.0214x + 763.3571$

Quadratic is better.

27. $(1, 450), (1.25, 375), (1.5, 330)$

(a) $\sum x_i = 3.75$

$\sum y_i = 1155$

$\sum x_i y_i = 1413.75$

$\sum x_i^2 = 4.8125$

Thus, $a = -240$, $b = 685$, and $y = -240x + 685$.

(b) When $x = 1.4$, $y = 349$.

(c) $y = 500$ when $x \approx 0.77$.

29. (a) Using a graphing utility, the least squares regression line is $y = 13.8x + 22.1$.

(b) If $x = 1.6$, $y \approx 44.18$ bushels per acre.

31. (a) $y = -0.48t + 19.74$

For 2010, $t = 40$ and $y \approx 0.54$ deaths/1000 births.

(b) $y = 0.0027t^2 - 0.51t + 19.03$

For 2010, $t = 40$ and $y \approx 2.95$ deaths/1000 births.

33. (a) (0, 0), (2, 15), (4, 30), (6, 50), (8, 65), (10, 70)

$$\sum x_i = 30$$
$$\sum y_i = 230$$
$$\sum x_i^2 = 220$$
$$\sum x_i^3 = 1800$$
$$\sum x_i^4 = 15,664$$
$$\sum x_i y_i = 1670$$
$$\sum x_i^2 y_i = 13,500$$

$$15,664a + 1800b + 220c = 13,500$$
$$1800a + 220b + 30c = 1670$$
$$220a + 30b + 6c = 230$$

Thus, $a = -\frac{25}{112}$, $b = \frac{541}{56}$, and $c = -\frac{25}{14}$, and we

have $y = -\frac{25}{112}x^2 + \frac{541}{56}x - \frac{25}{14}$.

(b) When $x = 5$, $y \approx 40.9$ mph.

35. Linear: $y = 3.7569x + 9.0347$

Quadratic: $y = 0.006316x^2 + 3.6252x + 9.4282$

37. Linear: $y = 0.9374x + 6.2582$

Quadratic: $y = -0.08715x^2 + 2.8159x + 0.3975$

39. Positive correlation

$r = 0.9981$

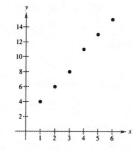

41. No correlation

$r = 0$

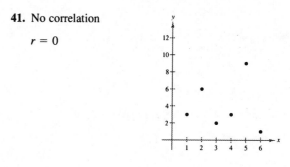

43. Let $t = 20$ correspond to age 20. Using a graphing utility, you obtain,

$$y = -49.95t^2 + 4442.6t - 41,941.$$

When $t = 28$, $y \approx \$43,291$.

45. True, the slope is negative.

47. True, the correlation is poor.

Section 7.8   Double Integrals and Area in the Plane

1. $\displaystyle\int_0^x (2x - y)\, dy = \left(2xy - \frac{y^2}{2}\right)\Big]_0^x = \frac{3x^2}{2}$

3. $\displaystyle\int_1^{2y} \frac{y}{x}\, dx = y \ln|x|\Big]_1^{2y} = y \ln|2y|$

5. $\displaystyle\int_0^{\sqrt{9-x^2}} x^2 y\, dy = \frac{x^2 y^2}{2}\Big]_0^{\sqrt{9-x^2}} = \frac{x^2(9 - x^2)}{2} = \frac{9x^2 - x^4}{2}$

7. $\displaystyle\int_{e^y}^y \frac{y \ln x}{x}\, dx = \frac{y(\ln x)^2}{2}\Big]_{e^y}^y = \frac{y}{2}[(\ln y)^2 - y^2]$

9. $\displaystyle\int_0^{x^3} ye^{-y/x}\,dy = -xye^{-y/x}\Big]_0^{x^3} + x\int_0^{x^3} e^{-y/x}\,dy$

$\qquad = -x^4e^{-x^2} - \Big[x^2e^{-y/x}\Big]_0^{x^3}$

$\qquad = -x^4e^{-x^2} - x^2e^{-x^2} + x^2$

$\qquad = x^2(1 - e^{-x^2} - x^2e^{-x^2})$

11. $\displaystyle\int_0^2\int_0^1 (x - y)\,dy\,dx = \int_0^2\Big[xy - \frac{y^2}{2}\Big]_0^1\,dx$

$\qquad = \int_0^2\Big(x - \frac{1}{2}\Big)\,dx$

$\qquad = \Big[\frac{x^2}{2} - \frac{1}{2}x\Big]_0^2$

$\qquad = 2 - 1 = 1$

13. $\displaystyle\int_0^4\int_0^3 xy\,dy\,dx = \int_0^4\Big[\frac{xy^2}{2}\Big]_0^3\,dx = \frac{9}{2}\int_0^4 x\,dx = \frac{9}{2}\Big[\frac{x^2}{2}\Big]_0^4 = 36$

15. $\displaystyle\int_0^1\int_0^{\sqrt{1-y^2}} (x + y)\,dx\,dy = \int_0^1\Big[\frac{x^2}{2} + xy\Big]_0^{\sqrt{1-y^2}}\,dy = \int_0^1\Big[\frac{1}{2}(1 - y^2) + y\sqrt{1 - y^2}\Big]\,dy = \Big[\frac{1}{2}\Big(y - \frac{y^3}{3}\Big) - \frac{1}{2}\Big(\frac{2}{3}\Big)(1 - y^2)^{3/2}\Big]_0^1$

$\qquad = \frac{1}{2}\Big[y - \frac{y^3}{3} - \frac{2}{3}(1 - y^2)^{3/2}\Big]_0^1 = \frac{2}{3}$

17. $\displaystyle\int_1^2\int_0^4 (x^2 - 2y^2 + 1)\,dx\,dy = \int_1^2\Big[\frac{x^3}{3} - 2xy^2 + x\Big]_0^4\,dy$

$\qquad = \int_1^2\Big(\frac{64}{3} - 8y^2 + 4\Big)\,dy$

$\qquad = \Big[\frac{76}{3}y - \frac{8y^3}{3}\Big]_1^2$

$\qquad = \Big[\frac{4}{3}(19y - 2y^3)\Big]_1^2 = \frac{20}{3}$

19. $\displaystyle\int_0^2\int_0^{\sqrt{1-y^2}} -5xy\,dx\,dy = -5\int_0^2\Big[\frac{x^2}{2}y\Big]_0^{\sqrt{1-y^2}}\,dy$

$\qquad = -5\int_0^2\frac{(1 - y^2)y}{2}\,dy$

$\qquad = -\frac{5}{2}\Big[\frac{y^2}{2} - \frac{y^4}{4}\Big]_0^2$

$\qquad = -\frac{5}{2}(2 - 4) = 5$

21. $\displaystyle\int_0^2\int_0^{4-x^2} x^3\,dy\,dx = \int_0^2 x^3y\Big]_0^{4-x^2}\,dx$

$\qquad = \int_0^2 (4x^3 - x^5)\,dx$

$\qquad = \Big[\Big(x^4 - \frac{x^6}{6}\Big)\Big]_0^2$

$\qquad = \frac{16}{3}$

23. Since (for fixed x)

$\qquad \displaystyle\lim_{b\to\infty}\Big[-2e^{-(x+y)/2}\Big]_0^b = 2e^{-x/2},$

we have the following.

$\qquad \displaystyle\int_0^\infty\int_0^\infty e^{-(x+y)/2}\,dy\,dx = \int_0^\infty 2e^{-x/2}\,dx$

$\qquad\qquad = \lim_{b\to\infty}\Big[-4e^{-x/2}\Big]_0^b = 4$

25. $\displaystyle\int_0^1\int_0^2 dy\,dx = \int_0^1 2\,dx = 2$

$\qquad \displaystyle\int_0^2\int_0^1 dx\,dy = \int_0^2 dy = 2$

27. $\displaystyle\int_0^1\int_{2y}^2 dx\,dy = \int_0^1 (2 - 2y)\,dy$

$\qquad = (2y - y^2)\Big]_0^1 = 1$

$\qquad \displaystyle\int_0^2\int_0^{x/2} dy\,dx = \int_0^2\frac{x}{2}\,dx$

$\qquad = \frac{x^2}{4}\Big]_0^2 = 1$

29. $\int_0^2 \int_{x/2}^1 dy\,dx = \int_0^2 \left(1 - \frac{x}{2}\right) dx = \left(x - \frac{x^2}{4}\right)\Big]_0^2 = 1$

$\int_0^1 \int_0^{2y} dx\,dy = \int_0^1 2y\,dy = y^2\Big]_0^1 = 1$

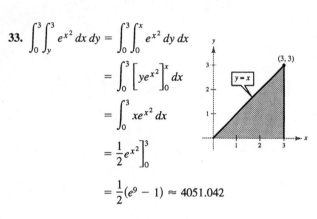

31. $\int_0^1 \int_{y^2}^{\sqrt[3]{y}} dx\,dy = \int_0^1 \left(\sqrt[3]{y} - y^2\right) dy$

$= \left[\left(\frac{3}{4}y^{4/3} - \frac{y^3}{3}\right)\right]_0^1$

$= \frac{5}{12}$

$\int_0^1 \int_{x^3}^{\sqrt{x}} dy\,dx = \int_0^1 \left(\sqrt{x} - x^3\right) dx$

$= \left[\left(\frac{2}{3}x^{3/2} - \frac{x^4}{4}\right)\right]_0^1$

$= \frac{5}{12}$

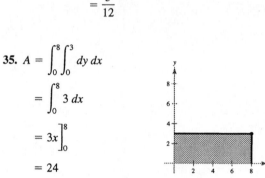

33. $\int_0^3 \int_y^3 e^{x^2} dx\,dy = \int_0^3 \int_0^x e^{x^2} dy\,dx$

$= \int_0^3 \left[ye^{x^2}\right]_0^x dx$

$= \int_0^3 xe^{x^2} dx$

$= \frac{1}{2}e^{x^2}\Big]_0^3$

$= \frac{1}{2}(e^9 - 1) \approx 4051.042$

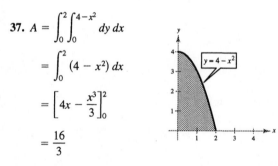

35. $A = \int_0^8 \int_0^3 dy\,dx$

$= \int_0^8 3\,dx$

$= 3x\Big]_0^8$

$= 24$

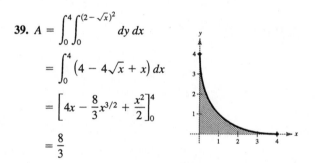

37. $A = \int_0^2 \int_0^{4-x^2} dy\,dx$

$= \int_0^2 (4 - x^2) dx$

$= \left[4x - \frac{x^3}{3}\right]_0^2$

$= \frac{16}{3}$

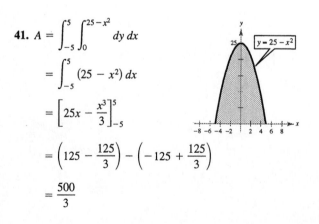

39. $A = \int_0^4 \int_0^{(2-\sqrt{x})^2} dy\,dx$

$= \int_0^4 \left(4 - 4\sqrt{x} + x\right) dx$

$= \left[4x - \frac{8}{3}x^{3/2} + \frac{x^2}{2}\right]_0^4$

$= \frac{8}{3}$

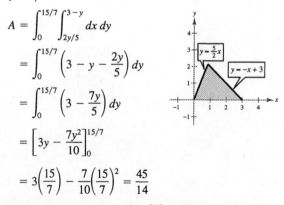

41. $A = \int_{-5}^5 \int_0^{25-x^2} dy\,dx$

$= \int_{-5}^5 (25 - x^2) dx$

$= \left[25x - \frac{x^3}{3}\right]_{-5}^5$

$= \left(125 - \frac{125}{3}\right) - \left(-125 + \frac{125}{3}\right)$

$= \frac{500}{3}$

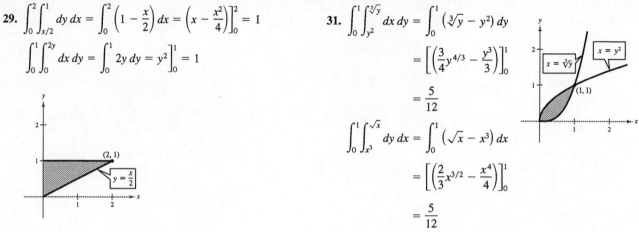

43. The point of intersection of the two graphs is found by equating $y = \frac{5}{2}x$ and $y = 3 - x$, which yields $x = \frac{6}{7}$ and $y = \frac{15}{7}$.

$A = \int_0^{15/7} \int_{2y/5}^{3-y} dx\,dy$

$= \int_0^{15/7} \left(3 - y - \frac{2y}{5}\right) dy$

$= \int_0^{15/7} \left(3 - \frac{7y}{5}\right) dy$

$= \left[3y - \frac{7y^2}{10}\right]_0^{15/7}$

$= 3\left(\frac{15}{7}\right) - \frac{7}{10}\left(\frac{15}{7}\right)^2 = \frac{45}{14}$

Note: Area of triangle is $\frac{1}{2}(3)\left(\frac{15}{7}\right) = \frac{45}{14}$.

45. $A = \int_0^2 \int_x^{2x} dy\, dx$

$= \int_0^2 (2x - x)\, dx$

$= \int_0^2 x\, dx$

$= \dfrac{x^2}{2}\Big]_0^2 = 2$

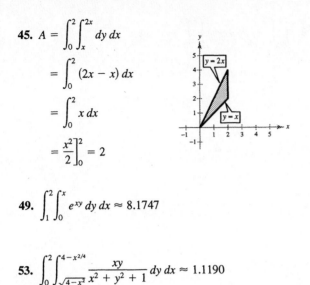

47. $\int_0^1 \int_0^2 e^{-x^2-y^2}\, dx\, dy \approx 0.65876$

49. $\int_1^2 \int_0^x e^{xy}\, dy\, dx \approx 8.1747$

51. $\int_0^1 \int_x^1 \sqrt{1-x^2}\, dy\, dx = \dfrac{\pi}{4} - \dfrac{1}{3} \approx 0.4521$

53. $\int_0^2 \int_{\sqrt{4-x^2}}^{4-x^2/4} \dfrac{xy}{x^2+y^2+1}\, dy\, dx \approx 1.1190$

55. False. Because $dA = dx\, dy = dy\, dx$, it doesn't matter what order the integration is performed.

Section 7.9   Applications of Double Integrals

1. $\int_0^2 \int_0^1 (3x + 4y)\, dy\, dx = \int_0^2 (3xy + 2y^2)\Big]_0^1 dx$

$= \int_0^2 (3x + 2)\, dx$

$= \left[\left(\dfrac{2}{3}x^2 + 2x\right)\right]_0^2$

$= 10$

3. $\int_0^1 \int_y^{\sqrt{y}} x^2y^2\, dx\, dy = \int_0^1 \dfrac{x^3y^2}{3}\Big]_y^{\sqrt{y}} dy$

$= \dfrac{1}{3}\int_0^1 (y^{7/2} - y^5)\, dy$

$= \dfrac{1}{3}\left[\dfrac{2}{9}y^{9/2} - \dfrac{1}{6}y^6\right]_0^1$

$= \dfrac{1}{54}$

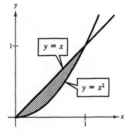

5. $\int_0^1 \int_0^{\sqrt{1-x^2}} y\, dy\, dx = \int_0^1 \dfrac{y^2}{2}\Big]_0^{\sqrt{1-x^2}} dx$

$= \dfrac{1}{2}\int_0^1 (1 - x^2)\, dx$

$= \dfrac{1}{2}\left(x - \dfrac{x^3}{3}\right)\Big]_0^1$

$= \dfrac{1}{3}$

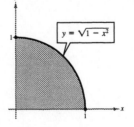

7. $\displaystyle\int_{-a}^{a}\int_{-\sqrt{a^2-x^2}}^{\sqrt{a^2-x^2}} dy\,dx = \int_{-a}^{a}\Big[y\Big]_{-\sqrt{a^2-x^2}}^{\sqrt{a^2-x^2}}\,dx$

$\displaystyle\qquad\qquad = \int_{-a}^{a} 2\sqrt{a^2-x^2}\,dx$

$\displaystyle\qquad\qquad = \pi a^2 \text{ (area of circle)}$

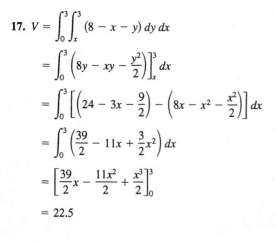

9. $\displaystyle\int_{0}^{3}\int_{0}^{5} xy\,dy\,dx = \int_{0}^{5}\int_{0}^{3} xy\,dx\,dy$

$\displaystyle\int_{0}^{3}\int_{0}^{5} xy\,dy\,dx = \int_{0}^{3}\frac{xy^2}{2}\Big]_{0}^{5}\,dx$

$\displaystyle\qquad\qquad = \int_{0}^{3}\frac{25}{2}x\,dx$

$\displaystyle\qquad\qquad = \frac{25}{4}x^2\Big]_{0}^{3}$

$\displaystyle\qquad\qquad = \frac{225}{4}$

11. $\displaystyle\int_{0}^{2}\int_{x}^{2x}\frac{y}{x^2+y^2}\,dy\,dx = \int_{0}^{2}\int_{y/2}^{y}\frac{y}{x^2+y^2}\,dx\,dy + \int_{2}^{4}\int_{y/2}^{2}\frac{y}{x^2+y^2}\,dx\,dy$

$\displaystyle\int_{0}^{2}\int_{x}^{2x}\frac{y}{x^2+y^2}\,dy\,dx = \int_{0}^{2}\Big[\frac{1}{2}\ln(x^2+y^2)\Big]_{x}^{2x}\,dx$

$\displaystyle\qquad\qquad = \frac{1}{2}\int_{0}^{2}\big[\ln(5x^2)-\ln(2x^2)\big]\,dx = \frac{1}{2}\int_{0}^{2}\ln\frac{5}{2}\,dx = \Big(\frac{1}{2}\ln\frac{5}{2}\Big)x\Big]_{0}^{2} = \ln\frac{5}{2}$

13. $\displaystyle\int_{0}^{1}\int_{y/2}^{1/2} e^{-x^2}\,dx\,dy = \int_{0}^{1/2}\int_{0}^{2x} e^{-x^2}\,dy\,dx$

$\displaystyle\qquad\qquad = \int_{0}^{1/2} 2xe^{-x^2}\,dx$

$\displaystyle\qquad\qquad = -e^{-x^2}\Big]_{0}^{1/2}$

$\displaystyle\qquad\qquad = 1 - e^{-1/4}$

$\displaystyle\qquad\qquad \approx 0.2212$

15. $\displaystyle V = \int_{0}^{2}\int_{0}^{4}\frac{y}{2}\,dx\,dy$

$\displaystyle\qquad = \int_{0}^{2}\frac{xy}{2}\Big]_{0}^{4}\,dy$

$\displaystyle\qquad = \int_{0}^{2} 2y\,dy$

$\displaystyle\qquad = y^2\Big]_{0}^{2}$

$\displaystyle\qquad = 4$

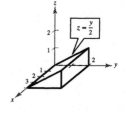

$0 \le x \le 4$

$0 \le y \le 2$

17. $\displaystyle V = \int_{0}^{3}\int_{x}^{3}(8-x-y)\,dy\,dx$

$\displaystyle\qquad = \int_{0}^{3}\Big(8y - xy - \frac{y^2}{2}\Big)\Big]_{x}^{3}\,dx$

$\displaystyle\qquad = \int_{0}^{3}\Big[\Big(24 - 3x - \frac{9}{2}\Big) - \Big(8x - x^2 - \frac{x^2}{2}\Big)\Big]\,dx$

$\displaystyle\qquad = \int_{0}^{3}\Big(\frac{39}{2} - 11x + \frac{3}{2}x^2\Big)\,dx$

$\displaystyle\qquad = \Big[\frac{39}{2}x - \frac{11x^2}{2} + \frac{x^3}{2}\Big]_{0}^{3}$

$\displaystyle\qquad = 22.5$

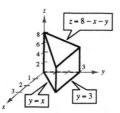

19. $V = \int_0^6 \int_0^{4-(2x/3)} \left(3 - \dfrac{x}{2} - \dfrac{3y}{4} \right) dy \, dx$

$= \int_0^6 \left[3y - \dfrac{xy}{2} - \dfrac{3y^2}{8} \right]_0^{4-(2x/3)} dx$

$= \int_0^6 \left(6 - 2x + \dfrac{x^2}{6} \right) dx$

$= \left[6x - x^2 + \dfrac{x^3}{18} \right]_0^6$

$= 12$

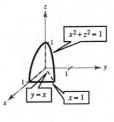

$2x + 3y + 4z = 12$

21. $V = \int_0^1 \int_0^y (1 - xy) \, dx \, dy$

$= \int_0^1 \left[x - \dfrac{x^2 y}{2} \right]_0^y dy$

$= \int_0^1 \left(y - \dfrac{y^3}{2} \right) dy$

$= \left[\dfrac{y^2}{2} - \dfrac{y^4}{8} \right]_0^1$

$= \dfrac{3}{8}$

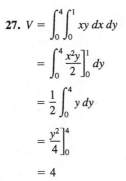

$z = 1 - xy$

$y = x$ $y = 1$

23. $V = 4 \int_0^1 \int_0^1 (4 - x^2 - y^2) \, dy \, dx$

$= 4 \int_0^1 \left[4y - x^2 y - \dfrac{y^3}{3} \right]_0^1 dx$

$= 4 \int_0^1 \left[4 - x^2 - \dfrac{1}{3} \right] dx$

$= 4 \int_0^1 \left(\dfrac{11}{3} - x^2 \right) dx$

$= 4 \left[\dfrac{11x}{3} - \dfrac{x^3}{3} \right]_0^1$

$= 4 \left(\dfrac{11}{3} - \dfrac{1}{3} \right)$

$= 4 \left(\dfrac{10}{3} \right)$

$= \dfrac{40}{3}$

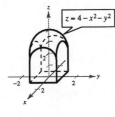

$z = 4 - x^2 - y^2$

$-1 \leq x \leq 1$

$-1 \leq y \leq 1$

25. $V = \int_0^1 \int_0^x \sqrt{1 - x^2} \, dy \, dx$

$= \int_0^1 x \sqrt{1 - x^2} \, dx$

$= -\dfrac{1}{3}(1 - x^2)^{3/2} \Big]_0^1$

$= \dfrac{1}{3}$

$x^2 + z^2 = 1$

$y = x$ $x = 1$

27. $V = \int_0^4 \int_0^1 xy \, dx \, dy$

$= \int_0^4 \dfrac{x^2 y}{2} \Big]_0^1 dy$

$= \dfrac{1}{2} \int_0^4 y \, dy$

$= \dfrac{y^2}{4} \Big]_0^4$

$= 4$

29. $V = \int_0^2 \int_0^4 x^2 \, dy \, dx$

$= \int_0^2 x^2 y \Big]_0^4 \, dx$

$= \int_0^2 4x^2 \, dx$

$= \frac{4x^3}{3} \Big]_0^2$

$= \frac{32}{3}$

31. Population $= \int_0^2 \int_0^2 \frac{120,000}{(2 + x + y)^3} \, dy \, dx$

$= \int_0^2 \left[-60,000(2 + x + y)^{-2} \right]_0^2 \, dx$

$= \int_0^2 \frac{24,000(3 + x)}{(4 + x)^2(2 + x^2)} \, dx$

$= 60,000 \left[\frac{1}{4 + x} - \frac{1}{2 + x} \right]_0^2$

$= 10,000 \text{ people}$

33. Average $= \frac{1}{8} \int_0^4 \int_0^2 x \, dy \, dx$

$= \frac{1}{8} \int_0^4 2x \, dx$

$= \frac{x^2}{8} \Big]_0^4$

$= 2$

35. Average $= \frac{1}{4} \int_0^2 \int_0^2 (x^2 + y^2) \, dx \, dy$

$= \frac{1}{4} \int_0^2 \left[\frac{x^3}{3} + xy^2 \right]_0^2 \, dy$

$= \frac{1}{4} \int_0^2 \left(\frac{8}{3} + 2y^2 \right) \, dy$

$= \left[\frac{1}{4} \left(\frac{8}{3}y + \frac{2}{3}y^3 \right) \right]_0^2$

$= \frac{8}{3}$

37. Average $= \frac{1}{1250} \int_{100}^{150} \int_{50}^{75} \left[(500 - 3p_1)p_1 + (750 - 2.4p_2)p_2 \right] dp_1 \, dp_2$

$= \frac{1}{1250} \int_{100}^{150} \int_{50}^{75} \left[-3p_1^2 + 500p_1 - 2.4p_2^2 + 750p_2 \right] dp_1 \, dp_2$

$= \frac{1}{1250} \int_{100}^{150} \left[-p_1^3 + 250p_1^2 - 2.4p_1p_2^2 + 750p_1p_2 \right]_{50}^{75} dp_2$

$= \frac{1}{1250} \int_{100}^{150} \left[484,375 - 60p_2^2 + 18,750p_2 \right] dp_2$

$= \frac{1}{1250} \left[484,375p_2 - 20p_2^3 + 9375p_2^2 \right]_{100}^{150}$

$= \$75,125$

Review Exercises for Chapter 7

1.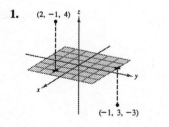

3. $d = \sqrt{(2 - 0)^2 + (5 - 0)^2 + (9 - 0)^2}$

$= \sqrt{4 + 25 + 81}$

$= \sqrt{110}$

5. Midpoint $= \left(\frac{2 - 4}{2}, \frac{6 + 2}{2}, \frac{4 + 8}{2} \right) = (-1, 4, 6)$

7. $(x - 0)^2 + (y - 1)^2 + (z - 0)^2 = 5^2$

$x^2 + (y - 1)^2 + z^2 = 25$

9. Center $= \left(\dfrac{3 + 5}{2}, \dfrac{4 + 8}{2}, \dfrac{0 + 2}{2} \right) = (4, 6, 1)$

Radius $= \sqrt{(4 - 3)^2 + (6 - 4)^2 + (1 - 0)^2} = \sqrt{6}$

Sphere: $(x - 4)^2 + (y - 6)^2 + (z - 1)^2 = 6$

11. $x^2 + 4x + 4 + y^2 - 2y + 1 + z^2 - 8z + 16 = -5 + 4 + 1 + 16$

$\qquad (x + 2)^2 + (y - 1)^2 + (z - 4)^2 = 16 = 4^2$

Center: $(-2, 1, 4)$

Radius: 4

13. Let $z = 0$.

$\qquad (x + 2)^2 + (y - 1)^2 + (0 - 3)^2 = 25$

$\qquad\qquad (x + 2)^2 + (y - 1)^2 = 16 = 4^2$

Circle of radius 4

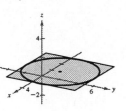

15. x-intercept: $(6, 0, 0)$

y-intercept: $(0, 3, 0)$

z-intercept: $(0, 0, 2)$

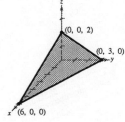

17. x-intercept: $(2, 0, 0)$

y-intercept: $(0, 4, 0)$

z-intercept: $(0, 0, -2)$

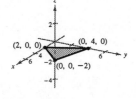

19. The graph is a sphere whose standard equation is

$\qquad (x - 1)^2 + (y + 2)^2 + (z - 3)^2 = 9.$

21. The graph is an ellipsoid.

23. The graph is an elliptic paraboloid.

25. The graph is the top half of a circular cone whose standard equation is

$\qquad x^2 + y^2 - z^2 = 0.$

27. $f(x, y) = xy^2$

(a) $f(2, 3) = 2(3)^2 = 18$

(b) $f(0, 1) = 0(1)^2 = 0$

(c) $f(-5, 7) = -5(7)^2 = -245$

(d) $f(-2, -4) = -2(-4)^2 = -32$

29. The domain is the set of all points inside or on the circle $x^2 + y^2 = 1$ and the range is $[0, 1]$.

31. The level curves are lines of slope $-\frac{2}{5}$.

$c = 0$: $10 - 2x - 5y = 0$
$$2x + 5y = 10$$

$c = 2$: $10 - 2x - 5y = 2$
$$2x + 5y = 8$$

$c = 4$: $10 - 2x - 5y = 4$
$$2x + 5y = 6$$

$c = 5$: $10 - 2x - 5y = 5$
$$2x + 5y = 5$$

$c = 10$: $10 - 2x - 5y = 10$
$$2x + 5y = 0$$

33. $z = (xy)^2$

$c = 1$: $(xy)^2 = 1$,

$c = 4$: $(xy)^2 = 4$,

$c = 9$: $(xy)^2 = 9$,

$c = 12$: $(xy)^2 = 12$,

$c = 16$: $(xy)^2 = 16$,

The level curves are hyperbolas.

$$y^2 = \frac{1}{x^2}, y = \pm\frac{1}{x}$$

$$y = \pm\frac{2}{x}$$

$$y = \pm\frac{3}{x}$$

$$y = \pm\frac{2\sqrt{3}}{x}$$

$$y = \pm\frac{4}{x}$$

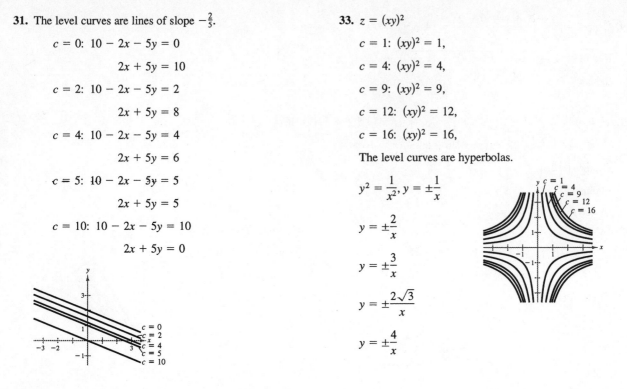

35. (a) The level curves represent lines of equal rainfall, and separate the four colors. As the color darkens from light green to dark green, the average yearly precipitation increases.

(b) The small eastern portion containing Davenport

(c) The northwestern portion containing Sioux City

37. Southwest

39. $P = \dfrac{MV}{T} = \dfrac{(2500)6}{6000} = \$2.50$

41. $f(x, y) = x^2y + 3xy + 2x - 5y$

$f_x = 2xy + 3y + 2$

$f_y = x^2 + 3x - 5$

43. $z = 6x^2\sqrt{y} + 3\sqrt{xy} - 7xy$

$z_x = 12x\sqrt{y} + \dfrac{3}{2}\sqrt{\dfrac{y}{x}} - 7y$

$z_y = 3\dfrac{x^2}{\sqrt{y}} + \dfrac{3}{2}\sqrt{\dfrac{x}{y}} - 7x$

45. $f(x, y) = \ln(2x + 3y)$

$f_x = \dfrac{2}{2x + 3y}$

$f_y = \dfrac{3}{2x + 3y}$

47. $f_x = 2xe^y - y^2e^x$

$f_y = x^2e^y - 2ye^x$

49. $\dfrac{\partial w}{\partial x} = yz^2$

$\dfrac{\partial w}{\partial y} = xz^2$

$\dfrac{\partial w}{\partial z} = 2xyz$

51. (a) $z_x = 3$

(b) $z_y = -4$

53. (a) $z_x = -2x$;   At $(1, 2, 3)$, $z_x = -2$.

(b) $z_y = -2y$;   At $(1, 2, 3)$, $z_y = -4$.

55. $f_x = 3x^2 - 4y^2$

$f_y = -8xy + 3y^2$

$f_{xx} = 6x$

$f_{yy} = -8x + 6y$

$f_{xy} = f_{yx} = -8y$

57. $f_x = \dfrac{-x}{\sqrt{64 - x^2 - y^2}}$

$f_y = \dfrac{-y}{\sqrt{64 - x^2 - y^2}}$

$f_{xx} = \dfrac{y^2 - 64}{(64 - x^2 - y^2)^{3/2}}$

$f_{yy} = \dfrac{x^2 - 64}{(64 - x^2 - y^2)^{3/2}}$

$f_{xy} = \dfrac{-xy}{(64 - x^2 - y^2)^{3/2}} = f_{yx}$

59. $C_x = 5x^{-2/3}y^{1/3} + 99,\quad C_x(250, 175) \approx 99.70$

$C_y = 5x^{1/3}y^{-2/3} + 139,\quad C_y(250, 175) \approx 140.04$

61. (a) $\dfrac{\partial A}{\partial w} = 101.4(0.425)w^{0.425 - 1}h^{0.725} = 43.095w^{-0.575}h^{0.725}$

$\dfrac{\partial A}{\partial h} = 101.4(0.725)w^{0.425}h^{0.725 - 1} = 73.515w^{0.425}h^{-0.275}$

(b) $\dfrac{\partial A}{\partial w}(180, 70) = 43.095(180)^{-0.575}(70)^{0.725} \approx 47.35$

The surface area increases approximately 47 cm^2 per pound for a human weighing 180 pounds and 70 inches tall.

63. The first partial derivatives of f, $f_x(x, y) = 2x + 2y$ and $f_y(x, y) = 2x + 2y$, are zero when $y = -x$. The points $(x, -x, 0)$ are relative minima.

Note: The Second-Partials Test fails since $d = 0$.

65. The first partial derivatives of f, $f_x = 2x + 6y + 6$ and $f_y = 6x + 6y$, are zero at $\left(\frac{3}{2}, -\frac{3}{2}\right)$. The second partials are $f_{xx} = 2, f_{xy} = 6$, and $f_{yy} = 6$. Since $f_{xx} > 0$ and $d = f_{xx}f_{yy} - (f_{xy})^2 < 0,\ \left(\frac{3}{2}, -\frac{3}{2}, \frac{25}{2}\right)$ is a saddle point. No relative extrema.

67. The first partial derivatives of f, $f_x = 3x^2 - y$ and $f_y = 2y - x$, are zero at $\left(\frac{1}{6}, \frac{1}{12}\right)$ and $(0, 0)$. The second partials are $f_{xx} = 6x, f_{yy} = 2$, and $f_{xy} = 1$.

At $(0, 0)$, $d < 0 \Longrightarrow$ saddle point at $(0, 0, 0)$.

At $\left(\frac{1}{6}, \frac{1}{12}\right)$, $f_{xx} > 0$ and $d > 0 \Longrightarrow$ relative minimum at $\left(\frac{1}{6}, \frac{1}{12}, \frac{-1}{432}\right)$.

69. The first partial derivatives of f, $f_x = 3x^2 - 3$ and $f_y = 3y^2 - 3$, are zero at $(\pm 1, \pm 1), (\pm 1, \mp 1)$. The second partials are $f_{xx} = 6x, f_{yy} = 6y$, and $f_{xy} = 0$.

$(1, 1)$: $f_{xx} > 0, d > 0$ Relative minimum: $(1, 1, -2)$

$(1, -1)$: $f_{xx} > 0, d < 0$ Saddle point: $(1, -1, 2)$

$(-1, 1)$: $f_{xx} < 0, d < 0$ Saddle point: $(-1, 1, 2)$

$(-1, -1)$: $f_{xx} < 0, d > 0$ Relative maximum: $(-1, -1, 6)$

71. (a) $R = x_1 p_1 + x_2 p_2$

$= x_1(100 - x_1) + x_2(200 - 0.5x_2)$

$= -x_1^2 - \tfrac{1}{2}x_2^2 + 100x_1 + 200x_2$

(b) $R_{x_1} = -2x_1 + 100 = 0 \Longrightarrow x_1 = 50$

$R_{x_2} = -x_2 + 200 = 0 \Longrightarrow x_2 = 200$

By the Second-Partials Test, $(50, 200)$ is a maximum.

(c) $R(50, 200) = \$22{,}500.00$

73. $F(x, y, \lambda) = x^2 y - \lambda(x + 2y - 2)$

$\left.\begin{array}{l} F_x(x, y, \lambda) = 2xy - \lambda = 0 \\ F_y(x, y, \lambda) = x^2 - 2\lambda = 0 \end{array}\right\}\quad 4xy = x^2$

$F_\lambda(x, y, \lambda) = -(x + 2y - 2) = 0,\quad y = \dfrac{2 - x}{2}$

Thus, $x = 0$ or $x = \frac{4}{3}$, and the corresponding y-values are $y = 1$ or $y = \frac{1}{3}$. This implies that the extrema occur at $(0, 1, 0)$ (relative minimum) and $\left(\frac{4}{3}, \frac{1}{3}, \frac{16}{27}\right)$ (relative minimum).

75. $F(x, y, z, \lambda) = xyz - \lambda(x + 2y + z - 4)$

$\left.\begin{array}{l} F_x = yz - \lambda = 0 \\ F_y = xz - 2\lambda = 0 \\ F_z = xy - \lambda = 0 \end{array}\right\}\ \ \begin{array}{l} xz = 2yz = 2xy \Longrightarrow x = 2y \\ \\ z = 2y \end{array}$

$F_\lambda = -(x + 2y + z - 4) = 0$

$2y + 2y + 2y - 4 = 0 \Longrightarrow y = \frac{2}{3}, x = \frac{4}{3}, z = \frac{4}{3}$

At $\left(\frac{4}{3}, \frac{2}{3}, \frac{4}{3}\right)$, the relative maximum value is $\frac{32}{27}$.

77. $F(x, y, z, \lambda, \mu) = x^2 + y^2 + z^2 - \lambda(x + z - 6) - \mu(y + z - 8)$

$$F_x = 2x - \lambda = 0 \qquad\qquad x = \frac{\lambda}{2}$$

$$F_y = 2y - \mu = 0 \qquad\qquad y = \frac{\mu}{2}$$

$$F_z = 2z - \lambda - \mu = 0 \qquad z = \frac{\lambda + \mu}{2} = x + y$$

$$F_\lambda = -(x + z - 6) = 0 \qquad x + z = 6 \Rightarrow 2x + y = 6$$
$$F_\mu = -(y + z - 8) = 0 \qquad y + z = 8 \Rightarrow x + 2y = 8$$

$x = \dfrac{4}{3}, y = \dfrac{10}{3}, z = \dfrac{14}{3}$

$f\left(\frac{4}{3}, \frac{10}{3}, \frac{14}{3}\right) = 34\frac{2}{3}$ is a relative minimum.

79. $F(x, y, z, \lambda, \mu) = xy - \lambda(x^2 + y^2 - 16) - \mu(x - 2z)$ 　　　Thus,

$$F_x = y - 2\lambda x - \mu \quad = 0 \qquad\qquad\qquad 2x^2 = 16$$
$$F_y = x - 2\lambda y \quad\quad\; = 0$$
$$F_z = 2\mu \quad\quad\quad\quad\; = 0 \quad \Rightarrow \mu = 0 \qquad x = \pm 2\sqrt{2} \text{ and } z = \frac{1}{2}x = \pm\sqrt{2}.$$
$$F_\lambda = -(x^2 + y^2 - 16) = 0 \qquad\qquad\quad \text{Maximum: } f\left(2\sqrt{2}, 2\sqrt{2}, \sqrt{2}\right) = 8$$
$$F_\mu = -(x - 2z) \quad\quad = 0$$

$y = 2\lambda x$

$x = 2\lambda y = 2\lambda(2\lambda x) \Rightarrow 1 = 4\lambda^2 \Rightarrow \lambda = \pm\dfrac{1}{2}$

$y = 2\lambda x \Rightarrow \dfrac{y}{x} = \dfrac{x}{y} \Rightarrow x^2 = y^2$

$x = 2\lambda y$

81. Maximize $f(x, y) = 4x + xy + 2y$, subject to the constraint $20x + 4y = 2000$.

$F(x, y, \lambda) = 4x + xy + 2y - \lambda(20x + 4y - 2000)$

$$F_x(x, y, \lambda) = 4 + y - 20\lambda = 0 \qquad 4 + y = 5(x + 2)$$
$$F_y(x, y, \lambda) = x + 2 - 4\lambda = 0 \qquad\qquad y = 5x + 6$$
$$F_\lambda(x, y, \lambda) = -(20x + 4y - 2000) = 0, \qquad y = 500 - 5x$$

Thus, $x = 49.4$ and $y = 5(49.4) + 6 = 253$ which implies that the maximum production level is $f(49.4, 253) \approx 13,202$.

83. (a) $\sum x_i = 1$ 　　　　　$a = \dfrac{4(15) - (1)(0)}{4(15) - (1)^2} = \dfrac{60}{59}$

$\qquad\;\; \sum y_i = 0$

$\qquad\;\; \sum x_i^2 = 15 \qquad\qquad b = \dfrac{1}{4}\left(0 - \dfrac{60}{59}(1)\right) = -\dfrac{15}{59}$

$\qquad\;\; \sum x_i y_i = 15 \qquad\qquad y = \dfrac{60}{59}x - \dfrac{15}{59}$

(b) $\left(\dfrac{60}{59}(-2) - \dfrac{15}{59} + 3\right)^2 + \left(\dfrac{60}{59}(-1) - \dfrac{15}{59} + 1\right)^2 + \left(\dfrac{60}{59}(1) - \dfrac{15}{59} - 2\right)^2 + \left(\dfrac{60}{59}(3) - \dfrac{15}{59} - 2\right)^2 \approx 2.746$

85. Using a graphing utility, $y = 14x + 19$. For 20 pounds per acre, $x = 0.2$ and $y = 14(0.2) + 19 = 21.8$ bushels per acre.

87. $\sum x_i = 6$

$\sum y_i = 50$

$\sum x_i^2 = 22$

$\sum x_i y_i = 100$

$\sum x_i^3 = 72$

$\sum x_i^2 y_i = 406$

$\sum x_i^4 = 274$

The system

$$274a + 74b + 22c = 406$$

$$74a + 22b + 6c = 100$$

$$22a + 6b + 5c = 50$$

has approximate solutions $a = 1.71$, $b = -2.57$, and $c = 5.56$. Therefore, the least squares quadratic is $y = 1.71x^2 - 2.57x + 5.56$.

89. $\displaystyle\int_0^1 \int_0^{1+x} (3x + 2y)\, dy\, dx = \int_0^1 \left[(3xy + y^2) \right]_0^{1+x} dx$

$$= \int_0^1 [3x(1 + x) + (1 + x)^2]\, dx$$

$$= \int_0^1 (4x^2 + 5x + 1)\, dx$$

$$= \left[\frac{4x^3}{3} + \frac{5x^2}{2} + x \right]_0^1$$

$$= \frac{4}{3} + \frac{5}{2} + 1$$

$$= \frac{29}{6}$$

91. $\displaystyle\int_1^2 \int_1^{2y} \frac{x}{y^2}\, dx\, dy = \int_1^2 \left[\frac{x^2}{2y^2} \right]_1^{2y} dy$

$$= \int_1^2 \left[\frac{4y^2}{2y^2} - \frac{1}{2y^2} \right] dy$$

$$= \int_1^2 \left(2 - \frac{1}{2} y^{-2} \right) dy$$

$$= \left[2y + \frac{1}{2y} \right]_1^2$$

$$= \left(4 + \frac{1}{4} \right) - \left(2 + \frac{1}{2} \right)$$

$$= \frac{7}{4}$$

93. $A = \displaystyle\int_{-2}^2 \int_5^{9-x^2} dy\, dx = \int_5^9 \int_{-\sqrt{9-y}}^{\sqrt{9-y}} dx\, dy$

$$\int_{-2}^2 \int_5^{9-x^2} dy\, dx = \int_{-2}^2 [(9 - x^2) - 5]\, dx$$

$$= \int_{-2}^2 (4 - x^2)\, dx$$

$$= \left[\left(4x - \frac{x^3}{3} \right) \right]_{-2}^2$$

$$= \left(8 - \frac{8}{3} \right) - \left(-8 + \frac{8}{3} \right)$$

$$= \frac{32}{3}$$

95. $A = \displaystyle\int_{-3}^6 \int_{1/3(x+3)}^{\sqrt{x+3}} dy\, dx$

$$= \int_{-3}^6 \left(\sqrt{x + 3} - \frac{1}{3}(x + 3) \right) dx$$

$$= \left[\frac{2}{3}(x + 3)^{3/2} - \frac{x^2}{6} - x \right]_{-3}^6$$

$$= (18 - 6 - 6) - \left(0 - \frac{3}{2} + 3 \right)$$

$$= \frac{9}{2}$$

97. $V = \displaystyle\int_0^4 \int_0^4 (xy)^2\, dy\, dx \frac{1}{2}$

$$= \int_0^4 \int_0^4 x^2 y^2\, dy\, dx$$

$$= \int_0^4 \frac{x^2 y^3}{3} \Big]_0^4 dx$$

$$= \int_0^4 \frac{64x^2}{3}\, dx$$

$$= \frac{64x^3}{9} \Big]_0^4$$

$$= \frac{4096}{9}$$

99. Average $= \dfrac{\displaystyle\int_0^{10} \int_0^{25-2.5x} (0.25 - 0.025x - 0.01y)\, dy\, dx}{\text{area}} = \dfrac{10^{5/12}}{125} = 0.083\overline{3}$ mile

Practice Test for Chapter 7

1. Find the distance between the points $(3, -7, 2)$ and $(5, 11, -6)$ and find the midpoint of the line segment joining the two points.

2. Find the standard form of the equation of the sphere whose center is $(1, -3, 0)$ and whose radius is $\sqrt{5}$.

3. Find the center and radius of the sphere whose equation is $x^2 + y^2 + z^2 - 4x + 2y + 8z = 0$.

4. Sketch the graph of the plane.

   (a) $3x + 8y + 6z = 24$ (b) $y = 2$

5. Identify the surface.

   (a) $\dfrac{x^2}{16} + \dfrac{y^2}{4} - \dfrac{z^2}{9} = 1$ (b) $z = \dfrac{x^2}{25} + y^2$

6. Find the domain of the function.

   (a) $f(x, y) = \ln(3 - x - y)$ (b) $f(x, y) = \dfrac{1}{x^2 + y^2}$

7. Find the first partial derivatives of $f(x, y) = 3x^2 + 9xy^2 + 4y^3 - 3x - 6y + 1$.

8. Find the first partial derivatives of $f(x, y) = \ln(x^2 + y^2 + 5)$.

9. Find the first partial derivatives of $f(x, y, z) = x^2 y^3 \sqrt{z}$.

10. Find the second partial derivatives of $z = \dfrac{x}{x^2 + y^2}$.

11. Find the relative extrema of $f(x, y) = 3x^2 + 4y^2 - 6x + 16y - 4$.

12. Find the relative extrema of $f(x, y) = 4xy - x^4 - y^4$.

13. Use Lagrange multipliers to find the minimum of $f(x, y) = xy$ subject to the constraint $4x - y = 16$.

14. Use Lagrange multipliers to find the minimum of $f(x, y) = x^2 - 16x + y^2 - 8y + 12$ subject to the constraint $x + y = 4$.

15. Find the least squares regression line for the points $(-3, 7)$, $(1, 5)$, $(8, -2)$, and $(4, 4)$.

16. Find the least squares regression quadratic for the points $(-5, 8)$, $(-1, 2)$, $(1, 3)$, and $(5, 5)$.

17. Evaluate $\displaystyle\int_0^3 \int_0^{\sqrt{x}} xy^3 \, dy \, dx$.

18. Evaluate $\displaystyle\int_{-1}^2 \int_0^{3y} (x^2 - 4xy) \, dx \, dy$.

19. Set up a double integral to find the area of the indicated region.

(a)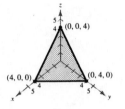

(b)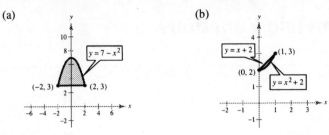

20. Find the volume of the solid bounded by the first octant and the plane $x + y + z = 4$.

Graphing Calculator Required

21. Use a graphing utility to find the least squares regression line and the correlation coefficient for the given data points.

(0, 20.4), (1, 21.3), (2, 22.9), (3, 23.4), (4, 24), (5, 24.2), (6, 24.9), (7, 25.3), (8, 26.2), (9, 27.5), (10, 29), (11, 30.3), and (12, 31.1)

22. Use a graphing calculator or a computer algebra system to approximate $\displaystyle\int_2^3 \int_0^{x+2} e^{x^2 y} \, dy \, dx$.

C H A P T E R   8
Trigonometric Functions

C H A P T E R 8
Trigonometric Functions

Section 8.1   Radian Measure of Angles

Solutions to Odd-Numbered Exercises

1. (a) Positive: $45° + 360° = 405°$

Negative: $45° - 360° = -315°$

(b) Positive: $-41° + 360° = 319°$

Negative: $-41° - 360° = -401°$

3. (a) Positive: $300° + 360° = 660°$

Negative: $300° - 360° = -60°$

(b) Positive: $740° - 2(360°) = 20°$

Negative: $740° - 3(360°) = -340°$

5. (a) Positive: $\dfrac{\pi}{9} + 2\pi = \dfrac{19\pi}{9}$

Negative: $\dfrac{\pi}{9} - 2\pi = -\dfrac{17\pi}{9}$

(b) Positive: $\dfrac{2\pi}{3} + 2\pi = \dfrac{8\pi}{3}$

Negative: $\dfrac{2\pi}{3} - 2\pi = -\dfrac{4\pi}{3}$

7. (a) Positive: $-\dfrac{9\pi}{4} + 2(2\pi) = \dfrac{7\pi}{4}$

Negative: $-\dfrac{9\pi}{4} + 2\pi = -\dfrac{\pi}{4}$

(b) Positive: $-\dfrac{2\pi}{15} + 2\pi = \dfrac{28\pi}{15}$

Negative: $-\dfrac{2\pi}{15} - 2\pi = -\dfrac{32\pi}{15}$

9. $30°\left(\dfrac{\pi \text{ radians}}{180°}\right) = \dfrac{\pi}{6}$ radians

11. $225°\left(\dfrac{\pi \text{ radians}}{180°}\right) = \dfrac{5\pi}{4}$ radians

13. $315°\left(\dfrac{\pi \text{ radians}}{180°}\right) = \dfrac{7\pi}{4}$ radians

15. $-30°\left(\dfrac{\pi \text{ radians}}{180°}\right) = -\dfrac{\pi}{6}$ radians

17. $-270°\left(\dfrac{\pi \text{ radians}}{180°}\right) = -\dfrac{3\pi}{2}$ radians

19. $390°\left(\dfrac{\pi \text{ radians}}{180°}\right) = \dfrac{13\pi}{6}$ radians

21. $\dfrac{3\pi}{2}\left(\dfrac{180°}{\pi}\right) = 270°$

23. $\dfrac{11\pi}{6}\left(\dfrac{180°}{\pi}\right) = 330°$

25. $-\dfrac{5\pi}{3}\left(\dfrac{180}{\pi}\right) = -300°$

27. $\dfrac{9\pi}{4}\left(\dfrac{180°}{\pi}\right) = 405°$

29. $\dfrac{19\pi}{6}\left(\dfrac{180°}{\pi}\right) = 570°$

31. $-270°\left(\dfrac{\pi \text{ radians}}{180°}\right) = -\dfrac{3\pi}{2}$ radians

33. $144°\left(\dfrac{\pi \text{ radians}}{180°}\right) = \dfrac{4\pi}{5}$ radians

35. The angle θ is $\theta = 90° - 30° = 60°$. Since the vertical side is $c/2$, we can use the Pythagorean Theorem to find the length of the hypotenuse as follows.

$$c^2 = \left(5\sqrt{3}\right)^2 + \left(\frac{c}{2}\right)^2$$

$$\frac{3c^2}{4} = 75$$

$$c^2 = 100$$

$$c = 10$$

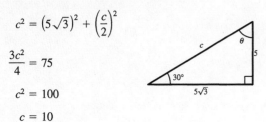

37. The angle θ is $\theta = 90° - 60° = 30°$. By the Pythagorean Theorem, the value of a is

$$a = \sqrt{8^2 - 4^2}$$

$$= \sqrt{64 - 16}$$

$$= \sqrt{48}$$

$$= 4\sqrt{3}.$$

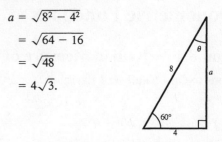

39. Since the triangle is isosceles, we have $\theta = 40°$.

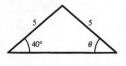

41. Since the large triangle is similar to the two smaller triangles, $\theta = 60°$. The hypotenuse of the large triangle is

$$\sqrt{2^2 + \left(2\sqrt{3}\right)^2} = \sqrt{16} = 4.$$

By similar triangles we have

$$\frac{s}{2} = \frac{2\sqrt{3}}{4} \Longrightarrow s = \sqrt{3}.$$

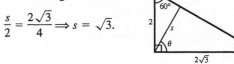

43. $h^2 + 3^2 = 6^2$

$\quad\quad h^2 = 27$

$\quad\quad\quad h = 3\sqrt{3}$

$A = \frac{1}{2}bh$

$\quad = \frac{1}{2}(6)\left(3\sqrt{3}\right)$

$\quad = 9\sqrt{3}$ square inches

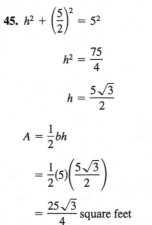

45. $h^2 + \left(\frac{5}{2}\right)^2 = 5^2$

$\quad\quad\quad h^2 = \frac{75}{4}$

$\quad\quad\quad h = \frac{5\sqrt{3}}{2}$

$A = \frac{1}{2}bh$

$\quad = \frac{1}{2}(5)\left(\frac{5\sqrt{3}}{2}\right)$

$\quad = \frac{25\sqrt{3}}{4}$ square feet

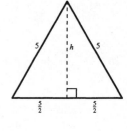

47. Using similar triangles, we have

$$\frac{h}{24} = \frac{6}{8}$$

which implies that $h = 18$ feet.

49.

r	8 ft	15 in.	85 cm	24 in.	$\frac{12,963}{\pi}$ mi
s	12 ft	24 in.	200.28 cm	96 in.	8642 mi
θ	1.5	1.6	$\frac{3\pi}{4}$	4	$\frac{2\pi}{3}$

51. (a) The radian measure is

$$75°\left(\frac{\pi \text{ radians}}{180°}\right) = \frac{5\pi}{12} \text{ radians}.$$

   (b) The distance moved is

$$S = \frac{5\pi}{12}(18.75) = 7.8125\pi \text{ in.} \approx 24.54 \text{ in.}$$

53. (a) Revolutions $= \dfrac{(3142 \text{ radians/minute})}{(2\pi \text{ radians/revolution})}$

$$\approx 500 \text{ revolutions/minute}$$

   (b) Time $= \dfrac{(10,000 \text{ revolutions})}{(500 \text{ revolutions/minute})} = 20 \text{ minutes}$

55. False. An obtuse angle is between 90° and 180°.

57. True. The angles would be 90°, 89°, and 1°.

Section 8.2   The Trigonometric Functions

1. Since $x = 3$ and $y = 4$, it follows that $r = \sqrt{3^2 + 4^2} = 5$. Therefore, we have the following.

$$\sin \theta = \tfrac{4}{5} \qquad \csc \theta = \tfrac{5}{4}$$
$$\cos \theta = \tfrac{3}{5} \qquad \sec \theta = \tfrac{5}{3}$$
$$\tan \theta = \tfrac{4}{3} \qquad \cot \theta = \tfrac{3}{4}$$

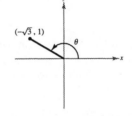

3. Since $x = -12$ and $y = -5$, it follows that $r = \sqrt{(-12)^2 + (-5)^2} = 13$. Therefore, we have the following.

$$\sin \theta = -\tfrac{5}{13} \qquad \csc \theta = -\tfrac{13}{5}$$
$$\cos \theta = -\tfrac{12}{13} \qquad \sec \theta = -\tfrac{13}{12}$$
$$\tan \theta = \tfrac{5}{12} \qquad \cot \theta = \tfrac{12}{5}$$

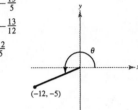

5. Since $x = -\sqrt{3}$ and $y = 1$, it follows that $r = \sqrt{3 + 1^2} = 2$. Therefore, we have the following.

$$\sin \theta = \frac{1}{2} \qquad \csc \theta = 2$$

$$\cos \theta = -\frac{\sqrt{3}}{2} \qquad \sec \theta = -\frac{2\sqrt{3}}{3}$$

$$\tan \theta = -\frac{\sqrt{3}}{3} \qquad \cot \theta = -\sqrt{3}$$

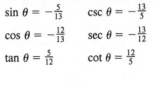

7. $\csc \theta = \dfrac{1}{\sin \theta} = \dfrac{1}{1/2} = 2$

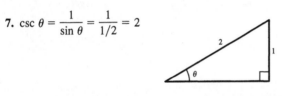

9. Since $x = 4$ and $r = 5$, the length of the opposite side is $y = \sqrt{5^2 - 4^2} = 3$. Therefore, we have

$$\cot \theta = \frac{x}{y} = \frac{4}{3}.$$

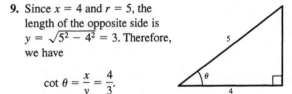

11. Since $x = 15$ and $y = 8$, the length of the hypotenuse is $r = \sqrt{15^2 + 8^2} = 17$. Therefore, we have the following.

$$\sec \theta = \frac{r}{x} = \frac{17}{15}$$

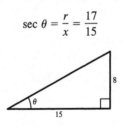

13. The length of the third side of the triangle is

$$x^2 = 3^2 - 1^2 = 8$$
$$x = 2\sqrt{2}.$$

Therefore, we have the following.

$$\sin \theta = \frac{1}{3} \qquad\qquad \csc \theta = 3$$

$$\cos \theta = \frac{2\sqrt{2}}{3} \qquad\qquad \sec \theta = \frac{3}{2\sqrt{2}} = \frac{3\sqrt{2}}{4}$$

$$\tan \theta = \frac{1}{2\sqrt{2}} = \frac{\sqrt{2}}{4} \qquad \cot \theta = 2\sqrt{2}$$

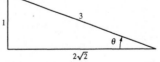

15. Since $x = 2$ and $r = 3$, the length of the opposite side is
$y = \sqrt{3^2 - 2^2} = \sqrt{5}$. Therefore, we have the following.

$$\sin \theta = \frac{\sqrt{5}}{3} \qquad \csc \theta = \frac{3}{\sqrt{5}} = \frac{3\sqrt{5}}{5}$$

$$\cos \theta = \frac{2}{3} \qquad \sec \theta = \frac{3}{2}$$

$$\tan \theta = \frac{\sqrt{5}}{2} \qquad \cot \theta = \frac{2}{\sqrt{5}} = \frac{2\sqrt{5}}{5}$$

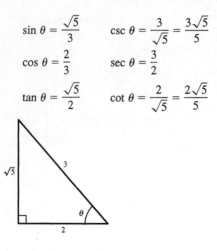

17. $\tan \theta = 3.5 = \frac{7}{2}$. Since $y = 7$ and $x = 2$, the length of the hypotenuse is $r = \sqrt{7^2 + 2^2} = \sqrt{53}$. Therefore, we have the following.

$$\sin \theta = \frac{7}{\sqrt{53}} = \frac{7\sqrt{53}}{53} \qquad \csc \theta = \frac{\sqrt{53}}{7}$$

$$\cos \theta = \frac{2}{\sqrt{53}} = \frac{2\sqrt{53}}{53} \qquad \sec \theta = \frac{\sqrt{53}}{2}$$

$$\tan \theta = \frac{7}{2} \qquad \cot \theta = \frac{2}{7}$$

19. Since the sine is negative and the cosine is positive, θ must lie in Quadrant IV.

21. Since the sine is positive and the secant is positive, θ must lie in Quadrant I.

23. Since the cosecant is positive and the tangent is negative, θ must lie in Quadrant II.

25. (a) $\sin 60° = \dfrac{\sqrt{3}}{2}$
$\cos 60° = \dfrac{1}{2}$
$\tan 60° = \sqrt{3}$

(b) $\sin\left(\dfrac{-2\pi}{3}\right) = -\dfrac{\sqrt{3}}{2}$
$\cos\left(\dfrac{-2\pi}{3}\right) = -\dfrac{1}{2}$
$\tan\left(\dfrac{-2\pi}{3}\right) = \sqrt{3}$

27. (a) $\sin\left(-\dfrac{\pi}{6}\right) = -\dfrac{1}{2}$
$\cos\left(-\dfrac{\pi}{6}\right) = \dfrac{\sqrt{3}}{2}$
$\tan\left(-\dfrac{\pi}{6}\right) = -\dfrac{\sqrt{3}}{3}$

(b) $\sin 150° = \dfrac{1}{2}$
$\cos 150° = -\dfrac{\sqrt{3}}{2}$
$\tan 150° = -\dfrac{\sqrt{3}}{3}$

29. (a) $\sin 225° = -\dfrac{\sqrt{2}}{2}$
$\cos 225° = -\dfrac{\sqrt{2}}{2}$
$\tan 225° = 1$

(b) $\sin(-225°) = \dfrac{\sqrt{2}}{2}$
$\cos(-225°) = -\dfrac{\sqrt{2}}{2}$
$\tan(-225°) = -1$

31. (a) $\sin 750° = \dfrac{1}{2}$
$\cos 750° = \dfrac{\sqrt{3}}{2}$
$\tan 750° = \dfrac{\sqrt{3}}{3}$

(b) $\sin 510° = \dfrac{1}{2}$
$\cos 510° = -\dfrac{\sqrt{3}}{2}$
$\tan 510° = -\dfrac{\sqrt{3}}{3}$

33. (a) $\sin 12° \approx 0.2079$

(b) $\csc 12° \approx 4.8097$

35. (a) $\tan\left(\dfrac{\pi}{9}\right) \approx 0.3640$

(b) $\tan\left(\dfrac{10\pi}{9}\right) \approx 0.3640$

37. (a) $\cos(-110°) \approx -0.3420$

(b) $\cos 250° \approx -0.3420$

39. (a) $\csc 2.62 = \dfrac{1}{\sin 2.62} \approx 2.0070$

(b) $\csc 150° = \dfrac{1}{\sin 150°} = 2.0000$

41. (a) $\theta = \dfrac{\pi}{6}$ or $\theta = \dfrac{5\pi}{6}$

(b) $\theta = \dfrac{7\pi}{6}$ or $\theta = \dfrac{11\pi}{6}$

43. (a) $\theta = \dfrac{\pi}{3}$ or $\theta = \dfrac{2\pi}{3}$

(b) $\theta = \dfrac{3\pi}{4}$ or $\theta = \dfrac{7\pi}{4}$

45. (a) $\theta = \dfrac{3\pi}{4}$ or $\theta = \dfrac{7\pi}{4}$
(b) $\theta = \dfrac{5\pi}{6}$ or $\theta = \dfrac{11\pi}{6}$

47. Solving for $\sin \theta$ produces the following.

$$\sin \theta = \pm\frac{\sqrt{2}}{2}$$

$$\theta = \frac{\pi}{4}, \frac{3\pi}{4}, \frac{5\pi}{4}, \frac{7\pi}{4}$$

49. $\tan\theta(\tan\theta - 1) = 0$

$\tan\theta = 0 \qquad$ or $\quad \tan\theta = 1$

$\theta = 0, \pi, 2\pi \qquad\qquad \theta = \dfrac{\pi}{4}, \dfrac{5\pi}{4}$

51. $\qquad \sin 2\theta - \cos\theta = 0$

$2\sin\theta\cos\theta - \cos\theta = 0$

$\cos\theta(2\sin\theta - 1) = 0$

$\cos\theta = 0 \qquad$ or $\quad 2\sin\theta = 1$

$\theta = \dfrac{\pi}{2}, \dfrac{3\pi}{2} \qquad\qquad \theta = \dfrac{\pi}{6}, \dfrac{5\pi}{6}$

53. Dividing both sides by $\cos\theta$ produces the following.

$\tan\theta = 1$

$\theta = \dfrac{\pi}{4}, \dfrac{5\pi}{4}$

55. Using the identity $\cos^2\theta = 1 - \sin^2\theta$ produces the following.

$1 - \sin^2\theta + \sin\theta = 1$

$\sin^2\theta - \sin\theta = 0$

$\sin\theta(\sin\theta - 1) = 0$

$\sin\theta = 0 \qquad$ or $\quad \sin\theta = 1$

$\theta = 0, \pi, 2\pi \qquad\qquad \theta = \dfrac{\pi}{2}$

57. Since

$\tan 30° = \dfrac{1}{\sqrt{3}} = \dfrac{y}{100}$

it follows that

$y = \dfrac{100}{\sqrt{3}} = \dfrac{100\sqrt{3}}{3}.$

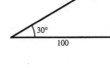

59. Since

$\cot 60° = \dfrac{1}{\sqrt{3}} = \dfrac{x}{25}$

it follows that

$x = \dfrac{25}{\sqrt{3}} = \dfrac{25\sqrt{3}}{3}.$

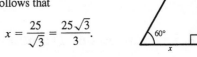

61. Since

$\sin 40° = \dfrac{10}{r} \approx 0.6428$

it follows that

$r = \dfrac{10}{0.6428} \approx 15.5572.$

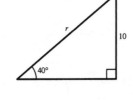

63. Let h be the height of the ladder. Then

$\sin 75° = \dfrac{h}{20} \approx 0.9659$

and $h = 20(0.9659) \approx 19.3185$ feet.

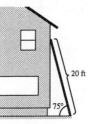

65. Let x be the distance from the shore. Then

$\cot 3° = \dfrac{x}{150} \approx 19.0811$

and

$x = 150(19.0811) \approx 2862.2$ feet.

67. (a) 10:00 P.M.: $T(0) = 98.6 + 4 \cos 0 = 98.6 + 4 = 102.6°$

(b) 4:00 A.M.: $T(6) = 98.6 + 4 \cos\left(\dfrac{6\pi}{36}\right) = 98.6 + 4\left(\dfrac{\sqrt{3}}{2}\right) \approx 102.1°$

(c) 10:00 A.M.: $T(12) = 98.6 + 4 \cos\left(\dfrac{12\pi}{36}\right) = 98.6 + 4\left(\dfrac{1}{2}\right) = 100.6°$

The temperature returns to normal when

$$98.6 = 98.6 + 4 \cos\left(\dfrac{\pi t}{36}\right)$$

$$0 = \cos\left(\dfrac{\pi t}{36}\right)$$

$$\dfrac{\pi}{2} = \dfrac{\pi t}{36}$$

$$t = 18 \text{ or } 4 \text{ P.M.}$$

69.

x	0	2	4	6	8	10
$f(x)$	0	2.7021	2.7756	1.2244	1.2979	4

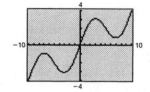

Section 8.3   Graphs of Trigonometric Functions

1. Period: $\dfrac{2\pi}{2} = \pi$

Amplitude: 2

3. Period: $\dfrac{2\pi}{1/2} = 4\pi$

Amplitude: $\dfrac{3}{2}$

5. Period: $\dfrac{2\pi}{\pi} = 2$

Amplitude: $\dfrac{1}{2}$

7. Period: $\dfrac{2\pi}{1} = 2\pi$

Amplitude: 2

9. Period: $\dfrac{2\pi}{10} = \dfrac{\pi}{5}$

Amplitude: 2

11. Period: $\dfrac{2\pi}{2/3} = 3\pi$

Amplitude: $\dfrac{1}{2}$

13. Period: $\dfrac{2\pi}{4\pi} = \dfrac{1}{2}$

Amplitude: 3

15. Period: $\dfrac{\pi}{2}$

17. Period: $\dfrac{2\pi}{5}$

19. Period: $\dfrac{\pi}{\pi/6} = 6$

21. The graph of this function has a period of π and matches graph (c).

23. The graph of this function has a period of 2 and matches graph (f).

25. The graph of this function has a period of 4π and matches graph (b).

27. $y = \sin \dfrac{x}{2}$

Period: 4π

Amplitude: 1

x-intercepts:

$(0, 0), (2\pi, 0), (4\pi, 0)$

Maximum: $(\pi, 1)$

Minimum: $(3\pi, -1)$

29. $y = 2 \cos \dfrac{2x}{3}$

Period: 3π

Amplitude: 2

x-intercepts:

$\left(\dfrac{3\pi}{4}, 0\right), \left(\dfrac{9\pi}{4}, 0\right), \left(\dfrac{15\pi}{4}, 0\right)$

Maxima: $(0, 2), (3\pi, 2)$

Minimum: $\left(\dfrac{3\pi}{2}, -2\right)$

31. $y = -2 \sin 6x$

Period: $\dfrac{\pi}{3}$

Amplitude: 2

x-intercepts: $(0, 0), \left(\dfrac{\pi}{6}, 0\right), \left(\dfrac{\pi}{3}, 0\right), \left(\dfrac{\pi}{2}, 0\right), \left(\dfrac{2\pi}{3}, 0\right)$

Maxima: $\left(\dfrac{\pi}{4}, 2\right), \left(\dfrac{7\pi}{4}, 2\right)$

Minima: $\left(\dfrac{\pi}{12}, -2\right), \left(\dfrac{5\pi}{12}, -2\right)$

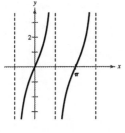

33. $y = \cos 2\pi x$

Period: 1

Amplitude: 1

x-intercepts: $\left(-\dfrac{3}{4}, 0\right), \left(-\dfrac{1}{4}, 0\right), \left(\dfrac{1}{4}, 0\right), \left(\dfrac{3}{4}, 0\right)$

Maxima: $(-1, 1), (0, 1), (1, 1)$

Minima: $\left(-\dfrac{1}{2}, -1\right), \left(\dfrac{1}{2}, 1\right)$

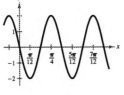

35. $y = 2 \tan x$

Period: π

x-intercepts: $(0, 0), (\pi, 0)$

Asymptotes: $x = -\dfrac{\pi}{2}, x = \dfrac{\pi}{2}, x = \dfrac{3\pi}{2}$

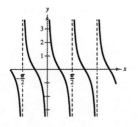

37. $y = -\sin \dfrac{2\pi x}{3}$

Period: 3

Amplitude: 1

x-intercepts: $(0, 0), \left(\dfrac{3}{2}, 0\right), (3, 0)$

Maximum: $\left(\dfrac{9}{4}, 1\right)$ Minimum: $\left(\dfrac{3}{4}, -1\right)$

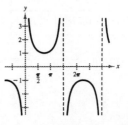

39. $y = \cot 2x$

Period: $\dfrac{\pi}{2}$

x-intercepts: $\left(-\dfrac{3\pi}{4}, 0\right), \left(-\dfrac{\pi}{4}, 0\right), \left(\dfrac{\pi}{4}, 0\right), \left(\dfrac{3\pi}{4}, 0\right)$

Asymptotes: $x = -\dfrac{\pi}{2}, x = 0, x = \dfrac{\pi}{2}$

41. $y = \csc \dfrac{2x}{3}$

Period: 3π

Asymptotes: $x = 0, x = \dfrac{3\pi}{2}, 3\pi$

Relative minimum: $\left(\dfrac{3\pi}{4}, 1\right)$

Relative maximum: $\left(\dfrac{9\pi}{4}, -1\right)$

43. $y = 2 \sec 2x$

Period: π

Asymptotes: $x = -\dfrac{\pi}{4}, x = \dfrac{\pi}{4}, x = \dfrac{3\pi}{4}, x = \dfrac{5\pi}{4}, x = \dfrac{7\pi}{4}$

Relative minima: $(0, 2), (\pi, 2)$

Relative maxima: $\left(\dfrac{\pi}{2}, -2\right), \left(\dfrac{3\pi}{2}, -2\right)$

45. $y = \csc 2\pi x$

Period: 1

Asymptotes: $x = -\dfrac{3}{2}, x = -1, x = -\dfrac{1}{2}, x = 0, x = \dfrac{1}{2},$
$x = 1, x = \dfrac{3}{2}$

Relative minima: $\left(-\dfrac{7}{4}, 1\right), \left(-\dfrac{3}{4}, 1\right), \left(\dfrac{1}{4}, 1\right), \left(\dfrac{5}{4}, 1\right)$

Relative maxima: $\left(-\dfrac{5}{4}, -1\right), \left(-\dfrac{1}{4}, -1\right), \left(\dfrac{3}{4}, -1\right), \left(\dfrac{7}{4}, -1\right)$

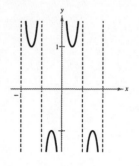

47.

x	-0.1	-0.01	-0.001	0.001	0.01	0.1
$f(x)$	-0.0997	-0.01	-0.001	0.001	0.01	0.0997

$$\lim_{x \to 0} \frac{1 - \cos 2x}{2x} = 0$$

49.

x	-0.1	-0.01	-0.001	0.001	0.01	0.1
$f(x)$	0.2	0.20	0.200	0.200	0.20	0.2

From this table, we estimate that $\displaystyle\lim_{x \to 0} \frac{\sin x}{5x} = \frac{1}{5}$.

51.

x	-0.1	-0.01	-0.001	0.001	0.01	0.1
$f(x)$	-0.1499	-0.0150	-0.0015	0.0015	0.0150	0.1499

From this table, we estimate that $\displaystyle\lim_{x \to 0} \frac{3(1 - \cos x)}{x} = 0$.

53.

x	-0.1	-0.01	-0.001	0.001	0.01	0.1
$f(x)$	2.027	2.0003	2	2	2.0003	2.027

From this table, we estimate that $\displaystyle\lim_{x \to 0} \frac{\tan 2x}{x} = 2$.

55.

x	-0.1	-0.01	-0.001	0.001	0.01	0.1
$f(x)$	-1.516	-0.1599	-0.016	0.016	0.1599	1.516

From this table, we estimate that $\displaystyle\lim_{x \to 0} \frac{1 - \cos^2 4x}{x} = 0$.

57. Amplitude $= 1 \Longrightarrow a = -1$

Period $= \pi \Longrightarrow bx + c = 2x \Longrightarrow b = 2, c = 0$

Vertical shift $\Longrightarrow d = 1$

$y = -\sin(2x) + 1$

59. (a) Period $= \dfrac{2\pi}{\pi/3} = 6$ seconds

(b) Cycles per minute $= \dfrac{60}{6} = 10$

(c)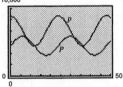

61. (a) Period $= \dfrac{2\pi}{880\pi} = \dfrac{1}{440}$

(b) Frequency $= \dfrac{1}{\text{period}} = 440$

(c)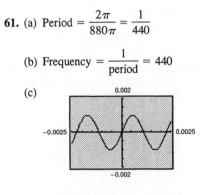

63. (a) $P = 8000 + 2500 \sin \dfrac{2\pi t}{24}$

$p = 12{,}000 + 4000 \cos \dfrac{2\pi t}{24}$

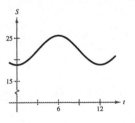

(b) As the population of the prey increases, the population of the predator increases as well. At some point, the predator eliminates the prey faster than the prey can reproduce, and the prey population decreases rapidly. As the prey becomes scarce, the predator population decreases, releasing the prey from predator pressure, and the cycle begins again.

65. The graph has a period of 12, an amplitude of 3.4, and oscillates about the line $y = 22.3$, as shown in the graph.

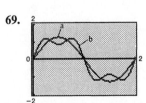

67. December 31, 2008 is the 8930th day.

Days in 1984:	165
5 leap years:	5(366)
18 non-leap years:	18(365)
Days in 2008:	365
Total:	8930 days

$P(8930) \approx 0.9977$

$E(8930) \approx -0.4339$

$I(8930) \approx -0.6182$

69.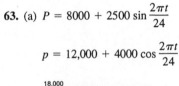

71. $\displaystyle\lim_{x \to 0} \dfrac{\sin x}{x} = 1$

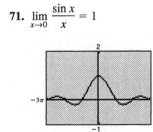

73. $\displaystyle\lim_{x \to 0} \dfrac{\sin 5x}{\sin 2x} = \dfrac{5}{2}$

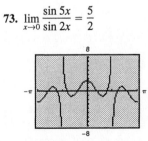

75. (a)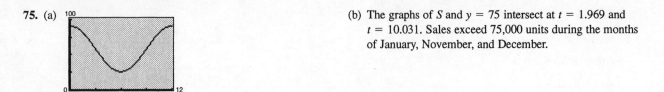

(b) The graphs of S and $y = 75$ intersect at $t = 1.969$ and $t = 10.031$. Sales exceed 75,000 units during the months of January, November, and December.

77. Answers will vary.

79. When the population of hares (prey) increases, the population of lynxes (predator) increases as well, because there is more food. At some point, the lynxes devour hares faster than the hares can reproduce and the hare population decreases. As food becomes scarce, the lynx population decreases as well. At some point, there are only a few lynx, so the hares can begin to increase again.

81. False, the period is $\dfrac{\pi}{4/3} = \dfrac{3\pi}{4}$.

83. False, $\tan\left(\dfrac{5\pi/4}{2}\right) \neq 1$.

Section 8.4   Derivatives of Trigonometric Functions

1. $y' = -3\cos x$

3. $y' = 2x + \sin x$

5. $f'(x) = \dfrac{2}{\sqrt{x}} - 3\sin x$

7. $f'(t) = -t^2\sin t + 2t\cos t$

9. $g'(t) = \dfrac{t(-\sin t) - (\cos t)(1)}{t^2}$

$\qquad = -\dfrac{t\sin t + \cos t}{t^2}$

11. $y' = \sec^2 x + 2x$

13. $y' = e^{x^2}(\sec x\tan x) + 2xe^{x^2}\sec x = e^{x^2}\sec x(\tan x + 2x)$

15. $y' = -\sin 3x(3) + 2\sin x\cos x = -3\sin 3x + 2\sin x\cos x$

17. $y' = (\cos \pi x)\pi = \pi\cos \pi x$

19. $y' = x\left(-\dfrac{1}{x^2}\right)\cos\left(\dfrac{1}{x}\right) + \left(\sin\dfrac{1}{x}\right)(1) = \sin\dfrac{1}{x} - \dfrac{1}{x}\cos\dfrac{1}{x}$

21. $y' = 12\sec^2 4x$

23. $y' = 4\tan(4x)\sec^2 4x(4) = 16\tan 4x\sec^2 4x$

Equivalently,

$\qquad y' = 16\tan 4x(1 + \tan^2 4x) = 16\tan 4x + 16\tan^3 4x.$

25. $y' = e^{2x}(2\cos 2x) + 2e^{2x}\sin 2x$

$\qquad = 2e^{2x}(\cos 2x + \sin 2x)$

27. $y = (\cos x)^2$

$\qquad y' = 2(\cos x)(-\sin x) = -2\cos x\sin x = -\sin 2x$

29. $y' = -2\cos x\sin x - 2\cos x\sin x$

$\qquad = -4\cos x\sin x = -2\sin 2x$

31. $y' = \dfrac{\cos x}{\sin x} = \cot x$

33. $y' = \dfrac{1}{\csc x^2 - \cot x^2}(-2x\csc x^2\cot x^2 + 2x\csc^2 x^2)$

$\qquad = \dfrac{2x\csc x^2(\csc x^2 - \cot x^2)}{\csc x^2 - \cot x^2}$

$\qquad = 2x\csc x^2$

35. $y' = \sec^2 x - 1 = \tan^2 x$

37. $y' = \dfrac{1}{\sin^2 x} 2 \sin x \cos x$

$\quad = \dfrac{2 \cos x}{\sin x}$

$\quad = 2 \cot x$

39. $y' = \sec^2 x, \qquad y'\left(-\dfrac{\pi}{4}\right) = \left(\dfrac{2}{\sqrt{2}}\right)^2 = 2$

$\qquad y - (-1) = 2\left[x - \left(-\dfrac{\pi}{4}\right)\right]$

$\qquad\qquad y = 2x - 1 + \dfrac{\pi}{2}$

41. $y' = 4 \cos 4x, \qquad y'(\pi) = 4$

$\qquad y - 0 = 4(x - \pi)$

$\qquad\quad y = 4x - 4\pi$

43. $y = \dfrac{\cos x}{\sin x} = \cot x$

$\qquad y' = -\csc^2 x, \qquad\qquad y'\left(\dfrac{3\pi}{4}\right) = -2$

$\qquad\qquad y + 1 = -2\left(x - \dfrac{3\pi}{4}\right)$

$\qquad\qquad\quad y = -2x - 1 + \dfrac{3\pi}{2}$

45. $y' = \dfrac{1}{\cot x}(-\csc^2 x), \qquad y'\left(\dfrac{\pi}{4}\right) = -2$

$\qquad y - 0 = -2\left(x - \dfrac{\pi}{4}\right)$

$\qquad\quad y = -2x + \dfrac{\pi}{2}$

47. $\qquad \sin x + \cos 2y = 1$

$\qquad \cos x - 2 \sin 2y \dfrac{dy}{dx} = 0$

$\qquad\qquad\qquad \dfrac{dy}{dx} = \dfrac{\cos x}{2 \sin 2y}$

$\qquad$ At $\left(\dfrac{\pi}{2}, \dfrac{\pi}{4}\right)$, we have $\dfrac{dy}{dx} = 0$.

49. Since $y' = 2 \cos x - 3 \sin x$ and $y'' = -2 \sin x + 3 \cos x$, it follows that $y'' + y = 0$.

51. Since $y' = -2 \sin 2x + 2 \cos 2x$ and $y'' = -4 \cos 2x - 4 \sin 2x$, it follows that $y'' + 4y = 0$.

53. Since

$$y' = \dfrac{5}{4} \cos\left(\dfrac{5x}{4}\right),$$

the slope of the tangent line at $(0, 0)$ is $\frac{5}{4}$. There is one complete cycle of the graph in the interval $[0, 2\pi]$.

55. Since $y' = 2 \cos 2x$, the slope of the tangent line at $(0, 0)$ is 2. There are two complete cycles of the graph in the interval $[0, 2\pi]$.

57. Since $y' = \cos x$, the slope of the tangent line at $(0, 0)$ is 1. There is one complete cycle of the graph in the interval $[0, 2\pi]$.

59. The first derivative is zero when

$$2 \cos x + 2 \cos 2x = 0$$
$$2[\cos x + 2 \cos^2 x - 1] = 0$$
$$2 \cos^2 x + \cos x - 1 = 0$$
$$(2 \cos x - 1)(\cos x + 1) = 0$$
$$\cos x = \dfrac{1}{2} \text{ or } \cos x = -1.$$

Critical numbers: $x = \dfrac{\pi}{3}, \dfrac{5\pi}{3}, \pi$

Relative maximum: $\left(\dfrac{\pi}{3}, \dfrac{3\sqrt{3}}{2}\right)$

Relative minimum: $\left(\dfrac{5\pi}{3}, -\dfrac{3\sqrt{3}}{2}\right)$

61. The first derivative is zero when

$$1 - 2 \cos x = 0$$
$$\cos x = \dfrac{1}{2}.$$

Critical numbers: $x = \dfrac{\pi}{3}, \dfrac{5\pi}{3}, \dfrac{7\pi}{3}, \dfrac{11\pi}{3}$

Relative minimum: $\left(\dfrac{\pi}{3}, \dfrac{\pi}{3} - \sqrt{3}\right)$

Relative maximum: $\left(\dfrac{5\pi}{3}, \dfrac{5\pi}{3} + \sqrt{3}\right)$

63. The first derivative is zero when

$$f'(x) = e^{-x}(-\sin x) - e^{-x}\cos x = 0$$

$$e^{-x}(\sin x + \cos x) = 0$$

$$\tan x = -1.$$

Critical numbers: $x = \dfrac{3\pi}{4}, \dfrac{7\pi}{4}$

Relative maximum: $\left(\dfrac{7\pi}{4}, 0.0029\right)$

Relative minimum: $\left(\dfrac{3\pi}{4}, -0.0670\right)$

67. (a)

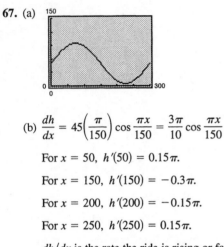

(b) $\dfrac{dh}{dx} = 45\left(\dfrac{\pi}{150}\right)\cos\dfrac{\pi x}{150} = \dfrac{3\pi}{10}\cos\dfrac{\pi x}{150}$

For $x = 50$, $h'(50) = 0.15\pi$.

For $x = 150$, $h'(150) = -0.3\pi$.

For $x = 200$, $h'(200) = -0.15\pi$.

For $x = 250$, $h'(250) = 0.15\pi$.

dh/dx is the rate the ride is rising or falling as it moves horizontally from its starting point.

69. $f'(x) = 2\sec x(\sec x \tan x) = 2\sec^2 x \tan x$

$g'(x) = 2\tan x \sec^2 x$

73. (a) $f(t) = t^2 \sin t$

$f'(t) = t^2 \cos t + 2t \sin t$

$= t(t \cos t + 2 \sin t)$

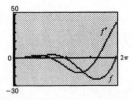

65. $h(t) = 0.20t + 0.03 \sin 2\pi t$

$h'(t) = 0.20 + 0.06\pi \cos 2\pi t$

$h''(t) = -0.12\pi^2 \sin 2\pi t = 0$

$t = 0, \frac{1}{2}$

(a) $h'(t)$ is maximum when $t = 0$ (midnight).

(b) $h'(t)$ is minimum when $t = \frac{1}{2}$ (noon).

(c) $\dfrac{3\pi}{10}\cos\dfrac{\pi x}{150} = 0 \Longrightarrow \dfrac{\pi x}{150} = \dfrac{\pi}{2} \Longrightarrow x = 75$ and $h(75) = 95$ ft

Lowest: $\dfrac{\pi x}{150} = \dfrac{3\pi}{2} \Longrightarrow x = 225$ and $h(225) = 5$ ft

(d) dh/dx is greatest when

$$\cos\dfrac{\pi x}{150} = \pm 1 \text{ or } \dfrac{\pi x}{150} = \pi \Longrightarrow x = 150 \text{ ft}.$$

71. $\theta = 0.2 \cos 8t$

(a) Maximum angular displacement is 0.2 or -0.2.

(b) $\dfrac{d\theta}{dt} = 0.2(8)(-\sin 8t) = -1.6 \sin 8t$

When $t = 3$,

$$\dfrac{d\theta}{dt} = -1.6 \sin 24 \approx 1.4489 \text{ radians/sec}.$$

(b) $f'(t) = 0$ when $t = 0$, $t \approx 2.289$, and $t \approx 5.087$.

(c)

Interval	$(0, 2.289)$	$(2.289, 5.087)$	$(5.087, 2\pi)$
Sign of f'	$+$	$-$	$+$
Conclusion	f is increasing.	f is decreasing.	f is increasing.

75. (a) $f(x) = \sin x - \frac{1}{3}\sin 3x + \frac{1}{5}\sin 5x$

$f'(x) = \cos x - \cos 3x + \cos 5x$

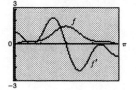

(b) $f'(x) = 0$ when $x \approx 0.524$,

$x \approx 1.571$, and $x \approx 2.618$.

(c)

Interval	$(0, 0.524)$	$(0.524, 1.571)$	$(1.571, 2.618)$	$(2.618, \pi)$
Sign of f'	$+$	$+$	$-$	$-$
Conclusion	f is increasing.	f is increasing.	f is decreasing.	f is decreasing.

77. (a) $f(x) = \sqrt{2x}\,\sin x$

$f'(x) = \sqrt{2x}\,\cos x + \dfrac{\sin x}{\sqrt{2x}}$

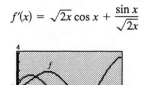

(b) $f'(x) = 0$ when $x \approx 1.837$ and $x \approx 4.816$.

(c)

Interval	$(0, 1.837)$	$(1.837, 4.816)$	$(4.816, \pi)$
Sign of f'	$+$	$-$	$+$
Conclusion	f is increasing.	f is decreasing.	f is increasing.

79. $f(x) = \dfrac{x}{\sin x}$

Relative maximum: $(4.49, -4.60)$

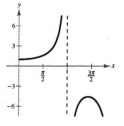

81. $f(x) = \ln x \cos x$

Relative maximum: $(1.27, 0.07)$

Relative minimum: $(3.38, -1.18)$

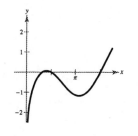

83. $f(x) = \sin(0.1x^2)$

Relative maximum: $(3.96, 1)$

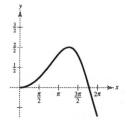

85. False. $y' = \frac{1}{2}(1 - x)^{-1/2}(-1)$

87. False. If $y = x\sin^3 x$, then $y' = \sin^3 x + 3x\sin^2 x(\cos x)$.

Section 8.5   Integrals of Trigonometric Functions

1. $\displaystyle\int (2 \sin x + 3 \cos x)\, dx = -2 \cos x + 3 \sin x + C$

3. $\displaystyle\int (1 - \csc t \cot t)\, dt = t + \csc t + C$

5. $\displaystyle\int (\csc^2 \theta - \cos \theta)\, d\theta = -\cot \theta - \sin \theta + C$

7. $\displaystyle\int \sin 2x\, dx = \frac{1}{2}\int 2 \sin 2x\, dx = -\frac{1}{2}\cos 2x + C$

9. $\displaystyle\int x \cos x^2\, dx = \frac{1}{2}\int 2x \cos x^2\, dx = \frac{1}{2}\sin x^2 + C$

11. $\displaystyle\int \sec^2 \frac{x}{2}\, dx = 2\int \frac{1}{2}\sec^2 \frac{x}{2}\, dx = 2 \tan \frac{x}{2} + C$

13. $\displaystyle\int \tan 3x\, dx = \frac{1}{3}\int 3 \tan 3x\, dx = -\frac{1}{3}\ln|\cos 3x| + C$

15. $\displaystyle\int \tan^3 x \sec^2 x\, dx = \frac{\tan^4 x}{4} + C$

17. $\displaystyle\int \cot \pi x\, dx = \frac{1}{\pi}\int \pi \cot \pi x\, dx = \frac{1}{\pi}\ln|\sin \pi x| + C$

19. $\displaystyle\int \csc 2x\, dx = \frac{1}{2}\int 2 \csc 2x\, dx$

$\displaystyle\qquad\qquad = \frac{1}{2}\ln|\csc 2x - \cot 2x| + C$

21. $\displaystyle\int \frac{\sec^2 2x}{\tan 2x}\, dx = \frac{1}{2}\int \frac{1}{\tan 2x}(2 \sec^2 2x)\, dx$

$\displaystyle\qquad\qquad = \frac{1}{2}\ln|\tan 2x| + C$

23. $\displaystyle\int \frac{\sec x \tan x}{\sec x - 1}\, dx = \ln|\sec x - 1| + C$

25. $\displaystyle\int \frac{\sin x}{1 + \cos x}\, dx = -\ln|1 + \cos x| + C$

27. $\displaystyle\int \frac{\csc^2 x}{\cot^3 x}\, dx = -\int \cot^{-3} x(-\csc^2 x)\, dx = -\frac{\cot^{-2} x}{-2} + C = \frac{1}{2}\tan^2 x + C$

29. $\displaystyle\int e^x \sin e^x\, dx = -\cos e^x + C$

31. $\displaystyle\int e^{\sin x} \cos x\, dx = e^{\sin x} + C$

33. $\displaystyle\int (\sin 2x + \cos 2x)^2\, dx = \int (\sin^2 2x + 2 \sin 2x \cos 2x + \cos^2 2x)\, dx = \int (1 + \sin 4x)\, dx = x - \frac{1}{4}\cos 4x + C$

35. Using integration by parts, we let $u = x$ and $dv = \cos x\, dx$. Then $du = dx$ and $v = \sin x$.

$\displaystyle\int x \cos x\, dx = x \sin x - \int \sin x\, dx$

$\displaystyle\qquad\qquad = x \sin x + \cos x + C$

37. Using integration by parts, we let $u = x$ and $dv = \sec^2 x\, dx$. Then $du = dx$ and $v = \tan x$.

$\displaystyle\int x \sec^2 x\, dx = x \tan x - \int \tan x\, dx$

$\displaystyle\qquad\qquad = x \tan x + \ln|\cos x| + C$

39. $\displaystyle\int_0^{\pi/4} \cos \frac{4x}{3}\, dx = \frac{3}{4}\sin \frac{4x}{3}\Big]_0^{\pi/4}$

$\displaystyle\qquad\qquad = \frac{3}{4}\left(\sin \frac{\pi}{3}\right)$

$\displaystyle\qquad\qquad = \frac{3\sqrt{3}}{8}$

$\displaystyle\qquad\qquad \approx 0.6495$

41. $\displaystyle\int_{\pi/2}^{2\pi/3} \sec^2 \frac{x}{2}\, dx = 2 \tan \frac{x}{2}\Big]_{\pi/2}^{2\pi/3}$

$\displaystyle\qquad\qquad = 2(\sqrt{3} - 1)$

$\displaystyle\qquad\qquad \approx 1.4641$

43. $\displaystyle\int_{\pi/12}^{\pi/4} \csc 2x \cot 2x\,dx = -\frac{1}{2}\csc 2x\Big]_{\pi/12}^{\pi/4}$

$$= -\frac{1}{2}\left[\csc\frac{\pi}{2} - \csc\frac{\pi}{6}\right]$$

$$= -\frac{1}{2}[1 - 2]$$

$$= \frac{1}{2}$$

45. $\displaystyle\int_0^1 \tan(1 - x)\,dx = \ln|\cos(1 - x)|\Big]_0^1$

$$= \ln(\cos 0) - \ln(\cos 1)$$

$$\approx 0.6156$$

47. $\displaystyle\int_0^{2\pi} \cos\frac{x}{4}\,dx = 4\sin\frac{x}{4}\Big]_0^{2\pi} = 4$

49. Area $= \displaystyle\int_0^\pi (x + \sin x)\,dx = \left[\frac{x^2}{2} - \cos x\right]_0^\pi = \frac{\pi^2}{2} + 2 \approx 6.9348$ square units

51. Area $= \displaystyle\int_0^\pi (\sin x + \cos 2x)\,dx$

$$= \left[-\cos x + \frac{1}{2}\sin 2x\right]_0^\pi$$

$$= 2 \text{ square units}$$

53. $V = \pi\displaystyle\int_0^{\pi/4} \sec^2 x\,dx$

$$= \pi\tan x\Big]_0^{\pi/4}$$

$$= \pi \text{ cubic units}$$

55. Trapezoidal Rule: $\dfrac{\pi/2}{8}\left[f(0) + 2f\left(\dfrac{\pi}{8}\right) + 2f\left(\dfrac{\pi}{4}\right) + 2f\left(\dfrac{3\pi}{8}\right) + f\left(\dfrac{\pi}{2}\right)\right] \approx 1.3655$

Simpson's Rule: $\dfrac{\pi/2}{12}\left[f(0) + 4f\left(\dfrac{\pi}{8}\right) + 2f\left(\dfrac{\pi}{4}\right) + 4f\left(\dfrac{3\pi}{8}\right) + f\left(\dfrac{\pi}{2}\right)\right] \approx 1.3708$

Graphing utility: 1.3708

57. (a) $\dfrac{1}{3}\displaystyle\int_0^3\left[217 + 13\cos\frac{\pi(t-3)}{6}\right]dt = \frac{1}{3}\left[217t + \frac{78}{\pi}\sin\frac{\pi(t-3)}{6}\right]_0^3 = \frac{1}{3}\left(651 + \frac{78}{\pi}\right) \approx 225.28$ million barrels

(b) $\dfrac{1}{3}\displaystyle\int_3^6\left[217 + 13\cos\frac{\pi(t-3)}{6}\right]dt = \frac{1}{3}\left[217t + \frac{78}{\pi}\sin\frac{\pi(t-3)}{6}\right]_3^6 = \frac{1}{3}\left(1302 + \frac{78}{\pi} - 651\right) \approx 225.28$ million barrels

(c) $\dfrac{1}{12}\displaystyle\int_0^{12}\left[217 + 13\cos\frac{\pi(t-3)}{6}\right]dt = \frac{1}{12}\left[217t + \frac{78}{\pi}\sin\frac{\pi(t-3)}{6}\right]_0^{12} = \frac{1}{12}\left(2604 - \frac{78}{\pi} + \frac{78}{\pi}\right) \approx 217$ million barrels

59. Total $= \displaystyle\int_0^{12}[2.34\sin(0.38t + 1.91) + 2.22]\,dt \approx 18.54$ inches

61. (a) $C = 0.1\displaystyle\int_8^{20} 12\sin\frac{\pi(t-8)}{12}\,dt = -\frac{14.4}{\pi}\cos\frac{\pi(t-8)}{12}\Big]_8^{20} = -\frac{14.4}{\pi}(-1 - 1) \approx \$9.17$

(b) $C = 0.1\displaystyle\int_{10}^{18}\left[12\sin\frac{\pi(t-8)}{12} - 6\right]dt = \left[-\frac{14.4}{\pi}\cos\frac{\pi(t-8)}{12} - 0.6t\right]_{10}^{18} \approx \$3.14$

Savings $\approx 9.17 - 3.14 = \$6.03$

63. $V = \displaystyle\int_0^2 1.75\sin\frac{\pi t}{2}\,dt = -\frac{3.5}{\pi}\cos\frac{\pi t}{2}\Big]_0^2 \approx 2.2282$ liters

Capacity increases by $2.2282 - 1.7189 = 0.5093$ liter.

65. $\displaystyle\int_0^{\pi/2}\sqrt{x}\sin x\,dx \approx 0.9777$

67. $\displaystyle\int_0^\pi \sqrt{1 + \cos^2 x}\,dx \approx 3.8202$

69. True. $\sin x$ is periodic with period 2π.

71. False.
$$\int \sin^2 2x \cos 2x\,dx = \frac{1}{6}\sin^3 2x + C$$

Section 8.6 L'Hôpital's Rule

1. Yes, $\dfrac{0}{0}$.

3. No, $\dfrac{4}{\infty} \to 0$.

5. Yes, $\dfrac{\infty}{\infty}$.

7.

x	-0.1	-0.01	-0.001	0	0.001	0.01	0.1
$f(x)$	-0.35	-0.335	-0.3335		-0.3332	-0.332	-0.32

$$\lim_{x\to 0}\frac{e^{-x}-1}{3x} = \frac{-1}{3}$$

9.

x	-0.1	-0.01	-0.001	0	0.001	0.01	0.1
$f(x)$	0.1997	0.2	0.2		0.2	0.2	0.1997

$$\lim_{x\to 0}\frac{\sin x}{5x} = \frac{1}{5}$$

11. $\displaystyle\lim_{x\to\infty}\frac{\ln x}{x^2} = 0$

13. $\displaystyle\lim_{x\to 1}\frac{\ln(2-x)}{x-1} = -1$

15. $\displaystyle\lim_{x\to 2}\frac{x^2-x-2}{x^2-5x+6} = -3$

17. $\displaystyle\lim_{x\to 0}\frac{e^{-x}-1}{x} = \lim_{x\to 0}\frac{-e^{-x}}{1} = -1$
(See Exercise 7.)

19. $\displaystyle\lim_{x\to 0}\frac{\sin x}{5x} = \lim_{x\to 0}\frac{\cos x}{5} = \frac{1}{5}$
(See Exercise 9.)

21. $\displaystyle\lim_{x\to\infty}\frac{\ln x}{x^2} = \lim_{x\to\infty}\frac{1/x}{2x} = 0$

23. $\displaystyle\lim_{x\to 2}\frac{x^2-x-2}{x^2-5x+6} = \lim_{x\to 2}\frac{2x-1}{2x-5} = \frac{3}{-1} = -3$
(See Exercise 13.)

25. $\displaystyle\lim_{x\to 0}\frac{2x+1-e^x}{x} = \lim_{x\to 0}\frac{2-e^x}{1} = 1$

27. $\displaystyle\lim_{x\to\infty}\frac{\ln x}{e^x} = \lim_{x\to\infty}\frac{1/x}{e^x} = 0$

29. $\displaystyle\lim_{x\to\infty}\frac{4x^2+2x-1}{3x^2-7} = \lim_{x\to\infty}\frac{8x+2}{6x} = \lim_{x\to\infty}\frac{8}{6} = \frac{4}{3}$

31. $\displaystyle\lim_{x\to\infty}\frac{1-x}{e^x} = \lim_{x\to\infty}\frac{-1}{e^x} = 0$

33. $\displaystyle\lim_{x\to 1}\frac{\ln x}{x^2-1} = \lim_{x\to 1}\frac{1/x}{2x} = \frac{1}{2}$

35. $\displaystyle\lim_{x\to 0}\frac{\sin 2x}{\sin 5x} = \lim_{x\to 0}\frac{2\cos 2x}{5\cos 5x} = \frac{2}{5}$

37. $\displaystyle\lim_{x\to\pi/2}\frac{2\cos x}{3\cos x} = \lim_{x\to\pi/2}\frac{-2\sin x}{-3\sin x} = \frac{2}{3}$

39. $\displaystyle\lim_{x\to 0}\frac{\sin x}{e^x-1} = \lim_{x\to 0}\frac{\cos x}{e^x} = 1$

41. $\displaystyle\lim_{x\to\pi}\frac{\cos(x/2)}{x-\pi} = \lim_{x\to\pi}\frac{-(1/2)\sin(x/2)}{1} = -\frac{1}{2}$

43. $\displaystyle\lim_{x\to\infty} \frac{x}{\sqrt{x+1}} = \lim_{x\to\infty} \frac{x/x}{\sqrt{(1/x)+(1/x^2)}}$

$\displaystyle\qquad = \lim_{x\to\infty} \frac{1}{\sqrt{(1/x)+(1/x)}}$

$\displaystyle\qquad = \infty$

45. $\displaystyle\lim_{x\to0} \frac{\sqrt{4-x^2}-2}{x} = \lim_{x\to0} \frac{(1/2)(4-x^2)^{-1/2}(-2x)}{1}$

$\displaystyle\qquad = \lim_{x\to0} \frac{-x}{\sqrt{4-x^2}}$

$\displaystyle\qquad = 0$

47. $\displaystyle\lim_{x\to\infty} \frac{e^{3x}}{x^3} = \lim_{x\to\infty} \frac{3e^{3x}}{3x^2} = \lim_{x\to\infty} \frac{3e^{3x}}{2x} = \lim_{x\to\infty} \frac{9e^{3x}}{2} = \infty$

49. $\displaystyle\lim_{x\to\infty} \frac{x^2+2x+1}{x^2+3} = \lim_{x\to\infty} \frac{2x+2}{2x} = \lim_{x\to\infty} \frac{2}{2} = 1$

51. $\displaystyle\lim_{x\to-1} \frac{x^3+3x^2-6x-8}{2x^3-3x^2-5x+6} = \frac{0}{6} = 0$

53. $\displaystyle\lim_{x\to3} \frac{\ln(x-2)}{x-2} = \frac{\ln(1)}{1} = 0$

55. $\displaystyle\lim_{x\to1} \frac{2\ln x}{e^x} = \frac{2(0)}{e} = 0$

57. $\displaystyle\lim_{x\to\infty} \frac{x^2}{e^{4x}} = \lim_{x\to\infty} \frac{2x}{4e^{4x}} = \lim_{x\to\infty} \frac{2}{16e^{4x}} = 0$

59. $\displaystyle\lim_{x\to\infty} \frac{(\ln x)^4}{x} = \lim_{x\to\infty} \frac{4(\ln x)^3(1/x)}{1}$

$\displaystyle\qquad = \lim_{x\to\infty} \frac{4(\ln x)^3}{x}$

$\displaystyle\qquad = \lim_{x\to\infty} \frac{12(\ln x)^2(1/x)}{1}$

$\displaystyle\qquad = \lim_{x\to\infty} \frac{12(\ln x)^2}{x}$

$\displaystyle\qquad = \lim_{x\to\infty} \frac{24(\ln x)(1/x)}{1}$

$\displaystyle\qquad = \lim_{x\to\infty} \frac{24\ln x}{x}$

$\displaystyle\qquad = \lim_{x\to\infty} \frac{24(1/x)}{1}$

$\displaystyle\qquad = 0$

61. $\displaystyle\lim_{x\to\infty} \frac{(\ln x)^n}{x^m} = 0$

(The log term eventually disappears.)

63.

x	10	10^2	10^3	10^4	10^5	10^6
$\dfrac{(\ln x)^5}{x}$	6.47	20.71	15.73	6.63	2.02	0.503

$\displaystyle\lim_{x\to\infty} \frac{(\ln x)^5}{x} = 0$

65. The limit of the denominator is not 0.

67. The limit of the numerator is not 0.

69. (a)

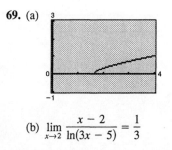

(b) $\displaystyle\lim_{x\to2} \frac{x-2}{\ln(3x-5)} = \frac{1}{3}$

71. (a)

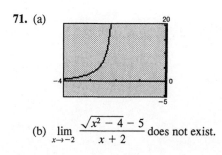

(b) $\displaystyle\lim_{x\to-2} \frac{\sqrt{x^2-4}-5}{x+2}$ does not exist.

73. $\displaystyle\lim_{x\to\infty} \frac{x}{\sqrt{x^2+1}} = \lim_{x\to\infty} \frac{\sqrt{x^2+1}}{x} = \lim_{x\to\infty} \frac{x}{\sqrt{x^2+1}}$

and repeated applications of L'Hôpital's Rule continue in this pattern.

$$\lim_{x\to\infty} \frac{x}{\sqrt{x^2+1}} = \lim_{x\to\infty} \frac{x/x}{\sqrt{(x^2/x^2)+(1/x^2)}}$$

$$= \lim_{x\to\infty} \frac{1}{\sqrt{1+(1/x^2)}}$$

$$= 1$$

75. (a)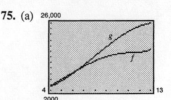

(b) CVS had a larger rate of growth.

(c) Ultimately, Rite Aid's rate of growth will be larger because of the e^t factor.

(d) The curves intersect at $t \approx 14.3$. From 2004 onward, Rite Aid will be larger.

77. Answers will vary.

79. False, $\displaystyle\lim_{x\to\infty} \frac{x}{1-x} = -1$.

81. False. For example, $\displaystyle\lim_{x\to\infty} \frac{x+1}{x} = 1$, but $x+1 \neq x$.

Review Exercises for Chapter 8

1. Positive coterminal angle: $\dfrac{7\pi}{4} + 2\pi = \dfrac{15\pi}{4}$

Negative coterminal angle: $\dfrac{7\pi}{4} - 2\pi = -\dfrac{\pi}{4}$

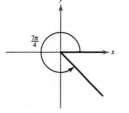

3. Positive coterminal angle: $\dfrac{3\pi}{2} + 2\pi = \dfrac{7\pi}{2}$

Negative coterminal angle: $\dfrac{3\pi}{2} - 2\pi = -\dfrac{\pi}{2}$

5. Positive coterminal angle: $135° + 360° = 495°$

Negative coterminal angle: $135° - 360° = -225°$

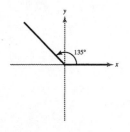

7. Positive coterminal angle: $-405° + 2(360°) = 315°$

Negative coterminal angle: $-405° + 360° = -45°$

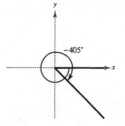

9. $210°\left(\dfrac{\pi\,\text{radians}}{180°}\right) = \dfrac{7\pi}{6}\,\text{radians}$

11. $-60°\left(\dfrac{\pi\,\text{radians}}{180°}\right) = -\dfrac{\pi}{3}\,\text{radians} \approx -\dfrac{\pi}{3}$

13. $-480°\left(\dfrac{\pi\,\text{radians}}{180°}\right) = -\dfrac{8\pi}{3}\,\text{radians}$

15. $-135°\left(\dfrac{\pi\,\text{radians}}{180°}\right) = \dfrac{3\pi}{4}\,\text{radians}$

17. $\dfrac{4\pi}{3}\left(\dfrac{180°}{\pi}\right) = 240°$

19. $-\dfrac{2\pi}{3}\left(\dfrac{180°}{\pi}\right) = -120°$

21. $b = \sqrt{8^2 - 4^2} = \sqrt{48} = 4\sqrt{3}$

$\theta = 90° - 30° = 60°$

23. $c = 5$, $\theta = 60°$ (equilateral triangle)

$a = \sqrt{5^2 - \left(\frac{5}{2}\right)^2} = \frac{5}{2}\sqrt{3}$

25. $h = \sqrt{16^2 - 4.4^2} = \sqrt{236.64} \approx 15.38$ feet

27. The reference angle is $\pi - \frac{2\pi}{3} = \frac{\pi}{3}$.

29. The reference angle is $\pi - \frac{5\pi}{6} = \frac{\pi}{6}$.

31. The reference angle is $240° - 180° = 60°$.

33. The reference angle is $420° - 360° = 60°$.

35. $\cos 45° = \dfrac{\sqrt{2}}{2}$

37. $\tan \dfrac{\pi}{3} = \sqrt{3}$

39. $\sin\left(\dfrac{5\pi}{3}\right) = -\dfrac{\sqrt{3}}{2}$

41. $\cot\left(\dfrac{-5\pi}{6}\right) = \sqrt{3}$

43. $\sec(-180°) = -1$

45. $\cos\left(\dfrac{-4\pi}{3}\right) = -\dfrac{1}{2}$

47. $\tan 33° \approx 0.6494$

49. $\sec \dfrac{12\pi}{5} \approx 3.2361$

51. $\sin\left(-\dfrac{\pi}{9}\right) \approx -0.3420$

53. $\cos 105° \approx -0.2588$

55. $\cos 70° = \dfrac{50}{r} \Longrightarrow r = \dfrac{50}{\cos 70°} \approx 146.19$

57. $\tan 20° = \dfrac{25}{x} \Longrightarrow x = \dfrac{25}{\tan 20°} \approx 68.69$

59. $2\cos x + 1 = 0$

$\cos x = -\dfrac{1}{2}$

$x = \dfrac{2\pi}{3} + 2k\pi,\ \dfrac{4\pi}{3} + 2k\pi$

61. $2\sin^2 x + 3\sin x + 1 = 0$

$(2\sin x + 1)(\sin x + 1) = 0$

$\sin x = -\dfrac{1}{2}$ or $\sin x = -1$

$x = \dfrac{7\pi}{6} + 2k\pi,\ \dfrac{11\pi}{6} + 2k\pi$

or $x = \dfrac{3\pi}{2} + 2k\pi$

63. $\sec^2 x - \sec x - 2 = 0$

$(\sec x - 2)(\sec x + 1) = 0$

$\sec x = 2$ or $\sec x = -1$

$\cos x = \dfrac{1}{2}$ or $\cos x = -1$

$x = \dfrac{\pi}{3} + 2k\pi$ or $\dfrac{5\pi}{3} + 2k\pi$

or $x = \pi + 2k\pi$

65. $h \approx 125\tan 33° \approx 81.18$ feet

67. $y = 2\cos 6x$

Period: $\dfrac{2\pi}{6} = \dfrac{\pi}{3}$

Amplitude: 2

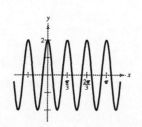

69. $y = \dfrac{1}{3}\tan x$

Period: π

71. Period: $\dfrac{2\pi}{2/5} = 5\pi$

Amplitude: 3

x-intercepts: $(0, 0), \left(\dfrac{5\pi}{2}, 0\right), (5\pi, 0)$

Maximum: $\left(\dfrac{5\pi}{4}, 3\right)$

Minimum: $\left(\dfrac{15\pi}{4}, -3\right)$

73. Period: 1

Asymptotes: $x = -\dfrac{1}{4}, x = \dfrac{1}{4}, x = \dfrac{3}{4}, x = \dfrac{5}{4}$

Maximum: $\left(\dfrac{1}{2}, -1\right), \left(\dfrac{3}{2}, -1\right)$

Minimum: $(0, 1), (1, 1)$

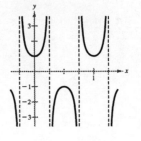

75. $S = 74 + \dfrac{3}{365}t - 40 \sin \dfrac{2\pi t}{365}$

77. $y = \sin 5\pi x$

$y' = 5\pi \cos 5\pi x$

79. $y' = -x \sec^2 x - \tan x$

81. $y' = \dfrac{x^2(-\sin x) - \cos x(2x)}{(x^2)^2} = \dfrac{-x \sin x - 2 \cos x}{x^3}$

83. $y' = 6 \sin 4x(4 \cos 4x) + 1$

$= 24 \sin 4x \cos 4x + 1$

$= 12 \sin 8x + 1$

85. $y' = 6 \csc^2 x(-\csc x \cot x)$

$= -6 \csc^3 x \cot x$

87. $y' = e^x(-\csc^2 x) + e^x \cot x$

$= e^x(\cot x - \csc^2 x)$

89. $y = 4 \sin 2x, \quad (\pi, 0)$

$y' = 8 \cos 2x, \quad y'(\pi) = 8$

$y - 0 = 8(x - \pi)$

$y = 8x - 8\pi$

91. $y = \dfrac{1}{4} \sin^2 2x, \quad \left(\dfrac{\pi}{4}, \dfrac{1}{4}\right)$

$y' = \dfrac{1}{4} 2 \sin 2x \cdot \cos 2x(2) = \sin 2x \cos 2x, \quad y'\left(\dfrac{\pi}{4}\right) = 0$

$y - \dfrac{1}{4} = 0\left(x - \dfrac{\pi}{4}\right)$

$y = \dfrac{1}{4}$

93. $y = e^x \tan 2x, \quad (0, 0)$

$y' = e^x(2 \sec^2 2x) + e^x \tan 2x, \quad y'(0) = 2$

$y - 0 = 2(x - 0)$

$y = 2x$

95. $f(x) = \dfrac{x}{2} + \cos x$

Relative maximum: $(0.523, 1.128)$

Relative minimum: $(2.616, 0.443)$

97. $f(x) = \sin x (\sin x + 1)$

Relative maxima: $\left(\dfrac{\pi}{2}, 2\right), \left(\dfrac{3\pi}{2}, 0\right)$

Relative minima: $\left(\dfrac{7\pi}{6}, -\dfrac{1}{4}\right), \left(\dfrac{11\pi}{6}, -\dfrac{1}{4}\right)$

99. $S = 74 + \dfrac{3}{365}t - 40 \sin \dfrac{2\pi t}{365}$

(a) Maximum is 116.25 at $t \approx 274.44$. So, maximum is 116.25 thousand on day 274.

(b) Minimum is 34.75 at $t \approx 90.56$. So, minimum is 34.75 thousand on day 90.

101. $\displaystyle\int (3 \sin x - 2 \cos x)\,dx = -3 \cos x - 2 \sin x + C$

103. $\displaystyle\int \sin^3 x \cos x \, dx = \dfrac{\sin^4 x}{4} + C$

105. $\displaystyle\int_0^\pi (1 + \sin x)\,dx = \Big[x - \cos x\Big]_0^\pi = [\pi - (-1)] - (-1) = \pi + 2\dfrac{1}{2}$

107. $\displaystyle\int_{-\pi/6}^{\pi/6} \sec^2 x \, dx = \tan x \Big]_{-\pi/6}^{\pi/6} = \dfrac{1}{\sqrt{3}} - \left(\dfrac{-1}{\sqrt{3}}\right) = \dfrac{2}{\sqrt{3}}$

109. $\displaystyle\int_{-\pi/3}^{\pi/6} 4 \sec x \tan x \, dx = 4 \sec x \Big]_{-\pi/3}^{\pi/3} = 0$

111. $\displaystyle\int_{-\pi/2}^{\pi/2} (2x + \cos x)\,dx = \Big[x^2 + \sin x\Big]_{\pi/4}^{\pi/2} = 2$

113. $\displaystyle\int_0^{\pi/2} \sin 2x \, dx = -\dfrac{1}{2}\cos 2x \Big]_0^{\pi/2} = -\dfrac{1}{2}(-1 - 1) = 1$

115. $\displaystyle\int_0^{\pi/2} (2 \sin x + \cos 3x)\,dx = \Big[-2 \cos x + \dfrac{1}{3}\sin 3x \Big]_0^{\pi/2}$

$= \left(-\dfrac{1}{3}\right) - (-2)$

$= \dfrac{5}{3}$

117. Average $= \dfrac{1}{12 - 0} \displaystyle\int_0^{12} \left[6.9 + \cos \dfrac{\pi(2t - 1)}{12} \right] dt$

$= \dfrac{1}{12}\left[6.9t + \dfrac{6}{\pi}\sin \dfrac{\pi(2t - 1)}{12} \right]_0^{12}$

$= 6.9$ quads

119. Total $= \displaystyle\int_0^{12} [2.77 \sin(0.39t + 1.91) + 2.52]\,dt = 21.11$ inches

121. $\displaystyle\lim_{x \to 1} \dfrac{3x - 1}{5x + 5} = \dfrac{2}{10} = \dfrac{1}{5}$

123. $\displaystyle\lim_{x \to 2} \dfrac{x^2 + 2x + 11}{x^3 + 4} = \dfrac{19}{12}$

125. $\displaystyle\lim_{x \to 1} \dfrac{x^3 - x^2 + 4x - 4}{x^3 - 6x^2 + 5x} = \lim_{x \to 1} \dfrac{3x^2 - 2x + 4}{3x^2 - 12x + 5} = \dfrac{5}{-4} = -\dfrac{5}{4}$

127. $\displaystyle\lim_{x \to 0} \dfrac{\sin \pi x}{\sin 2\pi x} = \lim_{x \to 0} \dfrac{\pi \cos \pi x}{2\pi \cos 2\pi x} = \dfrac{\pi}{2\pi} = \dfrac{1}{2}$

129. $\displaystyle\lim_{x \to 0} \dfrac{\cos^2 x - 1}{\sin x} = \lim_{x \to 0} \dfrac{-\sin^2 x}{\sin x} = 0$

131. $\displaystyle\lim_{x \to 0} \dfrac{\sin^2 x}{e^x} = \dfrac{0}{1} = 0$

133. $\displaystyle\lim_{x \to \infty} \dfrac{e^{5x}}{x^2} = \lim_{x \to \infty} \dfrac{5e^{5x}}{2x} = \lim_{x \to \infty} \dfrac{25e^{5x}}{2} = \infty$

135. $g(x)$ grows faster than $f(x)$.

Practice Test for Chapter 8

1. (a) Express $\dfrac{12\pi}{23}$ in degree measure.

   (b) Express $105°$ in radian measure.

2. Determine two coterminal angles (one positive and one negative) for the given angle.

   (a) $-220°$; give your answers in degrees.

   (b) $\dfrac{7\pi}{9}$; give your answers in radians.

3. Find the six trigonometric functions of the angle θ if it is in standard position and the terminal side passes through the point $(12, -5)$.

4. Solve for θ, $(0 \le \theta \le 2\pi)$: $\sin^2 \theta + \cos \theta = 1$

5. Sketch the graph of the given function.

   (a) $y = 3 \sin \dfrac{x}{4}$ (b) $y = \tan 2\pi x$

6. Find the derivative of $y = 3x - 3 \cos x$.

7. Find the derivative of $f(x) = x^2 \tan x$.

8. Find the derivative of $g(x) = \sin^3 x$.

9. Find the derivative of $y = \dfrac{\sec x}{x^2}$.

10. Find the derivative of $y = \sin 5x \cos 5x$.

11. Find the derivative of $y = \sqrt{\csc x}$.

12. Find the derivative of $y = \ln|\sec x + \tan x|$.

13. Find the derivative of $f(x) = \cot e^{2x}$.

14. Find $\dfrac{dy}{dx}$: $\sin(x^2 + y) = 3x$

15. Find $\dfrac{dy}{dx}$: $\tan x - \cot 3y = 4$

16. Evaluate $\displaystyle\int \cos 4x \, dx$.

17. Evaluate $\displaystyle\int \csc^2 \dfrac{x}{8} \, dx$.

18. Evaluate $\displaystyle\int x \tan x^2 \, dx$.

19. Evaluate $\displaystyle\int \sin^5 x \cos x \, dx$.

20. Evaluate $\displaystyle\int \dfrac{\cos^2 x}{\sin x} \, dx$.

21. Evaluate $\displaystyle\int e^{\tan x} \sec^2 x \, dx$.

22. Evaluate $\displaystyle\int \dfrac{\sin x}{1 + \cos x} \, dx$.

23. Evaluate $\displaystyle\int (\sec x - \tan x)^2 \, dx$.

24. Evaluate $\displaystyle\int x \cos x \, dx$.

25. Evaluate $\displaystyle\int_0^{\pi/4} (2x - \cos x) \, dx$.

26. Use L'Hôpital's Rule to find the limit: $\displaystyle\lim_{x \to \infty} \dfrac{e^{2x}}{x^2}$

27. Use L'Hôpital's Rule to find the limit: $\displaystyle\lim_{x \to 0} \dfrac{\sin 7x}{3x}$

28. Use L'Hôpital's Rule to find the limit: $\displaystyle\lim_{x \to \infty} \dfrac{5x^3 + 7x^2 - 8}{9x^3 + 2x + 4}$

Graphing Calculator Required

29. Use a graphing utility to graph $f(x) = \sin x + \cos 2x$. What are the minimum and maximum values that $f(x)$ takes on?

30. Use a graphing utility to find $\displaystyle\lim_{x \to \infty} \dfrac{x}{\sqrt{x^2 + 4}}$. Try using L'Hôpital's Rule on this limit—what happens?

CHAPTER 9
Probability and Calculus

C H A P T E R 9
Probability and Calculus

Section 9.1   Discrete Probability

Solutions to Odd-Numbered Exercises

1. (a) $S = \{HHH, HHT, HTH, THH, HTT, THT, TTH, TTT\}$

   (b) $A = \{HHH, HHT, HTH, THH\}$

   (c) $B = \{HTT, THT, TTH, TTT\}$

3. (a) $S = \{3, 6, 9, 12, 15, 18, 21, 24, 27, 30, 33, 36, 39, 42, 45, 48\}$

   (b) $A = \{12, 24, 36, 48\}$

   (c) $B = \{9, 36\}$

5. $1 - 0.29 - 0.47 = 0.24$

7. $1 - 0.9855 = 0.0145$

9.

Random variable	0	1	2
Frequency	1	2	1

11.

Random variable	0	1	2	3
Frequency	1	3	3	1

13. (a) $P(1 \le x \le 3) = P(1) + P(2) + P(3)$

$$= \tfrac{3}{20} + \tfrac{6}{20} + \tfrac{6}{20}$$

$$= \tfrac{15}{20}$$

$$= \tfrac{3}{4}$$

   (b) $P(x \ge 2) = P(2) + P(3) + P(4)$

$$= \tfrac{6}{20} + \tfrac{6}{20} + \tfrac{4}{20}$$

$$= \tfrac{16}{20}$$

$$= \tfrac{4}{5}$$

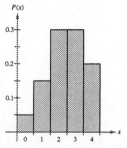

15. (a) $P(x \le 3) = 0.041 + 0.189 + 0.247 + 0.326$

$$= 0.803$$

   (b) $P(x > 3) = 0.159 + 0.038$

$$= 0.197$$

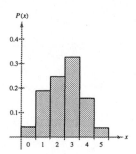

284

17. (a) $S = \{gggg, gggb, ggbg, gbgg, bggg, ggbb, gbbg, gbgb, bgbg, bbgg, bggb, gbbb, bgbb, bbgb, bbbg, bbbb\}$

(b)

x	0	1	2	3	4
$P(x)$	$\frac{1}{16}$	$\frac{4}{16}$	$\frac{6}{16}$	$\frac{4}{16}$	$\frac{1}{16}$

(c)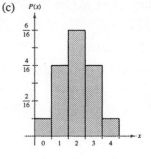

(d) Probability of at least one boy $= 1 -$ probability of all girls

$$P = 1 - \tfrac{1}{16} = \tfrac{15}{16}$$

19. There are 16 possibilities, as indicated in the following chart.

	RY	Ry	rY	ry
RY	RRYY	RRYy	RrYY	RrYy
Ry	RRYy	RRyy	RrYy	Rryy
rY	RrYY	RrYy	rrYY	rrYy
ry	RrYy	Rryy	rrYy	rryy

(a) Round, yellow seeds: can occur 9 ways

RRYY (1)

RRYy (2)

RrYY (2)

RrYy   (4)

∴ Probability $= \frac{9}{16}$

(b) Wrinkled, yellow seeds: can occur 3 ways

rrYY (1)

rrYy (2)

∴ Probability $= \frac{3}{16}$

(c) Round, green seeds: can occur 3 ways

RRyy (1)

Rryy (2)

∴ Probability $= \frac{3}{16}$

(d) Wrinkled, green seeds: can occur 1 way

rryy (1)

∴ Probability $= \frac{1}{16}$

21. $E(x) = 1\left(\frac{1}{16}\right) + 2\left(\frac{3}{16}\right) + 3\left(\frac{8}{16}\right) + 4\left(\frac{3}{16}\right) + 5\left(\frac{1}{16}\right) = \frac{48}{16} = 3$

$V(x) = (1-3)^2\left(\frac{1}{16}\right) + (2-3)^2\left(\frac{3}{16}\right) + (3-3)^2\left(\frac{8}{16}\right) + (4-3)^2\left(\frac{3}{16}\right) + (5-3)^2\left(\frac{1}{16}\right) = 14\left(\frac{1}{16}\right) = \frac{7}{8} = 0.875$

$\sigma = \sqrt{V(x)} = 0.9354$

23. $E(x) = -3\left(\frac{1}{5}\right) + (-1)\left(\frac{1}{5}\right) + 0\left(\frac{1}{5}\right) + 3\left(\frac{1}{5}\right) + 5\left(\frac{1}{5}\right) = \frac{4}{5}$

$V(x) = \left(-3 - \frac{4}{5}\right)^2\left(\frac{1}{5}\right) + \left(-1 - \frac{4}{5}\right)^2\left(\frac{1}{5}\right) + \left(0 - \frac{4}{5}\right)^2\left(\frac{1}{5}\right) + \left(3 - \frac{4}{5}\right)^2\left(\frac{1}{5}\right) + \left(5 - \frac{4}{5}\right)^2\left(\frac{1}{5}\right) = 8.16$

$\sigma = \sqrt{V(x)} \approx 2.857$

25. (a) $E(x) = 1\left(\frac{1}{4}\right) + 2\left(\frac{1}{4}\right) + 3\left(\frac{1}{4}\right) + 4\left(\frac{1}{4}\right) = \frac{10}{4} = 2.5$

$V(x) = (1 - 2.5)^2\left(\frac{1}{4}\right) + (2 - 2.5)^2\left(\frac{1}{4}\right) + (3 - 2.5)^2\left(\frac{1}{4}\right) + (4 - 2.5)^2\left(\frac{1}{4}\right) = 1.25$

(b) $E(x) = 2\left(\frac{1}{16}\right) + 3\left(\frac{2}{16}\right) + 4\left(\frac{3}{16}\right) + 5\left(\frac{4}{16}\right) + 6\left(\frac{3}{16}\right) + 7\left(\frac{2}{16}\right) + 8\left(\frac{1}{16}\right) = \frac{80}{16} = 5$

$V(x) = (2 - 5)^2\left(\frac{1}{16}\right) + (3 - 5)^2\left(\frac{2}{16}\right) + (4 - 5)^2\left(\frac{3}{16}\right) + (5 - 5)^2\left(\frac{4}{16}\right) + (6 - 5)^2\left(\frac{3}{16}\right) + (7 - 5)^2\left(\frac{2}{16}\right) + (8 - 5)^2\left(\frac{1}{16}\right)$

$= \frac{5}{2} = 2.5$

27. (a) $E(x) = 10(0.25) + 15(0.30) + 20(0.25) + 30(0.15) + 40(0.05) = 18.50$

$V(x) = (10 - 18.50)^2(0.25) + (15 - 18.50)^2(0.30) + (20 - 18.50)^2(0.25) + (30 - 18.50)^2(0.15) + (40 - 18.50)^2(0.05)$

$= 65.25$

$\sigma = \sqrt{V(x)} \approx 8.078$

(b) Expected revenue: $R = \$2.95(18.50)(1000) = \$54,575$

29. $E(x) = 0(0.995) + 30,000(0.0036) + 60,000(0.0011) + 90,000(0.0003) = 201$

Each customer should be charged $201.

31. $E(x) = 35\left(\frac{1}{38}\right) + (-1)\left(\frac{37}{38}\right) = -\frac{2}{38} \approx -\$0.05$ or $E(x) = 36\left(\frac{1}{38}\right) + 0\left(\frac{37}{38}\right) = \frac{36}{38}$

$\frac{36}{38} - 1 = -\$0.05$

33. City 1: Expected value $= 0.3(20) + 0.7(-4) = 3.2$ million

City 2: Expected value $= 0.2(50) + 0.8(-9) = 2.8$ million

The company should open the store in City 1.

35. (a)

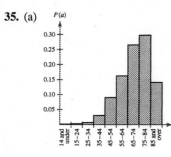

(b) 35–74 years: $.0307 + 0.0903 + 0.1625 + 0.2653 = 0.5488$ or 54.88%

(c) Over 45 years: $0.0903 + 0.1625 + 0.2653 + 0.2978 + 0.1408 = 0.9567$ or 95.67%

(d) Under 24 years: $0.0018 + 0.0036 = 0.0054$ or 0.54%

37. (a) Total number $= 14 + 26 + 7 + 2 + 1 = 50$

x	0	1	2	3	4
$P(x)$	$\frac{14}{50}$	$\frac{26}{50}$	$\frac{7}{50}$	$\frac{2}{50}$	$\frac{1}{50}$

(b)

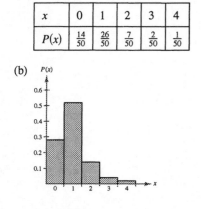

(c) $P(1 \leq x \leq 3) = \frac{26}{50} + \frac{7}{50} + \frac{2}{50} = \frac{35}{50}$

(d) $E(x) = 0 \cdot \frac{14}{50} + 1 \cdot \frac{26}{50} + 2 \cdot \frac{7}{50} + 3 \cdot \frac{2}{50} + 4 \cdot \frac{1}{50} = \frac{50}{50} = 1$

$V(x) = (0 - 1)^2 \frac{14}{50} + (1 - 1)^2 \frac{26}{50} + (2 - 1)^2 \frac{7}{50} + (3 - 1)^2 \frac{2}{50} + (4 - 1)^2 \frac{1}{50} \approx 0.76$

$\sigma = \sqrt{V(x)} \approx 0.87$

39. Mean $= 4.4$

Standard deviation $= 2.816$

Section 9.2   Continuous Random Variables

1. $\int_0^8 \frac{1}{8}\, dx = \frac{1}{8}x\Big]_0^8 = 1$ and $f(x) = \frac{1}{8} \geq 0$ on $[0, 8]$.

3. $\int_0^4 \frac{4-x}{8}\, dx = \frac{1}{8}\left(4x - \frac{x^2}{2}\right)\Big]_0^4 = 1$

and $f(x) = \frac{4-x}{8} \geq 0$ on $[0, 4]$.

5. $\int_0^1 6x(1-x)\, dx = \int_0^1 (6x - 6x^2)\, dx$

$\qquad\qquad = \left[3x^2 - 2x^3\right]_0^1 = 1$

and $f(x) = 6x(1-x) \geq 0$ on $[0, 1]$.

7. $\int_0^\infty \frac{1}{5}e^{-x/5}\, dx = \left[\lim_{b\to\infty} -e^{-x/5}\right]_0^b = 0 + 1 = 1$

and $f(x) = \frac{1}{5}e^{-x/5} \geq 0$ on $[0, \infty)$.

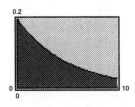

9. $\int_0^2 \frac{3}{8}x\sqrt{4-x^2}\, dx = -\frac{1}{2}\left(\frac{3}{8}\right)\left(\frac{2}{3}\right)(4-x^2)^{3/2}\Big]_0^2 = 1$

and $f(x) = \frac{3}{8}x\sqrt{4-x^2} \geq 0$ on $[0, 2]$.

11. $\int_0^3 \frac{4}{27}x^2(3-x)\, dx = \frac{4}{27}\int_0^3 (3x^2 - x^3)\, dx$

$\qquad\qquad = \frac{4}{27}\left[x^3 - \frac{x^4}{4}\right]_0^3$

$\qquad\qquad = \frac{4}{27}\left(27 - \frac{81}{4}\right)$

$\qquad\qquad = 1$

and $f(x) = \frac{4}{27}x^2(3-x) \geq 0$ on $[0, 3]$.

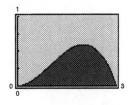

13. $\int_0^\infty \frac{1}{3}e^{-x/3}\, dx = \lim_{b\to\infty}\left[-e^{-x/3}\right]_0^b = 0 - (-1) = 1$ and $f(x) = \frac{1}{3}e^{-x/3} \geq 0$ on $[0, \infty)$.

15. $\int_1^4 kx\,dx = \left[\dfrac{kx^2}{2}\right]_1^4 = \dfrac{15}{2}k = 1 \Rightarrow k = \dfrac{2}{15}$

17. $\int_{-2}^2 k(4 - x^2)\,dx = k\left(4x - \dfrac{x^3}{3}\right)\Big]_{-2}^2 = \dfrac{32k}{3} = 1 \Rightarrow k = \dfrac{3}{32}$

19. $\int_0^\infty ke^{-x/2}\,dx = \lim_{b\to\infty}\left[-2ke^{-x/2}\right]_0^b = 2k = 1 \Rightarrow k = \dfrac{1}{2}$

21. $\int_a^b \dfrac{1}{10}\,dx = \dfrac{x}{10}\Big]_a^b = \dfrac{b-a}{10}$

   (a) $P(0 < x < 6) = \dfrac{6-0}{10} = \dfrac{3}{5}$

   (b) $P(4 < x < 6) = \dfrac{6-4}{10} = \dfrac{1}{5}$

   (c) $P(8 < x < 10) = \dfrac{10-8}{10} = \dfrac{1}{5}$

   (d) $P(x \ge 2) = P(2 < x < 10) = \dfrac{10-2}{10} = \dfrac{4}{5}$

23. $\int_a^b \dfrac{3}{16}\sqrt{x}\,dx = \left(\dfrac{3}{16}\right)\dfrac{2}{3}x^{3/2}\Big]_a^b = \dfrac{1}{8}\left[b\sqrt{b} - a\sqrt{a}\right]$

   (a) $P(0 < x < 2) = \dfrac{\sqrt{2}}{4} \approx 0.354$

   (b) $P(2 < x < 4) = 1 - \dfrac{\sqrt{2}}{4} \approx 0.646$

   (c) $P(1 < x < 3) = \dfrac{1}{8}\left(3\sqrt{3} - 1\right) \approx 0.525$

   (d) $P(x \le 3) = \dfrac{3\sqrt{3}}{8} \approx 0.650$

25. $\int_a^b \dfrac{1}{3}e^{-t/3}\,dt = e^{-t/3}\Big]_a^b = e^{-a/3} - e^{-b/3}$

   (a) $P(t < 2) = e^{-0/3} - e^{-2/3} \approx 0.4866$

   (b) $P(t \ge 2) = e^{-2/3} - 0 \approx 0.5134$

   (c) $P(1 < t < 4) = e^{-1/3} - e^{-4/3} \approx 0.4529$

   (d) $P(t = 3) = 0$

27. $P(a < t < b) = \int_a^b \dfrac{1}{30}\,dt = \dfrac{b-a}{30}$

   (a) $P(0 \le t \le 5) = \dfrac{5-0}{30} = \dfrac{1}{6}$

   (b) $P(18 < t < 30) = \dfrac{30-18}{30} = \dfrac{2}{5}$

29. $\int_a^b \dfrac{1}{3}e^{-t/3}\,dt = -e^{-t/3}\Big]_a^b = e^{-a/3} - e^{-b/3}$

   (a) $P(0 < t < 2) = e^{-0/3} - e^{-2/3} = 1 - e^{-2/3} \approx 0.487$

   (b) $P(2 < t < 4) = e^{-2/3} - e^{-4/3} \approx 0.250$

   (c) $P(t > 2) = 1 - P(0 < t < 2) = e^{-2/3} \approx 0.513$

31. $P(0 < t < 1) = \int_0^1 \dfrac{4}{3}e^{-4t/3}\,dt$

$$= -e^{-4t/3}\Big]_0^1$$

$$= 1 - e^{-4/3}$$

$$\approx 0.736$$

33. $\displaystyle\int_a^b \frac{1}{36} xe^{-x/e}\, dx = -\frac{1}{6} e^{-x/e}(x+6)\Big]_a^b$

    (a) $P(x < 6) = -\dfrac{1}{6} e^{-x/6}(x+6)\Big]_0^6 = -\dfrac{1}{6}(12e^{-1} - 6) \approx 0.264$

    (b) $P(6 < x < 12) = -\dfrac{1}{6} x^{-x/6}(x+6)\Big]_6^{12} = 2e^{-1} - 3e^{-2} \approx 0.330$

    (c) $P(x > 12) = 1 - P(x \le 12) = 1 - (1 - 3e^{-2}) \approx 0.406$

35. Note that $\displaystyle\int_0^{15} f(x)\, dx = 1.$

    (a) $P(0 \le x \le 10) = \displaystyle\int_0^{10} f(x)\, dx = 0.75$

        75% probability of receiving up to 10 inches of rain

    (c) $P(0 \le x < 5) = \displaystyle\int_0^5 f(x)\, dx = 0.25$

        25% probability of receiving less than 5 inches of rain

    (b) $P(10 \le x \le 15) = \displaystyle\int_{10}^{15} f(x)\, dx = 1 - 0.75 = 0.25$

        25% probability of receiving between 10 and 15 inches of rain

    (d) $P(12 \le x \le 15) \approx 0.0955$

        9.6% probability of receiving between 12 and 15 inches of rain

37. $f(x) = 4.7 + 0.61x - 4.15 \ln x - \dfrac{5.14}{x}, \quad 1 \le x \le 5$

    (a) $\displaystyle\int_1^4 f(x)\, dx \approx 0.987 \text{ or } 98.7\%$

    (b) $\displaystyle\int_2^3 f(x)\, dx \approx 0.366 \text{ or } 36.6\%$

    Note: $\displaystyle\int_1^5 f(x)\, dx \approx 1.05 \approx 100\%$

Section 9.3   Expected Value and Variance

1. (a) $\mu = \displaystyle\int_a^b xf(x)\, dx = \int_0^8 x\left(\frac{1}{8}\right) dx = \frac{x^2}{16}\Big]_0^8 = 4$

    (b) $\sigma^2 = \displaystyle\int_a^b x^2 f(x)\, dx - \mu^2 = \int_0^8 x^2\left(\frac{1}{8}\right) dx - (4)^2 = \frac{x^3}{24}\Big]_0^8 - 16 = \frac{64}{3} - 16 = \frac{16}{3}$

    (c) $\sigma = \dfrac{4}{\sqrt{3}} = \dfrac{4\sqrt{3}}{3}$

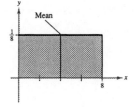

3. (a) $\mu = \displaystyle\int_a^b tf(t)\, dt = \int_0^6 t\left(\frac{t}{18}\right) dt = \frac{t^3}{54}\Big]_0^6 = 4$

    (b) $\mu^2 = \displaystyle\int_a^b t^2 f(t) - \mu^2 = \int_0^6 t^2\left(\frac{t}{18}\right) dt - 4^2 = \frac{t^4}{72}\Big]_0^6 - 4^2 = 18 - 16 = 2$

    (c) $\sigma = \sqrt{2}$

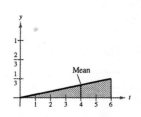

5. (a) $\mu = \int_a^b xf(x)\,dx = \int_0^1 x\left(\frac{5}{2}x^{3/2}\right)dx = \frac{5}{2}\int_0^1 x^{5/2}\,dx = \left(\frac{5}{2}\right)\frac{2}{7}x^{7/2}\Big]_0^1 = \frac{5}{7} \approx 0.714$

   (b) $\sigma^2 = \int_a^b x^2 f(x)\,dx - \mu^2 = \int_0^1 x^2\left(\frac{5}{2}x^{3/2}\right)dx - \left(\frac{5}{7}\right)^2 = \frac{5}{2}\int_0^1 x^{7/2}\,dx - \frac{25}{49} = \left(\frac{5}{2}\right)\frac{2}{9}x^{9/2}\Big]_0^1 - \frac{25}{49} = \frac{5}{9} - \frac{25}{49} = \frac{20}{441}$

   (c) $\sigma = \dfrac{2\sqrt{5}}{21}$

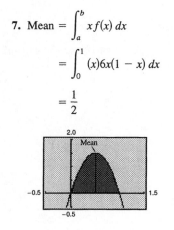

7. Mean $= \displaystyle\int_a^b xf(x)\,dx$

$\quad\quad = \displaystyle\int_0^1 (x)6x(1-x)\,dx$

$\quad\quad = \dfrac{1}{2}$

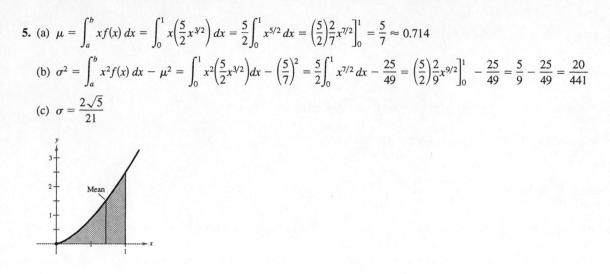

9. Mean $= \displaystyle\int_a^b xf(x)\,dx$

$\quad\quad = \displaystyle\int_0^3 x\,\frac{4}{3(x+1)^2}\,dx$

$\quad\quad = \dfrac{4}{3}\ln 4 - 1$

$\quad\quad \approx 0.848$

11. Median $= \displaystyle\int_0^m \frac{1}{9}e^{-t/9}\,dt = -e^{t/9}\Big]_0^m = 1 - e^{m/9} = \frac{1}{2}$

$e^{-m/9} = \dfrac{1}{2} \Longrightarrow -\dfrac{m}{9} = \ln\dfrac{1}{2}$

$\quad\quad\quad\quad m = -9\ln\dfrac{1}{2} \approx 6.238$

13. $f(x) = \frac{1}{10}$, $[0, 10]$ is a uniform density function.

Expected value (mean): $\dfrac{a+b}{2} = \dfrac{0+10}{2} = 5$

Variance: $\dfrac{(b-a)^2}{12} = \dfrac{(10-0)^2}{12} = \dfrac{100}{12} = \dfrac{25}{3}$

Standard deviation: $\dfrac{b-a}{\sqrt{12}} = \dfrac{10-0}{\sqrt{12}} \approx 2.887$

15. $f(x) = \frac{1}{8}e^{-x/8}$, $[0, \infty)$ is an exponential density function with $a = \frac{1}{8}$.

Expected value (mean): $\dfrac{1}{a} = 8$

Variance: $\dfrac{1}{a^2} = 64$

Standard deviation: $\dfrac{1}{a} = 8$

17. $f(x) = \dfrac{1}{11\sqrt{2\pi}}e^{-(x-100)^2/242}$,

$(-\infty, \infty)$ is a normal density function with $\mu = 100$ and $\sigma = 11$.

Expected value (mean): $\mu = 100$

Variance: $\sigma^2 = 121$

Standard deviation: $\sigma = 11$

19. Mean = 0

Standard deviation = 1

$P(0 \le x \le 0.85) \approx 0.3023$

21. Mean = 6

Standard deviation = 6

$P(x \ge 2.23) \approx 0.6896$

23. Mean = 8

Standard deviation = 2

$P(3 \le x \le 13) \approx 0.9876$

25. $\mu = 60, \sigma = 12$

(a) $P(x > 64) \approx 0.3694$

(b) $P(x > 70) \approx 0.2023$

(c) $P(x < 70) \approx 0.7977$

(d) $P(33 < x < 65) \approx 0.6493$

27. $f(t) = \dfrac{1}{10}$, $[0, 10]$ where $t = 0$ corresponds to 10:00 A.M.

(a) Mean $= \dfrac{10}{2} = 5$

The mean is 10:05 A.M.

Standard deviation $= \dfrac{10}{\sqrt{12}} \approx 2.887$ minutes

(b) $1 - \displaystyle\int_3^{10} \frac{1}{10}\, dx = 1 - \frac{7}{10} = \frac{3}{10} = 0.30$

29. (a) $f(t) = \dfrac{1}{2}e^{-t/2}$, since mean = 2.

(b) $P(0 < t < 1) = \displaystyle\int_0^1 \frac{1}{2}e^{-t/2}\, dt$

$= -e^{-t/2}\Big]_0^1$

$= 1 - e^{-1/2}$

≈ 0.3935

31. (a) Since $\mu = 5$, we have $f(t) = \dfrac{1}{5}e^{-t/5}$.

(b) $P(\mu - \sigma < t < \mu + \sigma) = P(0 < t < 10)$

$\displaystyle\int_0^{10} \frac{1}{5}e^{-t/5}\, dt = -e^{-t/5}\Big]_0^{10}$

$= 1 - e^{-2}$

≈ 0.865

$= 86.5\%$

33. (a) $\dfrac{174 - 150}{16} = \dfrac{3}{2} = 1.5$

Your score exceeded the national mean by 1.5 standard deviations.

(b) $P(x < 174) = 0.9332$

Thus, $0.9332 = 93.32\%$ of those who took the exam had scores lower than yours.

35. (a) $\mu = \displaystyle\int_0^6 \frac{1}{36}x^2(6 - x)\, dx = \frac{1}{36}\int_0^6 (6x^2 - x^3)\, dx = \frac{1}{36}\left[2x^3 - \frac{x^4}{4}\right]_0^6 = 3$

$\sigma^2 = \displaystyle\int_0^6 \frac{1}{36}x^3(6 - x)\, dx - (3)^2 = \frac{1}{36}\int_0^6 (6x^3 - x^4)\, dx - 9 = \frac{1}{36}\left[\frac{3x^4}{2} - \frac{x^5}{5}\right]_0^6 - 9 = \frac{54}{5} - 9 = \frac{9}{5}$

$\sigma = \sqrt{\dfrac{9}{5}} = \dfrac{3\sqrt{5}}{5} \approx 1.342$

(b) $\displaystyle\int_0^m \frac{1}{36}x(6 - x)\, dx = \frac{1}{36}\int_0^m (6x - x^2)\, dx = \frac{1}{36}\left[3x^2 - \frac{x^3}{3}\right]_0^m = \frac{1}{36}\left[3m^2 - \frac{m^3}{3}\right] = \frac{1}{2}$

$3m^2 - \dfrac{m^3}{3} = 18 \Longrightarrow 0 = (m - 3)(m^2 - 6m - 18)$

$m = 3$ or $m = \dfrac{6 \pm \sqrt{108}}{2} = \dfrac{6 \pm 6\sqrt{3}}{2} = 3 \pm 3\sqrt{3}$

In the interval $[0, 6]$, $m = 3$.

—CONTINUED—

35. —CONTINUED—

(c) $P(\mu - \sigma < x < \mu + \sigma) = P\left(3 - \dfrac{3\sqrt{5}}{5} < x < 3 + \dfrac{3\sqrt{5}}{5}\right)$

$\approx P(1.6584 < x < 4.3416)$

$= \displaystyle\int_{1.6584}^{4.3416} \dfrac{1}{36} x(6 - x)\, dx$

$= \dfrac{1}{36}\left[3x^2 - \dfrac{x^3}{3}\right]_{1.6584}^{4.3416}$

≈ 0.626

$= 62.6\%$

37. $\mu = \displaystyle\int_0^1 \dfrac{15}{4} x^2 \sqrt{1 - x}\, dx$

$= \dfrac{15}{4}\left(\dfrac{2}{-7}\right)\left[x^2(1 - x)^{3/2} - 2\left(\dfrac{2}{-5}\right)\left[x(1 - x)^{3/2} + \dfrac{2}{3}(1 - x)^{3/2}\right]\right]_0^1 = -\dfrac{15}{14}\left[0 - \dfrac{4}{5}\left(\dfrac{2}{3}\right)\right] = \dfrac{4}{7}$

$\sigma^2 = \displaystyle\int_0^1 \dfrac{15}{4} x^3 \sqrt{1 - x}\, dx - \left(\dfrac{4}{7}\right)^2$

$= \dfrac{15}{4}\left(\dfrac{2}{-9}\right)\left[\left[x^3(1 - x)^{3/2}\right]_0^1 - 3\int_0^1 x^2 \sqrt{1 - x}\, dx\right] - \dfrac{16}{49} = -\dfrac{5}{6}\left[0 - 3\left(\dfrac{4}{7}\right)\left(\dfrac{4}{15}\right)\right] - \dfrac{16}{49} = \dfrac{8}{21} - \dfrac{16}{49} = \dfrac{8}{147}$

39. Using a graphing utility:

(a) $\mu = \displaystyle\int_0^\infty x f(x)\, dx = \int_0^\infty \dfrac{1}{25} x^2 e^{-x/5}\, dx = 10$

(b) $P(x \le 4) \approx 0.1912$

41. Mean $= \dfrac{11}{2} =$ median

43. Mean $= \displaystyle\int_0^{1/2} x(4)(1 - 2x)\, dx = \dfrac{1}{6}$

Median $= \displaystyle\int_0^m 4(1 - 2x)\, dx = \dfrac{1}{2}$

$4x - 4x^2\Big]_0^m = \dfrac{1}{2}$

$4m - 4m^2 = \dfrac{1}{2} \Longrightarrow m \approx 0.1465$

$\left(m \approx 0.8536 \text{ is not in the interval } \left[0, \tfrac{1}{2}\right].\right)$

45. Mean $= 5$

Median $= 5 \ln 2 \approx 3.4657$

47. $\displaystyle\int_0^m f(x)\, dx = \int_0^m 0.28 e^{-0.28x}\, dx = 0.5$

$-e^{-0.28m} + 1 = 0.5$

$e^{-0.28m} = 0.5$

$m = \dfrac{1}{-0.28} \ln 0.5 \approx 2.4755$

49. (a) $\mu = \displaystyle\int_0^\infty \frac{1}{9}x^2 e^{-x/3}\, dx$ (Use integration by parts.)

$$= \lim_{b \to \infty} -3\left[\frac{x^2}{9}e^{-x/3} - 2\left(-\frac{x}{3} - 1\right)e^{-x/3}\right]_0^b$$

$$= 6$$

$$\sigma^2 = \int_0^\infty \frac{1}{9}x^3 e^{-x/3}\, dx - (6)^2 \quad \text{(Use integration by parts.)}$$

$$= \lim_{b \to \infty}\left[-\frac{x^3}{3}e^{-x/3}\right]_0^b + 9\int_0^\infty \frac{1}{9}x^2 e^{-x/3}\, dx - 36$$

$$= 0 + 9(6) - 36$$

$$= 18 \quad \text{(Use part (a).)}$$

$$\sigma = \sqrt{18} = 3\sqrt{2} \approx 4.243$$

(b) $P(x > 4) = 1 - P(x < 4) = 1 - \displaystyle\int_0^4 \frac{1}{9}xe^{-x/3}\, dx = 1 - \left[-\frac{1}{3}e^{-x/3}(x + 3)\right]_0^4 \approx 0.615$

51.

$$P(x < 12) = 0.05$$

$$P\left(z < \frac{12 - \mu}{0.15}\right) = 0.05$$

$$0.5000 - P\left(\frac{12 - \mu}{0.15} < z < 0\right) = 0.05$$

$$P\left(\frac{12 - \mu}{0.15} < z < 0\right) = 0.4500$$

$$\frac{12 - \mu}{0.15} \approx -1.645$$

$$\mu \approx 12.25$$

53. $\mu = 12.30, \; \sigma = 1.50$

$$f(x) = \frac{1}{1.50\sqrt{2\pi}}e^{-(x - 12.30)^2/4.5}$$

$$P(9 < x < 12) = \frac{1}{1.50\sqrt{2\pi}}\int_9^{12} e^{-(x - 12.30)^2/4.5}\, dx$$

$$\approx 0.4068$$

$$= 40.68\%$$

55. $f(x) = \dfrac{1}{\sigma\sqrt{2\pi}}e^{-(x - \mu)^2/2\sigma^2}, \quad \sigma = 4.8, \quad \mu = 20.8$

$$f(x) = \frac{1}{4.8\sqrt{2\pi}}e^{-(x - 20.8)^2/2(4.8)^2}$$

(a)

(b) $P(24 \le x \le 36) = \displaystyle\int_{24}^{36} f(x)\, dx \approx 0.252$ or 25.2%

Review Exercises for Chapter 9

1. The sample space consists of the twelve months of the year.

$S = \{$January, February, March, April, May, June, July, August, September, October, November, December$\}$

3. If the questions are numbered 1, 2, 3, and 4,

$S = \{123, 124, 134, 234\}$.

5. $S = \{0, 1, 2, 3\}$

7.

x	0	1	2	3
n(x)	1	3	3	1

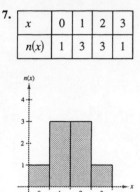

9. (a) $P(2 \le x \le 4) = P(2) + P(3) + P(4)$

$$= \frac{7}{18} + \frac{5}{18} + \frac{3}{18}$$

$$= \frac{15}{18} = \frac{5}{6}$$

(b) $P(x \ge 3) = P(3) + P(4) + P(5)$

$$= \frac{5}{18} + \frac{3}{18} + \frac{2}{18}$$

$$= \frac{10}{18} = \frac{5}{9}$$

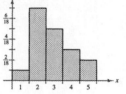

11.

x	2	3	4	5	6	7	8	9	10	11	12
n(x)	1	2	3	4	5	6	5	4	3	2	1

$n(S) = 36$

(a) $P(x = 8) = \frac{5}{36}$

(b) $P(x > 4) = 1 - P(x \le 4) = 1 - \frac{6}{36} = \frac{5}{6}$

(c) $P \text{ (doubles)} = \frac{6}{36} = \frac{1}{6}$

(d) $P \text{ (double sixes)} = \frac{1}{36}$

13. Mean $= 19.5$

15. (a) $E(x) = 10(0.10) + 15(0.20) + 20(0.50) + 30(0.15) + 40(0.05) = 20.5$

(b) $R = 20.5(1000)(0.75) = \$15,375$

17. $E(x) = \dfrac{24(1200) + 12(1500) + 35(2000) + 5(2200) + 4(3000)}{80} = \dfrac{139,800}{80} = 1747.50$

$V(x) = (1200 - 1747.50)^2(24) + (1500 - 1747.50)^2(12) + (2000 - 1747.50)^2(35)$

$\qquad + (2200 - 1747.50)^2(5) + (3000 - 1747.50)^2(4) = 218,243.7500$

$\sigma = \sqrt{V(x)} \approx 467.1657$

19. $E(x) = 0(0.10) + 1(0.28) + 2(0.39) + 3(0.17) + 4(0.04) + 5(0.02) = 1.83$

$V(x) = (0 - 1.83)^2(0.10) + (1 - 1.83)^2(0.28) + (2 - 1.83)^2(0.39) + (3 - 1.83)^2(0.17) + (4 - 1.83)^2(0.04)$

$\qquad + (5 - 1.83)^2(0.02) \approx 1.1611$

$\sigma = \sqrt{V(x)} \approx 1.0775$

21. $\displaystyle\int_0^4 \frac{1}{8}(4 - x)\, dx = \left[\frac{1}{2}x - \frac{x^2}{16}\right]_0^4 = 2 - 1 = 1$

23. $\displaystyle\int_1^9 \frac{1}{4\sqrt{x}}\, dx = \frac{1}{4}\left[2\sqrt{x}\right]_1^9 = 1$

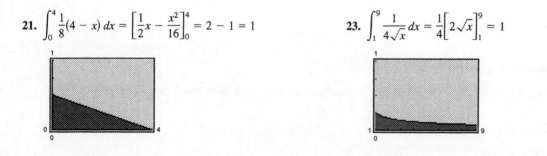

25. $P(0 < x < 2) = \int_0^2 \frac{1}{50}(10 - x)\, dx$

$$= \frac{1}{50}\left[10x - \frac{x^2}{2} \right]_0^2 = \frac{9}{25}$$

27. $P\left(0 < x < \frac{1}{2}\right) = \int_0^{1/2} \frac{2}{(x + 1)^2}\, dx$

$$= -\frac{2}{x + 1}\Big]_0^{1/2} = -\frac{4}{3} + 2 = \frac{2}{3}$$

29. (a) $P(t \le 10) = \int_0^{10} \frac{1}{20}\, dt = \frac{1}{20}t\Big]_0^{10} = \frac{1}{2}$

(b) $P(t \ge 15) = \int_{15}^{20} \frac{1}{20}\, dt = \frac{1}{20}t\Big]_{15}^{20} = 1 - \frac{3}{4} = \frac{1}{4}$

31. $P(t > 8) = \int_8^{13} \frac{1}{4\sqrt{t - 4}}\, dt = \frac{1}{2}\sqrt{t - 4}\Big]_8^{13} = \frac{1}{2}$

33. Mean $= \dfrac{5 - 0}{2} = 2.5$

35. Mean $= 6$

37. Mean $= \int_0^3 x\frac{2}{9}x(3 - x)\, dx = \frac{2}{9}\left[x^3 - \frac{x^4}{4} \right]_0^3 = \frac{3}{2}$

Variance $= \int_0^3 \left(x - \frac{3}{2}\right)^2 \frac{2}{9}x(3 - x)\, dx = \frac{9}{20}$

Standard deviation $= \sqrt{V(x)} = \dfrac{3}{2\sqrt{5}}$

39. Mean $= \int_0^\infty x\left(\frac{1}{2}e^{-x/2}\right) dx = 2$

Variance $= 4$

Standard deviation $= 2$

41. $\int_0^m 6x(1 - x)\, dx = \frac{1}{2}$

$3x^2 - 2x^3\Big]_0^m = \frac{1}{2}$

$3m^2 - 2m^3 = \frac{1}{2}$

$m = \frac{1}{2}$

43. $\int_0^m 0.25e^{-x/4}\, dx = \frac{1}{2}$

$1 - e^{-m/4} = \frac{1}{2}$

$m \approx 2.7726$

45. $f(t) = \frac{1}{3}e^{-1/3}\, dt, \quad 0 \le t < \infty$

(a) $P(t < 2) = \int_0^2 f(t)\, dt \approx 0.4866$

(b) $P(2 < t < 4) = \int_2^4 f(t)\, dt \approx 0.2498$

47. $f(x) = \dfrac{1}{\sigma\sqrt{2\pi}}e^{-(x - \mu)^2/2\sigma^2}$

$P(x \ge 50) = \int_{50}^\infty \frac{1}{3\sqrt{2\pi}}e^{-(x - 42)^2/[2(3)^2]}\, dx \approx 0.00383$

49. $\mu = 3.75, \omega = 0.5$

By Simpson's Rule with $n = 12$,

$P(3.5 < x < 4) \approx 0.3829.$

51. Draw a line dividing the area into two equal pieces.

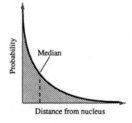

Practice Test for Chapter 9

1. A coin is tossed four times. What is the probability that at least two heads occur?

2. A card is chosen at random from a standard 52-card deck of playing cards. What is the probability that the card will be red and not a face card?

3. Find $E(x)$, $V(x)$, and σ for the given probability distribution.

x	-2	-1	0	3	4
$P(x)$	$\frac{2}{10}$	$\frac{1}{10}$	$\frac{4}{10}$	$\frac{2}{10}$	$\frac{1}{10}$

4. Find the constant k so that $f(x) = ke^{-x/4}$ is a probability density function over the interval $[0, \infty)$.

5. Find (a) $P(0 < x < 5)$ and (b) $P(x > 1)$ for the probability density function

    $$f(x) = \frac{x}{32}, \quad [0, 8].$$

6. Find (a) the mean, (b) the standard deviation, and (c) the median for the probability density function

    $$f(x) = \frac{3}{256}x(8 - x), \quad [0, 8].$$

7. Find (a) the mean, (b) the standard deviation, and (c) the median for the probability density function

    $$f(x) = \frac{6}{x^2}, \quad [2, 3].$$

8. Find the expected value, median, and standard deviation of the exponential density function $f(x) = 7e^{-7x}, \quad [0, \infty)$.

Technology Required

9. Find $P(1.67 < x < 3.24)$ using the standard normal probability density function.

10. The monthly revenue x (in thousands of dollars) of a given shop is normally distributed with $\mu = 20$ and $\sigma = 4$. Approximate $P(19 < x < 24)$.

CHAPTER 10
Series and Taylor Polynomials

CHAPTER 10
Series and Taylor Polynomials

Section 10.1 Sequences

Solutions to Odd-Numbered Exercises

1. 2, 4, 8, 16, 32

3. $\frac{1}{2}, \frac{2}{3}, \frac{3}{4}, \frac{4}{5}, \frac{5}{6}$

5. $3, \frac{9}{2}, \frac{27}{6}, \frac{81}{24}, \frac{243}{120}$

7. $-1, \frac{1}{4}, -\frac{1}{9}, \frac{1}{16}, -\frac{1}{25}$

9. This sequence converges since $\lim\limits_{n\to\infty} \dfrac{5}{n} = 0$.

11. This sequence converges since $\lim\limits_{n\to\infty} \dfrac{n+1}{n} = 1$.

13. This sequence converges since $\lim\limits_{n\to\infty} \dfrac{n^2 + 3n - 4}{2n^2 + n - 3} = \dfrac{1}{2}$.

15. This sequences diverges since
$$\lim_{n\to\infty} \frac{n^2 - 25}{n + 5} = \lim_{n\to\infty} (n - 5) = \infty.$$

17. This sequence converges since $\lim\limits_{n\to\infty} \dfrac{1 + (-1)^n}{n} = 0$.

19. This sequence diverges since
$$\lim_{n\to\infty} \frac{n!}{n} = \lim_{n\to\infty} (n - 1)! = \infty.$$

21. This sequence converges since $\lim\limits_{n\to\infty} \left(3 - \dfrac{1}{2^n} \right) = 3$.

23. This sequence converges since
$$\lim_{n\to\infty} \frac{3^n}{4^n} = \lim_{n\to\infty} \left(\frac{3}{4} \right)^n = 0.$$

25. This sequence diverges since
$$\lim_{n\to\infty} \frac{(n+1)!}{n!} = \lim_{n\to\infty} \frac{(n+1)n!}{n!}$$
$$= \lim_{n\to\infty} (n + 1) = \infty.$$

27. This sequence diverges since $\lim\limits_{n\to\infty} (-1)^n \dfrac{n}{n+1}$ does not exist.

29. The sequence $a_n = (-1)^n + 2$ oscillates between 1 and 3. Hence, $\lim\limits_{n\to\infty} a_n$ does not exist.

31. $a_n = 3n - 2$

33. $a_n = 5n - 6$

35. $a_n = \dfrac{n+1}{n+2}$

37. $a_n = \dfrac{(-1)^{n-1}}{2^{n-2}}$

39. $a_n = 1 + \dfrac{1}{n} = \dfrac{n+1}{n}$

41. $a_n = 2(-1)^n$

43. $a_n = (-1)^n \dfrac{x^n}{n}$

45. Since $a_n = 3n - 1$, the next two terms are $a_5 = 14$ and $a_6 = 17$.

47. Since $a_n = \dfrac{1}{3} + \dfrac{2n}{3}$, we have $a_5 = \dfrac{11}{3}$ and $a_6 = \dfrac{13}{3}$.

49. Since $a_n = 3\left(-\dfrac{1}{2}\right)^{n-1}$, the next two terms are $a_5 = \dfrac{3}{16}$ and $a_6 = -\dfrac{3}{32}$.

51. Since $a_n = 2(3^{n-1})$, the next two terms are 162 and 486.

53. Since $a_n = 20\left(\frac{1}{2}\right)^{n-1}$, the sequence is geometric.

55. Since $a_n = \frac{2}{3}n + 2$, the sequence is arithmetic.

57. One example is $a_n = \dfrac{3n+1}{4n}$.

59. $A_n = P\left(1 + \dfrac{r}{12}\right)^n = 9000\left(1 + \dfrac{0.06}{12}\right)^n = 9000(1.005)^n$

The first 10 terms are 9045.00, 9090.23, 9135.68, 9181.35, 9227.26, 9273.40, 9319.76, 9366.36, 9413.20, 9460.26.

61. $A_n = 2000(11)[(1.1)^n - 1]$

(a) $A_1 = \$2200$

$A_2 = \$4620$

$A_3 = \$7282$

$A_4 = \$10,210.20$

$A_5 = \$13,431.22$

$A_6 \approx \$16,974.34$

(b) $A_{20} \approx \$126,005.00$

(c) $A_{40} \approx \$973,703.62$

63. $S_6 = 130 + 70 + 40 = 240$

$S_7 = 240 + 130 + 70 = 440$

$S_8 = 440 + 240 + 130 = 810$

$S_9 = 810 + 440 + 240 = 1490$

$S_{10} = 1490 + 810 + 440 = 2740$

65. $a_n = \dfrac{2.4 + 0.16n^2}{1 + 0.024n^2}, \quad n = 0, 1, 2, \ldots, 15$

($n = 0$ corresponds to 1987)

(a) $a_0 = \dfrac{2.4 + 0.16(0^2)}{1 + 0.024(0^2)} = 2.4$

$a_1 = \dfrac{2.4 + 0.16(1)}{1 + 0.024(1)} = 2.5$

$a_2 \approx 2.77 \qquad a_9 \approx 5.22$

$a_3 \approx 3.16 \qquad a_{10} \approx 5.41$

$a_4 \approx 3.58 \qquad a_{11} \approx 5.57$

$a_5 = 4.0 \qquad a_{12} \approx 5.71$

$a_6 \approx 4.38 \qquad a_{13} \approx 5.82$

$a_7 \approx 4.71 \qquad a_{14} \approx 5.92$

$a_8 \approx 4.98 \qquad a_{15} = 6.0$

(b)

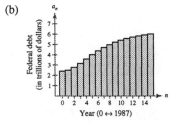

67. (a) $S_1 = 1 = 1^2$

$S_2 = 5 = 1^2 + 2^2$

$S_3 = 14 = 1^2 + 2^2 + 3^2$

$S_4 = 30 = 1^2 + 2^2 + 3^2 + 4^2$

$S_5 = 55 = 1^2 + 2^2 + 3^2 + 4^2 + 5^2$

(b) $S_{20} = 2870$

69. (a) $A_1 = 1.3 - 0.15(1.3) = 1.3(0.85)$

$A_2 = A_1 - 0.15A_1 = 0.85A_1 = 1.3(0.85)^2$

$A_3 = 1.3(0.85)^3$

$\vdots$

$A_n = 1.3(0.85)^n$

(b) $A_1 = \$1.105$ billion

$A_2 = \$0.939$ billion

$A_3 = \$0.798$ billion

$A_4 = \$0.679$ billion

(c) The sequence converges to 0.

$\lim_{n \to \infty} 1.3(0.85)^n = 0$

71. $A_1 = 16$

$A_2 = 16 + 16.10$

$A_3 = 16 + 16.10 + 16.20$

$\vdots$

$A_{100} = 16 + 16.10 + \cdots + [16 + (0.1)(99)]$

$\quad\quad\; = 16 + 16.10 + \cdots + 25.9$

$\quad\quad\; = 2095$

73. (a) $a_n = -0.265625n^3 + 5.32271n^2 - 9.7470n + 90.192$

($n = 1$ corresponds to 1994)

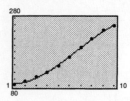

(b) For 2010, $n = 17$ and $a_{17} \approx \$157.7$ billion.

(This is not realistic because of the negative coefficient for n^3.)

75. $a_n = \left(1 + \dfrac{1}{n}\right)^n$

$a_1 = 2$

$a_{10} \approx 2.593742460$

$a_{100} \approx 2.704813829$

$a_{1000} \approx 2.716923932$

$a_{10,000} \approx 2.718145927$

Note: $e \approx 2.718281828$

Section 10.2   Series and Convergence

1. $S_1 = 1$

$S_2 = \frac{5}{4} = 1.25$

$S_3 = \frac{49}{36} \approx 1.361$

$S_4 = \frac{205}{144} \approx 1.424$

$S_5 = \frac{5269}{3600} \approx 1.464$

3. $S_1 = 3$

$S_2 = \frac{9}{2} = 4.5$

$S_3 = \frac{21}{4} = 5.25$

$S_4 = \frac{45}{8} = 5.625$

$S_5 = \frac{93}{16} = 5.8125$

5. This series diverges by the *n*th-Term Test since

$$\lim_{n\to\infty} \frac{n}{n+1} = 1 \neq 0.$$

7. This series diverges by the *n*th-Term Test since

$$\lim_{n\to\infty} \frac{n^2}{n^2+1} = 1 \neq 0.$$

9. This series diverges by the Test for Convergence of a Geometric Series since $|r| = \left|\frac{3}{2}\right| > 1$.

11. This series diverges by the Test for Convergence of a Geometric Series since $|r| = |1.055| > 1$.

13. This series converges by the Test for Convergence of a Geometric Series since $|r| = \left|\frac{3}{4}\right| < 1$.

15. This series converges by the Test for Convergence of a Geometric Series since $|r| = |0.9| < 1$.

17. Since $a = 1$ and $r = \frac{1}{2}$, we have

$$S = \frac{1}{1 - (1/2)} = 2.$$

19. Since $a = 1$ and $r = -\frac{1}{2}$, we have

$$S = \frac{1}{1 + (1/2)} = \frac{2}{3}.$$

21. Since $a = 2$ and $r = 1/\sqrt{2}$, we have

$$S = \frac{2}{1 - (1/\sqrt{2})}$$

$$= \frac{2\sqrt{2}}{\sqrt{2} - 1}\left(\frac{\sqrt{2} + 1}{\sqrt{2} + 1}\right)$$

$$= 4 + 2\sqrt{2}$$

$$\approx 6.828.$$

23. Since $a = 1$ and $r = 0.1$, we have

$$S = \frac{1}{1 - 0.1} = \frac{1}{0.9} = \frac{10}{9}.$$

25. Since $a = 2$ and $r = -\frac{1}{3}$, we have

$$S = \frac{2}{1 + (1/3)} = \frac{3}{2}.$$

27. $\displaystyle\sum_{n=0}^{\infty} \left(\frac{1}{2^n} - \frac{1}{3^n}\right) = \sum_{n=0}^{\infty} \left(\frac{1}{2}\right)^n - \sum_{n=0}^{\infty} \left(\frac{1}{3}\right)^n$

$$= \frac{1}{1 - (1/2)} - \frac{1}{1 - (1/3)}$$

$$= 2 - \frac{3}{2} = \frac{1}{2}$$

29. $\displaystyle\sum_{n=0}^{\infty} \left(\frac{1}{3^n} + \frac{1}{4^n}\right) = \sum_{n=0}^{\infty} \left(\frac{1}{3}\right)^n + \sum_{n=0}^{\infty} \left(\frac{1}{4}\right)^n$

$$= \frac{1}{1 - (1/3)} + \frac{1}{1 - (1/4)}$$

$$= \frac{3}{2} + \frac{4}{3} = \frac{17}{6}$$

31. This series diverges by the nth-Term Test since

$$\lim_{n \to \infty} \frac{n + 10}{10n + 1} = \frac{1}{10} \neq 0.$$

33. This series diverges by the nth-Term Test since

$$\lim_{n \to \infty} \frac{n! + 1}{n!} = 1 \neq 0.$$

35. This series diverges by the nth-Term Test since

$$\lim_{n \to \infty} \frac{3n - 1}{2n + 1} = \frac{3}{2} \neq 0.$$

37. This series diverges by the Test for Convergence of a Geometric Series since $r = 1.075 > 1$.

39. This series converges by the Test for Convergence of a Geometric Series since $r = \frac{1}{4} < 1$.

41. $0.66\overline{6} = \displaystyle\sum_{n=0}^{\infty} 0.6(0.1)^n = \frac{0.6}{1 - 0.1} = \frac{0.6}{0.9} = \frac{2}{3}$

43. $0.81\overline{81} = \displaystyle\sum_{n=0}^{\infty} 0.81(0.01)^n = \frac{0.81}{1 - 0.01} = \frac{0.81}{0.99} = \frac{9}{11}$

45. (a) $\displaystyle\sum_{i=0}^{n-1} 8000(0.9)^i = \frac{8000[1 - (0.9)^{(n-1)+1}]}{1 - 0.9}$

$$= 80,000(1 - 0.9^n)$$

(b) $\displaystyle\sum_{i=0}^{\infty} 8000(0.9)^i = \frac{8000}{1 - 0.9} = 80,000$

47. $D_1 = 16$

$D_2 = 0.64(16) + 0.64(16) = 32(0.64)$

$D_3 = 32(0.64)^2$

$\vdots$

$D = -16 + \displaystyle\sum_{n=0}^{\infty} 32(0.64)^n$

$$= -16 + \frac{32}{1 - 0.64}$$

$$= -16 + \frac{32}{0.36}$$

$$= \frac{2624}{36} \approx 72.89 \text{ feet}$$

49. $A = \displaystyle\sum_{n=1}^{60} 100\left(1 + \frac{0.10}{12}\right)^n$

$$= -100 + \sum_{n=0}^{60} 100\left(1 + \frac{0.10}{12}\right)^n$$

$$= -100 + \frac{100\left[1 - \left(1 + \frac{0.10}{12}\right)^{61}\right]}{1 - \left(1 + \frac{0.10}{12}\right)}$$

$$\approx \$7808.24$$

51. $A = \sum\limits_{n=0}^{\infty} 100(0.75)^n = \dfrac{100}{1 - 0.75} = \$400$ million

53. $A = \sum\limits_{n=0}^{19} 0.01(2)^n = \dfrac{0.01(1 - 2^{20})}{1 - 2} = \$10,485.75$

55. $\sum\limits_{n=1}^{\infty} n\left(\dfrac{1}{2}\right)^n = \dfrac{1}{2} + 2\left(\dfrac{1}{4}\right) + 3\left(\dfrac{1}{8}\right) + 4\left(\dfrac{1}{16}\right) + 5\left(\dfrac{1}{32}\right) + 6\left(\dfrac{1}{64}\right) + \cdots$

$$= \dfrac{1}{2} + \dfrac{1}{2} + \dfrac{3}{8} + \dfrac{1}{4} + \dfrac{5}{32} + \dfrac{3}{32} + \cdots$$

$$= 2$$

57. At factory: 500

1 mile away: (0.85)500

2 miles away: (0.85)²500

12 miles away: $(0.85)^{12}500 \approx 71.12$ ppm

59. (a) $\sum\limits_{i=1}^{10} 880 = \$8800$

(b) $\sum\limits_{i=1}^{168} 880 = (168)(880) = \$147,840$

Hence, $147,840 - 100,000 = \$47,840$ more.

61. $\sum\limits_{n=1}^{\infty} n^2\left(\dfrac{1}{2}\right)^n = 6$

63. $\sum\limits_{n=1}^{\infty} \dfrac{1}{(2n)!} \approx 0.5431$

65. $\sum\limits_{n=1}^{\infty} e^2\left(\dfrac{1}{e}\right)^n = \dfrac{e^2}{e - 1} \approx 4.3003$

67. False. For example, $\sum\limits_{n=1}^{\infty} 1/n$ diverges even though $\lim\limits_{n\to\infty} 1/n = 0$.

Section 10.3 *p*-Series and the Ratio Test

1. The series

$$\sum_{n=1}^{\infty} \dfrac{1}{n^2}$$

is a *p*-series with $p = 2$.

3. The series

$$\sum_{n=1}^{\infty} \dfrac{1}{3^n} = \sum_{n=0}^{\infty} \dfrac{1}{3}\left(\dfrac{1}{3}\right)^n$$

is *not* a *p*-series. This series is geometric with $r = \tfrac{1}{3}$.

5. The series

$$\sum_{n=1}^{\infty} \dfrac{1}{n^n} = 1 + \dfrac{1}{2^2} + \dfrac{1}{3^3} + \dfrac{1}{4^4} + \cdots$$

is *not* a *p*-series. The exponent changes with each term.

7. This series converges since $p = 3 > 1$.

9. This series diverges since $p = \tfrac{1}{3} < 1$.

11. This series converges since $p = 1.03 > 1$.

13. $1 + \dfrac{1}{\sqrt{2}} + \dfrac{1}{\sqrt{3}} + \dfrac{1}{\sqrt{4}} + \cdots = \sum\limits_{n=1}^{\infty} \dfrac{1}{\sqrt{n}} = \sum\limits_{n=1}^{\infty} \dfrac{1}{n^{1/2}}$

Therefore, this series diverges since $p = \tfrac{1}{2} < 1$.

15. $1 + \dfrac{1}{2\sqrt{2}} + \dfrac{1}{3\sqrt{3}} + \dfrac{1}{4\sqrt{4}} + \cdots = \sum\limits_{n=1}^{\infty} \dfrac{1}{n^{3/2}}$

Therefore, the series converges since $p = \tfrac{3}{2} > 1$.

17. Since $a_n = 3^n/n!$, we have

$$\lim_{n\to\infty} \left|\dfrac{a_{n+1}}{a_n}\right| = \lim_{n\to\infty} \left|\dfrac{3^{n+1}}{(n+1)!} \cdot \dfrac{n!}{3^n}\right|$$

$$= \lim_{n\to\infty} \dfrac{3}{n+1} = 0$$

and the series converges.

19. Since $a_n = n!/3^n$, we have

$$\lim_{n\to\infty} \left|\dfrac{a_{n+1}}{a_n}\right| = \lim_{n\to\infty} \left|\dfrac{(n+1)!}{3^{n+1}} \cdot \dfrac{3^n}{n!}\right|$$

$$= \lim_{n\to\infty} \dfrac{n+1}{3} = \infty$$

and the series diverges.

21. Since $a_n = n/4^n$, we have

$$\lim_{n \to \infty} \left| \frac{a_{n+1}}{a_n} \right| = \lim_{n \to \infty} \left| \frac{n+1}{4^{n+1}} \cdot \frac{4^n}{n} \right|$$

$$= \lim_{n \to \infty} \frac{n+1}{4n} = \frac{1}{4}$$

and the series converges.

23. Since $a_n = 2^n/n^5$, we have

$$\lim_{n \to \infty} \left| \frac{a_{n+1}}{a_n} \right| = \lim_{n \to \infty} \left| \frac{2^{n+1}}{(n+1)^5} \cdot \frac{n^5}{2^n} \right|$$

$$= \lim_{n \to \infty} \frac{2n^5}{(n+1)^5} = 2$$

and the series diverges.

25. Since

$$a_n = \frac{(-1)^n 2^n}{n!},$$

we have

$$\lim_{n \to \infty} \left| \frac{a_{n+1}}{a_n} \right| = \lim_{n \to \infty} \left| \frac{(-1)^{n+1} 2^{n+1}}{(n+1)!} \cdot \frac{n!}{(-1)^n 2^n} \right|$$

$$= \lim_{n \to \infty} \left| \frac{-2}{n+1} \right| = 0$$

and the series converges.

27. Since

$$a_n = \frac{4^n}{3^n + 1},$$

we have

$$\lim_{n \to \infty} \left| \frac{a_{n+1}}{a_n} \right| = \lim_{n \to \infty} \left| \frac{4^{n+1}}{3^{n+1} + 1} \cdot \frac{3^n + 1}{4^n} \right|$$

$$= \frac{4}{3} > 1$$

and the series diverges.

29. Since

$$a_n = \frac{n5^n}{n!},$$

we have

$$\lim_{n \to \infty} \left| \frac{a_{n+1}}{a_n} \right| = \lim_{n \to \infty} \left| \frac{(n+1)5^{n+1}}{(n+1)!} \cdot \frac{n!}{n5^n} \right|$$

$$= \lim_{n \to \infty} \left| \frac{5}{n} \right| = 0 < 1$$

and the series converges.

31. $\displaystyle\sum_{n=1}^{\infty} \frac{1}{n^3} \approx \frac{1}{1} + \frac{1}{2^3} + \frac{1}{3^3} + \frac{1}{4^3} = \frac{2035}{1728} \approx 1.1777$

The error is less than

$$\frac{1}{(p-1)N^{p-1}} = \frac{1}{(3-1)4^{3-1}} = \frac{1}{32}.$$

33. $\displaystyle\sum_{n=1}^{\infty} \frac{1}{n^{3/2}} \approx 1 + \frac{1}{2\sqrt{2}} + \frac{1}{3\sqrt{3}} + \cdots + \frac{1}{10\sqrt{10}} \approx 1.9953$

The error is less than

$$\frac{1}{(p-1)N^{p-1}} = \frac{1}{[(3/2)-1]10^{[(3/2)-1]}}$$

$$= \frac{2}{\sqrt{10}} \approx 0.6325.$$

35. $\displaystyle\lim_{n \to \infty} \left| \frac{a_{n+1}}{A_n} \right| = \lim_{n \to \infty} \frac{1/[(n+1)^{3/2}]}{1/n^{3/2}}$

$$= \lim_{n \to \infty} \left(\frac{n}{n+1} \right)^{3/2} = 1,$$

and the Ratio Test is inconclusive. (The series is a convergent *p*-series.)

37. $\displaystyle\sum_{n=1}^{\infty} \frac{2}{\sqrt[4]{n^3}} = \sum_{n=1}^{\infty} \frac{2}{n^{3/4}} = 2 + \frac{2}{2^{3/4}} + \cdots$

Diverges (*p*-series with $p = \frac{3}{4} < 1$)   Matches (a).

39. $\displaystyle\sum_{n=1}^{\infty} \frac{2}{n\sqrt{n}} = \sum_{n=1}^{\infty} \frac{2}{n^{3/2}} = 2 + \frac{2}{2^{3/2}} + \cdots$

Converges (*p*-series with $p = \frac{3}{2} > 1$)   Matches (b).

41. This series diverges by the *n*th-Term Test since

$$\lim_{n \to \infty} \frac{2n}{n+1} = 2 \neq 0.$$

43. This series converges by the *p*-series test since $p = \frac{4}{3} > 1$.

$$\sum_{n=1}^{\infty} \frac{1}{n\sqrt[3]{n}} \approx 3.6009$$

45. This series converges by the Geometric Series Test since $|r| = |-\frac{2}{3}| = \frac{2}{3} < 1.$

$$\sum_{n=0}^{\infty} \frac{(-1)^n 2^n}{3^n} = \frac{1}{1 - \left(-\frac{2}{3}\right)} = \frac{3}{5}$$

47. Since both series are convergent *p*-series, their difference is convergent.

$$\sum_{n=1}^{\infty} \left(\frac{1}{n^2} - \frac{1}{n^3}\right) \approx 0.4429$$

49. This series diverges by the Geometric Series Test since $r = \frac{5}{4} > 1.$

51. This series diverges by the Ratio Test since

$$a_n = \frac{n!}{3^{n-1}} \text{ and } \lim_{n\to\infty} \left|\frac{a_{n+1}}{a_n}\right| = \lim_{n\to\infty} \left|\frac{(n+1)!}{3^n} \cdot \frac{3^{n-1}}{n!}\right| = \lim_{n\to\infty} \frac{n+1}{3} = \infty.$$

53. The series diverges by the *n*th Term Test because $\lim_{n\to\infty} \frac{2^n}{2^{n+1}+1} = \frac{1}{2}.$

55. $\sum_{n=1}^{\infty} \frac{2^n}{5^{n-1}} = 5 \sum_{n=1}^{\infty} \left(\frac{2}{5}\right)^n$ converges by the Geometric Series Test since $r = \frac{2}{5} < 1.$

$$5\sum_{n=1}^{\infty} \left(\frac{2}{5}\right)^n = \frac{5(2/5)}{1 - 2/5} = \frac{2}{(3/5)} = \frac{10}{3}.$$

57. $\sum_{n=1}^{100} \frac{1}{n^2} = \frac{1}{1} + \frac{1}{2^2} + \frac{1}{3^2} + \frac{1}{4^2} + \cdots + \frac{1}{100^2} \approx 1.635$

$$\frac{\pi^2}{6} \approx 1.644934$$

59. (a) $a(n) = 0.1396n^2 + 0.309n + 12.32$

($t = 1$ corresponds to 1996)

$$\sum_{n=1}^{\infty} (0.1396n^2 + 0.309n + 12.32)\frac{1}{2}$$

(b) No, the Ratio Test is inconclusive. (Clearly, the series diverges.)

Section 10.4   Power Series and Taylor's Theorem

1. $\sum_{n=0}^{\infty} \left(\frac{x}{4}\right)^n = 1 + \frac{x}{4} + \left(\frac{x}{4}\right)^2 + \left(\frac{x}{4}\right)^3 + \left(\frac{x}{4}\right)^4 + \cdots$

3. $\sum_{n=0}^{\infty} \frac{(-1)^{n+1}(x+1)^n}{n!} = -1 + (x+1) - \frac{(x+1)^2}{2} + \frac{(x+1)^3}{6} - \frac{(x+1)^4}{24} + \cdots$

5. $\lim_{n\to\infty} \left|\frac{a_{n+1}x^{n+1}}{a_n x^n}\right| = \lim_{n\to\infty} \left|\frac{(x/2)^{n+1}}{(x/2)^n}\right| = \lim_{n\to\infty} \left|\frac{x}{2}\right| < 1 \implies |x| < 2$

Radius = 2

7. $\lim_{n\to\infty} \left|\frac{a_{n+1}x^{n+1}}{a_n x^n}\right| = \lim_{n\to\infty} \left|\frac{(-1)^{n+1}x^{n+1}/3(n+1)}{(-1)^n x^n/3n}\right| = \lim_{n\to\infty} \left|\frac{n}{n+1}x\right| = |x| < 1 \implies |x| < 1$

Radius = 1

9. $\lim_{n\to\infty} \left|\frac{a_{n+1}x^{n+1}}{a_n x^n}\right| = \lim_{n\to\infty} \left|\frac{(-1)^{n+1}x^{n+1}/(n+1)!}{(-1)^n x^n/n!}\right| = \lim_{n\to\infty} \left|\frac{x}{n+1}\right| = 0 \implies -\infty < x < \infty$

Radius = ∞

11. $\lim\limits_{n\to\infty}\left|\dfrac{a_{n+1}x^{n+1}}{a_n x^n}\right| = \lim\limits_{n\to\infty}\left|\dfrac{(n+1)!x^{n+1}/2^{n+1}}{n!x^n/2^n}\right| = \lim\limits_{n\to\infty}\left|\dfrac{(n+1)x}{2}\right| = \infty$

This series converges only at $x = 0$.

Radius $= 0$

13. $\lim\limits_{n\to\infty}\left|\dfrac{a_{n+1}x^{n+1}}{a_n x^n}\right| = \lim\limits_{n\to\infty}\left|\dfrac{(-1)^{n+2}x^{n+1}/4^{n+1}}{(-1)^{n+1}x^n/4^n}\right| = \lim\limits_{n\to\infty}\left|\dfrac{x}{4}\right| < 1 \Rightarrow |x| < 4$

Radius $= 4$

15. $\lim\limits_{n\to\infty}\left|\dfrac{a_{n+1}(x-5)^{n+1}}{a_n(x-5)^n}\right| = \lim\limits_{n\to\infty}\left|\dfrac{(-1)^{n+2}(x-5)^{n+1}/(n+1)5^{n+1}}{(-1)^{n+1}(x-5)^n/n5^n}\right| = \lim\limits_{n\to\infty}\left|\dfrac{x-5}{5}\cdot\dfrac{n+1}{n}\right| \Rightarrow |x-5| < 5 = \left|\dfrac{x-5}{5}\right| < 1$

Radius $= 5$

17. $\lim\limits_{n\to\infty}\left|\dfrac{a_{n+1}(x-1)^{n+1}}{a_n(x-1)^n}\right| = \lim\limits_{n\to\infty}\left|\dfrac{(-1)^{n+2}(x-1)^{n+2}/(n+2)}{(-1)^{n+1}(x-1)^{n+1}/(n+1)}\right|$

$\qquad\qquad = \lim\limits_{n\to\infty}\left|(x-1)\dfrac{n+1}{n+2}\right|$

$\qquad\qquad = \lim\limits_{n\to\infty}|x-1|$

$\qquad\qquad = |x-1| < 1$

$\Rightarrow |x-1| < 1$

Radius $= 1$

19. $\lim\limits_{n\to\infty}\left|\dfrac{a_{n+1}(x-c)^{n+1}}{a_n(x-c)^n}\right| = \lim\limits_{n\to\infty}\left|\dfrac{(x-c)^n/c^n}{(x-c)^{n-1}/c^{n-1}}\right|$

$\qquad\qquad = \lim\limits_{n\to\infty}\left|\dfrac{x-c}{c}\right| < 1$

$\Rightarrow |x-c| < c$

Radius $= c$

21. $\lim\limits_{n\to\infty}\left|\dfrac{a_{n+1}x^{n+1}}{a_n x^n}\right| = \lim\limits_{n\to\infty}\left|\dfrac{(n+1)(-2)^n x^n/(n+2)!}{n(-2)^{n-1}x^{n-1}/(n+1)!}\right|$

$\qquad\qquad = \lim\limits_{n\to\infty}\left|\dfrac{n+1}{n}(-2)x\dfrac{1}{n+2}\right|$

$\qquad\qquad = 0$

$\Rightarrow -\infty < x < \infty$

Radius $= \infty$

23. $\lim\limits_{n\to\infty}\left|\dfrac{a_{n+1}x^{n+1}}{a_n x^n}\right| = \lim\limits_{n\to\infty}\left|\dfrac{x^{2n+3}/(2n+3)!}{x^{2n+1}/(2n+1)!}\right| = \lim\limits_{n\to\infty}\left|\dfrac{x^2}{(2n+3)(2n+2)}\right| = 0$

$\Rightarrow -\infty < x < \infty$

Radius $= \infty$

25. $f(x) = e^x \qquad\qquad f(0) = 1$

$\quad f'(x) = e^x \qquad\qquad f'(0) = 1$

$\quad f''(x) = e^x \qquad\qquad f''(0) = 1$

$\qquad\qquad\vdots \qquad\qquad\qquad\qquad\vdots$

$\quad f^{(n)}(x) = e^x \qquad\qquad f^{(n)}(0) = 1$

The power series for f is

$$e^x = f(0) + f'(0)x + \frac{f''(0)x^2}{2!} + \cdots = 1 + x + \frac{x^2}{2!} + \frac{x^3}{3!} + \cdots = \sum_{n=0}^{\infty}\frac{x^n}{n!}.$$

$\lim\limits_{n\to\infty}\left|\dfrac{x^{n+1}/(n+1)!}{x^n/n!}\right| = \lim\limits_{n\to\infty}\left|\dfrac{x}{n+1}\right| = 0 \Rightarrow R = \infty$

27. $f(x) = e^{2x}$ $f(0) = 1$

$f'(x) = 2e^{2x}$ $f'(0) = 2$

$f''(x) = 4e^{2x}$ $f''(0) = 4$

$f'''(x) = 8e^{2x}$ $f'''(0) = 8$

$$\vdots$$

$$f^{(n)}(0) = 2^n$$

The power series for f is $e^{2x} = f(0) + f'(0)x + \dfrac{f''(0)x^2}{2!} + \cdots = 1 + 2x + \dfrac{4x^2}{2!} + \dfrac{8x^3}{3!} + \cdots = \displaystyle\sum_{n=0}^{\infty} \dfrac{(2x)^n}{n!}$.

$$\lim_{n\to\infty} \left| \frac{(2x)^{n+1}/(n+1)!}{(2x)^n/n!} \right| = \lim_{n\to\infty} \left| \frac{2x}{n+1} \right| = 0 \Rightarrow R = \infty$$

29. $f(x) = \dfrac{1}{x+1}$ $f(0) = 1$

$f'(x) = \dfrac{-1}{(x+1)^2}$ $f'(0) = -1$

$f''(x) = \dfrac{2}{(x+1)^3}$ $f''(0) = 2$

$f'''(x) = \dfrac{-6}{(x+1)^4}$ $f'''(0) = -6$

$$\vdots$$

$$f^{(n)}(0) = (-1)^n n!$$

The power series for f is

$$\frac{1}{x+1} = f(0) + f'(0)x + \frac{f''(0)x^2}{2!} + \cdots$$

$$= 1 - x + \frac{2x^2}{2!} - \frac{6x^3}{3!} + \cdots + \frac{(-1)^n n! x^n}{n!} + \cdots = 1 - x + x^2 - x^3 + \cdots = \sum_{n=0}^{\infty} (-1)^n x^n.$$

$$\lim_{n\to\infty} \left| \frac{(-1)^{n+1} x^{n+1}}{(-1)^n x^n} \right| = \lim_{n\to\infty} |x| < 1 \Rightarrow R = 1$$

31. $f(x) = \sqrt{x}$ $f(1) = 1$

$f'(x) = \dfrac{1}{2\sqrt{x}}$ $f'(1) = \dfrac{1}{2}$

$f''(x) = -\dfrac{1}{4x\sqrt{x}}$ $f''(1) = -\dfrac{1}{4}$

$f'''(x) = \dfrac{3}{8x^2\sqrt{x}}$ $f'''(1) = \dfrac{3}{8}$

$f^{(4)}(x) = -\dfrac{15}{16x^3\sqrt{x}}$ $f^{(4)}(1) = -\dfrac{15}{16}$

The general pattern (for $n \geq 2$) is

$$f^{(n)}(1) = (-1)^{n-1} \frac{1 \cdot 3 \cdots\cdots 5 \cdot (2n-3)}{2^n}.$$

The power series for $f(x) = \sqrt{x}$ is:

$$f(1) + f'(1)(x-1) + \frac{f''(1)(x-1)^2}{2!} + \cdots = 1 + \frac{1}{2}(x-1) - \frac{1}{8}(x-1)^2 + \frac{1}{16}(x-1)^3 + \cdots$$

$$= 1 + \frac{1}{2}(x-1) + \sum_{n=2}^{\infty} \frac{(-1)^{n+1} 1 \cdot 3 \cdot 5 \cdots\cdots (2n-3)}{2^n n!}(x-1)^n$$

$$\lim_{n\to\infty} \left| \frac{1 \cdot 3 \cdot 5 \cdots\cdots (2n-1)(x-1)^{n+1}/2^{n+1}(n+1)!}{1 \cdot 3 \cdot 5 \cdots\cdots (2n-3)(x-1)^n/2^n n!} \right| = \lim_{n\to\infty} \left| \frac{(2n-1)(x-1)}{2(n+1)} \right| = |x-1| < 1 \Rightarrow R = 1$$

33. $f(x) = (1 + x)^{-3}$ $f(0) = 1$

$f'(x) = -3(1 + x)^{-4}$ $f'(0) = -3$

$f''(x) = 12(1 + x)^{-5}$ $f''(0) = 12$

$f'''(x) = -60(1 + x)^{-6}$ $f'''(0) = -60$

In general, $f^{(n)}(0) = (-1)^n \dfrac{(n + 2)!}{2}$. The power series is

$$\frac{1}{(1 + x)^3} = f(0) + f'(0)x + \frac{f''(0)x^2}{2!} + \cdots = 1 - 3x + 6x^2 - 10x^3 + \cdots = \sum_{n=0}^{\infty} (-1)^n \frac{(n + 2)(n + 1)}{2} x^n.$$

$R = 1$

35. $f(x) = (1 + x)^{-1/2}$ $f(0) = 1$

$f'(x) = -\dfrac{1}{2}(1 + x)^{-3/2}$ $f'(0) = -\dfrac{1}{2}$

$f''(x) = \dfrac{3}{4}(1 + x)^{-5/2}$ $f''(0) = \dfrac{3}{4}$

$f'''(x) = -\dfrac{15}{8}(1 + x)^{-7/2}$ $f'''(0) = -\dfrac{15}{8}$

Thus, the general pattern is given by $f^{(n)}(0) = (-1)^n \left[\dfrac{1 \cdot 3 \cdot 5 \cdots (2n - 1)}{2^n} \right]$. The power series for f is

$$\frac{1}{\sqrt{1 + x}} = f(0) + f'(0)x + \frac{f''(0)x^2}{2!} + \frac{f'''(0)x^3}{3!} + \cdots$$

$$= 1 - \frac{1}{2}x + \frac{1 \cdot 3x^2}{2^2 2!} - \frac{1 \cdot 3 \cdot 5x^3}{2^3 3!} + \cdots$$

$$= 1 + \sum_{n=1}^{\infty} \frac{(-1)^n 1 \cdot 3 \cdot 5 \cdots (2n - 1)}{2^n n!} x^n.$$

$R = 1$

37. (a) $f(x) = \displaystyle\sum_{n=0}^{\infty} \left(\frac{x}{2}\right)^n = \sum_{n=0}^{\infty} \frac{x^n}{2^n}$ (b) $f'(x) = \displaystyle\sum_{n=1}^{\infty} \frac{nx^{n-1}}{2^n}$

$\displaystyle\lim_{n \to \infty} \left| \frac{(x/2)^{n+1}}{(x/2)^n} \right| = \left| \frac{x}{2} \right| < 1 \Longrightarrow |x| < 2$ $\displaystyle\lim_{n \to \infty} \left| \frac{(n + 1)x/2^{n+1}}{nx^{n-1}/2^n} \right| = \lim_{n \to \infty} \left| \frac{n + 1}{2n} x \right|$

$R = 2$ $= \left| \frac{x}{2} \right| < 1 \Longrightarrow R = 2$

(c) $f''(x) = \displaystyle\sum_{n=2}^{\infty} \frac{n(n - 1)x^{n-2}}{2^n}$ (d) $\displaystyle\int f(x)\, dx = \sum_{n=0}^{\infty} \frac{x^{n+1}}{(n + 1)2^n} + C$

$\displaystyle\lim_{n \to \infty} \left| \frac{(n + 1)nx^{n-1}/2^{n+1}}{n(n - 1)x^{n-2}/2^n} \right| = \lim_{n \to \infty} \left| \frac{n + 1}{n - 1} \cdot \frac{x}{2} \right|$ $\displaystyle\lim_{n \to \infty} \left| \frac{x^{n+2}/(n + 2)2^{n+1}}{x^{n+1}/(n + 1)2^n} \right| = \lim_{n \to \infty} \left| \frac{n + 1}{n + 2} \cdot \frac{x}{2} \right|$

$= \left| \frac{x}{2} \right| 1 \Longrightarrow R = 2$ $= \left| \frac{x}{2} \right| < 1 \Longrightarrow R = 2$

39. (a) $f(x) = \sum_{n=0}^{\infty} \frac{(x+1)^{n+1}}{n+1}$

$$\lim_{n\to\infty} \left| \frac{(x+1)^{n+2}/(n+2)}{(x+1)^{n+1}/(n+1)} \right| = \lim_{n\to\infty} \left| \frac{n+1}{n+2}(x+1) \right|$$

$$= |x+1| < 1 \Rightarrow R = 1$$

(b) $f'(x) = \sum_{n=0}^{\infty} (x+1)^n$

$$\lim_{n\to\infty} \left| \frac{(x+1)^{n+1}}{(x+1)^n} \right| = |x+1|1 \Rightarrow R = 1$$

(c) $f''(x) = \sum_{n=1}^{\infty} n(x+1)^{n-1}$

$$\lim_{n\to\infty} \left| \frac{(n+1)(x+1)^n}{n(x+1)^{n-1}} \right| = \lim_{n\to\infty} \left| \frac{n+1}{n}(x+1) \right|$$

$$= |x+1| < 1 \Rightarrow R = 1$$

(d) $\int f(x)\,dx = \sum_{n=0}^{\infty} \frac{(x+1)^{n+2}}{(n+2)(n+1)} + C$

$$\lim_{n\to\infty} \left| \frac{(x+1)^{n+3}/(n+3)(n+2)}{(x+1)^{n+2}/(n+2)(n+1)} \right| = |x+1| < 1$$

$$\Rightarrow R = 1$$

41. Since the power series for e^x is

$$e^x = \sum_{n=0}^{\infty} \frac{x^n}{n!}$$

it follows that the power series for e^{x^3} is

$$e^{x^3} = \sum_{n=0}^{\infty} \frac{(x^3)^n}{n!} = \sum_{n=0}^{\infty} \frac{x^{3n}}{n!}.$$

43. $3x^2e^x = \frac{d}{dx}[e^{x^3}]$

$$= \sum_{n=1}^{\infty} \frac{3nx^{3n-1}}{n!}$$

$$= 3 \sum_{n=1}^{\infty} \frac{x^{3n-1}}{(n-1)!}$$

$$= 3 \sum_{n=0}^{\infty} \frac{x^{3n+2}}{n!}$$

45. Since the power series for $1/(1+x)$ is

$$f(x) = \frac{1}{1+x} = \sum_{n=0}^{\infty} (-1)^n x^n$$

it follows that the power series for $1/(1+x^4)$ is

$$f(x^4) = \frac{1}{1+x^4} = \sum_{n=0}^{\infty} (-1)^n x^{4n}.$$

47. $\frac{1}{1+x^2} = \sum_{n=0}^{\infty} (-1)^n x^{2n}$

$$\frac{2x}{1+x^2} = \sum_{n=0}^{\infty} (-1)^n(2x)x^{2n} = 2\sum_{n=0}^{\infty} (-1)^n x^{2n+1}$$

$$\ln(1+x^2) = \int \frac{2x}{1+x^2}\,dx$$

$$= 2\sum_{n=0}^{\infty} \frac{(-1)^n x^{2n+2}}{2n+2} = \sum_{n=0}^{\infty} \frac{(-1)^n x^{2n+2}}{n+1}$$

49. $\frac{1}{x} = \sum_{n=0}^{\infty} (-1)^n(x-1)^n$

$$\ln x = \int \frac{1}{x}\,dx = \sum_{n=0}^{\infty} \frac{(-1)^n(x-1)^{n+1}}{n+1}$$

51. $-\frac{1}{x} = \sum_{n=0}^{\infty} (-1)^{n+1}(x-1)^n$

Differentiating, $\frac{1}{x^2} = \sum_{n=1}^{\infty} (-1)^{n+1}n(x-1)^{n-1}.$

53. $e^x = 1 + x + \frac{x^2}{2!} + \frac{x^3}{3!} + \frac{x^4}{4!} + \frac{x^5}{5!} + \cdots$

$$e^{1/2} = 1 + \frac{1}{2} + \frac{1}{2!(2^2)} + \frac{1}{3!(2^3)} + \frac{1}{4!(2^4)} + \frac{1}{5!(2^5)} + \cdots$$

Since $1/[6!(2^6)] \approx 0.00002$, the first six terms are sufficient to approximate $e^{1/2}$ to four decimal places.

$$e^{1/2} \approx 1 + \frac{1}{2} + \frac{1}{8} + \frac{1}{48} + \frac{1}{384} + \frac{1}{3840} \approx 1.6487$$

55. $f(0.5) = \sum_{n=1}^{\infty} \dfrac{(-1)^{n+1}(0.5-1)^n}{n}$

$\qquad = \sum_{n=1}^{\infty} \dfrac{(-1)^{n+1}(-1)^n(1/2)^n}{n}$

$\qquad = \sum_{n=1}^{\infty} \dfrac{(-1)^{2n+1}}{2^n n}$

$\qquad = -\sum_{n=1}^{\infty} \dfrac{1}{2^n n}$

$\qquad \approx -0.6931$

57. $f(0.1) = \sum_{n=1}^{\infty} \dfrac{(-1)^{n+1}(0.1-1)^n}{n}$

$\qquad = \sum_{n=1}^{\infty} \dfrac{(-1)^{n+1}(-0.9)^n}{n}$

$\qquad = \sum_{n=1}^{\infty} \dfrac{(-1)^{2n+1}(0.9)^n}{n}$

$\qquad = -\sum_{n=1}^{\infty} \dfrac{(0.9)^n}{n}$

$\qquad \approx -2.3018$

Section 10.5   Taylor Polynomials

1. $e^x = \sum_{n=0}^{\infty} \dfrac{1}{n!} x^n$

(a) $S_1(x) = 1 + x$

(b) $S_2(x) = 1 + x + \dfrac{x^2}{2}$

(c) $S_3(x) = 1 + x + \dfrac{x^2}{2} + \dfrac{x^3}{6}$

(d) $S_4(x) = 1 + x + \dfrac{x^2}{2} + \dfrac{x^3}{6} + \dfrac{x^4}{24}$

3. $\sqrt{x+1} = 1 + \dfrac{x}{2^1 \cdot 1!} - \dfrac{x^2}{2^2 \cdot 2!} + \dfrac{3x^3}{2^3 \cdot 3!} - \dfrac{3 \cdot 5x^4}{2^4 \cdot 4!}$

(a) $S_1(x) = 1 + \dfrac{x}{2}$

(b) $S_2(x) = 1 + \dfrac{x}{2} - \dfrac{x^2}{8}$

(c) $S_3(x) = 1 + \dfrac{x}{2} - \dfrac{x^2}{8} + \dfrac{3x^3}{48} = 1 + \dfrac{x}{2} - \dfrac{x^2}{8} + \dfrac{x^3}{16}$

(d) $S_4(x) = 1 + \dfrac{x}{2} - \dfrac{x^2}{8} + \dfrac{3x^3}{48} - \dfrac{15x^4}{384}$

$\qquad = 1 + \dfrac{x}{2} - \dfrac{x^2}{8} + \dfrac{x^3}{16} - \dfrac{5x^4}{128}$

5. $f(x) = \dfrac{x}{x+1} = 1 - \dfrac{1}{x+1}$

(a) $S_1(x) = x$ (b) $S_2(x) = x - x^2$ (c) $S_3(x) = x - x^2 + x^3$ (d) $S_4(x) = x - x^2 + x^3 - x^4$

7. $S_1(x) = 1 + \dfrac{x}{2}$

$S_2(x) = 1 + \dfrac{x}{2} + \dfrac{x^2}{8}$

$S_3(x) = 1 + \dfrac{x}{2} + \dfrac{x^2}{8} + \dfrac{x^3}{48}$

$S_4(x) = 1 + \dfrac{x}{2} + \dfrac{x^2}{8} + \dfrac{x^3}{48} + \dfrac{x^4}{384}$

x	0	0.25	0.50	0.75	1.0
$f(x)$	1.0000	1.1331	1.2840	1.4550	1.6487
$S_1(x)$	1.0000	1.1250	1.2500	1.3750	1.5000
$S_2(x)$	1.0000	1.1328	1.2813	1.4453	1.6250
$S_3(x)$	1.0000	1.1331	1.2839	1.4541	1.6458
$S_4(x)$	1.0000	1.1331	1.2840	1.4549	1.6484

9. $\dfrac{1}{1+x^2} = \sum_{n=0}^{\infty} (-1)^n x^{2n}$

(a) $S_2(x) = 1 - x^2$

(c) $S_6(x) = 1 - x^2 + x^4 - x^6$

(b) $S_4(x) = 1 - x^2 + x^4$

(d) $S_8(x) = 1 - x^2 + x^4 - x^6 + x^8$

11. $S_4(x) = 1 - x^2 + x^4$

13. $y = -\frac{1}{2}x^2 + 1$ is a parabola through $(0, 1)$; matches (d).

15. $y = e^{-1/2}[(x+1) + 1]$ is a line; matches (a).

17. $S_6(x) = 1 - x + \dfrac{x^2}{2} - \dfrac{x^3}{6} + \dfrac{x^4}{24} - \dfrac{x^5}{120} + \dfrac{x^6}{720}$

$f\left(\dfrac{1}{2}\right) \approx 1 - \dfrac{1}{2} + \dfrac{1}{8} - \dfrac{1}{48} + \dfrac{1}{384} - \dfrac{1}{3840} + \dfrac{1}{46,080}$

≈ 0.607

19. $f(x) = \ln x,\, c = 2$

$S_6(x) = \ln 2 + \dfrac{1}{2}(x - 2) - \dfrac{1}{8}(x - 2)^2 + \dfrac{1}{24}(x - 2)^3 - \dfrac{1}{64}(x - 2)^4 + \dfrac{1}{160}(x - 2)^5 - \dfrac{1}{384}(x - 2)^6$

$f\left(\dfrac{3}{2}\right) \approx \ln 2 + \dfrac{1}{2}\left(-\dfrac{1}{2}\right) - \dfrac{1}{8}\left(\dfrac{1}{4}\right) + \dfrac{1}{24}\left(-\dfrac{1}{8}\right) - \dfrac{1}{64}\left(\dfrac{1}{16}\right) - \dfrac{1}{160}\left(\dfrac{1}{32}\right) - \dfrac{1}{384}\left(\dfrac{1}{64}\right) = 0.4055$

21. $S_6(x) = 1 - x^2 + \dfrac{x^4}{2} - \dfrac{x^6}{6}$

$\displaystyle\int_0^1 e^{-x^2}\, dx \approx \int_0^1 \left(1 - x^2 + \dfrac{x^4}{2} - \dfrac{x^6}{6}\right) dx$

$= \left[x - \dfrac{x^3}{3} + \dfrac{x^5}{10} - \dfrac{x^7}{42}\right]_0^1 \approx 0.74286$

23. $S_6(x) = 1 - \dfrac{1}{2}x^2 + \dfrac{3}{8}x^4 - \dfrac{5}{16}x^6$

$\displaystyle\int_0^{1/2} \dfrac{1}{\sqrt{1 + x^2}}\, dx \approx \int_0^{1/2} \left(1 - \dfrac{1}{2}x^2 + \dfrac{3}{8}x^4 - \dfrac{5}{16}x^6\right) dx$

$= \left[x - \dfrac{x^3}{6} + \dfrac{3x^5}{40} - \dfrac{5x^7}{112}\right]_0^{1/2}$

$= \dfrac{1}{2} - \dfrac{1}{48} + \dfrac{3}{1280} - \dfrac{5}{14,336} \approx 0.481$

25. Since the $(n + 1)$ derivative of $f(x) = e^x$ is e^x, the maximum value of $|f^{n+1}(x)|$ on the interval $[0, 2]$ is $e^2 < 8$. Therefore, the nth remainder is bounded by

$|R_n| \leq \left| \dfrac{8}{(n + 1)!}(x - 1)^{n+1} \right|, \quad 0 \leq x \leq 2$

$|R_n| \leq \dfrac{8}{(n + 1)!}(1)$

with $n = 7$, $\dfrac{8}{(7 + 1)!} = 1.98 \times 10^{-4} < 0.001$.

Thus, $n = 7$ will approximate e^x with an error less than 0.001.

27. $|R_5| \leq \dfrac{f^{(6)}(z)}{6!}x^6 = \dfrac{e^{-z}}{6!}x^6$

Since $e^{-z} \leq 1$ in the interval $[0, 1]$, it follows that

$R_5 \leq \dfrac{1}{6!} \approx 0.00139$.

29. (a) $\displaystyle\sum_{n=0}^{\infty} P(n) = \sum_{n=0}^{\infty} \left(\dfrac{1}{2}\right)^{n+1} = \sum_{n=0}^{\infty} \dfrac{1}{2}\left(\dfrac{1}{2}\right)^n = \dfrac{1/2}{1 - (1/2)} = 1$

(b) Expected value $= \displaystyle\sum_{n=0}^{\infty} nP(n) = \sum_{n=0}^{\infty} n\left(\dfrac{1}{2}\right)^{n+1} = 1$

(See Example 5.)

(c) The expected daily profit is $\$10(1) = \$10$.

Section 10.6   Newton's Method

1. $x_2 = x_1 - \dfrac{f(x_1)}{f'(x_1)} = 2.2 - \dfrac{(2.2)^2 - 5}{2(2.2)} \approx 2.2364$

3. $f'(x) = 3x^2 + 1$

n	x_n	$f(x_n)$	$f'(x_n)$	$\dfrac{f(x_n)}{f'(x_n)}$	$x_n - \dfrac{f(x_n)}{f'(x_n)}$
1	0.5000	-0.3750	1.7500	-0.2143	0.7143
2	0.7143	0.0787	2.5306	0.0311	0.6832
3	0.6832	0.0021	2.4002	0.0009	0.6823

Approximation: $x \approx 0.682$

5. $f'(x) = \dfrac{5}{2\sqrt{x-1}} - 2$

n	x_n	$f(x_n)$	$f'(x_n)$	$\dfrac{f(x_n)}{f'(x_n)}$	$x_n - \dfrac{f(x_n)}{f'(x_n)}$
1	1.2	-0.1639	3.5902	-0.0457	1.2457
2	1.2457	-0.0131	3.0440	-0.0043	1.2500
3	1.2500	-0.000094	3.0003	-0.00003	1.25

Approximation: $x \approx 1.25$ (exact!)

7. $f'(x) = \dfrac{1}{x} + 1$

n	x_n	$f(x_n)$	$f'(x_n)$	$\dfrac{f(x_n)}{f'(x_n)}$	$x_n - \dfrac{f(x_n)}{f'(x_n)}$
1	0.6000	0.0892	2.1667	0.4120	0.5588
2	0.5588	-0.0231	2.7895	-0.0083	0.5671
3	0.5671	-0.0002	3.7634	-0.0001	0.5672

Approximation: $x \approx 0.567$

9. $f'(x) = -2xe^{-x^2} - 2x = -2x(e^{-x^2} + 1)$

n	x_n	$f(x_n)$	$f'(x_n)$	$\dfrac{f(x_n)}{f'(x_n)}$	$x_n - \dfrac{f(x_n)}{f'(x_n)}$
1	0.8000	-0.1127	-2.4437	0.0461	0.7539
2	0.7539	-0.0019	-2.3619	0.0008	0.7531

Approximations: $x \approx \pm 0.753$

11. $f'(x) = 3x^2 - 27$

n	x_n	$f(x_n)$	$f'(x_n)$	$\dfrac{f(x_n)}{f'(x_n)}$	$x_n - \dfrac{f(x_n)}{f'(x_n)}$
1	-5.0000	-17.0000	48.0000	-0.3542	-4.6458
2	-4.6458	-1.8371	37.7513	-0.0487	-4.5972
3	-4.5972	-0.0329	36.4019	-0.0009	-4.5963

n	x_n	$f(x_n)$	$f'(x_n)$	$\dfrac{f(x_n)}{f'(x_n)}$	$x_n - \dfrac{f(x_n)}{f'(x_n)}$
1	-1.0000	-1.0000	-24.0000	0.0417	-1.0417
2	-1.0417	-0.0053	-23.7448	0.0002	-1.0419

n	x_n	$f(x_n)$	$f'(x_n)$	$\dfrac{f(x_n)}{f'(x_n)}$	$x_n - \dfrac{f(x_n)}{f'(x_n)}$
1	6.0000	27.0000	81.0000	0.3333	5.6667
2	5.6667	1.9630	69.3333	0.0283	5.6384
3	5.6384	0.0136	68.3731	0.0002	5.6382

Approximations: $x \approx -4.596, -1.042, 5.638$

13. Let $h(x) = f(x) - g(x) = 4 - x - \ln x$. Then $h'(x) = -1 - (1/x)$. Starting with $x_1 = 3$, you obtain

$x_1 = 3$

$x_2 = 2.926$

$x_3 = 2.92627$

$x \approx 2.926.$

15. Let $3 - x = 1/(x^2 + 1)$ and define

$$h(x) = \frac{1}{x^2 + 1} + x - 3, \text{ then } h'(x) = -\frac{2x}{(x^2 + 1)^2} + 1.$$

n	x_n	$h(x_n)$	$h'(x_n)$	$\dfrac{h(x_n)}{h'(x_n)}$	$x_n - \dfrac{h(x_n)}{h'(x_n)}$
1	3.0000	0.1000	0.9400	0.1064	2.8936
2	2.8936	0.0003	0.9341	0.0003	2.8933

Approximation: $x \approx 2.893$

17. From the graph we see that the function has one zero and it is in the interval $(11, 12)$. Let $x_1 = 12$.

$$f(x) = \frac{1}{4}x^3 - 3x^2 + \frac{3}{4}x - 2$$

$$f'(x) = \frac{3}{4}x^2 - 6x + \frac{3}{4}$$

$$x_{n+1} = x_n - \frac{f(x_n)}{f'(x_n)}$$

$$= x_n - \frac{(1/4)x_n^3 - 3x_n^2 + (3/4)x_n - 2}{(3/4)x_n^2 - 6x_n + (3/4)}$$

$$= x_n - \frac{x_n^3 - 12x_n^2 + 3x_n - 8}{3x_n^2 - 24x_n + 3}$$

$$= \frac{2x_n^3 - 12x_n^2 + 8}{2x_n^2 - 24x_n + 3}$$

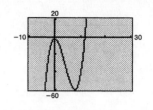

x	1	2	3	4
x_n	12.0000	11.8095	11.8033	11.8033

Zero: $x \approx 11.8033$

19. $f(x) = -x^4 + 5x^2 - 5$

$x \approx \pm 1.9021, \pm 1.1756$

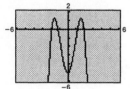

21. $f(x) = x^3 - 3.9x^2 + 4.79x - 1.881$

$x = 0.9, 1.1, 1.9$

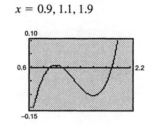

23. $f(x) = 3\sqrt{x - 1} - x$

$1.1459, 7.8541$

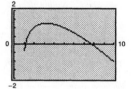

25. $f(x) = x^2 - \ln x - \frac{3}{2}$

$x \approx 0.2359, 1.3385$

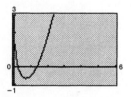

27. $f(x) = x^3 - \cos x$

0.8655

29. Newton's Method fails because $f'(x_1) = 0$.

31. Newton's Method fails because

$$\lim_{x \to \infty} x_n = \begin{cases} 1 = x_1 = x_3 = \dots \\ 0 = x_2 = x_4 = \dots \end{cases}.$$

Therefore, the limit does not exist.

33. Let $f(x) = x^2 - a$, then $f'(x) = 2x$.

$$x_{n+1} = x_n - \frac{x_n^2 - a}{2x_n} = \frac{x_n^2 + a}{2x_n}$$

35. $x_{i+1} = \frac{x_i^2 + 7}{2x_i}$

i	1	2	3	4	5
x_i	2.0000	2.7500	2.6477	2.6458	2.6458

Approximation: $\sqrt{7} \approx 2.646$

37. Let $f(x) = x^4 - 6$, then $f'(x) = 4x^3$.

$$x_{i+1} = x_i - \frac{x_i^4 - 6}{4x_i^3} = \frac{3x_i^4 + 6}{4x_i^3}$$

i	1	2	3	4	5
x_i	2.0000	1.6875	1.5778	1.5652	1.5651

Approximation: $\sqrt[4]{6} \approx 1.565$

39. Let $f(x) = (1/x) - a$, then $f'(x) = -1/x^2$.

$$x_{n+1} = x_n - \frac{(1/x_n) - a}{-1/(x_n^2)} = x_n + (x_n - ax_n^2) = x_n(2 - ax_n)$$

41. The time is given by

$$T = \frac{\sqrt{x^2 + 4}}{3} + \frac{\sqrt{x^2 - 6x + 10}}{4}.$$

To minimize the time, we set dT/dx equal to zero and solve for x. This produces the equation

$$7x^4 - 42x^3 + 43x^2 + 216x - 324 = 0.$$

Let $f(x) = 7x^4 - 42x^3 + 43x^2 + 216x - 324$. Since $f(1) = -100$ and $f(2) = 56$, the solution is in the interval $(1, 2)$.

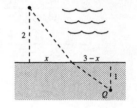

n	x_n	$f(x_n)$	$f'(x_n)$	$\dfrac{f(x_n)}{f'(x_n)}$	$x_n - \dfrac{f(x_n)}{f'(x_n)}$
1	1.7000	19.5887	135.6240	0.1444	1.5556
2	1.5556	−1.0414	150.2782	−0.0069	1.5625
3	1.5629	−0.0092	149.5693	−0.0001	1.5626

Approximation: $x \approx 1.563$ miles

43. To maximize C, we set dC/dt equal to zero and solve for t. This produces

$$C' = -\frac{3t^4 - 2t^3 + 300t + 50}{(50 + t^3)^2} = 0.$$

Let $f(t) = 3t^4 + 2t^3 - 300t - 50$. Since $f(4) = -354$ and $f(5) = 575$, the solution is in the interval $(4, 5)$.

n	t_n	$f(t_n)$	$f'(t_n)$	$\dfrac{f(t_n)}{f'(t_n)}$	$t_n - \dfrac{f(t_n)}{f'(t_n)}$
1	4.5000	12.4375	915.0000	0.0136	4.4864
2	4.4864	0.0658	904.3822	0.0001	4.4863

Approximation: $t \approx 4.486$ hours

45. $P(t) = A(t) - C(t) = 100{,}000e^{0.8\sqrt{t}}e^{-0.10t} - \displaystyle\int_0^t 1000e^{-0.10u}\,du = 100{,}000e^{0.8\sqrt{t}-0.10t} - 1000\int_0^t e^{-0.10u}\,du$

$P'(t) = 100{,}000\left(\dfrac{0.4}{\sqrt{t}} - 0.10\right)e^{0.8\sqrt{t}-0.10t} - 1000e^{-0.10t} = 1000e^{-0.10t}\left[100\left(\dfrac{0.4}{\sqrt{t}} - 0.10\right)e^{0.8\sqrt{t}} - 1\right] = 0$

Let $f(t) = \left(\dfrac{40}{\sqrt{t}} - 10\right)e^{0.8\sqrt{t}} - 1$, then:

$$f'(t) = \left(\frac{40}{\sqrt{t}} - 10\right)\left(\frac{0.4}{\sqrt{t}}e^{0.8\sqrt{t}}\right) + e^{0.8\sqrt{t}}\left(-\frac{20}{t\sqrt{t}}\right) = e^{0.8\sqrt{t}}\left(\frac{16}{t} - \frac{4}{\sqrt{t}} - \frac{20}{t\sqrt{t}}\right)$$

$$t_{n+1} = t_n - \frac{f(t_n)}{f'(t_n)}$$

n	1	2	3	4
t_n	16	15.8696	15.8686	15.8686

The timber should be harvested in 15.8686 years.

47. False. let $f(x) = \dfrac{x^2 - 1}{x - 1}$.

49. True

Review Exercises for Chapter 10

1. $a_n = \left(-\dfrac{1}{3}\right)^n$: $-\dfrac{1}{3}, \dfrac{1}{9}, -\dfrac{1}{27}, \dfrac{1}{81}, -\dfrac{1}{243}$

3. $a_n = \dfrac{4^n}{n!}$: $4, 8, 10\frac{2}{3}, 10\frac{2}{3}, \dfrac{128}{15}$

5. The sequence converges since
$$\lim_{n \to \infty} \frac{2n + 3}{n^2} = 0.$$

7. The sequence diverges since
$$\lim_{n \to \infty} \frac{n^3}{n^2 + 1} = \infty.$$

9. The sequence converges since
$$\lim_{n \to \infty} \left(5 + \frac{1}{3^n}\right) = 5 + 0 = 5.$$

11. The sequence converges since
$$\lim_{n \to \infty} \frac{1}{n^{4/3}} = 0.$$

13. $a_n = \dfrac{n}{3n}$ or $a_n = \dfrac{1}{3}$, $n \neq 0$

15. $a_n = (-1)^{n-1}\left(\dfrac{2^{n-1}}{3^n}\right)$, $n = 1, 2, 3, \ldots$

OR

$a_n = (-1)^n\left(\dfrac{2^n}{3^{n+1}}\right)$, $n = 0, 1, 2, \ldots$

17. (a) $a_1 = 15{,}000$

$a_2 = 15{,}000 + 10{,}000$

$\vdots$

$a_n = 15{,}000 + 10{,}000(n - 1)$

(b) $a_1 + a_2 + a_3 + a_4 + a_5 = \displaystyle\sum_{n=1}^{15} [15{,}000 + 10{,}000(n - 1)]$

$= (15{,}000)5 + 100{,}000$

$= \$175{,}000$

19. Beginning: 1
Year 1: $1 + 0.07 = 1.07$
Year 2: $1.07 + 1.07(0.07) = 1.1449$
Year 3: 1.2250
Year 4: 1.3108
Year 5: 1.4026
Year 6: 1.5007
Year 7: 1.6058
Year 8: 1.7182
Year 9: 1.8385
Year 10: 1.9672

21. $S_0 = 1$

$S_1 = 1 + \dfrac{3}{2} = \dfrac{5}{2} = 2.5$

$S_2 = 1 + \dfrac{3}{2} + \dfrac{9}{4} = \dfrac{19}{4} = 4.75$

$S_3 = 1 + \dfrac{3}{2} + \dfrac{9}{4} + \dfrac{27}{8} = \dfrac{65}{8} = 8.125$

$S_4 = 1 + \dfrac{3}{2} + \dfrac{9}{4} + \dfrac{27}{8} + \dfrac{81}{16} = \dfrac{211}{16} = 13.1875$

23. $S_1 = \dfrac{1}{2!} = \dfrac{1}{2} = 0.5$

$S_2 = \dfrac{1}{2!} - \dfrac{1}{4!} = \dfrac{11}{24} \approx 0.4583$

$S_3 = \dfrac{1}{2!} - \dfrac{1}{4!} + \dfrac{1}{6!} = \dfrac{331}{720} \approx 0.4597$

$S_4 = \dfrac{1}{2!} - \dfrac{1}{4!} + \dfrac{1}{6!} - \dfrac{1}{8!} = \dfrac{18{,}535}{40{,}320} \approx 0.4597$

$S_5 = \dfrac{1}{2!} - \dfrac{1}{4!} + \dfrac{1}{6!} - \dfrac{1}{8!} + \dfrac{1}{10!} = \dfrac{1{,}668{,}151}{3{,}628{,}800} \approx 0.4597$

25. This series diverges by the nth-Term Test since
$$\lim_{n \to \infty} \frac{n^2 + 1}{n(n + 1)} = 1 \neq 0.$$

27. This geometric series converges since $r = 0.25 < 1$.

29. $\lim\limits_{n\to\infty} \dfrac{2n}{n+5} = 2 \neq 0$, and the series diverges.

31. $\lim\limits_{n\to\infty} \left(\dfrac{5}{4}\right)^n = \infty \neq 0$, and the series diverges.

33. $S_N = \sum\limits_{k=0}^{N} \left(\dfrac{1}{5}\right)^k = \dfrac{1-(1/5)^{N+1}}{1-(1/5)} = \dfrac{5}{4}\left[1-\left(\dfrac{1}{5}\right)^{N+1}\right]$

35. $S_N = \sum\limits_{k=0}^{N} \left(\dfrac{1}{2^k}\right) + \sum\limits_{k=0}^{N} \dfrac{1}{4^k}$

$$= \dfrac{1-(1/2)^{N+1}}{1-(1/2)} + \dfrac{1-(1/4)^N}{1-(1/4)}$$

$$= 2\left[1-\left(\dfrac{1}{2}\right)^{N+1}\right] + \dfrac{4}{3}\left[1-\left(\dfrac{1}{4}\right)^{N+1}\right]$$

37. $\sum\limits_{n=0}^{\infty} \dfrac{1}{4}(4^n)$ diverges since $\lim\limits_{n\to\infty} a_n \neq 0$.

39. $\sum\limits_{n=0}^{\infty} [(0.5)^n + (0.2)^n] = \dfrac{1}{1-0.5} + \dfrac{1}{1-0.2}$

$$= 2 + \dfrac{5}{4} = \dfrac{13}{4}$$

41. (a) $D_1 = 8$

$D_2 = 0.7(8) + 0.7(8) = 16(0.7)$

$D_3 = 16(0.7)^2$

$D = -8 + 16 + 16(0.7) + 16(0.7)^2 + \cdots$

(b) $D = -8 + \sum\limits_{n=0}^{\infty} 16(0.7)^n$

$$= -8 + \dfrac{16}{1-0.7}$$

$$= -8 + \dfrac{160}{3} = \dfrac{136}{3} \text{ feet}$$

43. $500{,}000\left(1+\dfrac{i}{4}\right)^{4(25)} = 1{,}000{,}000$

$$\left(1+\dfrac{i}{4}\right)^{100} = 2$$

$$1+\dfrac{i}{4} = 2^{1/100}$$

$$i = 4[2^{0.01}-1] = 0.02782 \text{ or } 2.782\%$$

45. $\sum\limits_{n=1}^{\infty} \dfrac{1}{n^4}$ converges by the p-series test since $p = 4 > 1$.

47. $\sum\limits_{n=1}^{\infty} \dfrac{1}{n\sqrt[4]{n}} = \sum\limits_{n=1}^{\infty} \dfrac{1}{n^{5/4}}$ converges by the p-series test since $p = \dfrac{5}{4} > 1$.

49. $6, 5, 4\frac{2}{3}, \ldots$

Matches (a).

51. $10, 3, \ldots$

Matches (d).

53. $\sum\limits_{n=1}^{\infty} \dfrac{1}{n^6} \approx 1 + \dfrac{1}{2^6} + \dfrac{1}{3^6} + \dfrac{1}{4^6} \approx 1.0172$

error $< \dfrac{1}{(5)4^5} \approx 1.9531 \times 10^{-4}$

55. $\sum\limits_{n=1}^{\infty} \dfrac{1}{n^{5/4}} \approx 1 + \dfrac{1}{2^{5/4}} + \cdots + \dfrac{1}{6^{5/4}} \approx 2.09074$

error $< \dfrac{1}{(1/4)(6)^{1/4}} < 2.5558$

57. This series converges by the Ratio Test since

$$\lim\limits_{n\to\infty} \left|\dfrac{(n+1)4^{n+1}}{(n+1)!} \cdot \dfrac{n!}{n4^n}\right| = \lim\limits_{n\to\infty} \dfrac{4}{n} = 0 < 1.$$

59. This series diverges by the Ratio Test since

$$\lim\limits_{n\to\infty} \left|\dfrac{(-1)^{n+1}3^{n+1}}{n+1} \cdot \dfrac{n}{(-1)^n 3^n}\right| = \lim\limits_{n\to\infty} \left|3\dfrac{n}{n+1}\right|$$

$$= 3 > 1.$$

61. This series converges by the Ratio Test since

$$\lim\limits_{n\to\infty} \left|\dfrac{(n+1)2^{n+1}}{(n+1)!} \cdot \dfrac{n!}{n2^n}\right| = \lim\limits_{n\to\infty} \dfrac{2}{n} = 0 < 1.$$

63. $\displaystyle\lim_{n\to\infty}\left|\frac{(-1)^{n+1}(x-2)^{n+1}/(n+2)^2}{(-1)^n(x-2)^n/(n+1)^2}\right| = \lim_{n\to\infty}\left|\frac{(n+1)^2}{(n+2)^2}(x-2)\right| = |x-2| < 1 \Rightarrow R = 1$

65. $\displaystyle\lim_{n\to\infty}\left|\frac{(n+1)!(x-3)^{n+1}}{n!(x-3)^n}\right| = \lim_{n\to\infty}|(n+1)(x-3)| = \infty \Rightarrow R = 0$

67. $f(x) = e^{-0.5x}$ $\qquad f(0) = 1$

$\quad f'(x) = -\dfrac{1}{2}e^{-1/2x}$ $\qquad f'(0) = -\dfrac{1}{2}$

$\quad f''(x) = \dfrac{1}{4}e^{-1/2x}$ $\qquad f''(0) = \dfrac{1}{4}$

$\qquad\qquad\qquad\vdots$

$\qquad\qquad\qquad f^{(n)}(0) = \left(-\dfrac{1}{2}\right)^n$

$e^{-0.5x} = 1 - \dfrac{1}{2}x + \dfrac{1}{4}\cdot\dfrac{x^2}{2} - \dfrac{1}{8}\cdot\dfrac{x^3}{3!} + \cdots$

$\qquad = \displaystyle\sum_{n=0}^{\infty}\left(-\dfrac{1}{2}\right)^n\dfrac{x^n}{n!}$

69. $f(x) = \dfrac{1}{x}$ $\qquad f(-1) = -1$

$\qquad\qquad\qquad\qquad f'(-1) = -1$

$\quad f'(x) = -\dfrac{1}{x^2}$ $\qquad f''(-1) = -2$

$\qquad\qquad\qquad\qquad f'''(-1) = -6$

$\quad f''(x) = \dfrac{2}{x^3}$ $\qquad\qquad\vdots$

$\qquad\qquad\qquad\qquad f^{(n)}(-1) = -(n!)$

$\quad f'''(x) = -\dfrac{6}{x^4}$

The power series for f is

$\dfrac{1}{x} = f(-1) + f'(-1)(x+1) + \dfrac{f''(-1)(x+1)^2}{2!} + \cdots$

$\quad = -1 - (x+1) - \dfrac{2(x+1)^2}{2!} - \dfrac{6(x+1)^3}{3!} - \cdots$

$\quad = -[1 + (x+1) + (x+1)^2 + (x+1)^3 + \cdots]$

$\quad = -\displaystyle\sum_{n=0}^{\infty}(x+1)^n.$

71. $\ln(x+2) = \ln\left[2\left(\dfrac{x}{2}+1\right)\right]$

$\qquad\qquad = \ln 2 + \ln\left(\dfrac{x}{2}+1\right)$

$\qquad\qquad = \ln 2 + \dfrac{1}{2}x - \dfrac{1}{8}x^2 + \dfrac{1}{24}x^3 - \dfrac{1}{64}x^4 + \cdots$

$\qquad\qquad = \ln 2 + \displaystyle\sum_{n=1}^{\infty}(-1)^{n+1}\dfrac{(x/2)^2}{n}$

73. $(1+x^2)^2 = 1 + 2x^2 + \dfrac{2(1)x^4}{2!} + \cdots$

$\qquad\qquad = 1 + 2x^2 + x^4 + \cdots$

75. $f(x) = x^2e^x$

$\qquad = x^2\left[1 + x + \dfrac{x^2}{2!} + \cdots\right]$

$\qquad = \displaystyle\sum_{n=0}^{\infty}\dfrac{x^{n+2}}{n!}$

77. $f(x) = \dfrac{x^2}{x+1}$

$\qquad = x^2[1 - x + x^2 - x^3 + \cdots]$

$\qquad = \displaystyle\sum_{n=0}^{\infty}(-1)^nx^{n+2}$

79. $\dfrac{1}{(x+3)^2} \approx \dfrac{1}{9} - \dfrac{2}{27}x + \dfrac{1}{27}x^2 - \dfrac{4}{243}x^3 + \dfrac{5}{729}x^4 - \dfrac{2}{729}x^5 + \dfrac{7}{6561}x^6$

81. $\ln(x+2) \approx \ln 3 + \dfrac{1}{3}(x-1) - \dfrac{1}{18}(x-1)^2 + \dfrac{1}{81}(x-1)^3 - \dfrac{1}{324}(x-1)^4 + \dfrac{1}{1215}(x-1)^5 - \dfrac{1}{4374}(x-1)^6$

83. $f(1.25) \approx 4.770479903$

85. $f(1.5) \approx 0.9162835738$

87. error $= R_n = \dfrac{f^{(n+1)}(z)}{(n+1)!}(x - c)^{n+1}$

$$f(x) = \frac{2}{x} \implies f^{(6)}(x) = \frac{1440}{x^7} \le 1440 \text{ on } \left[1, \frac{3}{2}\right].$$

$$R_n \le \frac{1440}{6!}(x - 1)^6 < 2\left(\frac{1}{2}\right)^6 = \frac{1}{32}$$

89. $\sqrt{1 + x^3} = 1 + \dfrac{x^3}{2} - \dfrac{x^6}{8} + \cdots$

$$\int_0^{0.3} \sqrt{1 + x^3}\, dx \approx \left[x + \frac{x^4}{8} - \frac{x^7}{56}\right]_0^{0.3}$$

$$= 0.3 + \frac{(0.3)^4}{8} - \frac{(0.3)^7}{56} \approx 0.301$$

91. $\ln(x^2 + 1) = x^2 - \dfrac{1}{2}x^4 + \dfrac{1}{3}x^6 - \cdots$

$$\int_0^{0.75} \ln(x^2 + 1)\, dx \approx \int_0^{0.75} \left(x^2 - \frac{1}{2}x^4 + \frac{1}{3}x^6\right) dx = \left[\frac{x^3}{3} - \frac{x^5}{10} + \frac{x^7}{21}\right]_0^{0.75} \approx 0.12325$$

93. Expected value $= \displaystyle\sum_{n=0}^{\infty} nP(n) = \sum_{n=0}^{\infty} 2n\left(\frac{1}{3}\right)^{n+1}$

$$= 2\left[0\left(\frac{1}{3}\right) + 1\left(\frac{1}{3}\right)^2 + 2\left(\frac{1}{3}\right)^3 + \cdots\right]$$

Since $(1 - x)^{-2} = 1 + 2x + 3x^2 + 4x^3 + \cdots$ (binomial series),

$$\left(1 - \frac{1}{3}\right)^{-2} = 1 + 2\left(\frac{1}{3}\right) + 3\left(\frac{1}{3}\right)^2 + 4\left(\frac{1}{3}\right)^3 + \cdots = \frac{9}{4}$$

and

$$\text{Expected value} = 2\left[1\left(\frac{1}{3}\right)^2 + 2\left(\frac{1}{3}\right)^3 + 3\left(\frac{1}{3}\right)^4 + \cdots\right]$$

$$= 2\left(\frac{1}{3}\right)^2\left[1 + 2\left(\frac{1}{3}\right) + 3\left(\frac{1}{3}\right)^2 + \cdots\right]$$

$$= 2\left(\frac{1}{9}\right)\left(\frac{9}{4}\right) = \frac{1}{2}$$

Expected production cost $= (23.00)\left(\dfrac{1}{2}\right) = \$11.50.$

95. $f(x) = 2x^3 + 3x - 1$

$f'(x) = 6x^2 + 3$

$x_{n+1} = x_n - \dfrac{f(x_n)}{f'(x_n)} = x_n - \dfrac{2x_n^3 + 3x_n - 1}{6x_n^3 + 3}$

$x_1 = 0$ (initial guess)

$x_2 = \dfrac{1}{3}$

$x_3 = 0.31\overline{31}$

$x_4 = 0.3129$

$x \approx 0.313$

97. $f(x) = \ln 3x + x$

$f'(x) = \dfrac{1}{x} + 1$

$x_{n+1} = x_n - \dfrac{f(x_n)}{f'(x_n)}$

$x_1 = 0.5$ (initial guess)

$x_2 = 0.1982$

$x_3 = 0.2514$

$x_4 = 0.2576$

$x \approx 0.258$

99. Let

$$h(x) = f(x) - g(x) = x^5 - (x + 3)$$

$$h'(x) = 5x^4 - 1.$$

$x_1 = 1.5$ (initial guess)

$x_2 = 1.37275$

$x_3 = 1.34279$

$x_4 = 1.3413$

$x \approx 1.341$

101. Let

$$h(x) = f(x) - g(x) = x^3 - e^{-x}$$

$$h'(x) = 3x^2 + e^{-x}.$$

$x_1 = 1$ (initial guess)

$x_2 = 0.8123$

$x_3 = 0.7743$

$x_4 = 0.7729$

$x \approx 0.773$

Practice Test for Chapter 10

1. Find the general term of the sequence $\frac{1}{2}, \frac{2}{5}, \frac{3}{10}, \frac{4}{17}, \frac{5}{26}, \ldots \ldots$

2. Find the general term of the sequence $5, -7, 9, -11, 13, \ldots \ldots$

3. Determine the convergence or divergence of the sequence whose general term is $a_n = \dfrac{n^2}{3n^2 + 4}$.

4. Determine the convergence or divergence of the sequence whose general term is $a_n = \dfrac{4n}{\sqrt{n^2 + 1}}$.

5. Find the sum of the series $\displaystyle\sum_{n=0}^{\infty} \left(\frac{1}{5^n} - \frac{1}{7^n} \right)$.

6. Determine the convergence or divergence of the series $\displaystyle\sum_{n=1}^{\infty} \frac{3^n}{n!}$.

7. Determine the convergence or divergence of the series $\displaystyle\sum_{n=1}^{\infty} \frac{1}{n\sqrt[3]{n}}$.

8. Determine the convergence or divergence of the series $\displaystyle\sum_{n=1}^{\infty} \frac{n}{2n + 3}$.

9. Determine the convergence or divergence of the series $\displaystyle\sum_{n=0}^{\infty} \frac{(-1)^n 6^n}{5^n}$.

10. Determine the convergence or divergence of the series $\displaystyle\sum_{n=1}^{\infty} \frac{\sqrt[3]{n}}{\sqrt{n}}$.

11. Determine the convergence or divergence of the series $\displaystyle\sum_{n=1}^{\infty} \frac{5^n n!}{(n + 1)!}$.

12. Determine the convergence or divergence of the series $\displaystyle\sum_{n=0}^{\infty} 4(0.27)^n$.

13. Determine the convergence or divergence of the series $\displaystyle\sum_{n=1}^{\infty} \left(1 + \frac{1}{3^n} \right)$.

14. Find the radius of convergence of the power series $\displaystyle\sum_{n=0}^{\infty} \frac{(-1)^n (x - 3)^n}{(n + 4)^2}$.

15. Find the radius of convergence of the power series $\displaystyle\sum_{n=0}^{\infty} \frac{x^n}{(n + 1)!}$.

16. Apply Taylor's Theorem to find the power series (centered at 0) for $f(x) = e^{-4x}$.

17. Apply Taylor's Theorem to find the power series (centered at 1) for $f(x) = \dfrac{1}{\sqrt[3]{x}}$.

18. Use the ninth-degree Taylor polynomial for e^{x^3} to approximate the value of $\displaystyle\int_0^{0.213} e^{x^3}dx$.

19. Use Newton's Method to approximate the zero of the function $f(x) = x^3 + x - 3$. (Make your approximation good to three decimal places.)

20. Use Newton's Method to approximate $\sqrt[4]{10}$ to three decimal places.

Graphing Calculator Required

21. Use SUM SEQ to evaluate the following sums.

   (a) $\displaystyle\sum_{n=0}^{10} \frac{4}{2^n}$ (b) $\displaystyle\sum_{n=1}^{8} 3n!$

22. Graph the function $y = e^{2x}$ as well as the Taylor polynomials of degree 2, 4, and 6 on the same set of axes.

23. Use a program similar to the one on page 703 in the textbook to approximate the real roots of $3x^4 - 2x^3 + 5x^2 + 6x - 10 = 0$.

APPENDICES

APPENDIX A
Alternative Introduction to the Fundamental Theorem of Calculus

Solutions to Odd-Numbered Exercises

1. Left Riemann sum: 0.518
Right Riemann sum: 0.768

3. Left Riemann sum: 0.746
Right Riemann sum: 0.646

5. Left Riemann sum: 0.859
Right Riemann sum: 0.659

7. Midpoint Rule: 0.673

9. (a)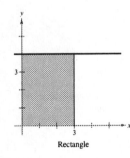

(b) Answers will vary.

(c) Answers will vary.

(d) Answers will vary.

(e)

n	5	10	50	100
Left sum, S_L	1.6	1.8	1.96	1.98
Right sum, S_R	2.4	2.2	2.04	2.02

11. $\displaystyle\int_0^5 3\, dx$

13. $\displaystyle\int_{-4}^4 (4 - |x|)\, dx = \int_{-4}^0 (4 + x)\, dx + \int_0^4 (4 - x)\, dx$

15. $\displaystyle\int_{-2}^2 (4 - x^2)\, dx$

17. $\displaystyle\int_0^2 \sqrt{x + 1}\, dx$

19. $A = 12$

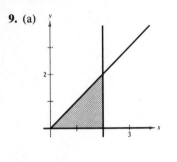
Rectangle

21. $A = 8$

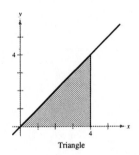
Triangle

23. $A = 14$

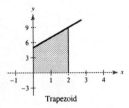
Trapezoid

25. $A = 1$

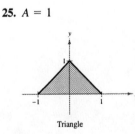
Triangle

27. $A = \dfrac{9\pi}{2}$

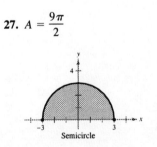
Semicircle

29. $S = 1 + 2 + 3 + \cdots + (n - 1) + n$

$\quad S = n + (n - 1) + \cdots + 3 + 2 + 1$

Adding,

$\quad 2S = (n + 1) + (n + 1) + \cdots + (n + 1)$ (n times)

$\quad 2S = n(n + 1)$

$\quad S = \dfrac{n(n + 1)}{2}$

31. $\displaystyle\sum_{i=1}^{n} f(x_i)\,\Delta x > \int_{1}^{5} f(x)\,dx$

A P P E N D I X   C
Differential Equations

Section C.1   Solutions of Differential Equations

Solutions to Odd-Numbered Exercises

1. $y' = 3x^2$

3. $y' = -2e^{-2x}$ and $y' + 2y = -2e^{-2x} + 2(e^{-2x}) = 0$

5. $y' = 6x^2$ and $y' - \dfrac{3}{x}y = 6x^2 - \dfrac{3}{x}(2x^3) = 0$

7. $y'' = 2$ and $x^2y'' - 2y = x^2(2) - 2(x^2) = 0$

9. $y' = 4e^{2x}$

$\quad y'' = 8e^{2x}$ and $y'' - y' - 2y = 8e^{2x} - 4e^{2x} - 2(2x^{2x}) = 0$

11. By differentiation, we have

$\quad \dfrac{dy}{dx} = -\dfrac{1}{x^2}.$

13. By differentiating, we have

$\quad \dfrac{dy}{dx} = 4Ce^{4x} = 4y.$

15. Since $\dfrac{dy}{dt} = -\left(\dfrac{1}{3}\right)Ce^{-t/3}$, we have $3\dfrac{dy}{dt} + y - 7 = 3\left(-\dfrac{1}{3}Ce^{-t/3}\right) + (Ce^{-t/3} + 7) - 7 = 0.$

17. Since $y' = 2Cx - 3$, we have $xy' - 3x - 2y = x(2Cx - 3) - 3x - 2(Cx^2 - 3x) = 0.$

19. Since $y' = 2x + 2 - (C/x^2)$, we have $xy' + y = \left(2x^2 + 2x - \dfrac{C}{x}\right) + \left(x^2 + 2x + \dfrac{C}{x}\right) = 3x^2 + 4x = x(3x + 4).$

21. Since $y' = \tfrac{1}{2}C_1e^{x/2} - 2C_2e^{-2x}$, we have $y'' = \tfrac{1}{4}C_1e^{x/2} + 4C_2e^{-2x}$, and it follows that

$\quad 2y'' + 3y' - 2y = \tfrac{1}{2}C_1e^{x/2} + 8C_2e^{-2x} + \tfrac{3}{2}C_1e^{x/2} - 6C_2e^{-2x} - 2C_1e^{x/2} - 2C_2e^{-2x} = 0.$

23. Since $y' = 4bx^3/(4 - a) + aCx^{a-1}$, we have

$\quad y' - \dfrac{ay}{x} = \left[\dfrac{4bx^3}{4 - a} + aCx^{a-1}\right] - \dfrac{a}{x}\left[\dfrac{bx^4}{4 - a} + Cx^a\right]$

$\quad\qquad = \dfrac{4bx^3}{4 - a} + aCx^{a-1} - \dfrac{abx^3}{4 - a} - aCx^{a-1} = \dfrac{bx^3(4 - a)}{4 - a} = bx^3.$

25. Since $y' = -2(1 + Ce^{x^2})^{-2}2xCe^{x^2} = -\dfrac{4xCe^{x^2}}{(1 + Ce^{x^2})^2}$, we have

$$y' + 2xy = -\frac{4xCe^{x^2}}{(1 + Ce^{x^2})^2} + \frac{4x}{1 + Ce^{x^2}} = \frac{-4xCe^{x^2} + 4x + 4xCe^{x^2}}{(1 + Ce^{x^2})^2} = x\left(\frac{2}{1 + Ce^{x^2}}\right)^2 = xy^2$$

27. Since $y' = \ln x + 1 + C$, we have $x(y' - 1) - (y - 4) = x(\ln x + 1 + C - 1) - (x \ln x + Cx + 4 - 4) = 0$.

29. By implicit differentiation, we have $2x + 2yy' = Cy'$, which implies that $2x = y'(C - 2y)$ and

$$y' = \frac{2x}{C - 2y} = \frac{2xy}{Cy - 2y^2} = \frac{2xy}{(x^2 + y^2) - 2y^2} = \frac{2xy}{x^2 - y^2}.$$

31. $\quad 2x + y + xy' = 0 \Rightarrow y' = \dfrac{-2x - y}{x} = -2 - \dfrac{y}{x}$

$$y'' = -\frac{y'}{x} + \frac{y}{x^2} = \frac{2}{x} + \frac{2y}{x^2}$$

$$x^2y'' - 2(x + y) = x^2\left(\frac{2}{x} + \frac{2y}{x^2}\right) - 2(x + y) = 0$$

33. $\quad y' = -2e^{-2x}$

$\quad y'' = 4e^{-2x}$

$\quad y''' = -8e^{-2x}$

$\quad y^{(4)} = 16e^{-2x}$

Therefore, we have $y^{(4)} - 16y = 16e^{-2x} - 16(e^{-2x}) = 0$.

35. $\quad y = 4x^{-1}$

$\quad y' = -4x^{-2}$

$\quad y'' = 8x^{-3}$

$\quad y''' = -24x^{-4}$

$\quad y^{(4)} = 96x^{-5}$

Therefore, we have $y^{(4)} - 16y = 96x^{-5} - 16(4x^{-1}) \neq 0$ and y is not a solution of the given differential equation.

37. $\quad y = \frac{2}{9}xe^{-2x}$

$\quad y' = -\frac{4}{9}xe^{-2x} + \frac{2}{9}e^{-2x}$

$\quad y'' = \frac{8}{9}xe^{-2x} - \frac{8}{9}e^{-2x}$

$\quad y''' = -\frac{16}{9}xe^{-2x} + \frac{24}{9}e^{-2x}$

Therefore, $y''' - 3y' + 2y = \left(-\frac{16}{9}xe^{-2x} + \frac{24}{9}e^{-2x}\right) - 3\left(-\frac{4}{9}xe^{-2x} + \frac{2}{9}e^{-2x}\right) + 2\left(\frac{2}{9}xe^{-2x}\right) = 2e^{-2x}$.

This is *not* a solution to $y''' - 3y' + 2y = 0$.

39. $\quad y = xe^x$

$\quad y' = xe^x + e^x$

$\quad y'' = xe^x + 2e^x$

$\quad y''' = xe^x + 3e^x$

Therefore, $y''' - 3y' + 2y = (xe^x + 3e^x) - 3(xe^x + e^x) + 2(xe^x) = 0$.

This *is* a solution to $y''' - 3y' + 2y = 0$.

41. Since $y' = -2Ce^{-2x} = -2y$, it follows that $y' + 2y = 0$. To find the particular solution, we use the fact that $y = 3$ when $x = 0$. That is, $3 = Ce^0 = C$. Thus, $C = 3$ and the particular solution is $y = 3e^{-2x}$.

43. Since $y' = C_2(1/x)$ and $y'' = -C_2(1/x^2)$, it follows that $xy'' + y' = 0$. To find the particular solution, we use the fact that $y = 5$ and $y' = 1/2$ when $x = 1$. That is,

$$\frac{1}{2} = C_2\frac{1}{1} \qquad \Rightarrow \qquad C_2 = \frac{1}{2}$$

$$5 = C_1 + \frac{1}{2}(0) \qquad \Rightarrow \qquad C_1 = 5$$

Thus, the particular solution is

$$y = 5 + \frac{1}{2}\ln|x| = 5 + \ln\sqrt{|x|}.$$

45. Since $y' = 4C_1e^{4x} - 3C_2e^{-3x}$, and $y'' = 16C_1e^{4x} + 9C_2e^{-3x}$, it follows that $y'' - y' - 12y = 0$. To find the particular solution, we use the fact that $y = 5$ and $y' = 6$ when $x = 0$. That is,

$$C_1 + C_2 = 5$$

$$4C_1 - 3C_2 = 6$$

which implies that $C_1 = 3$ and $C_2 = 2$. The particular solution is

$$y = 3e^{4x} + 2e^{-3x}$$

47. Since

$$y' = e^{2x/3}\left(\tfrac{2}{3}C_1 + \tfrac{2}{3}C_2x + C_2\right)$$

$$y'' = e^{2x/3}\left(\tfrac{4}{9}C_1 + \tfrac{4}{9}C_2x + \tfrac{4}{3}C_2\right)$$

it follows that $9y'' - 12y' + 4y = 0$. To find the particular solution, we use the fact that $y = 4$ when $x = 0$, and $y = 0$ when $x = 3$. That is,

$$4 = e^0[C_1 + C_2(0)] \quad \Longrightarrow \quad C_1 = 4$$

$$0 = e^2[4 + C_2(3)] \quad \Longrightarrow \quad C_2 = -\tfrac{4}{3}$$

Therefore, the particular solution is $y = e^{2x/3}\left(4 - \tfrac{4}{3}x\right) = \tfrac{4}{3}e^{2x/3}(3 - x)$.

49. When $C = 1$, the graph is a parabola $y = x^2$.

When $C = 2$, the graph is a parabola $y = 2x^2$.

When $C = 4$, the graph is a parabola $y = 4x^2$.

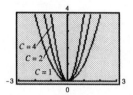

51. When $C = 0$, the graph is a straight line.

When $C = 1$, the graph is a parabola opening upward with a vertex at $(-2, 0)$.

When $C = -1$, the graph is a parabola opening downward with a vertex at $(-2, 0)$.

When $C = 2$, the graph is a parabola opening upward with a vertex at $(-2, 0)$.

When $C = -2$, the graph is a parabola opening downward with a vertex at $(-2, 0)$.

53. $y = \displaystyle\int 3x^2\, dx = x^3 + C$

55. $y = \displaystyle\int \frac{x + 3}{x}\, dx = \int\left(1 + \frac{3}{x}\right) dx = x + 3\ln|x| + C$

57. By using partial fractions:

$$\frac{1}{x^2 - 1} = \frac{A}{x - 1} + \frac{B}{x + 1}$$

$$1 = A(x + 1) + B(x - 1)$$

When $x = 1$: $1 = 2A \Longrightarrow A = \tfrac{1}{2}$.

When $x = -1$: $1 = -2B \Longrightarrow B = -\tfrac{1}{2}$.

We have $y = \displaystyle\int \frac{1}{x^2 - 1}\, dx = \frac{1}{2}\int\left[\frac{1}{x - 1} - \frac{1}{x + 1}\right] dx = \frac{1}{2}[\ln|x - 1| - \ln|x + 1|] + C = \frac{1}{2}\ln\left|\frac{x - 1}{x + 1}\right| + C$.

59. Letting $u = x - 3$, we have the following.

$$y = \int x\sqrt{x - 3}\, dx = \int (u + 3)u^{1/2}\, du = \int (u^{3/2} + 3u^{1/2})\, du$$

$$= \tfrac{2}{5}u^{5/2} + 2u^{3/2} + C = \tfrac{2}{5}u^{3/2}(u + 5) + C = \tfrac{2}{5}(x - 3)^{3/2}(x + 2) + C$$

61. Since $y = 4$ when $x = 4$, we have $4^2 = C4^3$ which implies that $C = \frac{1}{4}$ and the particular solutions is $y^2 = x^3/4$.

63. Since $y = 3$ when $x = 0$, we have $3 = Ce^0$ which implies that $C = 3$ and the particular solution is $y = 3e^x$.

65. (a) Since $N = 100$ when $t = 0$, it follows that $C = 650$. Therefore, the population function is $N = 750 - 650e^{-kt}$. Moreover, since $N = 160$ when $t = 2$, it follows that

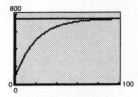

$$160 = 750 - 650e^{-2k}$$

$$e^{-2k} = \frac{59}{65}$$

$$k = -\frac{1}{2}\ln\frac{59}{65} \approx -0.0484$$

Thus, the population function is $N = 750 - 650e^{-0.0484t}$.

(b) See accompanying graph.

(c) When $t = 4$, $N \approx 214$.

67. $x = 30,000 - Ce^{-kt}$. Since the product is new, $x = 0$ when $t = 0 \Rightarrow C = 30,000$. When $t = 1$, $x = 2000 \Rightarrow$

$$2000 = 30,000 - 30,000e^{-k}$$

$$\frac{1}{15} = 1 - e^{-k}$$

$$e^{-k} = \frac{14}{15}$$

$$k = -\ln\left(\frac{14}{15}\right) \approx 0.06899$$

$$x = 30,000 - 30,000e^{-0.06899t}$$

Year, t	2	4	6	8	10
Units, x	3867	7235	10,169	12,725	14,951

69. Since $\dfrac{ds}{dh} = -\dfrac{13}{\ln 3}\left(\dfrac{1/2}{h/2}\right) = -\dfrac{13}{\ln 3}\dfrac{1}{h}$ and $-13/\ln 3$ is a constant, we can conclude that the equation is a solution to

$$\dfrac{ds}{dh} = \dfrac{k}{h} \text{ where } k = -\dfrac{13}{\ln 3}.$$

71. $R = 0.07$

C cannot be determined.

73. False. From Example 1, $y = e^x$ is a solution of $y'' - y = 0$, but $y = e^x + 1$ is not.

Section C.2 Separation of Variables

1. Yes, $\dfrac{dy}{dx} = \dfrac{x}{y+3}$

$$(y+3)\,dy = x\,dx$$

3. Yes, $\dfrac{dy}{dx} = \dfrac{1}{x} + 1$

$$dy = \left(\dfrac{1}{x} + 1\right)dx$$

5. No, the variables cannot be separated.

7. $\dfrac{dy}{dx} = 2x$

$$\int dy = \int 2x\,dx$$

$$y = x^2 + C$$

9. $3y^2\dfrac{dy}{dx} = 1$

$$\int 3y^2\,dy = \int dx$$

$$y^3 = x + C$$

$$y = \sqrt[3]{x + C}$$

11. $(y+1)\dfrac{dy}{dx} = 2x$

$$\int (y+1)\,dy = \int 2x\,dx$$

$$\dfrac{(y+1)^2}{2} = x^2 + C_1$$

$$C = 2x^2 - (y+1)^2$$

13. $\dfrac{dy}{dx} = xy$

$$\int \dfrac{1}{y}\,dy = \int x\,dx$$

$$\ln|y| = \dfrac{1}{2}x^2 + C_1$$

$$y = e^{(x^2/2)+C_1} = e^{C_1}e^{x^2/2} = Ce^{x^2/2}$$

15. $\dfrac{dy}{dt} = \dfrac{e^t}{4y}$

$$\int 4y\,dy = \int e^t\,dt$$

$$2y^2 = e^t + C_1$$

$$y^2 = \dfrac{1}{2}e^t + C$$

17. $\dfrac{dy}{dx} = \sqrt{1-y}$

$$\int (1-y)^{-1/2}\,dy = \int dx$$

$$-2(1-y)^{1/2} = x + C_1$$

$$\sqrt{1-y} = \dfrac{-x}{2} + C$$

$$1 - y = \left(C - \dfrac{x}{2}\right)^2$$

$$y = 1 - \left(C - \dfrac{x}{2}\right)^2$$

19. $(2+x)\dfrac{dy}{dx} = 2y$

$$\int \dfrac{1}{2y}\,dy = \int \dfrac{1}{2+x}\,dx$$

$$\dfrac{1}{2}\ln|y| = \ln|C_1(2+x)|$$

$$\sqrt{y} = C_1(2+x)$$

$$y = C(2+x)^2$$

21. $xy' = y$

$$x\dfrac{dy}{dx} = y$$

$$\int \dfrac{1}{y}\,dy = \int \dfrac{1}{x}\,dx$$

$$\ln|y| = \ln|x| + \ln|C|$$

$$\ln|y| = \ln|Cx|$$

$$y = Cx$$

23. $y' = \dfrac{dy}{dx} = \dfrac{x}{y} - \dfrac{x}{1+y} = x\left(\dfrac{1}{y+y^2}\right)$

$$\int (y + y^2)\,dy = \int x\,dx$$

$$\dfrac{y^2}{2} + \dfrac{y^3}{3} = \dfrac{x^2}{2} + C_1$$

$$3y^2 + 2y^3 = 3x^2 + C$$

25. $e^x(y' + 1) = 1$

$$\left(\dfrac{dy}{dx} + 1\right) = e^{-x}$$

$$\dfrac{dy}{dx} = e^{-x} - 1$$

$$\int dy = \int (e^{-x} - 1)\,dx$$

$$y = -e^{-x} - x + C$$

27. $y\dfrac{dy}{dx} = e^x$

$$\int y\,dy = \int e^x\,dx$$

$$\dfrac{y^2}{2} = e^x + C$$

When $x = 0$, $y = 4$. Therefore, $C = 7$ and the particular solution is $y^2 = 2e^x + 14$.

29. $\dfrac{dy}{dx} = -x(y + 4)$

$$\int \dfrac{1}{y+4}\,dy = \int -x\,dx$$

$$\ln|y + 4| = -\dfrac{x^2}{2} + C_1$$

$$|y + 4| = e^{-x^2/2+C_1}$$

$$y = -4 + Ce^{-x^2/2}$$

When $x = 0$, $y = -5 \Rightarrow C = -1$ and $y = -4 - e^{-x^2/2}$.

31. $dP = 6P \, dt$

$$\int \frac{1}{P} \, dP = \int 6 \, dt$$

$$\ln|P| = 6t + C_1$$

$$P = Ce^{6t}$$

When $t = 0$, $P = 5$. Therefore, $C = 5$ and the particular solution is $P = 5e^{6t}$.

33. $\dfrac{dy}{dx} = \dfrac{6x}{5y}$

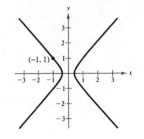

$$\int 5y \, dy = \int 6x \, dx$$

$$\frac{5}{2}y^2 = 3x^2 + C_1 \implies 5y^2 = 6x^2 + C$$

When $x = -1$, $y = 1 \implies 5 = 6 + C \implies C = -1$.
Therefore, $5y^2 = 6x^2 - 1$ or $6x^2 - 5y^2 = 1$ (hyperbola).

35. $\dfrac{dv}{dt} = 3.456 - 0.1v$

$$\int \frac{dv}{3.456 - 0.1v} = \int dt$$

$$-10 \ln|3.456 - 0.1v| = t + C_1$$

$$(3.456 - 0.1v)^{-10} = C_2 e^t$$

$$3.456 - 0.1v = Ce^{-0.1t}$$

$$v = -10Ce^{-0.1t} + 34.56$$

When $t = 0$, $v = 0$. Therefore, $C = 3.456$ and the solution is $v = 34.56(1 - e^{-0.1t})$.

37.   From the differential equation, we have $T = Ce^{kt} + T_0$. We have $T_0 = 90$, and when $t = 0$, $T = 1500$. Thus, $1500 = Ce^0 + 90$ which implies that $C = 1410$. When $t = 1$, $T = 1120$ which implies that

$$1120 = 1410e^k + 90$$

$$k = \ln \frac{1030}{1410} = \ln \frac{103}{141}.$$

Therefore, $T = 1410e^{[\ln(103/141)]t} + 90$. When $t = 5$, we have $T \approx 383.298°$.

39.   From the differential equation, we have $T = Ce^{kt} + T_0$. Since $T_0 = 0$ and $T = 70$ when $t = 0$, we have $T = 70e^{kt}$.
When $t = 1$, $T = 48$ and we have $48 = 70e^k$ which implies that $k = \ln \frac{48}{70} = \ln \frac{24}{35}$. Therefore, $T = 70e^{[\ln(24/35)]t}$.

(a) When $t = 6$, we have $T \approx 7.277°$.

(b) When $T = 10$, we have

$$10 = 70e^{[\ln(24/35)]t}$$

$$\ln\left(\frac{1}{7}\right) = \ln\left(\frac{24}{35}\right)t$$

$$t = \frac{\ln(1/7)}{\ln(24/35)} \approx 5.158 \text{ hours.}$$

41. $\displaystyle\int \frac{dN}{30 - N} = \int k \, dt$

$$-\ln|30 - N| = kt + C_1$$

$$30 - N = e^{-(kt + C_1)}$$

$$30 - N = C_2 e^{-kt}$$

$$N = 30 + Ce^{-kt}$$

43. $\dfrac{dy}{dx} = k\dfrac{y}{x}$

$$\int \frac{dy}{y} = \int \frac{k}{x} \, dx$$

$$\ln|y| = k\ln|x| + C_1 = \ln|x|^k + \ln C = \ln(C|x|^k)$$

$$y = Cx^k$$

Section C.3   First-Order Linear Differential Equations

1. $x^3 - 2x^2y' + 3y = 0$

$-2x^2y' + 3y = -x^3$

$y' + \dfrac{-3}{2x^2}y = \dfrac{x}{2}$

3. $xy' + y = xe^x$

$y' + \dfrac{1}{x}y = e^x$

5. $y + 1 = (x - 1)y'$

$(1 - x)y' + y = -1$

$y' + \dfrac{1}{1 - x}y = \dfrac{1}{x - 1}$

7. For this linear differential equation, we have $P(x) = 3x$ and $Q(x) = 6x$. Therefore, the integrating factor is $u(x) = e^{\int 3x\, dx} = e^{3x}$ and the general solution is

$$y = \frac{1}{u(x)}\int Q(x)u(x)\, dx = e^{-3x}\int 6e^{3x}\, dx = e^{-3x}(2e^{3x} + C) = 2 + Ce^{-3x}.$$

9. For this linear differential equation, we have $P(x) = 1$ and $Q(x) = e^{-x}$. Therefore, the integrating factor is $u(x) = e^{\int dx} = e^x$ and the general solution is

$$y = \frac{1}{u(x)}\int Q(x)u(x)\, dx = e^{-x}\int e^{-x}e^x\, dx = e^{-x}(x + C).$$

11. For this linear differential equation, we have $P(x) = 1/x$ and $Q(x) = 3x + 4$. Therefore, the integrating factor is

$$u(x) = e^{\int 1/x\, dx} = e^{\ln x} = x$$

and the general solution is

$$y = \frac{1}{u(x)}\int Q(x)u(x)\, dx = \frac{1}{x}\int (3x + 4)x\, dx = \frac{1}{x}(x^3 + 2x^2 + C) = x^2 + 2x + \frac{C}{x}.$$

13. For this linear differential equation, we have $P(x) = 5x$ and $Q(x) = x$. Therefore, the integrating factor is

$$u(x) = e^{\int 5x\, dx} = e^{(5/2)x^2}$$

and the general solution is

$$y = \frac{1}{u(x)}\int Q(x)u(x)\, dx = \frac{1}{e^{(5/2)x^2}}\int xe^{(5/2)x^2}\, dx = \frac{1}{e^{(5/2)x^2}}\left(\frac{1}{5}e^{(5/2)x^2} + C\right) = \frac{1}{5} + Ce^{-(5/2)x^2}.$$

15. For this linear differential equation

$$y' + y\left(\frac{1}{x - 1}\right) = x + 1$$

we have $P(x) = 1/(x - 1)$ and $Q(x) = x + 1$. Therefore, the integrating factor is $u(x) = e^{\int 1/(x-1)\, dx} = e^{\ln(x-1)} = x - 1$ and the general solution is

$$y = \frac{1}{u(x)}\int Q(x)u(x)\, dx = \frac{1}{x - 1}\int (x + 1)(x - 1)\, dx = \frac{1}{x - 1}\left(\frac{x^3}{3} - x + C_1\right) = \frac{x^3 - 3x + C}{3(x - 1)}.$$

17. For this linear differential equation,

$$y' + \frac{2}{x^3}y = \frac{1}{x^3}e^{1/x^2}$$

we have $P(x) = 2/x^3$ and $Q(x) = (1/x^3)e^{1/x^2}$. Therefore, the integrating factor is

$$u(x) = e^{\int (2/x^3)\, dx} = e^{-1/x^2}$$

and the general solution is

$$y = \frac{1}{u(x)}\int Q(x)u(x)\, dx = e^{1/x^2}\int \frac{1}{x^3}e^{1/x^2}e^{-1/x^2}\, dx = e^{1/x^2}\int \frac{1}{x^3}\, dx = e^{1/x^2}\left(-\frac{1}{2x^2} + C\right).$$

19. Separation of Variables: $\dfrac{dy}{dx} = 4 - y$

$$\int \frac{dy}{4 - y} = \int dx$$

$$-\ln|4 - y| = x + C_1$$

$$4 - y = e^{-(x + C_1)}$$

$$4 - y = C_2 e^{-x}$$

$$y = 4 - C_2 e^x = 4 + Ce^{-x}$$

First-Order Linear: $P(x) = 1, \qquad Q(x) = 4$

$$u(x) = e^{\int 1\,dx} = e^x$$

$$y = \frac{1}{e^x} \int 4e^x\,dx = \frac{1}{e^x}[4e^x + C] = 4 + Ce^{-x}$$

21. Separation of Variables: $\dfrac{dy}{dx} = 2x(1 + y)$

$$\int \frac{dy}{1 + y} = \int 2x\,dx$$

$$\ln|1 + y| = x^2 + C_1$$

$$1 + y = e^{x^2 + C_1}$$

$$y = Ce^{x^2} - 1$$

First-Order Linear: $P(x) = -2x, \qquad Q(x) = 2x$

$$u(x) = e^{\int -2x\,dx} = e^{-x^2}$$

$$y = \frac{1}{e^{-x^2}} \int 2xe^{-x^2}\,dx = e^{x^2}[-e^{-x^2} + C] = Ce^{x^2} - 1$$

23. $y' - 2x = 0$ matches (c) because

$$y = x^2 + C \implies y' = 2x \implies y - 2x = 0.$$

25. $y' - 2xy = 0$ matches (a) because

$$y = Ce^{x^2} \implies y' = Ce^{x^2}(2x) = 2xy.$$

27. Since $P(x) = 1$ and $Q(x) = 6e^x$, the integrating factor is

$$u(x) = e^{\int dx} = e^x$$

and the general solution is

$$y = \frac{1}{e^x} \int 6e^x(e^x)\,dx = \frac{1}{e^x}[3e^{2x} + C] = 3e^x + Ce^{-x}.$$

Since $y = 3$ when $x = 0$, it follows that $C = 0$, and the particular solution is $y = 3e^x$.

29. Since $P(x) = 1/x$ and $Q(x) = 0$, the integrating factor is $u(x) = e^{\int 1/x\,dx} = e^{\ln x} = x$ and the general solution is

$$y = \frac{1}{x} \int 0\,dx = \frac{C}{x}.$$

Since $y = 2$ when $x = 2$, it follows that $C = 4$, and the particular solution is $y = 4/x$ or $xy = 4$.

31. Since $P(x) = 3x^2$ and $Q(x) = 3x^2$, the integrating factor is $u(x) = e^{\int 3x^2\,dx} = e^{x^3}$ and the general solution is

$$y = e^{-x^3} \int 3x^2 e^{x^3}\,dx = e^{-x^3}(e^{x^3} + C) = 1 + Ce^{-x^3}.$$

Since $y = 6$ when $x = 0$, it follows that $C = 5$ and the particular solution is $y = 1 + 5e^{-x^3}$.

33. Since $P(x) = -2/x$ and $Q(x) = -x$, the integrating factor is

$$u(x) = e^{\int -2/x\, dx} = e^{-2 \ln x} = \frac{1}{x^2}$$

and the general solution is

$$y = x^2 \int (-x)\left(\frac{1}{x^2}\right) dx = x^2(-\ln|x| + C).$$

Since $y = 5$ when $x = 1$, it follows that $C = 5$, and the particular solution is $y = x^2(5 - \ln|x|)$.

35. Since $P(t) = 0.2$ and $Q(t) = 20 + 0.2t$, the integrating factor is $u(t) = e^{\int 0.2\, dt} = e^{t/5}$ and the general solution is

$$S = e^{-t/5} \int e^{t/5}\left(20 + \frac{t}{5}\right) dt.$$

Using integration by parts, the integral is

$$S = e^{-t/5}(100e^{t/5} + te^{t/5} - 5e^{t/5} + C) = 100 + t - 5 + Ce^{-t/5} = 95 + t + Ce^{-t/5}.$$

Since $S = 0$ when $t = 0$, it follows that $C = -95$, and the particular solution is $S = t + 95(1 - e^{-t/5})$. During the first 10 years, the sales are as follows.

t	0	1	2	3	4	5	6	7	8	9	10
s	0	18.22	33.32	45.86	56.31	65.05	72.39	78.57	83.82	88.30	92.14

37. $\dfrac{dp}{dx}\left(1 - \dfrac{400}{3x}\right) = \dfrac{p}{x}$

$\dfrac{dp}{dx}\left(x - \dfrac{400}{3}\right) = p$

$\displaystyle\int \dfrac{dp}{p} = \int \dfrac{dx}{x - (400/3)}$

$\ln|p| = \ln\left|x - \dfrac{400}{3}\right| + \ln|C|$

$p = C\left(x - \dfrac{400}{3}\right)$

$340 = C\left(20 - \dfrac{400}{3}\right) \Rightarrow C = -3$

$p = -3\left(x - \dfrac{400}{3}\right) = 400 - 3x$

39.

$D(t) = S(t)$

$480 + 5p(t) - 2p'(t) = 300 + 8p(t) + p'(t)$

$180 = 3p'(t) + 3p(t)$

$60 = p'(t) + p(t)$

$P(t) = 1,$

$u(t) = e^{\int 1\, dt} = e^t$

$p(t) = \dfrac{1}{e^t} \int 60e^t\, dt = \dfrac{1}{e^t}[60e^t + C]$

$= 60 + Ce^{-t}$

$p(0) = 60 + C = 75 \Rightarrow C = 15$

$p(t) = 60 + 15e^{-t} = 15(4 + e^{-t})$

41. (a) Since $P(t) = -r$ and $Q(t) = Pt$, the integrating factor is $u(t) = e^{\int -r\, dt} = e^{-rt}$ and the general solution is

$$A = e^{rt} \int Pte^{-rt}\, dt = Pe^{rt}\left(-\frac{t}{r}e^{-rt} - \frac{1}{r^2}e^{-rt} + C_1\right) = \frac{P}{r^2}(-rt - 1 + Ce^{rt}).$$

Since $A = 0$ when $t = 0$, it follows that $C = 1$, and the particular solution is

$$A = \frac{P}{r^2}(e^{rt} - rt - 1).$$

(b) When $t = 10$, $P = 500{,}000$ and $r = 0.09$, $A \approx \$34{,}543{,}402$.

43. $\dfrac{dv}{dt} + \dfrac{k}{m}v = g$

$u(t) = e^{\int (k/m)\,dt} = e^{kt/m}$

$v = \dfrac{1}{u(t)}\displaystyle\int Q(t)u(t)\,dt = \dfrac{1}{e^{kt/m}}\int -\,ge^{kt/m}\,dt = \dfrac{1}{e^{kt/m}}\left(\dfrac{gm}{k}e^{kt/m} + C\right) = \dfrac{gm}{k} + Ce^{kt/m}$

45. Answers will vary.

Section C.4   Applications of Differential Equations

1. The general solution is $y = Ce^{kx}$. Since $y = 1$ when $x = 0$, it follows that $C = 1$. Thus, $y = e^{kx}$. Since $y = 2$ when $x = 3$, it follows that $2 = e^{3k}$ which implies that

$$k = \frac{\ln 2}{3} \approx 0.2310.$$

Thus, the particular solution is $y \approx e^{0.2310x}$.

3. The general solution is $y = Ce^{kx}$. Since $y = 4$ when $x = 0$, it follows that $C = 4$. Thus, $y = 4e^{kx}$. Since $y = 1$ when $x = 4$, it follows that $\frac{1}{4} = e^{4k}$ which implies that

$$k = \tfrac{1}{4}\ln\tfrac{1}{4} \approx -0.3466.$$

Thus, the particular solution is $y \approx 4e^{-0.3466x}$.

5. The general solution is $y = Ce^{kx}$. Since $y = 2$ when $x = 2$ and $y = 4$ when $x = 3$, it follows that $2 = Ce^{2k}$ and $4 = Ce^{3k}$. By equating C-values from these two equations, we have the following.

$$2e^{-2k} = 4e^{-3k}$$

$$\tfrac{1}{2} = e^{-k} \implies k = \ln 2 \approx 0.6931$$

This implies that

$$C = 2e^{-2\ln 2} = 2e^{\ln(1/4)} = 2\left(\tfrac{1}{4}\right) = \tfrac{1}{2}.$$

Thus, the particular solution is

$$y = \tfrac{1}{2}e^{x\ln 2} \approx \tfrac{1}{2}e^{0.6931x}.$$

7. The general solution is $y = Ae^{kt}$ with $A = 2000$. Since $y = 2983.65$ when $t = 5$, we have

$$2983.65 = 2000e^{5k}$$

$$k = \frac{\ln(1.491825)}{5} \approx 0.08.$$

Thus, the particular solution is $y = 2000e^{0.08t}$. When $t = 10$, $y = 2000e^{0.08(10)} \approx \$4451.08$.

9. $\dfrac{dS}{dt} = k(L - S)$

$\displaystyle\int \dfrac{dS}{L - S} = \int k\,dt$

$-\ln|L - S| = kt + C_1$

$L - S = e^{-kt-C_1}$

$S = L + Ce^{-kt}$

Since $S = 0$ when $t = 0$, we have $0 = L + C \implies C = -L$. Thus, $S = L(1 - e^{-kt})$.

11. The general solution is $y = Ce^{20kx}(20 - y)$. Since $y = 1$ when $x = 0$, it follows that $C = \frac{1}{19}$. Thus,

$$y = \frac{1}{19}e^{20kx}(20 - y).$$

Since $y = 10$ when $x = 5$, it follows that

$$19 = e^{100k}$$

$$20k = \frac{\ln 19}{5} \approx 0.5889.$$

Thus, the particular solution is

$$y = \frac{1}{19}e^{0.5889x}(20 - y)$$

$$y(19 + e^{0.5889x}) = 20e^{0.5889x}$$

$$y = \frac{20e^{0.5889x}}{19 + e^{0.5889x}} = \frac{20}{1 + 19e^{-0.5889x}}.$$

13. The general solution is $y = Ce^{5000kx}(5000 - y)$. Since $y = 250$ when $x = 0$, it follows that $C = \frac{1}{19}$. Thus,

$$y = \frac{1}{19}e^{5000kx}(5000 - y).$$

Since $y = 2000$ when $x = 25$, it follows that

$$\frac{38}{3} = e^{125,000k}$$

$$5000k = \frac{\ln(38/3)}{25} \approx 0.10156.$$

Thus, the particular solution is

$$y = \frac{1}{19}e^{0.10156x}(5000 - y)$$

$$y(19 + e^{0.10156x}) = 5000e^{0.10156x}$$

$$y = \frac{5000e^{0.10156x}}{19 + e^{0.10156x}} = \frac{5000}{1 + 19e^{-0.10156x}}.$$

15.

$$\frac{dN}{dt} = kN(500 - N)$$

$$\int \frac{dN}{N(500 - N)} = \int k\,dt$$

$$\frac{1}{500}\int \left[\frac{1}{N} + \frac{1}{500 - N}\right]dN = \int k\,dt$$

$$\ln|N| - \ln|500 - N| = 500(kt + C_1)$$

$$\frac{N}{500 - N} = e^{500kt + C_2} = Ce^{500kt}$$

$$N = \frac{500Ce^{500kt}}{1 + Ce^{500kt}}$$

When $t = 0$, $N = 100$. Thus, $100 = \frac{500C}{1 + C} \Rightarrow C = 0.25$. Thus,

$$N = \frac{125e^{500kt}}{1 + 0.25e^{500kt}}.$$

When $t = 4$, $N = 200$. Thus,

$$200 = \frac{125e^{2000k}}{1 + 0.25e^{2000k}} \Rightarrow k = \frac{\ln(8/3)}{2000} \approx 0.00049.$$

Therefore,

$$N = \frac{125e^{0.2452t}}{1 + 0.25e^{0.2452t}} = \frac{500}{1 + 4e^{-0.2452t}}.$$

17. The differential equation is given by the following.

$$\frac{dP}{dn} = kP(L - P)$$

$$\int \frac{1}{P(L - P)}\,dP = \int k\,dn$$

$$\frac{1}{L}[\ln|P| - \ln|L - P|] = kn + C_1$$

$$\frac{P}{L - P} = Ce^{Lkn}$$

$$P = \frac{CLe^{Lkn}}{1 + Ce^{Lkn}} = \frac{CL}{e^{-Lkn} + C}$$

19. The general solution is $y = \frac{-1}{kt + C}$. Since $y = 45$ when $t = 0$, it follows that $45 = \frac{-1}{C}$ and $C = \frac{-1}{45}$. Therefore,

$$y = -\frac{1}{kt - (1/45)} = \frac{45}{1 - 45kt}$$

Since $y = 4$ when $t = 2$, we have $4 = \frac{45}{1 - 45k(2)} \Rightarrow k = -\frac{45}{360}$. Thus,

$$y = \frac{45}{1 + (41/8)t} = \frac{360}{8 + 41t}.$$

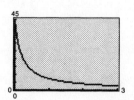

21. Since $y = 100$ when $t = 0$, it follows that $100 = 500e^{-C}$, which implies that $C = \ln 5$. Therefore, we have $y = 500e^{(-\ln 5)e^{-kt}}$. Since $y = 150$ when $t = 2$, it follows that

$$150 = 500e^{(-\ln 5)e^{-2k}}$$

$$e^{-2k} = \frac{\ln 0.3}{\ln 0.2}$$

$$k = -\frac{1}{2}\ln\frac{\ln 0.3}{\ln 0.2} \approx 0.1452.$$

Therefore, y is given by $y = 500e^{-1.6904e^{-0.1451t}}$.

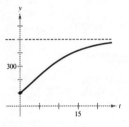

23. From Example 3, the general solution is

$$y = 60e^{-Ce^{-kt}}.$$

Since $y = 8$ when $t = 0$,

$$8 = 60e^{-C} \implies C = \ln\frac{15}{2} \approx 2.0149.$$

Since $y = 15$ when $t = 3$,

$$15 = 60e^{-2.0149e^{-3k}}$$

$$\frac{1}{4} + e^{-2.0149e^{-3k}}$$

$$\ln\frac{1}{4} = -2.0149e^{-3k}$$

$$k = -\frac{1}{3}\ln\left(\frac{\ln 1/4}{-2.0149}\right) \approx 0.1246.$$

Thus,

$$y = 60e^{-2.0149e^{-0.1246t}}.$$

When $t = 10$, $y \approx 34$ beavers.

25. Following Example 4, the differential equation is

$$\frac{dy}{dt} = ky(1-y)(2-y)$$

and its general solution is

$$\frac{y(2-y)}{(1-y)^2} = Ce^{2kt}$$

$$y = \frac{1}{2} \text{ when } t = 0 \implies \frac{(1/2)(3/2)}{(1/2)^2} = C \implies C = 3$$

$$y = 0.75 = \frac{3}{4} \text{ when } t = 4 \implies \frac{(3/4)(5/4)}{(1/4)^2} = 15$$

$$= 3e^{2k(4)} \implies k = \frac{1}{8}\ln 5 \approx 0.2012.$$

Hence, the particular solution is

$$\frac{y(2-y)}{(1-y)^2} = 3e^{0.4024t}.$$

Using a symbolic algebra utility or graphing utility, you find that when $t = 10$,

$$\frac{y(2-y)}{(1-y)^2} = 3e^{0.4024(10)}$$

and $y \approx 0.92$, or 92%.

27. (a) $\dfrac{dy}{dt} = ky$

$$\int\frac{dy}{y} = \int k\,dt$$

$$\ln y = kt + C_1$$

$$y = e^{kt+C_1} = Ce^{kt}$$

(b) $y(0) = 20 \implies C = 20$

$$y(1) = 16 = 20e^k \implies k = \ln\frac{16}{20} = \ln\left(\frac{4}{5}\right)$$

$$y = 20e^{t\ln(4/5)}$$

When 75% has changed:

$$5 = 20e^{t\ln(4/5)}$$

$$\frac{1}{4} = e^{t\ln(4/5)}$$

$$t = \frac{\ln(1/4)}{\ln(4/5)} \approx 6.2 \text{ hours}$$

29. Since $Q' + \frac{1}{20}Q = \frac{5}{2}$ is a first-order linear differential equation with $P(x) = \frac{1}{20}$ and $R(x) = \frac{5}{2}$, we have the integrating factor $u(t) = e^{\int (1/20)\,dt} = e^{(1/20)t}$, and the general solution is

$$Q = e^{-0.05t}\int\frac{5}{2}e^{0.05t}\,dt = e^{-0.05t}(50e^{0.05t} + C) = 50 + Ce^{-0.05t}.$$

Since $Q = 0$ when $t = 0$, we have $C = -50$ and $Q = 50(1 - e^{-0.05t})$. Finally, when $t = 30$, we have $Q \approx 38.843$ lbs/gal.

31. The general solution is $y = Ce^{kt}$. Since $y = 0.60C$ when $t = 1$, we have $0.60C = Ce^{k} \Rightarrow k = \ln 0.60 \approx -0.5108$. Thus, $y = Ce^{-0.5108t}$. When $y = 0.20C$, we have

$$0.20C = Ce^{-0.5108t}$$

$$\ln 0.20 = -0.5108t$$

$$t \approx 3.15 \text{ hours.}$$

33. $\displaystyle\int \frac{1}{kP + N}\, dp = \int dt$

$\dfrac{1}{k} \ln|kP + N| = t + C_1$

$kP + N = C_2 e^{kt}$

$P = Ce^{kt} - \dfrac{N}{k}$

35. $\displaystyle\int \frac{1}{rA + P}\, dA = \int dt$

$\dfrac{1}{r} \ln|rA + P| = t + C_1$

$rA + P = Ce^{rt}$

$A = \dfrac{1}{r}(Ce^{rt} - P)$

Since $A = 0$ when $t = 0$, it follows that $C = P$. Therefore, we have

$$A = \frac{P}{r}(e^{rt} - 1).$$

37. Since $A = 120,000,000$ when $t = 8$ and $r = 0.1625$, we have

$$P = \frac{(0.1625)(120,000,000)}{e^{(0.1625)(8)} - 1} \approx \$7,305,295.15.$$

39. (a) $\displaystyle\int \frac{dC}{C} = \int -\frac{R}{V}\, dt$

$\ln|C| = -\dfrac{R}{V}t + K_1$

$C = Ke^{-Rt/V}$

Since $C = C_0$ when $t = 0$, it follows that $K = C_0$ and the function is $C = C_0 e^{-Rt/V}$.

(b) Finally, as $t \to \infty$, we have

$$\lim_{t \to \infty} C = \lim_{t \to \infty} C_0 e^{-Rt/V} = 0.$$

41. (a) $\displaystyle\int \frac{1}{Q - RC}\, dC = \int \frac{1}{V}\, dt$

$-\dfrac{1}{R} \ln|Q - RC| = \dfrac{t}{V} + K_1$

$Q - RC = e^{-R[(t/V) + K_1]}$

$C = \dfrac{1}{R}(Q - e^{-R[(t/V) + K_1]})$

$\quad = \dfrac{1}{R}(Q - Ke^{-Rt/V})$

Since $C = 0$ when $t = 0$, it follows that $K = Q$ and we have $C = \dfrac{Q}{R}(1 - e^{-Rt/V})$.

(b) As $t \to \infty$, the limit of C is Q/R.

Practice Test Solutions for Chapter 0

1. Rational   (Sec. 0.1)

2. (Sec. 0.1)

    (a) Satisfies   (b) Does not satisfy

    (c) Satisfies   (d) Satisfies

3. $x \geq 3$ (Sec. 0.1)

4. $-1 < x < 7$ (Sec. 0.1)

5. $\sqrt{19} > \frac{13}{3}$ (Sec. 0.1)

6. (Sec. 0.2)

    (a) $d = 10$

    (b) Midpoint: 2

7. $-\frac{11}{3} \leq x \leq 3$ (Sec. 0.2)

8. $x < -5$ or $x > \frac{33}{5}$ (Sec. 0.2)

9. $-\frac{25}{2} < x < \frac{55}{2}$ (Sec. 0.2)

10. $|x - 1| \leq 4$ (Sec. 0.2)

11. $3x^5$ (Sec. 0.3)

12. 1   (Sec. 0.3)

13. $2xy\sqrt[3]{4x}$ (Sec. 0.3)

14. $\frac{1}{4}(x + 1)^{-1/3}(x + 7)$ (Sec. 0.3)

15. $x < 5$ (Sec. 0.3)

16. $(3x + 2)(x - 7)$ (Sec. 0.4)

17. $(5x + 9)(5x - 9)$ (Sec. 0.4)

18. $(x + 2)(x^2 - 2x + 4)$ (Sec. 0.4)

19. $-3 \pm \sqrt{11}$ (Sec. 0.4)

20. $-1, 2, 3$ (Sec. 0.4)

21. $\dfrac{-3}{(x - 1)(x + 3)}$ (Sec. 0.5)

22. $\dfrac{x + 13}{2\sqrt{x + 5}}$ (Sec. 0.5)

23. $\dfrac{1}{\sqrt{x}(x + 2)^{3/2}}$ (Sec. 0.5)

24. $\dfrac{3y\sqrt{y^2 + 9}}{y^2 + 9}$ (Sec. 0.5)

25. $-\dfrac{1}{2(\sqrt{x} - \sqrt{x + 7})}$ (Sec. 0.5)

26. $-1, 2, 4$ (Sec. 0.4)

Practice Test Solutions for Chapter 1

1. $d = \sqrt{82}$ (Sec. 1.1)

2. Midpoint: $(1, 3)$ (Sec. 1.1)

3. Collinear   (Sec. 1.1)

4. $x = \pm 3\sqrt{5}$ (Sec. 1.1)

5. x-intercepts: $(\pm 2, 0)$ (Sec. 1.2)

    y-intercept: $(0, 4)$

6. x-intercepts: $(2, 0)$ (Sec. 1.2)

    No y-intercept

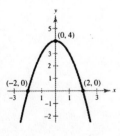

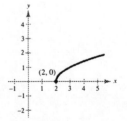

7. *x*-intercept: (3, 0)   (Sec. 1.2)

  y-intercept: (0, 3)

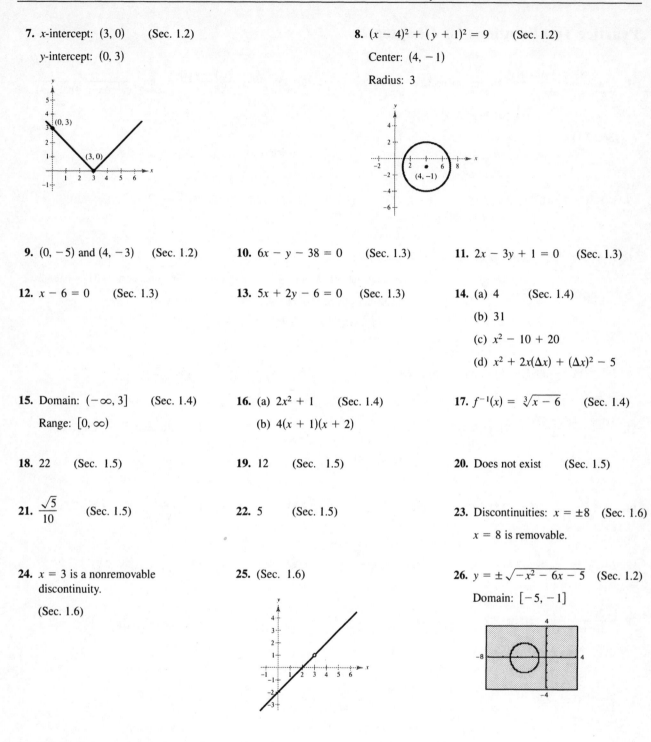

8. $(x - 4)^2 + (y + 1)^2 = 9$ (Sec. 1.2)

  Center: $(4, -1)$

  Radius: 3

9. $(0, -5)$ and $(4, -3)$ (Sec. 1.2)

10. $6x - y - 38 = 0$ (Sec. 1.3)

11. $2x - 3y + 1 = 0$ (Sec. 1.3)

12. $x - 6 = 0$ (Sec. 1.3)

13. $5x + 2y - 6 = 0$ (Sec. 1.3)

14. (a) 4   (Sec. 1.4)

  (b) 31

  (c) $x^2 - 10 + 20$

  (d) $x^2 + 2x(\Delta x) + (\Delta x)^2 - 5$

15. Domain: $(-\infty, 3]$ (Sec. 1.4)

  Range: $[0, \infty)$

16. (a) $2x^2 + 1$ (Sec. 1.4)

  (b) $4(x + 1)(x + 2)$

17. $f^{-1}(x) = \sqrt[3]{x - 6}$ (Sec. 1.4)

18. 22   (Sec. 1.5)

19. 12   (Sec. 1.5)

20. Does not exist   (Sec. 1.5)

21. $\dfrac{\sqrt{5}}{10}$ (Sec. 1.5)

22. 5   (Sec. 1.5)

23. Discontinuities: $x = \pm 8$ (Sec. 1.6)

  $x = 8$ is removable.

24. $x = 3$ is a nonremovable discontinuity.

  (Sec. 1.6)

25. (Sec. 1.6)

26. $y = \pm \sqrt{-x^2 - 6x - 5}$ (Sec. 1.2)

  Domain: $[-5, -1]$

27. The graph does **not** show that the function does not exist at $x = 3$ on many graphing utilities.

  $\displaystyle\lim_{x \to 3} f(x) = 6$ (Sec. 1.5)

Practice Test Solutions for Chapter 2

1. $\lim\limits_{\Delta x \to 0} \dfrac{f(x + \Delta x) - f(x)}{\Delta x} = \lim\limits_{\Delta x \to 0} (4x + 2\Delta x + 3)$

$\qquad\qquad\qquad\qquad\qquad\qquad = 4x + 3$

(Sec. 2.1)

2. $\lim\limits_{\Delta x \to 0} \dfrac{f(x + \Delta x) - f(x)}{\Delta x} = \lim\limits_{\Delta x \to 0} \dfrac{-1}{(x + \Delta x + 4)(x + 4)}$

$\qquad\qquad\qquad\qquad\qquad\qquad = -\dfrac{1}{(x - 4)^2}$

(Sec. 2.1)

3. $x - 4y + 2 = 0$ (Sec. 2.1)

4. $15x^2 - 12x + 15$ (Sec. 2.2)

5. $\dfrac{4x - 2}{x^3}$ (Sec. 2.2)

6. $\dfrac{2}{3\sqrt[3]{x}} + \dfrac{3}{5\sqrt[5]{x^2}}$ (Sec. 2.2)

7. (Sec. 2.3)

Average rate of change: 4

Instantaneous rates of change:

$\qquad f'(0) = 0, \quad f'(2) = 12$

8. (Sec. 2.3)

Marginal cost: $4.31 - 0.0002x$

9. $5x^4 + 28x^3 - 39x^2 - 56x + 36$ (Sec. 2.4)

10. $-\dfrac{x^2 + 14x + 8}{(x^2 - 8)^2}$ (Sec. 2.4)

11. $\dfrac{3x^4 + 14x^3 - 45x^2}{(x + 5)^2}$ (Sec. 2.4)

12. $-\dfrac{3x^2 + 4x + 1}{2\sqrt{x}(x^2 + 4x - 1)^2}$

(Sec. 2.4)

13. $72(6x - 5)^{11}$ (Sec. 2.5)

14. $-\dfrac{12}{\sqrt{4 - 3x}}$ (Sec. 2.5)

15. $\dfrac{18x}{(x^2 + 1)^4}$ (Sec. 2.5)

16. $\dfrac{\sqrt{10x}}{x(x + 2)^{3/2}}$ (Sec. 2.5)

17. $24x - 54$ (Sec. 2.6)

18. $-\dfrac{15}{16(3 - x)^{7/2}}$ (Sec. 2.6)

19. $-\dfrac{x^4}{y^4}$ (Sec. 2.7)

20. $-\dfrac{2(xy^3 + 1)}{3(x^2y^2 - 1)}$ (Sec. 2.7)

21. $\dfrac{8\sqrt{xy + 4} + y}{10\sqrt{xy + 4} - x} = \dfrac{41y - 32x}{50y - 41x}$

(Sec. 2.7)

22. $-\dfrac{8x^2}{y^2(x^3 - 4)^2}$ (Sec. 2.7)

23. $\dfrac{5}{12}$ (Sec. 2.8)

24. $\dfrac{dA}{dt} = 2\pi r \dfrac{dr}{dt}$ (Sec. 2.8)

$\qquad \dfrac{dr}{dt} = \dfrac{5}{4\pi}$

25. (Sec. 2.8)

$\qquad V = \dfrac{4}{3}\pi h^3$

$\qquad \dfrac{dV}{dt} = 4\pi h^2 \dfrac{dh}{dt}$

$\qquad \dfrac{dh}{dt} = \dfrac{1}{8\pi}$

26. (Sec. 2.4)

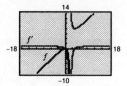

Horizontal Tangents at $x = 0$ and $x = 4$.
$f'(0) = f'(4) = 0$

27. (Sec. 2.7)

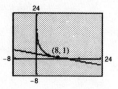

Tangent line: $y = -\dfrac{1}{4}x + 3$

Practice Test Solutions for Chapter 3

1. Increasing: $(-\infty, 0), (4, \infty)$

Decreasing: $(0, 4)$

(Sec. 3.1)

2. Increasing: $\left(-\infty, \frac{2}{3}\right)$

Decreasing: $\left(\frac{2}{3}, 1\right)$

(Sec. 3.1)

3. Relative minimum: $(2, -45)$

(Sec. 3.2)

4. Relative minimum: $(-3, 0)$

(Sec. 3.2)

5. Maximum: $(5, 0)$ (Sec. 3.2)

Minimum: $(2, -9)$

6. No inflection points   (Sec. 3.6)

7. Points of inflection: (Sec. 3.6)

$$\left(-\frac{1}{\sqrt{3}}, \frac{1}{4}\right), \left(\frac{1}{\sqrt{3}}, \frac{1}{4}\right)$$

8. $S = x + \dfrac{600}{x}$ (Sec. 3.4)

First number: $10\sqrt{6}$

Second number: $\dfrac{10\sqrt{6}}{3}$

9. $A = 3xy = 3x\left(\dfrac{3000 - 6x}{4}\right)$

$3x = 750$ feet, $y = 375$ feet

(Sec. 3.4)

10. $x \approx 13{,}333$ units   (Sec. 3.5)

11. $p = \$14{,}088$ (Sec. 3.5)

12. -1 (Sec. 3.5)

13. $-\infty$ (Sec. 3.6)

14. -2 (Sec. 3.6)

15. (Sec. 3.7)

Intercept: $(0, 0)$

Vertical asymptotes: $x = \pm 3$

Horizontal asymptote: $y = 1$

Relative maximum: $(0, 0)$

No inflection points

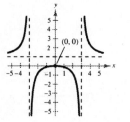

16. (Sec. 3.7)

Intercepts: $(-2, 0), \left(0, \frac{2}{5}\right)$

Horizontal asymptote: $y = 0$

Relative maximum: $\left(1, \frac{1}{2}\right)$

Relative minimum: $\left(-5, -\frac{1}{10}\right)$

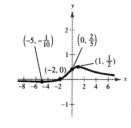

17. Intercept: $(0, -1)$ (Sec. 3.7)

No relative extrema

Inflection point: $(-1, -2)$

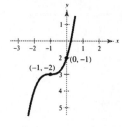

18. (Sec. 3.7)

Intercepts: $(0, 4), (2, 0)$

Relative minimum: $(2, 0)$

No inflection points

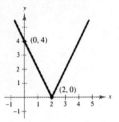

19. (Sec. 3.7)

Intercepts: $(2, 0), \left(0, \sqrt[3]{4}\right)$

Relative minimum: $(2, 0)$

No inflection points

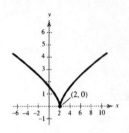

20. $\sqrt[3]{65} \approx 4.0208$ (Sec. 3.8)

21.

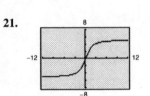

(Sec. 3.7)

Horizontal asymptotes at $y = \pm 5$.

No relative extrema.

22.

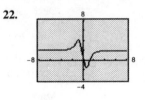

Yes, the graph crosses the horizontal asymptote $y = 2$.

(Sec. 3.7)

Practice Test Solutions for Chapter 4

1. (a) 81 (Sec. 4.1)

(b) $\frac{1}{32}$

(c) 1

2. (a) $x = 2$ (Sec. 4.1)

(b) $x = 32$

(c) $x = 5$

3. (a)

(b)

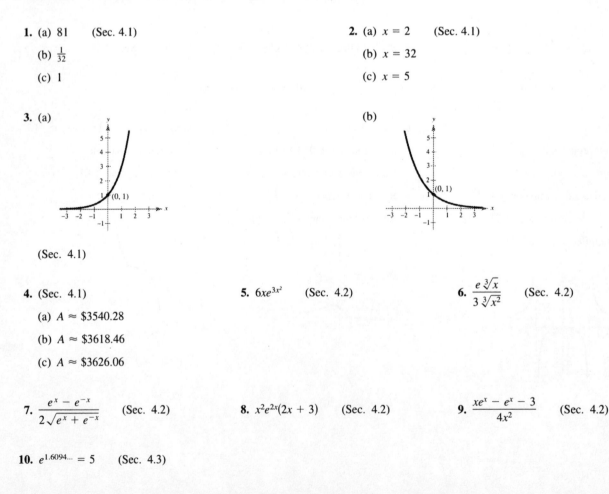

(Sec. 4.1)

4. (Sec. 4.1)

(a) $A \approx \$3540.28$

(b) $A \approx \$3618.46$

(c) $A \approx \$3626.06$

5. $6xe^{3x^2}$ (Sec. 4.2)

6. $\dfrac{e \sqrt[3]{x}}{3 \sqrt[3]{x^2}}$ (Sec. 4.2)

7. $\dfrac{e^x - e^{-x}}{2\sqrt{e^x + e^{-x}}}$ (Sec. 4.2)

8. $x^2 e^{2x}(2x + 3)$ (Sec. 4.2)

9. $\dfrac{xe^x - e^x - 3}{4x^2}$ (Sec. 4.2)

10. $e^{1.6094\ldots} = 5$ (Sec. 4.3)

11. (a)

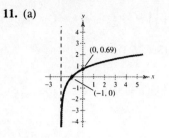

(b)

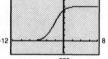

(Sec. 4.3)

12. (a) $\ln\left(\dfrac{3x+1}{2x-5}\right)$ (Sec. 4.3)

(b) $\ln\left(\dfrac{x^4}{y^3\sqrt{z}}\right)$

13. (a) $x = e^{17}$ (Sec. 4.3)

(b) $x = \dfrac{\ln 2}{3\ln 5}$

14. $\dfrac{6}{6x-7}$ (Sec. 4.4)

15. $\dfrac{4x+15}{x(2x+5)}$ (Sec. 4.4)

16. $\dfrac{1}{x(x+3)}$ (Sec. 4.4)

17. $x^3(1 + 4\ln x)$ (Sec. 4.4)

18. $\dfrac{1}{2x\sqrt{\ln x + 1}}$ (Sec. 4.4)

19. (a) $y = 7e^{-0.7611t}$ (Sec. 4.5)

(b) $y = 0.1501e^{0.4970t}$

20. $t \approx 5.776$ years   (Sec. 4.5)

21.

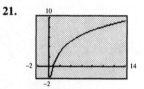

(Sec. 4.5)

The graphs are the same.

22.

(Sec. 4.1)

$\lim_{t\to\infty} f(t) = 600$

$\lim_{x\to-\infty} f(t) = 0$

Practice Test Solutions for Chapter 5

1. $x^3 - 4x^2 + 5x + C$ (Sec. 5.1)

2. (Sec. 5.1)

$$\frac{x^4}{4} + \frac{7x^3}{3} - 2x^2 - 28x + C$$

3. $\dfrac{x^2}{2} - 9x - \dfrac{1}{x} + C$ (Sec. 5.1)

4. $-\dfrac{1}{5}(1 - x^4)^{5/4} + C$ (Sec. 5.2)

5. $\dfrac{9}{14}(7x)^{2/3} + C$ (Sec. 5.2)

6. $-\dfrac{2}{33}(6 - 11x)^{3/2} + C$ (Sec. 5.2)

7. $\dfrac{4}{5}x^{5/4} + \dfrac{6}{7}x^{7/6} + C$ (Sec. 5.1)

8. $-\dfrac{1}{3x^3} + \dfrac{1}{4x^4} + C$ (Sec. 5.1)

9. $x - x^3 + \dfrac{3}{5}x^5 - \dfrac{1}{7}x^7 + C$

(Sec. 5.1)

10. $-\dfrac{5}{12(1 + 3x^2)^2} + C$

(Sec. 5.2)

11. $\left(\dfrac{1}{7}\right)e^{7x} + C$ (Sec. 5.3)

12. $\left(\dfrac{1}{8}\right)e^{4x^2} + C$ (Sec. 5.3)

13. $\left(\dfrac{1}{16}\right)(1 + 4e^x)^4 + C$

(Sec. 5.3)

14. $\left(\dfrac{1}{2}\right)e^{2x} + 4e^x + 4x + C$

(Sec. 5.3)

15. $\left(\dfrac{1}{2}\right)e^{2x} - 4x - e^{-x} + C$

(Sec. 5.3)

16. $\ln|x + 6| + C$ (Sec. 5.3)

17. $-\left(\dfrac{1}{3}\right)\ln|8 - x^3| + C$

(Sec. 5.3)

18. $\dfrac{1}{3}\ln(1 + 3e^x) + C$ (Sec. 5.3)

19. $\dfrac{(\ln x)^7}{7} + C$ (Sec. 5.3)

20. $\dfrac{x^2}{2} + x + 6\ln|x - 1| + C$ (Sec. 5.3)

(Use long division first)

21. -3 (Sec. 5.4)

22. $\dfrac{381}{7}$ (Sec. 5.4)

23. 2 (Sec. 5.4)

24. $A = 36$ (Sec. 5.5)

25. $A = \dfrac{1}{2}$ (Sec. 5.5)

26. $A = \dfrac{2}{3}$ (Sec. 5.5)

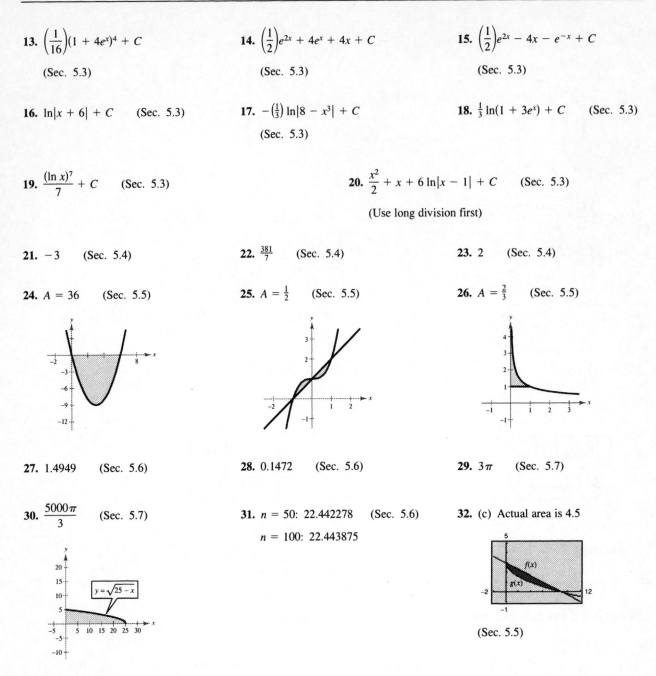

27. 1.4949 (Sec. 5.6)

28. 0.1472 (Sec. 5.6)

29. 3π (Sec. 5.7)

30. $\dfrac{5000\pi}{3}$ (Sec. 5.7)

31. $n = 50$: 22.442278 (Sec. 5.6)

$n = 100$: 22.443875

32. (c) Actual area is 4.5

(Sec. 5.5)

Practice Test Solutions for Chapter 6

1. $\dfrac{2}{5}(x + 3)^{3/2}(x - 2) + C$

(Sec. 6.1)

2. $-\dfrac{x - 1}{(x - 2)^2} + C$ (Sec. 6.1)

3. $\dfrac{2}{3}\ln\left|3\sqrt{x} + 1\right| + C$ (Sec. 6.1)

4. $\dfrac{(\ln 7x)^2}{2} + C$ (Sec. 6.1)

5. $\dfrac{1}{4}e^{2x}(2x - 1) + C$ (Sec. 6.2)

6. $\dfrac{x^4}{16}[4(\ln x) - 1] + C$ (Sec. 6.2)

7. $\dfrac{2}{35}(x - 6)^{3/2}(5x^2 + 24x + 96) + C$

(Sec. 6.2)

8. $\dfrac{1}{32}e^{4x}(8x^2 - 4x + 1) + C$

(Sec. 6.2)

9. $\ln\left|\dfrac{x + 3}{x - 2}\right| + C$ (Sec. 6.3)

10. $\ln \left| \dfrac{x^3}{(x + 4)^2} \right| + C$ (Sec. 6.3)   **11.** $5 \ln |x + 2| + \dfrac{7}{x + 2} + C$ **12.** $\dfrac{3}{2}x^2 + \ln \dfrac{|x|}{(x + 2)^2} + C$

(Sec. 6.3)   (Sec. 6.3)

13. $-\dfrac{\sqrt{16 - x^2}}{16x} + C$ (Sec. 6.4)   **14.** $x[(\ln x)^3 - 3(\ln x)^2 + 6(\ln x) - 6] + C$ (Sec. 6.4)

15. $1200x - 20,000 \ln(1 + e^{0.06x}) + C$ (Sec. 6.4)   **16.** (a) 15.567   (Sec. 6.5)

(b) 15.505

17. (a) 1.191   (Sec. 6.5)   **18.** Convergent; 6   (Sec. 6.6)

(b) 1.196

19. Divergent   (Sec. 6.6)   **20.** Divergent   (Sec. 6.6)

21. $n = 50$: 1.652674   (Sec. 6.5)   **22.** $n = 100$: 8.935335   (Sec. 6.5 and 6.6)

$n = 100$: 1.652674 $n = 1000$: 2.288003

$n = 10,000$: 1.636421

Converges $\left(\text{Actual answer is } \dfrac{\pi}{2} \right)$

Practice Test Solutions for Chapter 7

1. (a) $d = 14\sqrt{2}$ (Sec. 7.1)   **2.** $(x - 1)^2 + (y + 3)^2 + z^2 = 5$ **3.** Center: $(2, -1, -4)$ (Sec. 7.1)

(b) Midpoint: $(4, 2, -2)$ (Sec. 7.1)   Radius: $\sqrt{21}$

4. (Sec. 7.2)

(a) x-intercept: $(8, 0, 0)$ (b) $y = 2$

y-intercept: $(0, 3, 0)$ Parallel to xz-plane

z-intercept: $(0, 0, 4)$

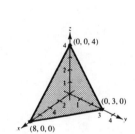

5. (Sec. 7.2)   **6.** (a) Domain: $x + y < 3$ (Sec. 7.3)

(a) Hyperboloid of one sheet   (b) Domain: all points in the xy-plane except the origin

(b) Elliptic paraboloid

7. $f_x(x, y) = 6x + 9y^2 - 3$ (Sec. 7.4)   **8.** $f_x(x, y) = \dfrac{2x}{x^2 + y^2 + 5}$ (Sec. 7.4)

$f_y(x, y) = 18xy + 12y^2 - 6$ $f_y(x, y) = \dfrac{2y}{x^2 + y^2 + 5}$

9. $\dfrac{\partial w}{\partial x} = 2xy^3 \sqrt{z}$ (Sec. 7.4)

$\dfrac{\partial w}{\partial y} = 3x^2y^2 \sqrt{z}$

$\dfrac{\partial w}{\partial z} = \dfrac{x^2y^3}{2\sqrt{z}}$

10. $\dfrac{\partial^2 z}{\partial x^2} = 2x\left(\dfrac{x^2 - 3y^2}{(x^2 + y^2)^3}\right)$ (Sec. 7.4)

$\dfrac{\partial^2 z}{\partial y \partial x} = 2y\left(\dfrac{3x^2 - y^2}{(x^2 + y^2)^3}\right)$

$\dfrac{\partial^2 z}{\partial x \partial y} = 2y\left(\dfrac{3x^2 - y^2}{(x^2 + y^2)^3}\right)$

$\dfrac{\partial^2 z}{\partial y^2} = 2x\left(\dfrac{3y^2 - x^2}{(x^2 + y^2)^3}\right)$

11. Relative minimum: $(1, -2, -23)$ (Sec. 7.5)

12. Saddle point: $(0, 0, 0)$ (Sec. 7.5)

Relative maxima: $(1, 1, 2),\ (-1, -1, 2)$

13. $f(2, -8) = -16$ (Sec. 7.6)   **14.** $f(4, 0) = -36$ (Sec. 7.6)   **15.** $y = \frac{1}{65}(-51x + 355)$

(Sec. 7.7)

16. $y = \frac{1}{6}x^2 - \frac{7}{26}x + \frac{7}{3}$ (Sec. 7.7)   **17.** $\frac{81}{16}$ (Sec. 7.8)   **18.** $-\frac{135}{4}$ (Sec. 7.8)

19. (a) $A = \displaystyle\int_{-2}^{2}\int_{3}^{7-x^2} dy\, dx = \int_{3}^{7}\int_{-\sqrt{7-y}}^{\sqrt{7-y}} dx\, dy$ (Sec. 7.8)

(b) $A = \displaystyle\int_{0}^{1}\int_{x^2+2}^{x+2} dy\, dx = \int_{2}^{3}\int_{y-2}^{\sqrt{y-2}} dx\, dy$

20. $V = \displaystyle\int_{0}^{4}\int_{0}^{4-x} (4 - x - y)\, dy\, dx = \dfrac{32}{3}$ (Sec. 7.9)

21. $y \approx 0.832t + 20.432$ (Sec. 7.7)

$r \approx 0.983$

22. 1.028531×10^{17} (Sec. 7.8)

Practice Test Solutions for Chapter 8

1. (a) $93.913°$ (Sec. 8.1)

(b) $\dfrac{7\pi}{12}$

2. (a) $140°,\ -580°$ (Sec. 8.1)

(b) $\dfrac{25\pi}{9},\ -\dfrac{11\pi}{9}$

3. $\sin\theta = \dfrac{y}{r} = -\dfrac{5}{13}$ $\csc\theta = \dfrac{r}{y} = -\dfrac{13}{5}$

$\cos\theta = \dfrac{x}{r} = \dfrac{12}{13}$ $\sec\theta = \dfrac{r}{x} = \dfrac{13}{12}$

$\tan\theta = \dfrac{y}{x} = -\dfrac{5}{12}$ $\cot\theta = \dfrac{x}{y} = -\dfrac{12}{5}$

(Sec. 8.2)

4. $\theta = 0,\ \dfrac{\pi}{2}, \dfrac{3\pi}{2}$ (Sec. 8.2)

5. (a) Period: 8π (Sec. 8.3)

Amplitude: 3

(b) Period: $\frac{1}{2}$

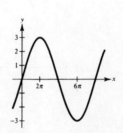

6. $3(1 + \sin x)$ (Sec. 8.4)

7. $x(x \sec^2 x + 2 \tan x)$
(Sec. 8.4)

8. $3 \sin^2 x \cos x$ (Sec. 8.4)

9. $\dfrac{\sec x(x \tan x - 2)}{x^3}$ (Sec. 8.4)

10. $5 \cos 10x$ (Sec. 8.4)

11. $-\dfrac{1}{2}\sqrt{\csc x}\,\cot x$ (Sec. 8.4)

12. $\sec x$ (Sec. 8.4)

13. $-2e^{2x}\csc^2 e^{2x}$ (Sec. 8.4)

14. $3 \sec(x^2 + y) - 2x$ (Sec. 8.4)

15. $-\dfrac{1}{3}\sin^2 3y \sec^2 x$ (Sec. 8.4)

16. $\dfrac{1}{4}\sin 4x + C$ (Sec. 8.5)

17. $-8 \cot \dfrac{x}{8} + C$ (Sec. 8.5)

18. $-\dfrac{1}{2}\ln|\cos x^2| + C$
(Sec. 8.5)

19. $\dfrac{\sin^6 x}{6} + C$ (Sec. 8.5)

20. $\ln|\csc x - \cot x| + \cos x + C$
(Sec. 8.5)

21. $e^{\tan x} + C$ (Sec. 8.5)

22. $-\ln|1 + \cos x| + C$
(Sec. 8.5)

23. $2 \tan x - 2 \sec x - x + C$
(Sec. 8.5)

24. $x \sin x + \cos x + C$
(Sec. 8.5)

25. $\dfrac{\pi^2 - 8\sqrt{2}}{16}$ (Sec. 8.5)

26. ∞ (Sec. 8.6)

27. $\frac{7}{3}$ (Sec. 8.6)

28. $\frac{5}{9}$ (Sec. 8.6)

29.

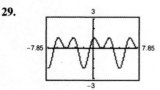

Minimum: -2

Maximum: 1.125

(Sec. 8.3)

30.

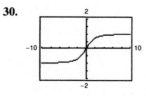

The limit is 1. L'Hôpital's Rule fails.

(Sec. 8.6)

Practice Test Solutions for Chapter 9

1. $\frac{11}{16}$ (Sec. 9.1)

2. $\frac{5}{13}$ (Sec. 9.1)

3. $E(x) = \frac{1}{2}$ (Sec. 9.1)

$V(x) = 4.05$

$\sigma \approx 2.012$

4. $k = \dfrac{1}{4}$ (Sec. 9.1)

5. (a) $\dfrac{25}{64}$ (Sec. 9.2)

(b) $\dfrac{63}{64}$

6. (a) 4   (Sec. 9.3)

(b) $\dfrac{4\sqrt{5}}{5}$

(c) 4

7. (a) $6 \ln\left(\frac{3}{2}\right) \approx 2.433$ (Sec. 9.3)

(b) $\sqrt{6 - 36\left(\ln\frac{3}{2}\right)^2} \approx 0.286$

(c) $\frac{12}{5}$

8. $\mu = \frac{1}{7}$ (Sec. 9.3)

Median: $\frac{\ln 2}{7}$

$\sigma = \frac{1}{7}$

9. 0.0469   (Sec. 9.3)

10. $P(19 < x < 24) = 0.4401$ (Sec. 9.3)

Practice Test Solutions for Chapter 10

1. $a_n = \frac{n}{n^2 + 1}$ (Sec. 10.1)

2. $a_n = (-1)^{n-1}(2n + 3)$

(Sec. 10.1)

3. Converges to $\frac{1}{3}$ (Sec. 10.1)

4. Converges to 4   (Sec. 10.1)

5. $\frac{1}{12}$ (Sec. 10.2)

6. Converges by the Ratio Test

(Sec. 10.3)

7. Converges since it is a p-series with $p = \frac{4}{3} > 1$.

(Sec. 10.3)

8. Diverges by the nth-Term Test   (Sec. 10.2)

9. Diverges since it is a geometric series with

$|r| = \left|-\frac{6}{5}\right| = \frac{6}{5} > 1$.

(Sec. 10.2)

10. Diverges since it is a p-series with $p = \frac{1}{6} < 1$.

(Sec. 10.3)

11. Diverges by the Ratio Test   (Sec. 10.3)

12. Converges since it is a geometric series with

$|r| = |0.27| = 0.27 < 1$.

(Sec. 10.2)

13. Diverges by the nth-Term Test

(Sec. 10.2)

14. $R = 1$ (Sec. 10.4)

15. $R = \lim_{n \to \infty} (n + 2) = \infty$

(Sec. 10.4)

16. $e^{-4x} = \sum_{n=0}^{\infty} \frac{(-4x)^n}{n!}$ (Sec. 10.5)

17. $\frac{1}{\sqrt[3]{x}} = 1 + \sum_{n=1}^{\infty} \frac{(-1)^n 1 \cdot 4 \cdot 7 \cdots (3n - 2)(x - 1)^n}{3^n n!}$

(Sec. 10.5)

18. 0.214   (Sec. 10.5)

19. $x \approx 1.213$ (Sec. 10.6)

20. $\sqrt[4]{10} \approx 1.778$ (Sec. 10.6)

21. (Sec. 10.2)

(a) 7.9961

(b) 138,699

22.

(Sec. 10.5)

23. -1.2090 and 0.9021

(Sec. 10.6)